ASCENT®
CENTER FOR TECHNICAL KNOWLEDGE

Autodesk® Advance Steel 2025 Fundamentals

Learning Guide
Imperial Units - Edition 1.0

ASCENT - Center for Technical Knowledge®
Autodesk® Advance Steel 2025
Fundamentals
Imperial Units - Edition 1.0

Prepared and produced by:

ASCENT Center for Technical Knowledge
630 Peter Jefferson Parkway, Suite 175
Charlottesville, VA 22911

866-527-2368
www.ASCENTed.com

Lead Contributor: Veredith Keller

ASCENT - Center for Technical Knowledge (a division of Rand Worldwide Inc.) is a leading developer of professional learning materials and knowledge products for engineering software applications. ASCENT specializes in designing targeted content that facilitates application-based learning with hands-on software experience. For over 25 years, ASCENT has helped users become more productive through tailored custom learning solutions.

We welcome any comments you may have regarding this guide, or any of our products. To contact us please email: feedback@ASCENTed.com.

© 2024 ASCENT - Center for Technical Knowledge

Contents

Chapter 3: Creating Connections   3-1

Chapter 4: Additional Model Objects   4-1

Chapter 5: Model Verifications   5-1

Chapter 6: Creating Fabrication Drawings   6-1

Preface

The Autodesk® Advance Steel software is a powerful 3D modeling application that streamlines the fabrication process using a 3D model. The 3D model is used to create fabrication drawings, Bill of Materials (BOM) lists, and files for Numerical Control (NC) machines.

Since structural steel projects are extremely complex, the Autodesk Advance Steel software is also complex. The objective of the *Autodesk® Advance Steel 2025: Fundamentals* guide is to enable you to create full 3D project models with a high level of detail and set them up in fabrication drawings. This guide focuses on the basic tools that the majority of users need. You begin by learning the user interface, basic 3D viewing tools, and the standard AutoCAD® tools that are routinely used. Specific Autodesk Advance Steel objects, including structural columns, beams, bracing, plates, bolts, anchors, welds, and additional 3D objects are also covered. You will also learn about the powerful model verification tools. Finally, you will learn to edit and generate all of the required documentation files that enable your design to accurately and effectively communicate the final design.

Topics Covered

- Understand the process of 3D modeling and extracting 2D documentation from a model in the Autodesk Advance Steel software.
- Navigate the Autodesk Advance Steel interface.
- Work with 3D viewing tools.
- Review helpful AutoCAD tools.
- Work with the User Coordinate System (UCS).
- Use the Autodesk Advance Steel Modify commands.
- Add structural grids.
- Create levels.
- Model columns and beams and add bracing.
- Create connections using the Connection Vault.
- Create special parts.
- Verify models using Clash Checking tools.
- Modify a drawing prototype.
- Work within the Drawing Style Manager.

- Create custom connections.

- Create plates and add bolts, anchors, and welds.

- Add grating and cladding.

- Model ladders, stairs, and railings.

- Create concrete objects such as footings.

- Number objects.

- Extract 2D drawings from the model using Drawing Styles and Drawing Processes.

- Review and modify 2D drawings using the Document Manager.

- Modify 2D details with parametric dimensions.

- Revise models and drawings.

- Create BOM lists.

- Export data to .NC and .DXF files.

Prerequisites

- Access to the 2025.0 version of the software, to ensure compatibility with this guide. Future software updates that are released by Autodesk may include changes that are not reflected in this guide. The practices and files included with this guide are not compatible with prior versions (e.g., 2024).

Note on Software Setup

This guide assumes a standard installation of the software using the default preferences during installation. Lectures and practices use the standard software templates and default options for the Content Libraries.

Note on Learning Guide Content

ASCENT's learning guides are intended to teach the technical aspects of using the software and do not focus on professional design principles and standards. The practices aim to demonstrate the capabilities and flexibility of the software rather than following specific design codes or standards.

In This Guide

The following highlights the key features of this guide.

Feature	Description
Practice Files	The Practice Files page includes a link to the practice files and instructions on how to download and install them. The practice files are required to complete the practices in this guide.
Chapters	A chapter consists of the following: Learning Objectives, Instructional Content, Practices, Chapter Review Questions, and Command Summary. • **Learning Objectives** define the skills you can acquire by learning the content provided in the chapter. • **Instructional Content**, which begins right after Learning Objectives, refers to the descriptive and procedural information related to various topics. Each main topic introduces a product feature, discusses various aspects of that feature, and provides step-by-step procedures on how to use that feature. Where relevant, examples, figures, helpful hints, and notes are provided. • **Practice** for a topic follows the instructional content. Practices enable you to use the software to perform a hands-on review of a topic. It is required that you download the practice files (using the link found on the Practice Files page) prior to starting the first practice. • **Chapter Review Questions**, located close to the end of a chapter, enable you to test your knowledge of the key concepts discussed in the chapter. • **Command Summary** concludes a chapter. It contains a list of the software commands that are used throughout the chapter and provides information on where the command can be found in the software.
Appendices	Appendices provide additional information to the main course content. It could be in the form of instructional content, practices, tables, projects, or skills assessment.

Practice Files

To download the practice files for this guide, use the following steps:

1. Type the URL *exactly as shown below* into the address bar of your Internet browser to access the Course File Download page.

   Note: If you are using the ebook, you do not have to type the URL. Instead, you can access the page by clicking the URL below.

   https://www.ascented.com/getfile/id/miltoniaPF

2. On the Course File Download page, click the **DOWNLOAD NOW** button, as shown below, to download the .ZIP file that contains the practice files.

3. Once the download is complete, unzip the file and extract its contents.

   The recommended practice files folder location is:
   C:\Autodesk Advance Steel 2025 Fundamentals Practice Files

   Note: It is recommended that you do not change the location of the practice files folder. Doing so may cause errors when completing the practices.

Stay Informed!

To receive information about upcoming events, promotional offers, and complimentary webcasts, visit:

www.ASCENTed.com/updates

Introduction to the Autodesk Advance Steel Software

The Autodesk® Advance Steel software is a program designed for steel fabricators that enables you to create a 3D model of steel parts and connections, and then extract 2D shop drawings and database files for Bills of Materials (BOMs) and Numerical Control (NC) machines. The software is based on the AutoCAD® software, but includes many additional tools and palettes designed specifically for steel fabrication.

Learning Objectives

- Describe the concepts and workflow of Autodesk Advance Steel.
- Navigate the user interface.
- Use navigation commands to display the model in 2D and 3D views.
- Review the AutoCAD tools that are helpful in the Autodesk Advance Steel software.
- Understand and modify the User Coordinate System (UCS).
- Move, copy, and mirror objects using **Advance Copy**.
- Trim and extend objects using the Autodesk Advance Steel commands.

1.1 Introduction to Autodesk Advance Steel

The Autodesk Advance Steel software expands on the features and functionality of the AutoCAD software to create 3D models of detail-heavy steel structures, as shown in Figure 1–1. It includes beams, columns, plates, and bolts, along with miscellaneous steel objects, such as stairs, railings, and ladders. The documentation for fabrication shop drawings is created automatically from the 3D model, as shown in Figure 1–1.

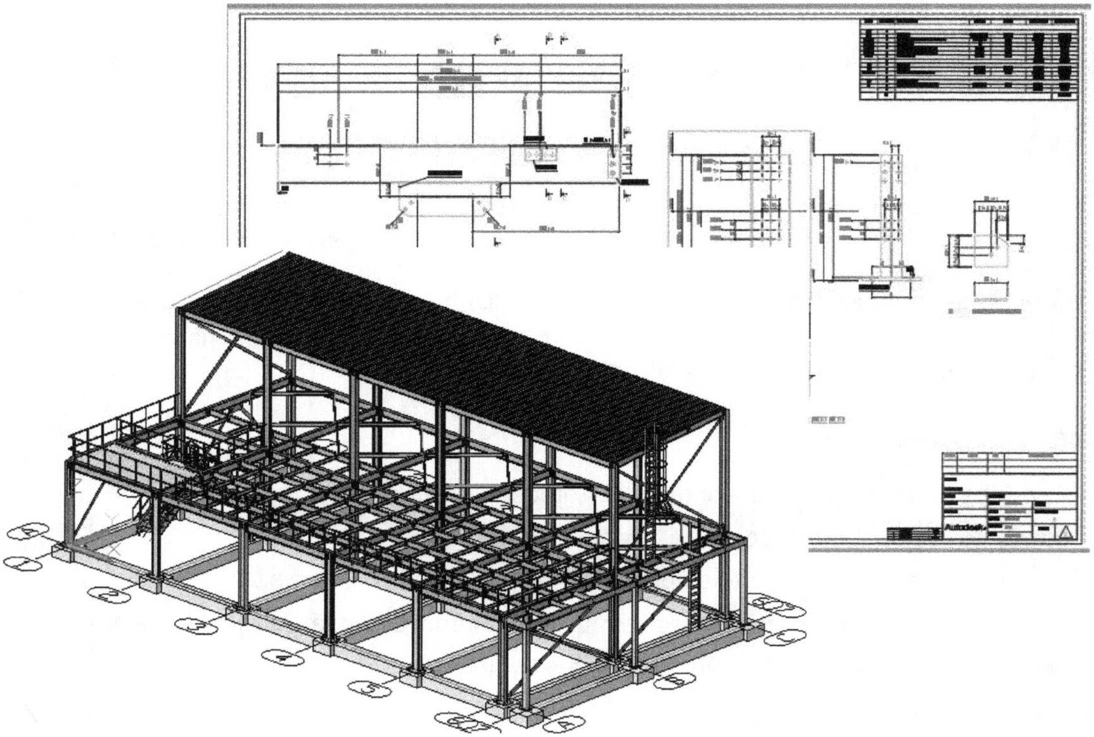

Figure 1–1

A lot of the work that you do in the Autodesk Advance Steel software is done using macros that are a series of standard AutoCAD commands, in addition to specific Autodesk Advance Steel commands and components to create the model and documentation. Therefore, most of the work done in Autodesk Advance Steel uses specific tools and tool palettes, rather than the standard AutoCAD commands.

- Autodesk Advance Steel projects are created from one of the specific Autodesk Advance Steel templates (i.e., ASTemplate.dwt or mm_ASTemplate.dwg) or a customized template created from one of these.

- The main drawing file contains the 3D model. Additional folders and files are created automatically for details and databases (e.g., for BOMs and NC files) as they are added, as shown in Figure 1–2.

Figure 1–2

1.2 Overview of the Interface

The Autodesk Advance Steel interface is designed for intuitive and efficient access to commands and special macros built specifically for Autodesk Advance Steel modeling. The interface includes the ribbon, Quick Access Toolbar, and Status Bar, which are common to most Autodesk® software. Similar to the AutoCAD software, Autodesk Advance Steel includes the Command Line, drawing windows, and layout tabs. It also includes tools that are specific to the Autodesk Advance Steel software, including the *Advance Tool Palette*, the *Connection Vault*, and the *Project Explorer*. The interface is shown in Figure 1–3. Shortcut menus and Autodesk Advance Steel dialog boxes are also an important part of using the software.

Note: The Drawing Window color has been changed to white for printing clarity.

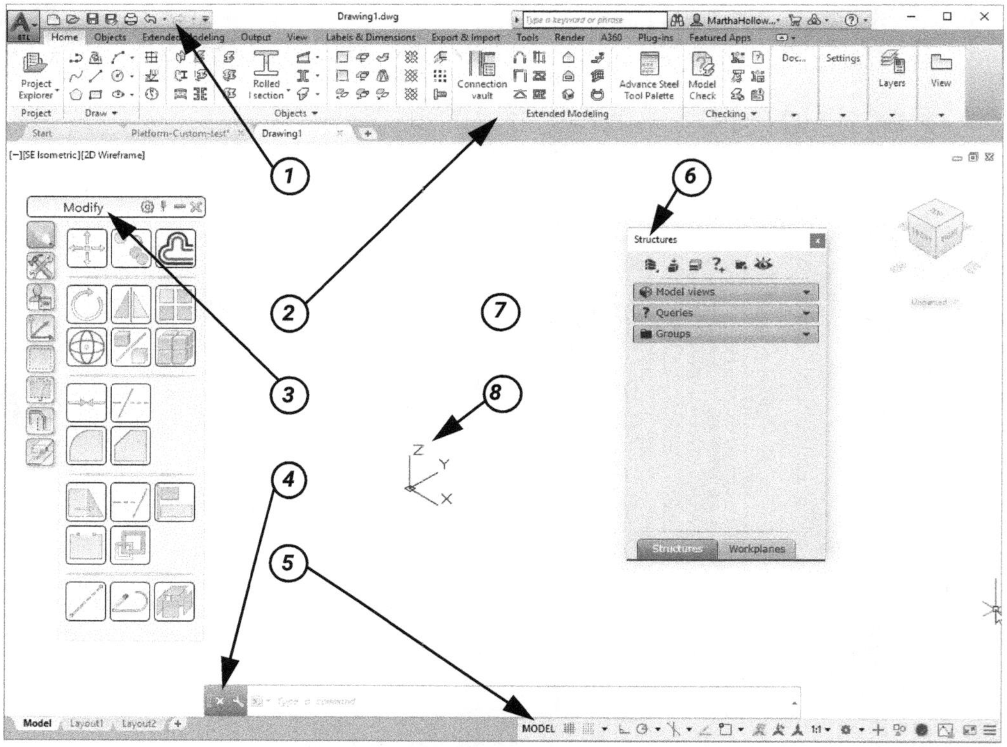

Figure 1–3

1. Quick Access Toolbar	5. Status Bar
2. Ribbon	6. Project Explorer
3. Advance Tool Palette	7. Drawing Window
4. Command Line	8. UCS Icon

1. Quick Access Toolbar

The Quick Access Toolbar (shown in Figure 1–4) includes several frequently used commands, including **New**, **Open**, **Save**, **Print**, **Undo**, and **Redo**. These commands are also available in the Application Menu (![STL icon]).

Figure 1–4

*Note: The **Undo** and **Redo** commands often do not work as expected in the Autodesk Advance Steel software because many of the commands are actually macros that run multiple commands. This means that you might have to undo several processes to fully undo a single command.*

2. Ribbon

Instead of the standard AutoCAD tools, the Autodesk Advance Steel ribbon includes specific tools used by the program, as shown in part of the *Home* tab in Figure 1–5.

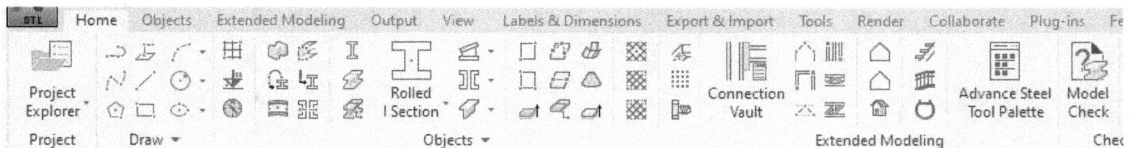

Figure 1–5

* Hover the cursor over a button to display the name of the tool.

3. Advance Tool Palette

The Advance Tool Palette is unique to the Autodesk Advance Steel software and holds many important tools, including some of the standard AutoCAD modify tools (shown in Figure 1–6) and Advance Steel modify tools (shown in Figure 1–7).

- Note that AutoCAD modification tools are not included on any of the ribbon tabs.

- To open the *Advance Tool Palette*, in the *Home* tab>*Extended Modeling* panel, click ▦ (Advance Steel Tool Palette).

Categories ———

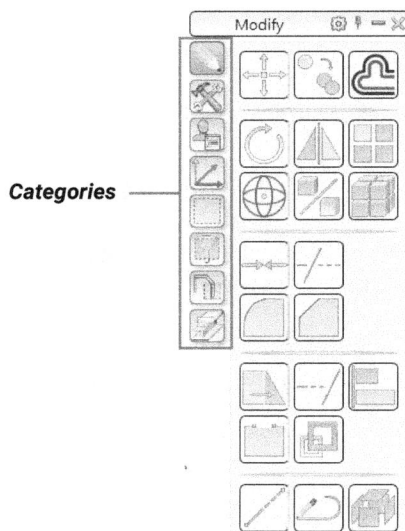

Figure 1–6

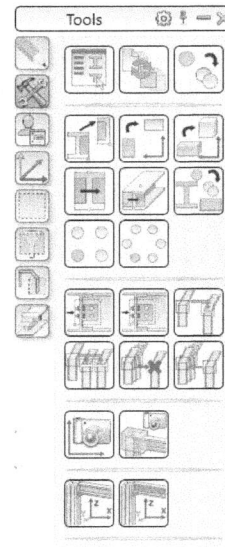

Figure 1–7

- Click on the buttons in the left column (called Categories) to access the different tools on the right.

- The *Advance Tool Palette* can float or be pinned into place. You can also minimize it (as shown in Figure 1–8), hide it, and modify the Theme Settings.

Theme Settings   *Minimize/Maximize*

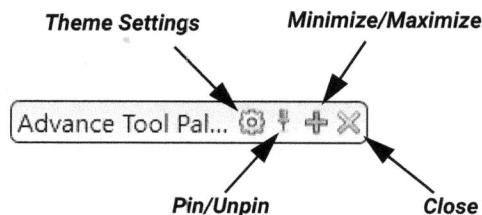

Pin/Unpin   *Close*

Figure 1–8

Advance Tool Palette Categories

	Modify	Includes standard AutoCAD modification commands, such as Move, Trim, and Fillet. These tools are most often used with standard AutoCAD objects (such as lines and polylines) or individual Autodesk Advance Steel objects that are not connected to other objects.
	Tools	Includes tools specifically created to use with Autodesk Advance Steel objects and objects that are connected together. It includes modification tools and ways to create groups.
	Custom connections	Includes tools to create individual custom connectors, including plates and bolts. Tools can also group and reuse custom connections.
	UCS	Includes tools that define the location and orientation of the User Coordinate System (UCS). Having a defined UCS is critical for many Autodesk Advance Steel commands.
	Selection	Includes tools that enable you to search, display, and mark objects by certain criteria.
	Selection filters	Includes tools that enable you to select specific types of elements, including all beams, or just curved or concrete beams, among other objects including slabs and bolts.
	Quick views	Includes tools that enable you to create views based on certain objects to help you modify the view by toggling all objects on, or selected objects off, etc.
	Features	Includes tools for modifying plates and beams (including miters, corner cuts, and copes), as well as cutting holes in objects.

- Commands found in this palette are referenced like this:

  In the *Advance Tool Palette>* (Tools) category, click (Advance Copy).

- Note that many of these tools are actually macros of multiple commands.

- Other tool palettes (including the Connection Vault) are similar and can be docked on top of each other.

4. Command Line

The command line is the same in the Autodesk Advance Steel software as it is in the AutoCAD software. Autodesk Advance Steel commands are much more complex (as seen in Figure 1–9), so they are rarely typed.

```
Command: _astm4crbeambyclass I
Please locate start point of system axis:_
Please locate end point of system axis:_

>._  - ASTORCRBEAMBYCLASS Please locate start point of system axis:_
```

Figure 1–9

- The prompts for Autodesk Advance Steel commands can be complex. If you are having trouble following a process, expand the command line to display multiple lines of prompts.

5. Status Bar

The Status Bar (shown in Figure 1–10) is essentially the same as in the AutoCAD software. Important tools found here include **Ortho**, **Object Snaps**, and **Isolate Objects**.

Figure 1–10

6. Project Explorer

The Project Explorer enables you to create *Levels* and *Model* views, and access other tools. This palette can remain floating on the screen (as shown in Figure 1–11) or be docked.

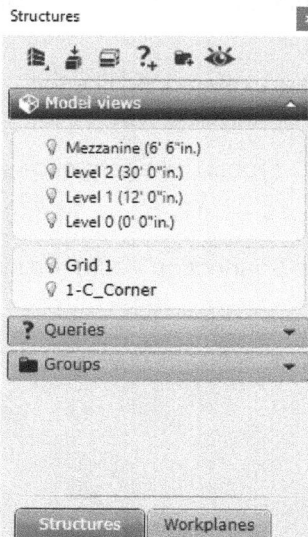

Figure 1–11

- To display the *Project Explorer*, in the *Home* tab>*Project* panel, click 🗔 (Project Explorer).

- If you have toggled off objects, you can toggle them all back on by clicking 👁 (Show All Elements).

7. Drawing Window

The drawing window is the area of the screen in which the drawing displays. Several drawing windows can be open at the same time. They can be resized, minimized, and maximized.

The drawing's File tabs (shown in Figure 1–12) are located near the top of the drawing window. They provide a quick way of switching between open drawings, creating new drawings, or closing drawings. The *Start* tab is always the first tab and persists in the File tabs bar. Clicking the start tab displays the *Start* window.

| Start | Drawing1* | × | Drawing2 | × | + |

Figure 1–12

- The *Model* and *Layout* tabs display at the bottom of the drawing window. Most of your work in the Autodesk Advance Steel software is done in the *Model* tab. Layouts are automatically created when you run the documentation tools.

8. UCS Icon

In the drawing window, the UCS icon indicates the current drawing planes. This is an important part of the Autodesk Advance Steel software as it controls the orientation of elements, such as plates and stairs. The style of the UCS icon changes with the visual style, as shown in Figure 1–13.

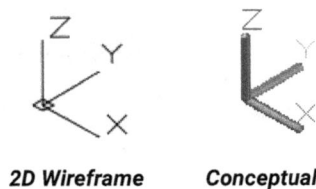

2D Wireframe   **Conceptual**

Figure 1–13

- To toggle the UCS icon on and off, in the *View* tab>*Viewport Tools* panel, click ⌐ (UCS Icon).

- If you change the UCS and want to return to the base location, in the command line, type **UCS**, press <Enter>, then type **W** (for World) and press <Enter>. Alternatively, in the *Advance Tool Palette*> ◩ (UCS) category, click 🗔 (UCS World).

Shortcut Menus

When you right-click, a shortcut menu usually displays next to the cursor. The menu that displays depends on what you are doing in the software and where you right-click. For example, when no objects are selected and you right-click in the drawing window, the menu in Figure 1–14 displays. When you have an Autodesk Advance Steel object selected, the menu in Figure 1–15 displays.

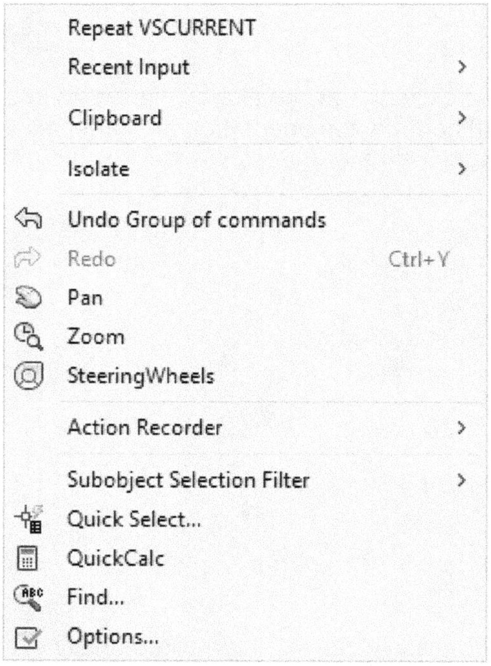

| Repeat VSCURRENT |
| Recent Input > |
| Clipboard > |
| Isolate > |
| Undo Group of commands |
| Redo   Ctrl+Y |
| Pan |
| Zoom |
| SteeringWheels |
| Action Recorder > |
| Subobject Selection Filter > |
| Quick Select... |
| QuickCalc |
| Find... |
| Options... |

Figure 1–14

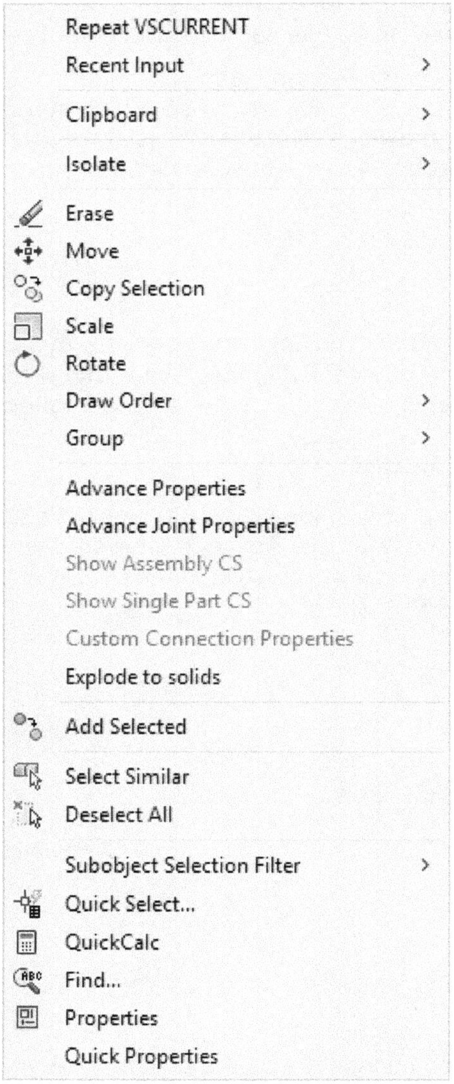

| Repeat VSCURRENT |
| Recent Input > |
| Clipboard > |
| Isolate > |
| Erase |
| Move |
| Copy Selection |
| Scale |
| Rotate |
| Draw Order > |
| Group > |
| Advance Properties |
| Advance Joint Properties |
| Show Assembly CS |
| Show Single Part CS |
| Custom Connection Properties |
| Explode to solids |
| Add Selected |
| Select Similar |
| Deselect All |
| Subobject Selection Filter > |
| Quick Select... |
| QuickCalc |
| Find... |
| Properties |
| Quick Properties |

Figure 1–15

Autodesk Advance Steel Dialog Boxes

Two dialog boxes are critical parts of using the Autodesk Advance Steel software: **Advance Properties** (as shown for a beam in Figure 1–16) and **Advance Joint Properties**. These dialog boxes are accessed through the shortcut menu when you have certain objects selected. The critical part of understanding these dialog boxes is that they are live (i.e., any changes that you make in the dialog box are instantly and automatically applied to the model).

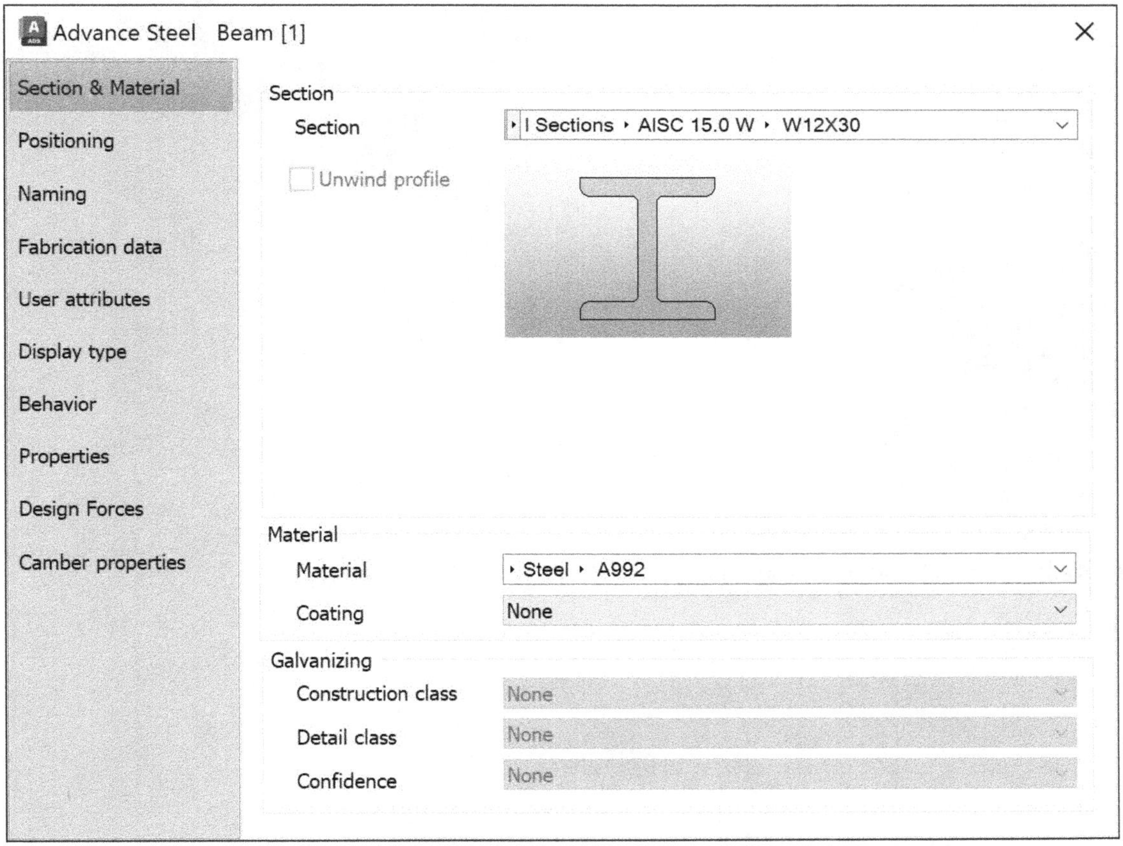

Figure 1–16

- **Advance Properties:** Displays the options for selected objects, such as beams, columns, or individual bolts.

- **Advance Joint Properties:** Displays the options for a full connection, such as the Base plate shown in Figure 1–17. This includes options for the plate, stiffeners, holes, and bolts as a group.

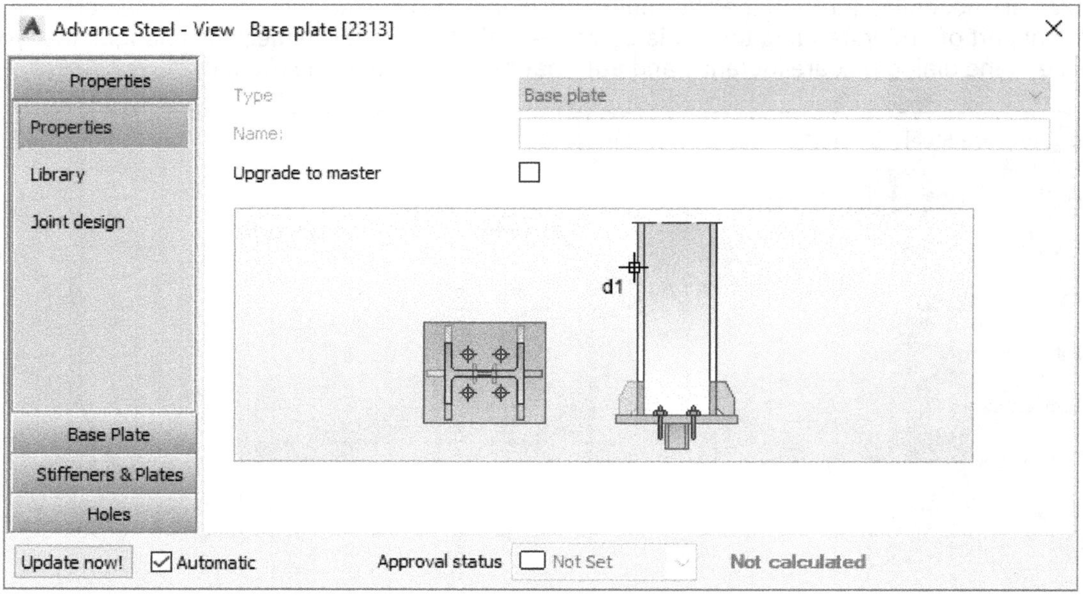

Figure 1–17

- Select the tabs on the left to display information on the right.

1.3 Viewing the Model

In Autodesk Advance Steel you are primarily working in 3D, and so you need to be able to view objects from all directions. There are several basic tools that enable you to do so: preset 3D views, the ViewCube, and Visual Styles, as shown in Figure 1–18.

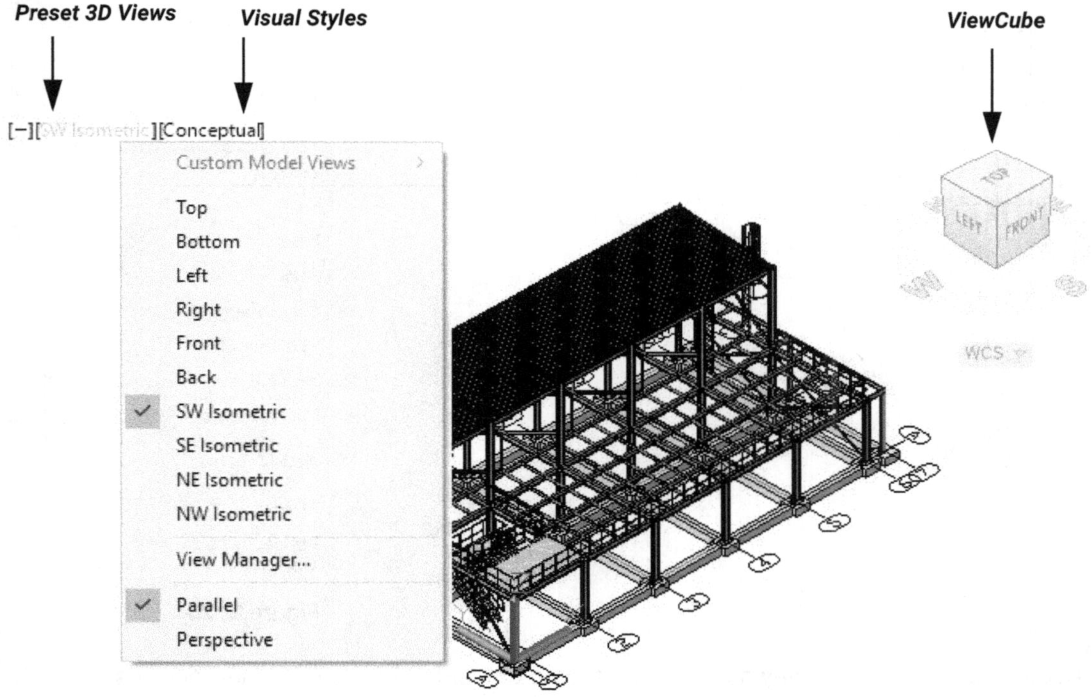

Preset 3D Views   **Visual Styles**   **ViewCube**

[−][SW Isometric][Conceptual]

Custom Model Views	>
Top	
Bottom	
Left	
Right	
Front	
Back	
✓ SW Isometric	
SE Isometric	
NE Isometric	
NW Isometric	
View Manager...	
✓ Parallel	
Perspective	

Figure 1–18

Accessing Preset 3D Views

There are several preset 3D views (shown in Figure 1–19) that enable you to quickly change the viewing angle. These presets include both orthographic and isometric views, and can be accessed in the top left corner of the drawing window, as shown in Figure 1–20.

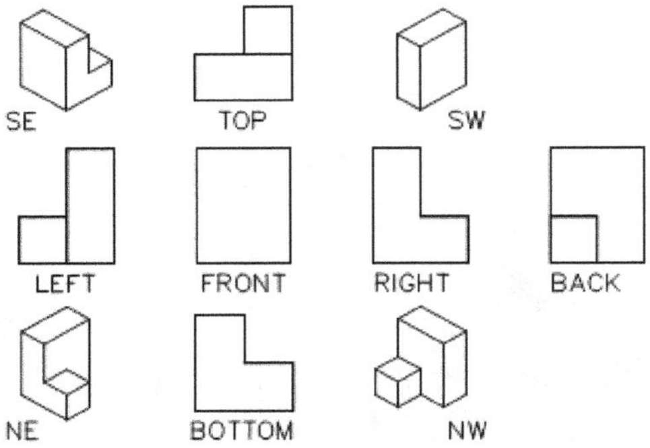

Figure 1–19

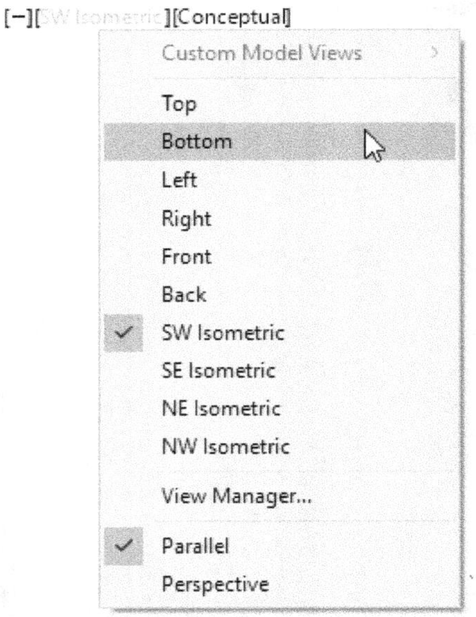

Figure 1–20

- Orthographic views display as if you are facing one side of a part. Isometric views typically display three sides, as if you are facing a corner.

- Orthographic views change the active drawing plane (UCS) of the view, while isometric views do not. To return to the flat drawing plane, select the **Top** view before continuing with a non-orthographic 3D view.

Using the ViewCube

The ViewCube provides visual clues as to where you are in a 3D drawing and makes it easier to navigate to standard views, such as **Top, Front, Right, Left, Corner**, and directional views. Move the cursor over one of the highlighted options and select it. You can also click and drag on the ViewCube to rotate the box, which rotates the model. The ViewCube is shown in Figure 1–21.

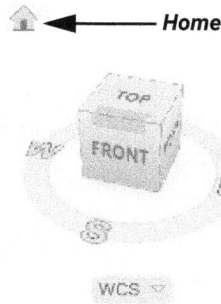

Figure 1–21

- ⌂ (Home) displays when you hover the cursor over the ViewCube. Click it to return to the view defined as **Home**.

- To toggle the ViewCube on and off, in the *View* tab>*Viewport Tools* panel, click

  ⬚ (ViewCube).

  *Note: To change the default **Home** view, set the view you want, right-click on the ViewCube, and select **Set Current View as Home**.*

Orbiting in 3D

The best tools for navigating a model in 3D are the mouse and keyboard. You can zoom in and out using the mouse wheel, and can pan by holding the mouse wheel and moving the mouse. Both methods are useful in 2D and 3D. However, in 3D you also need to view the model from all sides. Hold <Shift> and the mouse wheel to orbit the objects in your drawing, as shown in Figure 1–22.

Figure 1–22

- When you orbit, the target (what you are viewing) stays stationary while the camera (your viewpoint) moves.

- You can also hold <Ctrl> and the mouse wheel to swivel. This is similar to panning the camera as you drag the mouse. The target of the view changes.

- If you select objects before you start orbiting, only those objects display as you move around the drawing. This can be useful in complex drawings, because limiting the number of objects results in a smoother rotation of the view.

Using Visual Styles

While viewing a model, setting a visual style can help you gain a clearer understanding of the model. Visual styles control how elements display in a view. You can add and modify objects and orbit in any of the visual styles. Three useful visual styles include the 2D Wireframe, Conceptual, and X-ray styles, as shown in Figure 1–23.

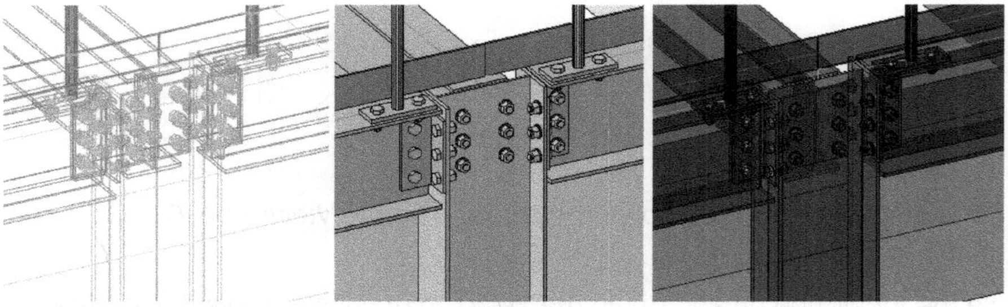

2D Wireframe   **Conceptua**   **X-ray**

Figure 1–23

- The visual styles list is available in the upper left corner of the drawing window, next to the 3D view presets. **Shaded** or **Shaded with Edges** visual styles include a fast 3D graphics system, as shown in Figure 1–24. By default, the command **FASTSHADEDMODE** is ON.

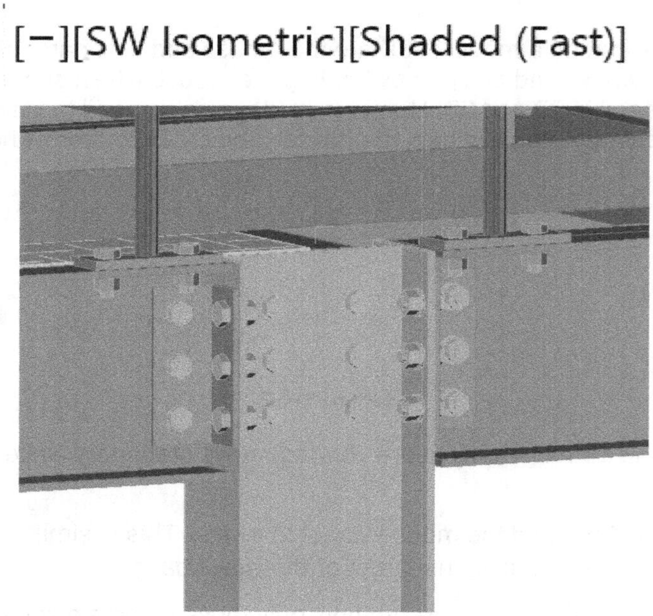

[−][SW Isometric][Shaded (Fast)]

Figure 1–24

Practice 1a
Open a Project and View the Model

Practice Objectives

- Review typical Autodesk Advance Steel project layouts.
- Open an Autodesk Advance Steel drawing.
- Review the user interface.
- View the 3D model.

In this practice, you will review the folder structure of a typical Autodesk Advance Steel project. You will open a drawing and review the user interface. You will then use the 3D viewing tools to display the model you will be creating in the practices, shown in Figure 1–25.

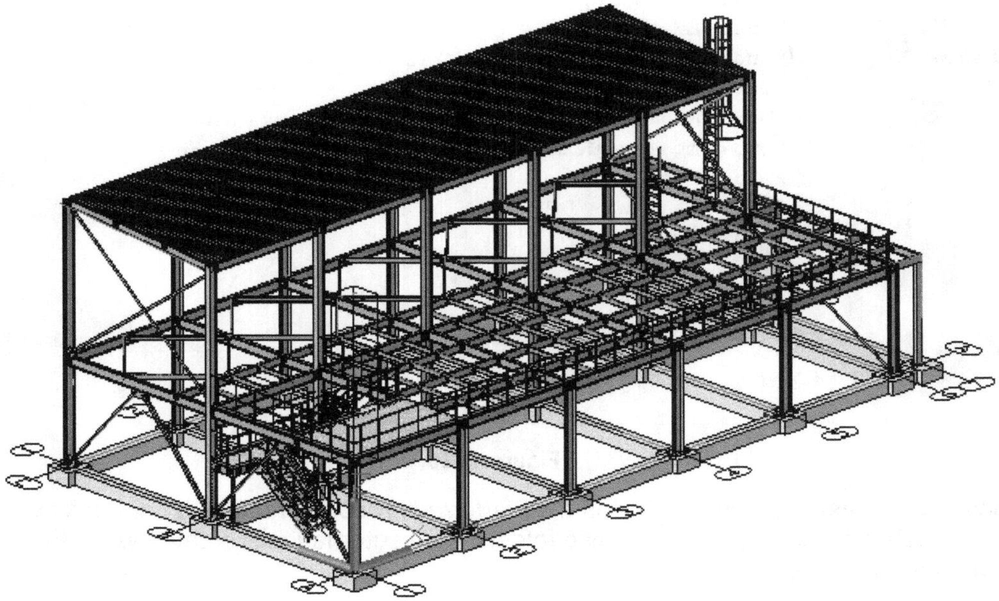

Figure 1–25

Task 1: Review files and open an Autodesk Advance Steel model.

1. In the Quick Access Toolbar, click 📂 (Open).
2. In the *Open* dialog box, navigate to the practice files folder.

3. Click once on **Platform-Introduction.dwg** to display the preview as shown in Figure 1–26. This is the primary model you will be working on.

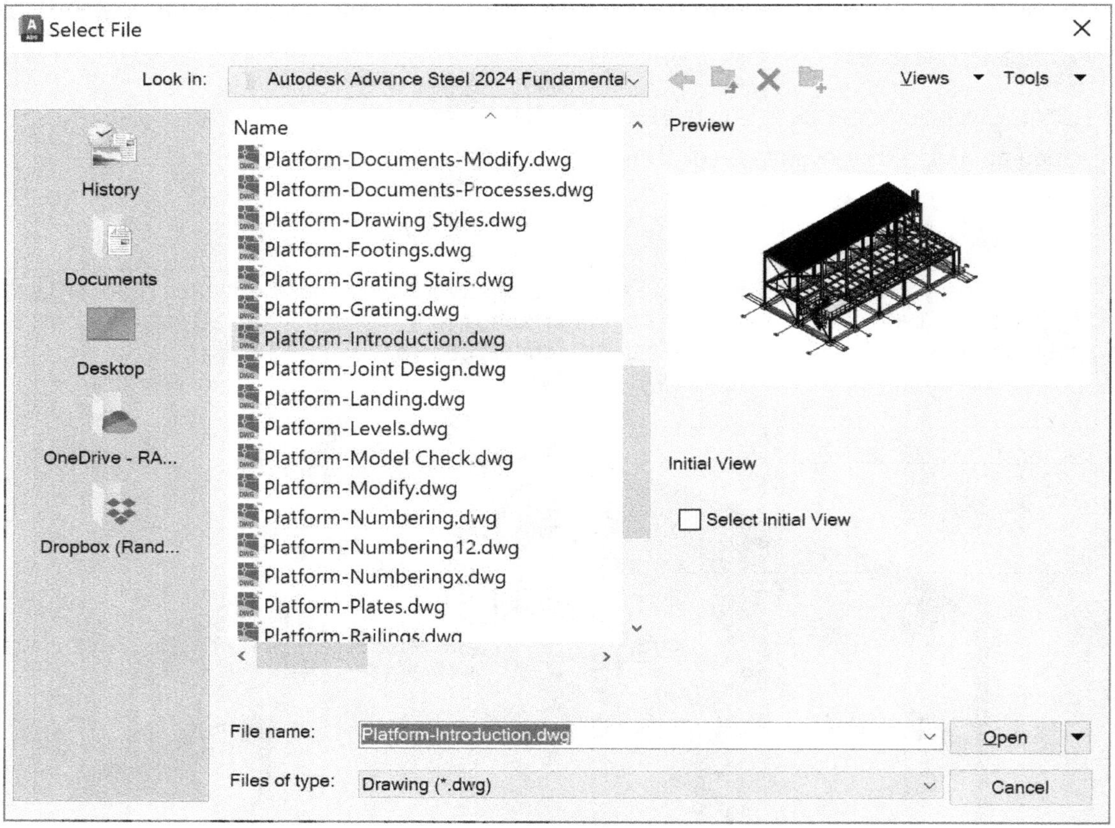

Figure 1–26

4. Scroll up the list and double-click on the *Platform-Introduction* folder. It contains two folders: *Databases* and *Details*. These folders were automatically created when the documentation files were processed.

5. Open the *Details* folder. Note that there are a number of detail drawings, but that they do not preview, as shown in Figure 1–27. These files are not typically opened directly, but rather are accessed through the *Document Manager* when you are in a model.

Figure 1–27

6. Return to the main folder and open **Platform-Introduction.dwg**.

Task 2: Review the user interface.

1. Review the different tabs of the ribbon. Note that many tools on the *Home* tab>*Objects* panel are also found on the *Objects* tab, as shown in Figure 1–28.

Home tab>Objects panel

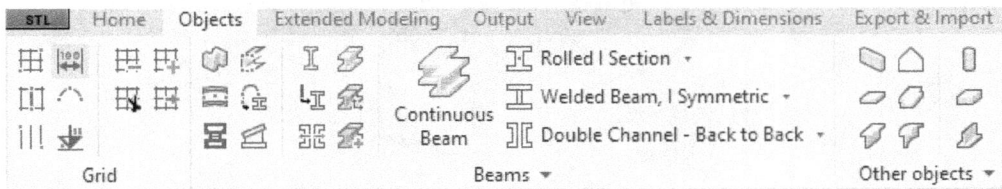

Objects tab

Figure 1–28

2. In the *Home* tab>*Extended Modeling* panel, click ▭ (Advance Steel Tool Palette) to toggle it off, and click it again to toggle it on.

3. In the *Advance Tool Palette*, click through the various categories to review the available tools.

4. In the ⬚ (Quick views) category, click ⬚ (All Visible). Additional elements display in the view, as shown in Figure 1–29.

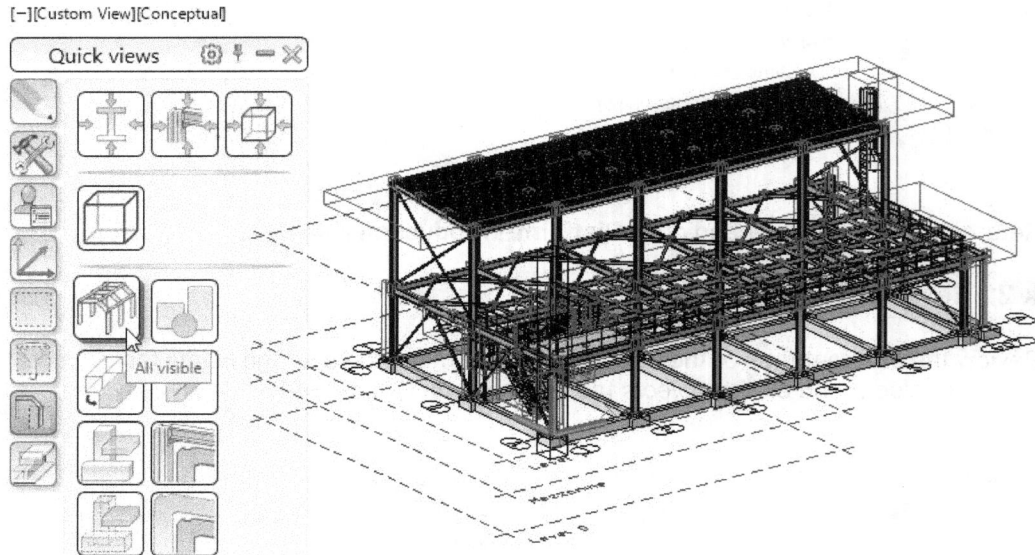

Figure 1–29

5. Select the object labeled **Level 0**. Hover the cursor over the object to display information about the layer it is on, as shown in Figure 1–30. Objects are automatically placed on the appropriate layer in the Autodesk Advance Steel software.

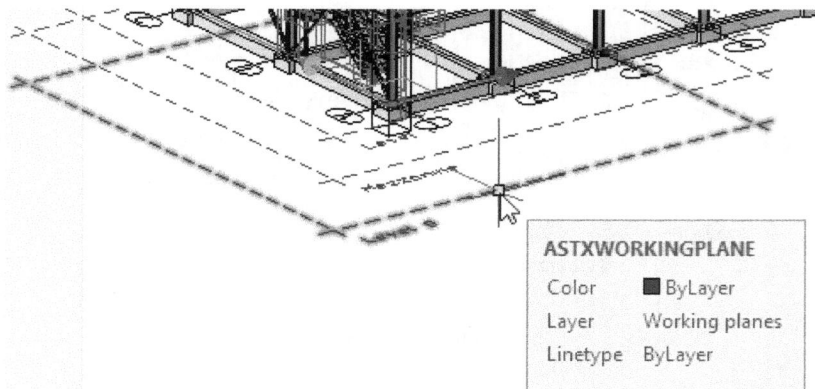

Figure 1–30

6. Right-click in the drawing window and select **Select Similar**.

7. In the *Advance Tool Palette*> ⬚ (Quick views) category, click ⬚ (Selected Objects off).

8. In the *Advance Tool Palette*> ⬚ (Selection filters) category, click ⬚ (Model view boxes).

9. In the *Advance Tool Palette*> ⬚ (Quick views) category, click ⬚ (Selected Objects off).

10. In the model, double-click on a column. The **Advance Properties** command displays, as shown in Figure 1–31. Click through the tabs on the left and note the different options.

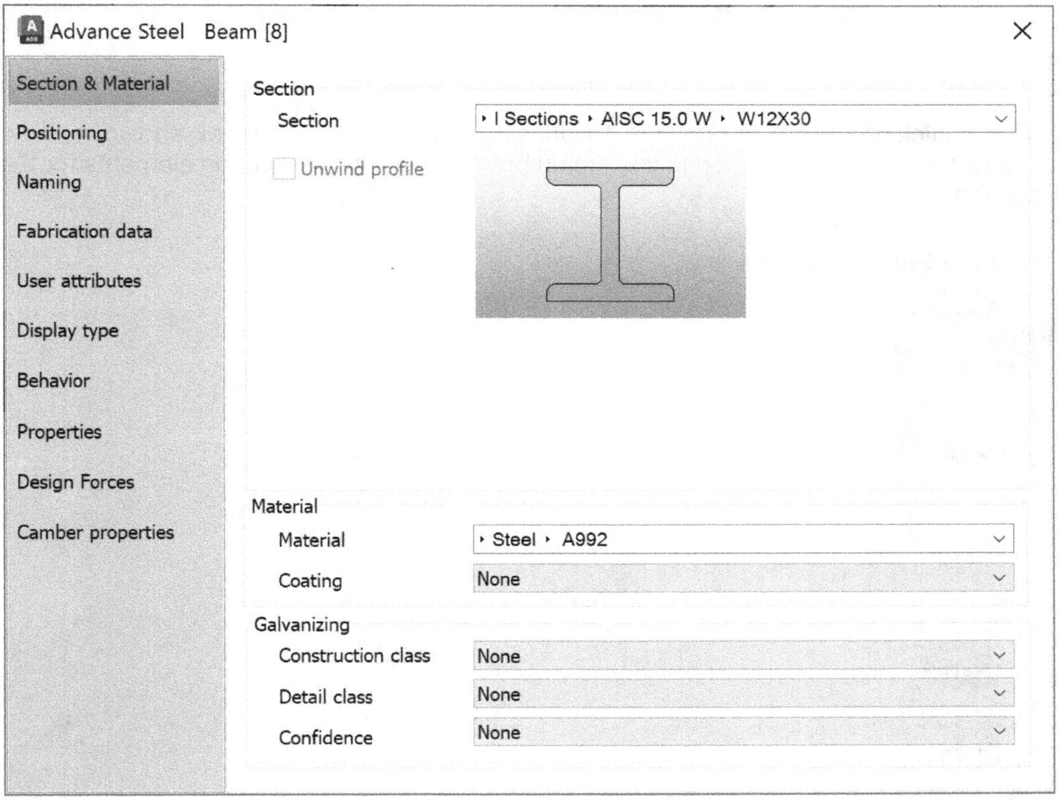

Advance Steel Beam [8]	✕

Section & Material | Section |
Positioning | Section | ▸ I Sections ▸ AISC 15.0 W ▸ W12X30 |
Naming | ☐ Unwind profile |
Fabrication data |
User attributes |
Display type |
Behavior |
Properties |
Design Forces |
Camber properties |

Material
Material | ▸ Steel ▸ A992
Coating | None

Galvanizing
Construction class | None
Detail class | None
Confidence | None

Figure 1–31

*Note: Columns are made from beams, but they have the Model Role of **Column**, and are typically placed on the **Column** layer when created with the appropriate Advance Steel command.*

11. Close the dialog box.

12. Zoom in on the base of column A2. Note that there is a box around the base plate on the **Connection boxes** layer, as shown in Figure 1–32.

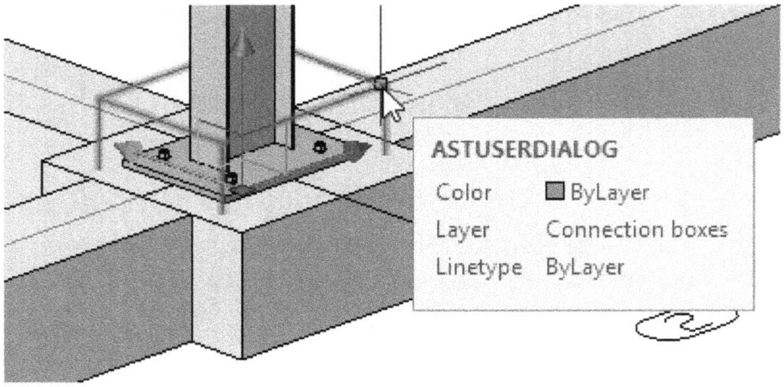

Figure 1–32

13. Double-click on the box to open the *Advance Joint Properties* dialog box, shown in Figure 1–33. The information in this dialog box controls the connection elements for the column.

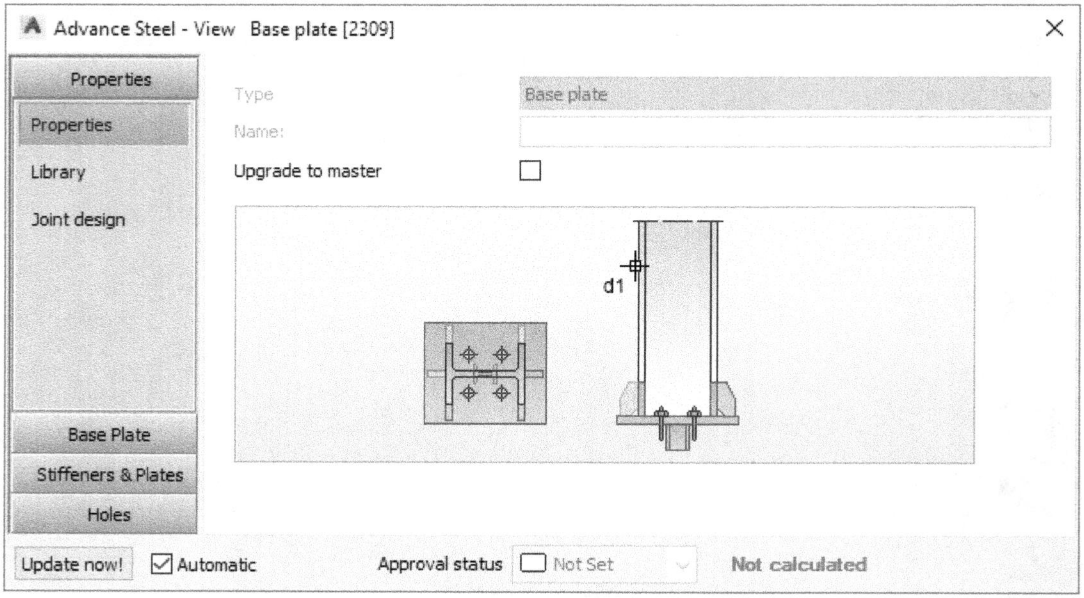

Figure 1–33

14. Close the dialog box.

15. In the *Advance Tool Palette>*  (Selection filters) category, click  (Joint boxes). In the *Advance Tool Palette>*  (Quick views) category, click  (Selected Objects off).

16. Double-click on the base plate. Note that the *Advance Properties* dialog box for the plate displays, but now you cannot make many changes to the plate because it is part of a connection object.

17. Close the dialog box.

18. Select the base plate again. Right-click (ensuring that you do not touch the UCS gizmo) and select **Advance Properties**, as shown in Figure 1–34. The same dialog box displays as when you double-clicked on the plate.

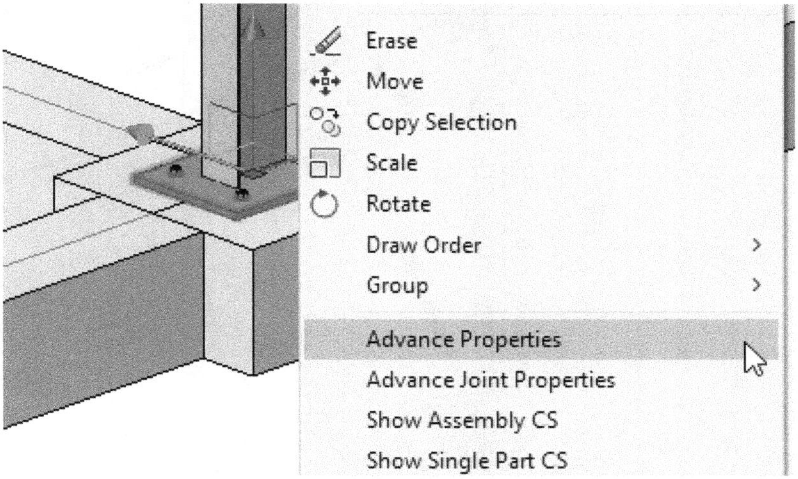

Figure 1–34

19. Select the base plate again. Right-click on the base plate and select **Advance Joint Properties**. This opens the *Full Connection* dialog box. It also turns the joint box on for this connection only.

20. Close the dialog box.

21. In the command line, type **Z** <Enter> **A**<Enter> to return to the full model view.

22. Save the drawing.

Task 3: View the 3D model.

1. Locate the UCS icon in the lower left of the model at column A1, as shown in Figure 1–35. Note the direction of the X, Y, and Z-axes.

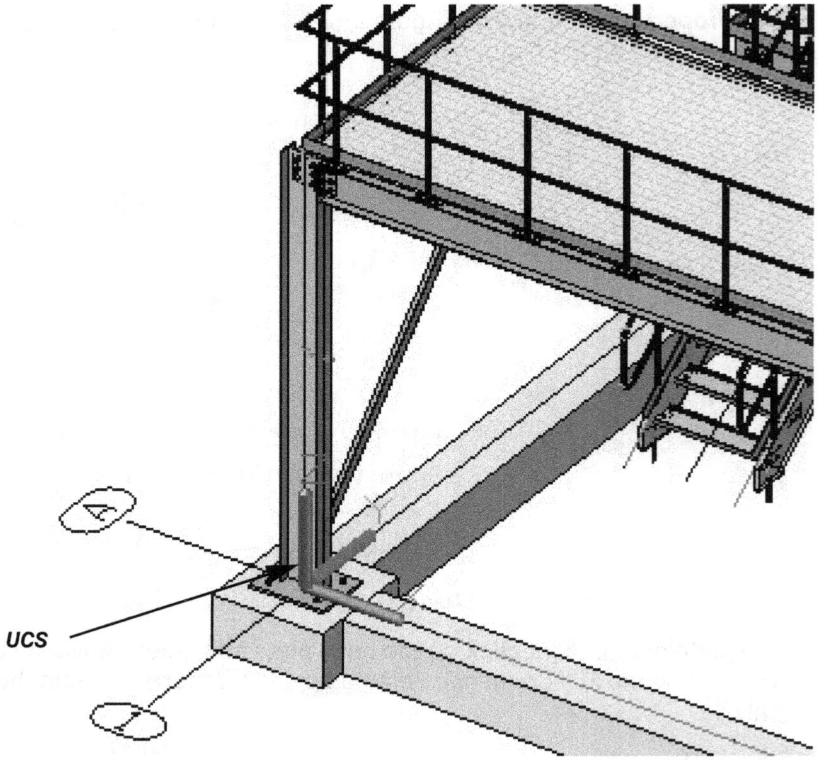

UCS

Figure 1–35

2. Click on various parts of the ViewCube to rotate the model. Note that the UCS does not change.

3. In the upper left of the drawing window, test the four preset isometric views. Note that the UCS still does not change.

4. Set the view to the **Left** preset. Note that the UCS now changes, as shown in Figure 1–36. The **Top**, **Bottom**, **Left**, **Right**, **Front**, and **Back** views change the UCS to that orientation.

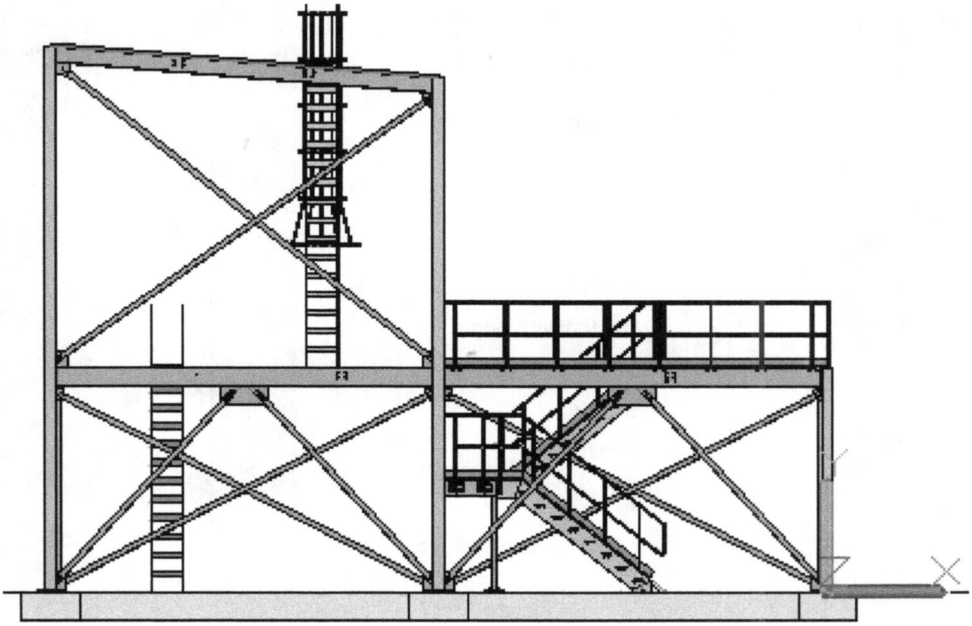

Figure 1–36

5. Hold <Shift> and the mouse wheel and rotate the model in the view. Even though you were in what looked like an elevation view, note that you are still in a 3D view.

6. In the *Advance Tool Palette>* (UCS) category, click  (UCS World), or in the command line, type **UCS** <Enter>, **W** <Enter>.

7. Zoom in on the UCS icon.

8. In the upper left corner of the view window, change the *Visual Style* to **2D Wireframe**, as shown in Figure 1–37. Note that the UCS icon changes with the rest of the model.

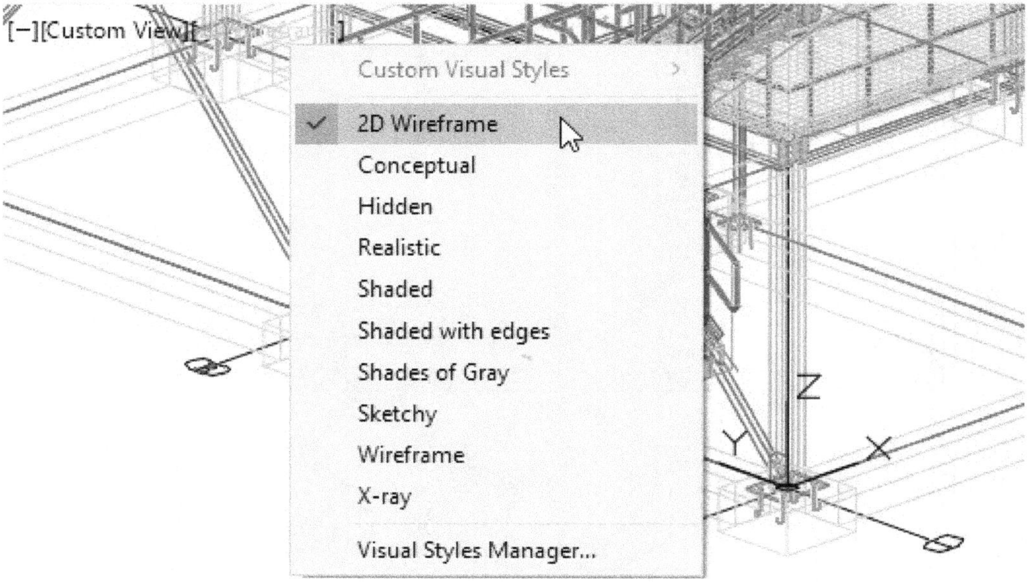

Figure 1–37

9. Test out other Visual Styles. Which one gives you the best view of the j-bolts in the footing?

10. Set the *Visual Style* to **Conceptual**.

11. Set the *Visual Style* to **Shaded** or **Shaded with edges**.

    Note: (Fast) is shown in the viewport control to indicate that the modern 3D graphics systems is being used.

12. Zoom out to display the entire model.

End of practice

1.4 Helpful AutoCAD Tools

Most of the tools that you use in the Autodesk Advance Steel software are customized macros that create complex objects and connections. Since the software is based on the AutoCAD software, there are some useful tools that you can use as you are drawing and modifying the Autodesk Advance Steel objects, including object snaps, Ortho, Polar Tracking (shown in Figure 1−38), basic modify tools such as Move and Copy, grips on objects, and the 3D Gizmo.

Object Snaps

Polar Tracking

Node

Figure 1−38

- **Warning:** One tool that does not work as expected in the Autodesk Advance Steel software is the **Undo** command. As many Autodesk Advance Steel commands are actually macros of many commands, pressing **Undo** once often does not do much. You might need to use the *Undo* drop-down list (shown in Figure 1−39) to undo numerous steps.

Figure 1−39

Status Bar Tools

Several tools on the Status Bar (shown in part in Figure 1–40) can help you as you are drawing and modifying objects, including the Ortho, Polar Tracking, and Object Snap Tracking tools. You can also toggle on the drawing view grid to help you visualize the space better. Object Snaps are probably the most critical of the tools.

Figure 1–40

Object Snaps

There are a lot of objects that you can snap to in Autodesk Advance Steel models, so it is important to use only the ones you need. For example, there are a lot of endpoints on an I-beam, but only one node at the end of each beam. It is safer to use object snap overrides (as shown in Figure 1–41) more frequently than setting the objects snaps (as shown in Figure 1–42).

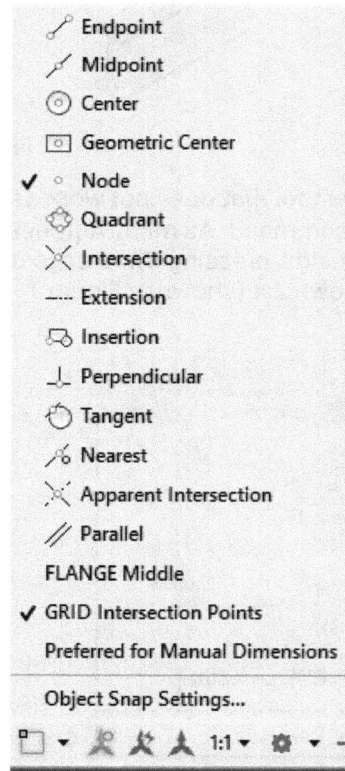

| Figure 1–41 | Figure 1–42 |

- The **Node** object snap and **GRID Intersection Points** object snap can be preset. Toggle off object snaps if they get in the way of what you are trying to do.

Basic Modify Tools

Basic AutoCAD modification tools (such as **Move**, **Copy**, and **Rotate**) are accessed in the

Advance Tool Palette> (Modify) category, as shown in Figure 1–43. These tools can be used if you are working with standard AutoCAD objects (such as lines) or when you are manipulating individual Autodesk Advance Steel objects (such as columns or beams).

Note: These AutoCAD tools can be used as long as there are no joints connecting objects.

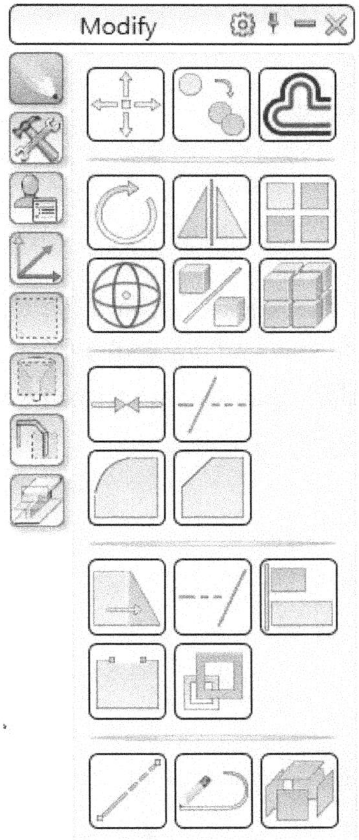

Figure 1–43

- Once you start making connections so that columns and beams work together with plates and bolts, you need to start using the Autodesk Advance Steel modify tools.

Hint: AutoCAD Break vs. Autodesk Advance Steel Split Beam

Not all of the standard AutoCAD tools will work with Autodesk Advance Steel objects. For example, the **Break** command does not work on beams. Instead, use the Autodesk Advance Steel command **Split Beam** to break a beam into separate parts, as shown in Figure 1–44.

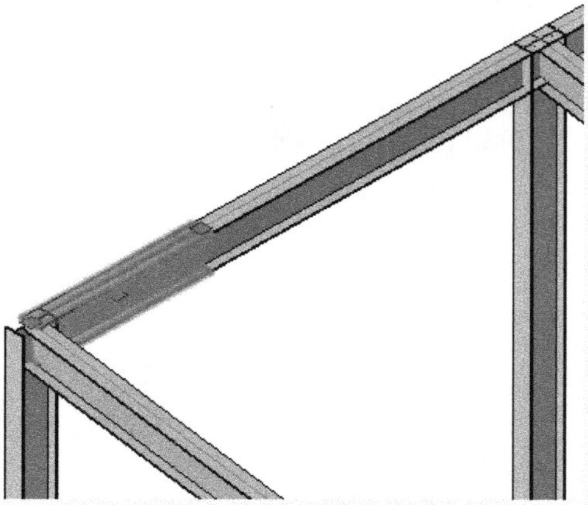

Figure 1–44

1. In the *Objects* tab>*Beams* panel, click 🍂 (Split Beam).

2. Select the beam you want to split and press <Enter>.

3. Select the point or points along the beam where you want it to split.

4. Press <Enter>.

Using Grips and the 3D Gizmo

You can use grips to modify objects, such as the length of beams or size of plates. They work in 2D and in 3D with added power when you use the 3D Gizmo, as shown in Figure 1−45.

Figure 1−45

- There are a few ways to edit a selected object using its grips. You can select the grip and right-click or hover your cursor over a grip to get editing options, as shown in Figure 1−46.

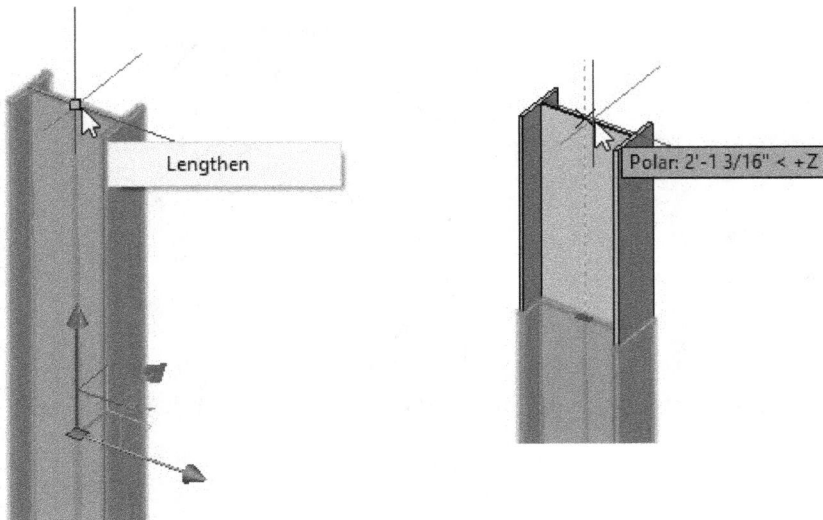

Figure 1−46

- 3D Gizmos can be used with both 2D and 3D objects. The Gizmo is primarily used to move and rotate, but can also scale objects, as shown in Figure 1–47.

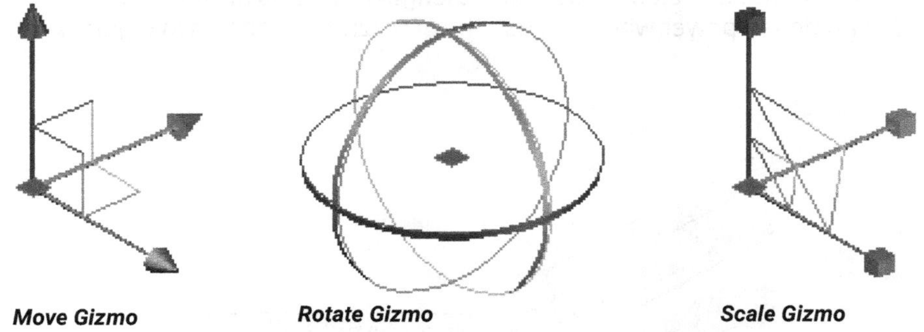

Move Gizmo   **Rotate Gizmo**   **Scale Gizmo**

Figure 1–47

- When you select objects in a 3D view, the UCS icon turns into the Gizmo tool. This tool enables you to limit the movement of objects along an axis or plane and to rotate in 3D space. When you hover the cursor over a Gizmo, it automatically jumps to a location or vertex grip. It can also be used as a grip without being moved to another grip.

- The color-coded icon displays in the current UCS. Red indicates the X-axis, green the Y-axis, and blue the Z-axis.

- As you move the cursor over one of the axes on the Gizmo tool, the active axis highlights in gold and displays a guideline in the axis color. If you move the cursor over a plane, the plane highlights in gold, as shown in Figure 1–48.

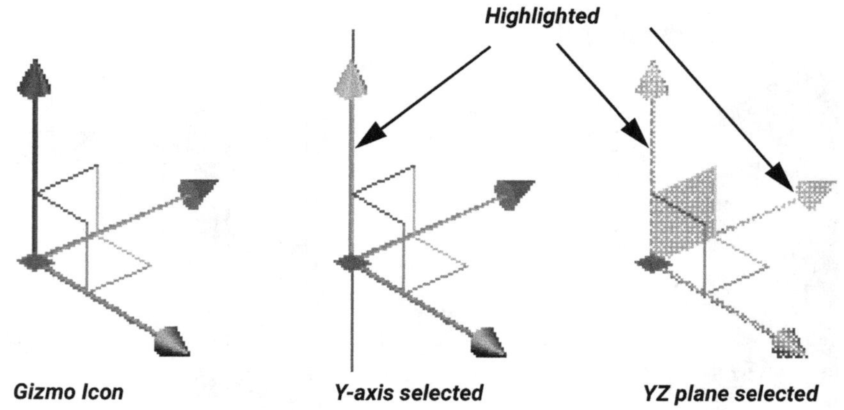

Highlighted

Gizmo Icon   **Y-axis selected**   **YZ plane selected**

Figure 1–48

How To: Use the 3D Gizmo

1. In a 3D view, select the objects you want to modify.

2. Move the Gizmo tool to a base point, if required.

3. Select an axis or plane on the Gizmo tool.

4. You can right-click on the Gizmo to display a menu and pick move, rotate, or scale.

5. When a Gizmo operation is in progress and you need to switch to a different Gizmo, press <Spacebar> or <Enter> to cycle through the options.

   - To move an object, move the cursor along the axis or in the plane. You can type coordinates or select a point to end the command.

   - To rotate an object, move the cursor along the axis. Type an angle value or select a point to finish the command.

   - To scale an object, select the center triangular plane between the three axes in the Gizmo tool, and move the cursor towards or away from the center of the Gizmo.

Practice 1b
Helpful AutoCAD Tools

Practice Objectives

- Use Object Snap, Ortho, and other AutoCAD drawing aids.
- Use grips and the 3D Gizmo to copy, move, and modify objects.
- Understand the difference between the AutoCAD Break command and the Autodesk Advance Steel Split Beam command.

In this practice, you will use grips and Ortho to shorten grid lines and lengthen beams. You will use the AutoCAD Copy command and Node object snap to copy columns and beams. You will then test the difference between the AutoCAD Break command and the Autodesk Advance Steel Split Beam command. Finally, you will use the 3D gizmo to copy beams in 3D and then use grips and node snaps to change the height on one end of the beams. The final model is shown in Figure 1−49.

Figure 1−49

Task 1: Use grips and object snaps to modify and copy objects.

1. In the practice files folder, open **Platform-Simple.dwg**.

2. In the Status Bar, ensure that ⌐ (Ortho) is toggled on and ▯ (Object Snaps) are toggled off.

3. Select the AB Grid object. Click on the grip for Grid A and move it in **30'-0"**, as shown in Figure 1–50.

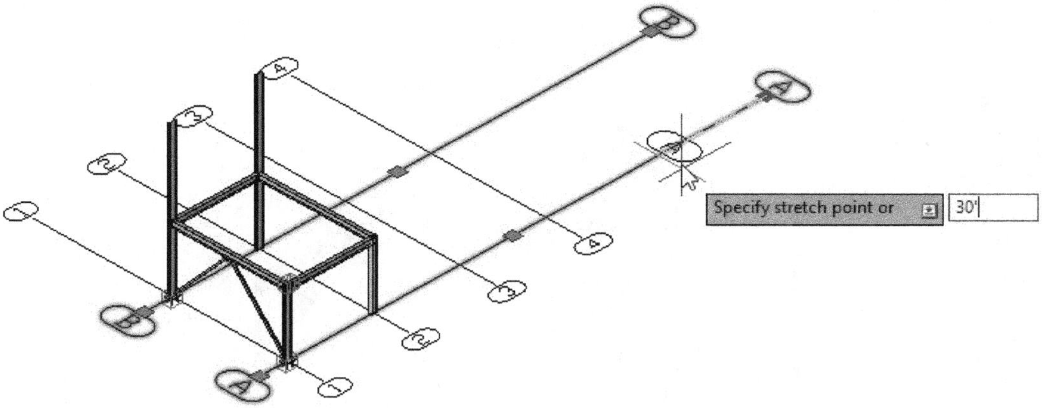

Figure 1–50

4. Repeat the process with Grid B.

5. Move Grids 1-4 in **20'-0"**, as shown in Figure 1–51.

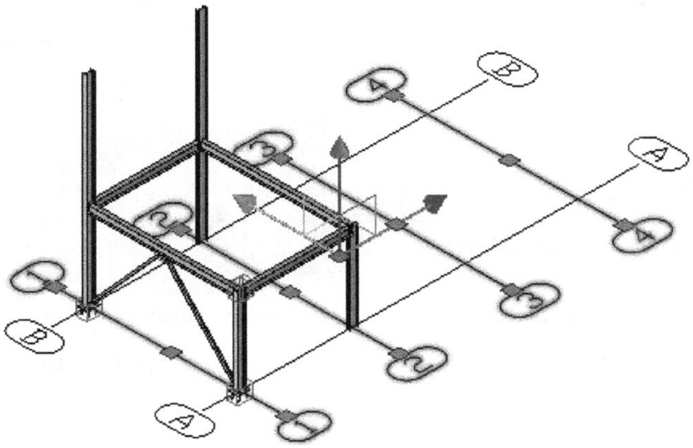

Figure 1–51

6. Toggle on Object Snaps and set the Object Snaps to **Node and GRID Intersection Points**. Ensure that no other object snaps are on.

7. In the *Advance Tool Palette>* ⬜ (Modify) category, click 🔲 (Copy). Select the columns and beam on Grid 2 and copy them to Grids 3 and 4 using the node object snap at the end of the grid lines.

8. Use grips to extend the beams along Grid A and B to Grid 4, as shown in Figure 1–52.

   • You can use the Node object snap for the beam on Grid A with the short columns, but you need to use the Perpendicular object snap override with the taller column on Grid B.

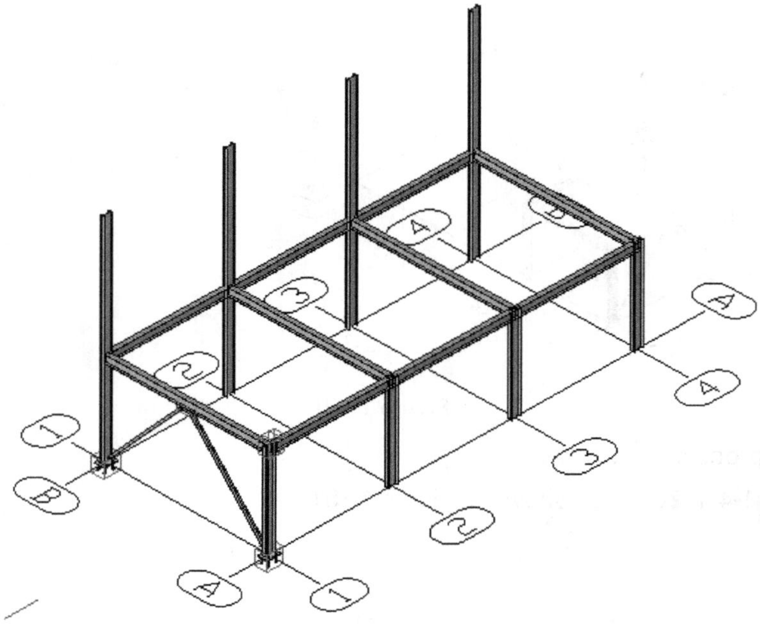

Figure 1–52

9. Save the drawing.

Task 2: Test Break vs. Split Beam.

1. Extend the Command Line so that at least 3 historical prompts display.

2. Zoom in on the top of Column A2 and select the beam. Note that the beam is all one piece.

3. In the *Advance Tool Palette>* ⬜ (Modify) category, click 🔲 (Break).

4. Select the beam. In the Command Line, note that the object cannot be broken, as shown in Figure 1−53. Press <Esc>.

```
X    Command:
     Command: _Break
     Select object: Object can't be broken
     🗂▾ BREAK Select object:                                    ▲
```

Figure 1−53

5. In the *Objects* tab>*Beams* panel, click 🪚 (Split Beams).

6. Select the beam and press <Enter>.

7. Select the Node object snap at the top of the beam, then pan over to Column A3 and select the Node there as well, and then press <Enter>. The beams are now separated, as shown in Figure 1−54.

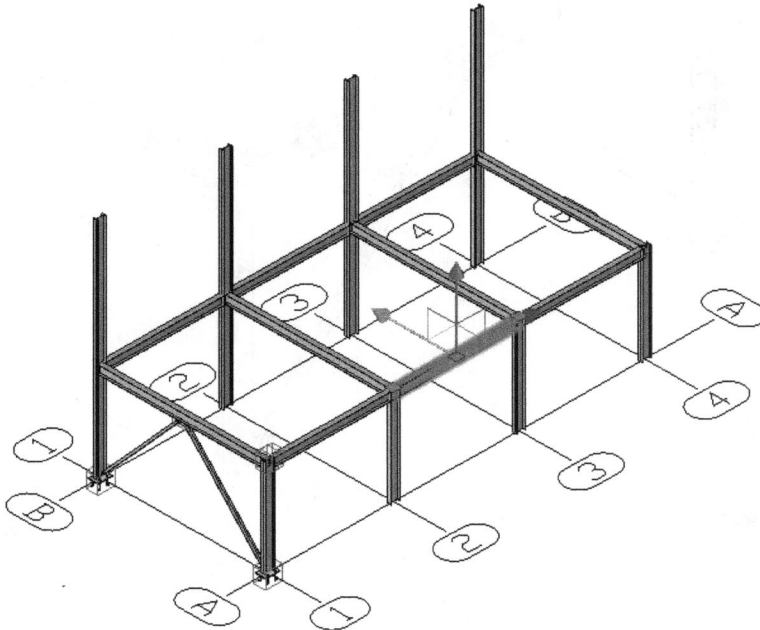

Figure 1−54

8. Repeat the process for the beam on Grid B.

9. Save the drawing.

Task 3: Copy objects in 3D.

1. Select one of the beams. Right-click and select **Select Similar**. Note that all of the beams and braces are selected.

2. Hold <Shift> and clear the selection of the two diagonal braces.

3. Move the cursor over the Z-axis (blue) of the 3D Gizmo and select it.

4. In the Command Line, type **C** (for Copy) and press <Enter> or click the down arrow and choose the **Copy** option, as shown in Figure 1–55.

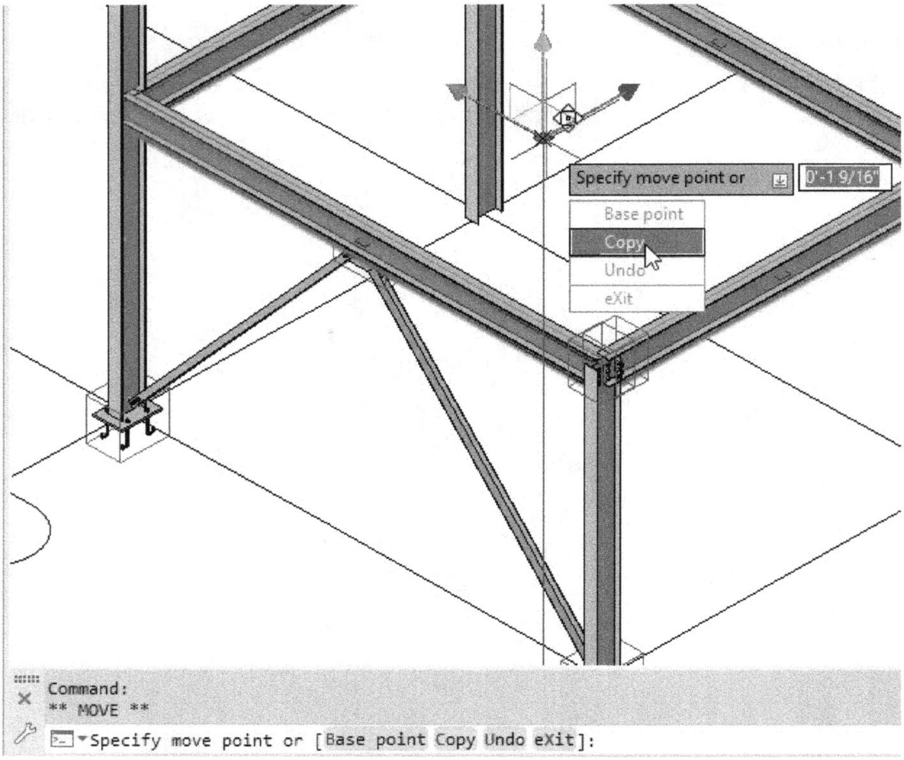

Figure 1–55

Note: Ortho should still be on for this step.

5. Move the cursor up, enter **12'**, and press <Esc> to finish the command, as shown in Figure 1–56.

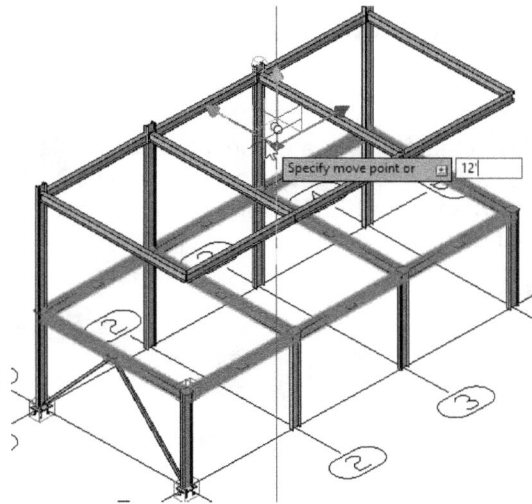

Figure 1–56

6. Use grips to extend the shorter columns up **12'-0"** by selecting the beam, hovering over the top grip, and selecting **Lengthen**. Drag the cursor in the up direction and type **12'**.

7. Use the AutoCAD **Move** command to move the back beams **3'-5 1/4"** up to the top of the columns.

8. Use grips and the Node object snap to angle the columns as shown in Figure 1–57.

Figure 1–57

9. Zoom out to display the entire model.

10. Save the drawing.

End of practice

1.5 Working with the User Coordinate System (UCS)

In the AutoCAD software, 2D objects are created on a single flat plane, which is usually the XY plane. In Autodesk Advance Steel, most of your work occurs in 3D and you need to be able to specify the XY plane (as shown in Figure 1–58) for a number of commands, such as when you draw plates or bracing members.

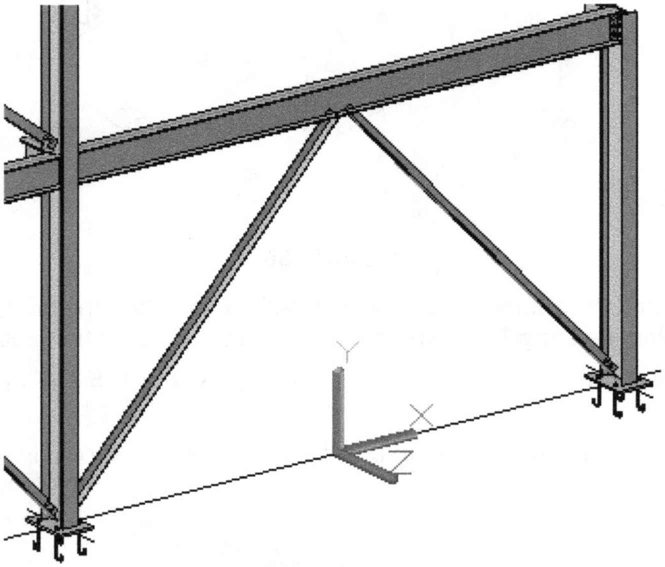

Figure 1–58

- There are three axes: the X-axis, Y-axis, and Z-axis. Three planes are also automatically created by the intersections of these axes. They are the XY plane, the YZ plane, and the XZ plane. Together these three axes and their planes form a user coordinate system, or UCS.

- The UCS is a user-defined working plane with X,Y coordinates that can be positioned at any location or orientation in space.

- Do not confuse the UCS position with the viewing direction. The position from which you view your drawing, known as the viewpoint, determines how you see your drawing. The UCS determines where you are drawing. It sets the position of the working plane.

- The POLAR, OTRACK, and ORTHO commands work with dynamic input in the Z-axis direction.

UCS Commands

UCS commands are found in the *Advance Tool Palette>* ![icon] (UCS) category, as shown in Figure 1–59. Some of these are the same tools as found in AutoCAD, and a few are specific to the Autodesk Advance Steel software.

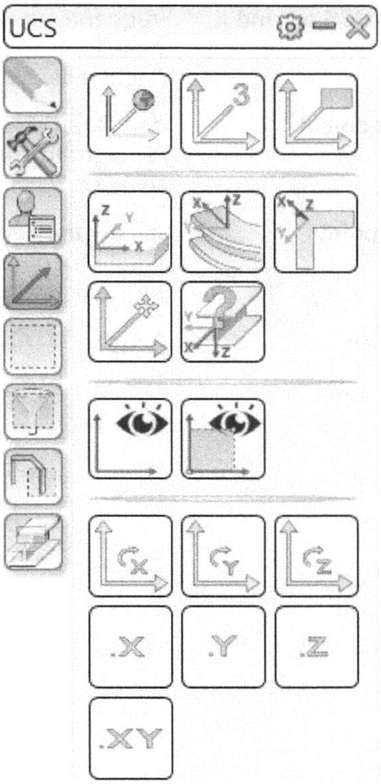

Figure 1–59

- Most of these tools can also be accessed using commands typed into the command line, as shown in Figure 1–60. Type **UCS** and then the method you want to use. **World** is the default for the UCS command.

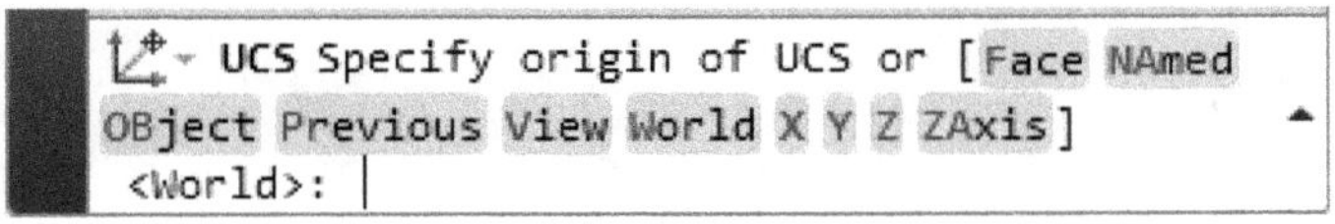

Figure 1–60

Commonly Used UCS Commands

	UCS World	Returns the UCS to the **Home** position at 0,0,0 in the drawing.
	Move UCS	Specify a new origin point. The axes are not modified
	Rotate UCS around X, Y, Z	Frequently used after moving the UCS. Click on these tools to rotate the other axes around the X, Y, or Z axis.
	UCS at object	Identify an object and then click on the line which you want to specify as the Z-axis (Autodesk Advance Steel only).
	UCS 3 points	Specify the new origin point, the positive direction of the X-axis, and the positive portion of the Y axis.
	UCS View	Orients the UCS to the current view.

1.6 Using the Autodesk Advance Steel Modify Commands

While you can use the standard AutoCAD modify commands for individual objects in the Autodesk Advance Steel software, there are times when these tools are not going to work, as shown in Figure 1–61. This is especially true once you start working with connections or if you want to copy features that have been cut into columns. In these cases, it is best to use the

Advance Copy command found in the *Advance Tool Palette>* ⚒ (Tools) category, as shown in Figure 1–62. There is also a specific tool to Trim or Extend Autodesk Advance Steel objects, such as beams and columns.

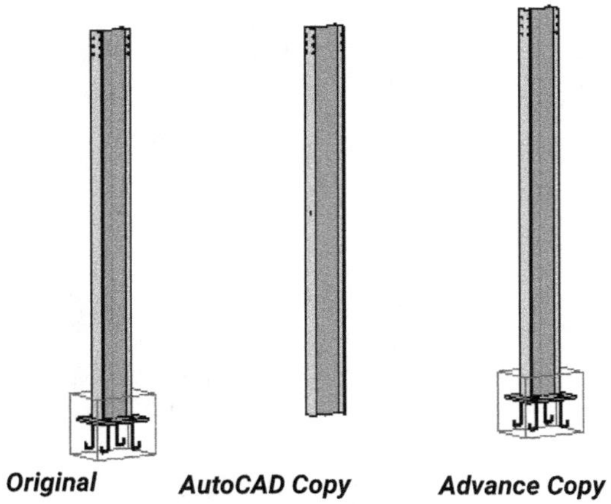

| Original | AutoCAD Copy | Advance Copy |

Figure 1–61

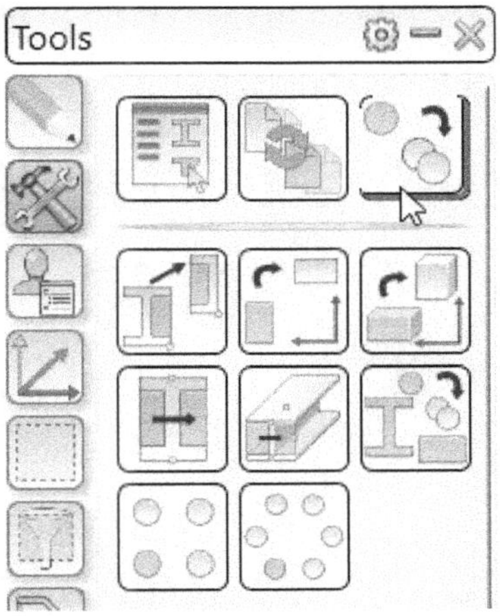

Figure 1–62

- The **Advance Copy** command includes options for **Copy**, **Move**, **Array**, **Polar array**, **Mirror**, **Rotate**, **Align**, and **Adapt**. You can also access some of these options directly using the other buttons in the *Tools* category.

How To: Transform Elements Using Advance Copy

1. In the *Advance Tool Palette>* ⚒ (Tools) category, click ◔ (Advance Copy).

2. In the *Transform elements* dialog box (shown in Figure 1–63), click ⬚ (Select objects).

3. In the drawing window, select the base objects (such as columns or beams) you want to use, and press <Enter> to return to the dialog box.

   - If you want to include the relationships between connected objects (such as a base plate on the bottom of a column), select **Include additional connections**, as shown in Figure 1–63.

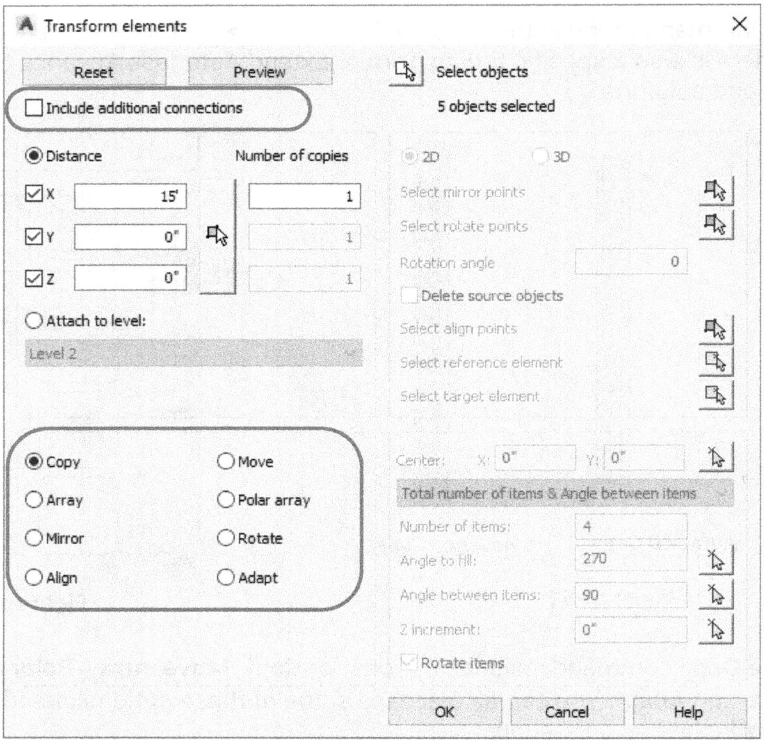

Figure 1–63

4. Select the type of command you want to use, as shown in the red box in Figure 1–63, above.

5. Depending on the type of command you select, various parts of the dialog box are available.

   Click ⬚ to specify distances or other points.

6. Click **Preview**.

7. Review the model to ensure that everything is placed as expected. When you are finished, in the *Preview* dialog box, click **OK**. If you need to make changes, click **Modify** to return to the dialog box or **Cancel** to exit the command.

- It is safer to preview the function, rather than just hoping that it works. Remember that commands in the Autodesk Advance Steel software are often macros of many commands, and trying to undo an operation might take many steps.

Using Advance Trim/Extend

The standard AutoCAD **Trim** and **Extend** commands do not work with Autodesk Advance Steel objects, but you can use the **Advance Trim/Extend** command, as shown in Figure 1–64.

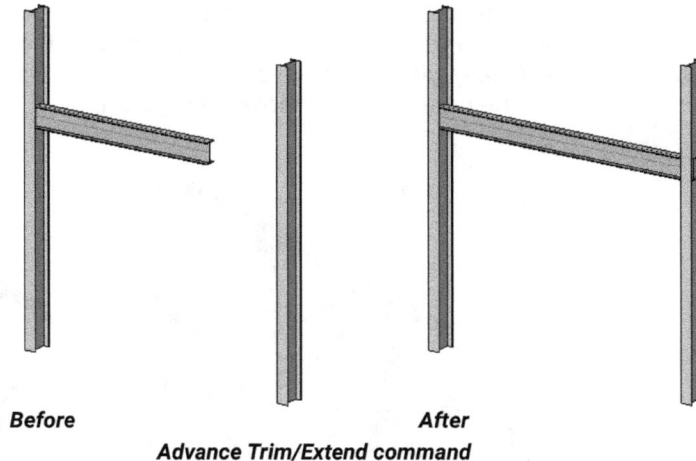

Before   *After*
Advance Trim/Extend command

Figure 1–64

How To: Trim or Extend Autodesk Advance Steel Objects

1. In the *Advance Tool Palette>* (Tools) category, scroll down and select (Advance Trim/Extend).

2. Select the **Trim**, **Extend**, or **Auto** operation mode (the default is **Auto**).

3. Select one of the following *Select* options:

   - **System:** The system line of a beam. This is the default.
   - **Center:** The center line of a beam.
   - **Face:** The face of a beam.
   - **Line:** An AutoCAD object.

4. Select the boundary object and press <Enter>

5. Select the object to be trimmed or extended.

6. Continue selecting other objects, or press <Enter> to finish the command.

Practice 1c
Use the Autodesk Advance Steel Modify Commands

Practice Objective

- Use Advance Copy to copy, move, and mirror Autodesk Advance Steel objects.

In this practice, you will use the **Advance Copy** command to copy and mirror columns, beams, and braces, with their connections. You will copy beams in 3D space and then extend columns to meet the new beam locations. Finally, you will move beams to a new location and use grips to modify the connecting beams. The final model is shown in Figure 1–65.

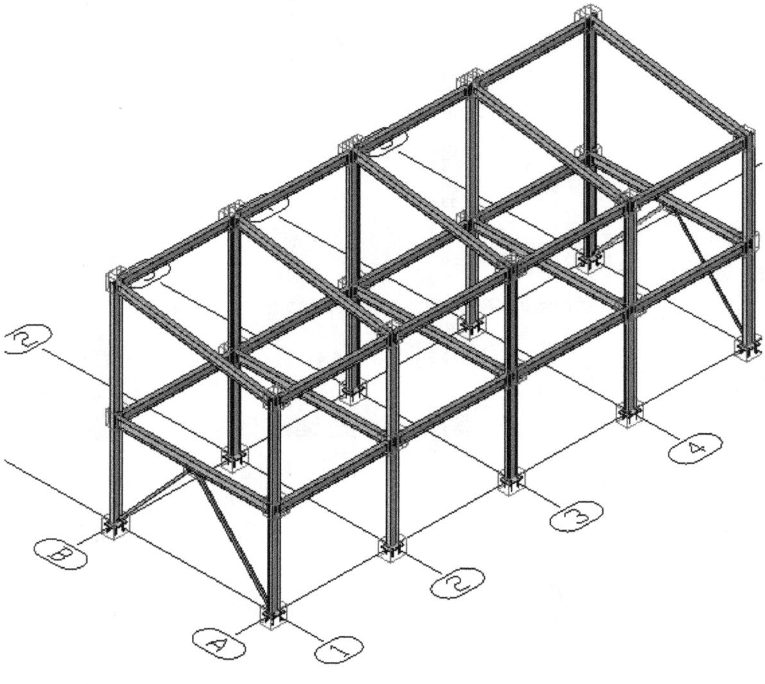

Figure 1–65

Task 1: Use Advance Copy to copy and mirror objects and their connections.

1. In the practice files folder, open **Platform-Modify.dwg**.

2. Investigate the existing objects in the model. Note that they all have connections where they intersect, as shown in Figure 1–66.

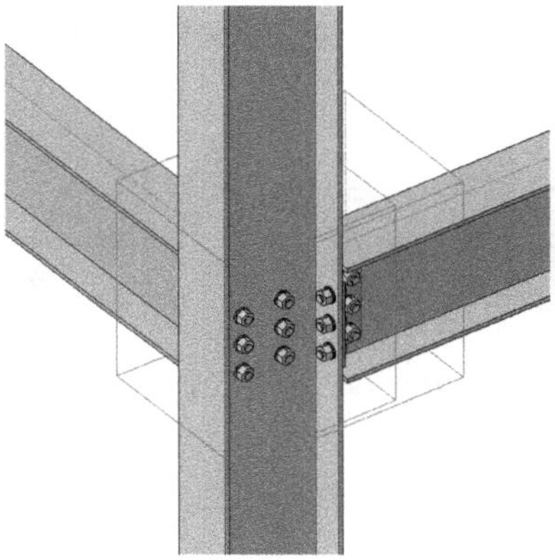

Figure 1–66

3. In the *Advance Tool Palette*> (Tools) category, click (Advance Copy).

4. In the *Transform elements* dialog box, click (Select objects).

5. In the drawing window, select the two columns on Grid 2 and the three connecting beams, as shown in Figure 1–67, and press <Enter>.

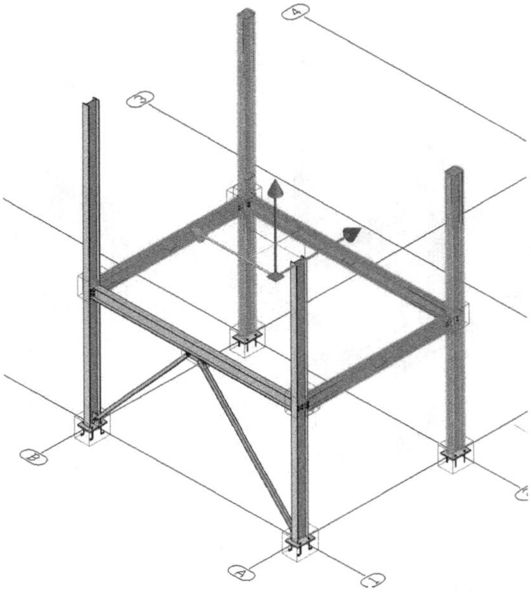

Figure 1–67

6. In the *Transform elements* dialog box, select **Include additional connections**.

7. Ensure that the **Copy** option is selected.

8. In the *Distance* area, set the *X* distance to **15'** and the *Number of copies* to **2**, as shown in Figure 1−68.

Figure 1−68

9. Click **Preview**. The copies and appropriate connections display as shown in Figure 1−69.

Figure 1−69

10. In the *Preview* dialog box, click **OK**.

11. Start the **Advance Copy** command again.

12. In the dialog box, click ![icon] (Select objects) and select the columns, beams, and braces that are highlighted in Figure 1–70.

13. In the dialog box, select the **Mirror** option. In the *Mirror* area, ensure that **2D** is selected, and then click ![icon] (Select mirror points).

14. Select the two end node points of Grid 3.

15. In the dialog box, click **Preview**. The mirrored objects should display as shown in Figure 1–70. Click **OK**.

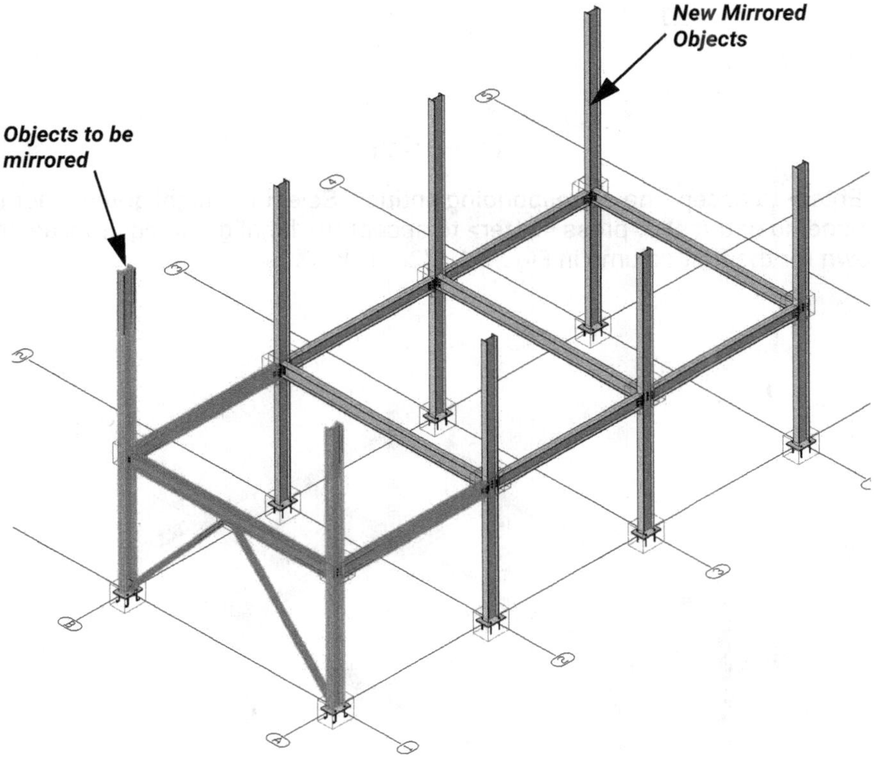

Figure 1–70

16. Save the drawing.

Task 2: Copy and extend objects in 3D space.

1. Start the **Advance Copy** command.
2. Select all of the beams in the model, but not the braces and columns.
3. Set the Z distance to **12'-0"** (as shown in Figure 1–71) and click **Preview**.

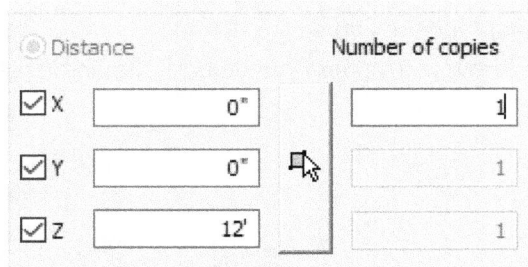

⊙ Distance		Number of copies	
☑ X	0"		1
☑ Y	0"		1
☑ Z	12'		1

Figure 1–71

4. Press <Enter> to accept the corresponding entities. Select the highlighted columns (corresponding entities) or press <Enter> to accept the highlighted column, as prompted and shown for the first column in Figure 1–72. Click **OK**.

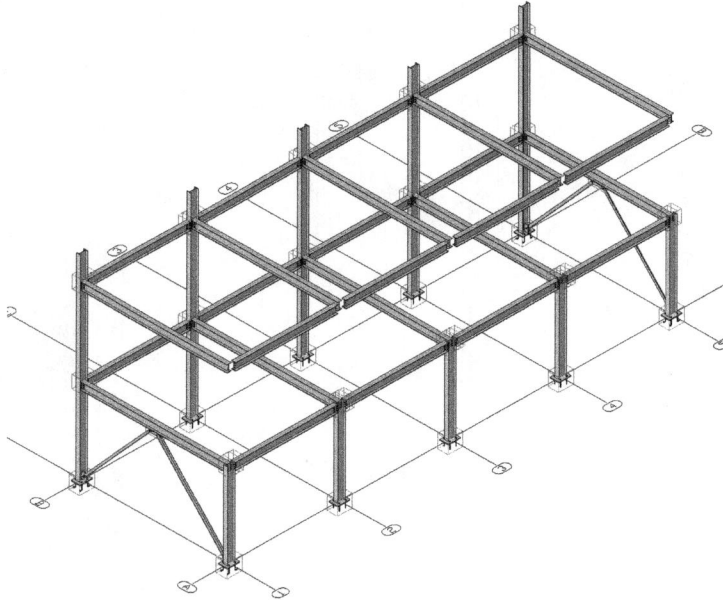

Figure 1–72

5. In the *Advance Tool Palette>* (Tools) category, scroll down and click (Advance Trim/Extend).

6. At the *Please select operation mode* prompt, select **Auto**.

7. At the *Please select option* prompt, select **System**.

8. Select the beams along Grid A on the second level. Press <Enter>, select the corresponding columns, and then press <Enter>. The columns are extended, as shown in Figure 1−73.

Figure 1−73

9. Press <Esc> to finish the command.

10. Use **Advance Copy** to move the top beams along Grid B in the Z direction 3' 5 1/4" to the top of 16'-0" columns.

11. Use the grips to modify the crossing beams so that they are at an angle, as shown in Figure 1–74.

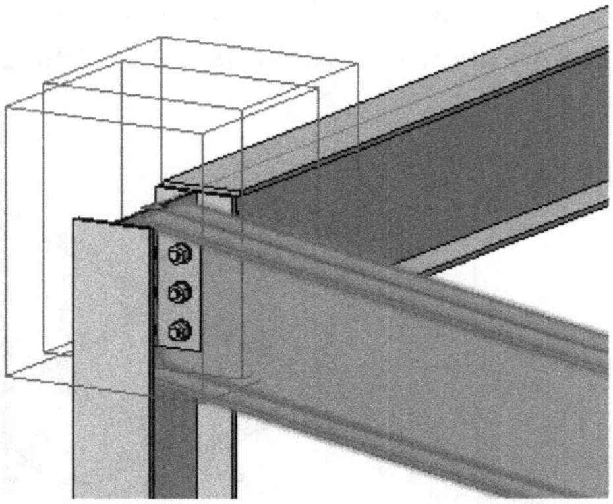

Figure 1–74

12. Save the drawing.

End of practice

Chapter Review Questions

1. Which of the following describes the Autodesk Advance Steel workflow?

   a. Start from an existing project and draw in 2D and 3D.

   b. Start from a template, draw the objects in 3D, and automatically create 2D drawings from the model.

   c. Start from an existing 2D drawing and extrude 3D elements.

   d. Start from a template and create 2D working documents.

2. When you are working in a 3D view, you need to change to a 2D view before you can add any objects.

   a. True

   b. False

3. Where do you find Autodesk Advance Steel specific tools in the User Interface? (Select all that apply.)

   a. Ribbon

   b. Advance Tool Palette

   c. Quick Access Toolbar

   d. Shortcut menu

4. In AutoCAD you can use many different object snaps as drawing aids. Which one of the following object snaps do you typically leave active in the Autodesk Advance Steel software?

   a. Endpoint

   b. Midpoint

   c. Node

   d. Center

5. Which of the following commands duplicates both the selected objects and any connections (such as clip angles)?

   a. Copy

   b. Advance Copy

   c. Array

   d. Advance Trim/Extend

Command Summary

Button	Command	Location
General Commands		
	Advance Copy	• **Advance Tool Palette:** *Tools* category
	Advance Steel Properties	• **Advance Tool Palette:** *Tools* category • **Double-click:** On an Advance Steel object • **Shortcut Menu:** Select objects, Advance Properties
	Advance Steel Tool Palette	• **Ribbon:** *Home* tab>*Extend Modeling* panel
	Advance Trim/Extend	• **Advance Tool Palette:** *Tools* category
	All Visible	• **Advance Tool Palette:** *Quick views* category
	Break	• **Advance Tool Palette:** *Modify* category
	Copy	• **Advance Tool Palette:** *Modify* category
	Open	• **Quick Access Toolbar** • **Application Menu:** Open>Drawing
	Project Explorer	• **Ribbon:** *Home* tab>*Project* panel
	Selected Objects off	• **Advance Tool Palette:** *Quick views* category
	Split Beam	• **Ribbon:** *Objects* tab>*Beam Tools* panel
	ViewCube	• **Ribbon:** *View* tab>*Viewport Tools* panel
UCS Commands		
	UCS Icon	• **Ribbon:** *View* tab>*Viewport Tools* panel
	UCS World	• **Advance Tool Palette:** *UCS* category

Button	Command	Location
	Move UCS	• **Advance Tool Palette:** *UCS* category
	Rotate UCS around X, Y, Z	• **Advance Tool Palette:** *UCS* category
	UCS at object	• **Advance Tool Palette:** *UCS* category
	UCS 3 points	• **Advance Tool Palette:** *UCS* category
	UCS View	• **Advance Tool Palette:** *UCS* category

Building Models

Building a 3D model is the first task in the Autodesk® Advance Steel software, before you can move to applying connections and creating 2D fabrication drawings. By starting a project based on the Autodesk Advance Steel template, you are prepared to add columns, beams, and bracing using the tools and properties that show the power and precision of the software. If you have or are working with users of the Autodesk® Revit® software, you can import steel building models into the Autodesk Advance Steel software to save you from having to re-enter work that has already been done.

Learning Objectives

- Start Autodesk Advance Steel projects based on a template.
- Specify project information.
- Place and modify structural grid groups.
- Create levels for object placement and viewing.
- Add columns and beams with *Advance Properties* modifications.
- Add bracing.
- Import and export Autodesk Revit models.

2.1 Starting Autodesk Advance Steel Projects

Autodesk Advance Steel projects need to be started using one of the supplied templates, or using a custom template built on the original templates. These templates include many default element presets and properties that are used when you add columns (shown in Figure 2–1), connections, and other objects, and settings to automatically organize elements into specific layers.

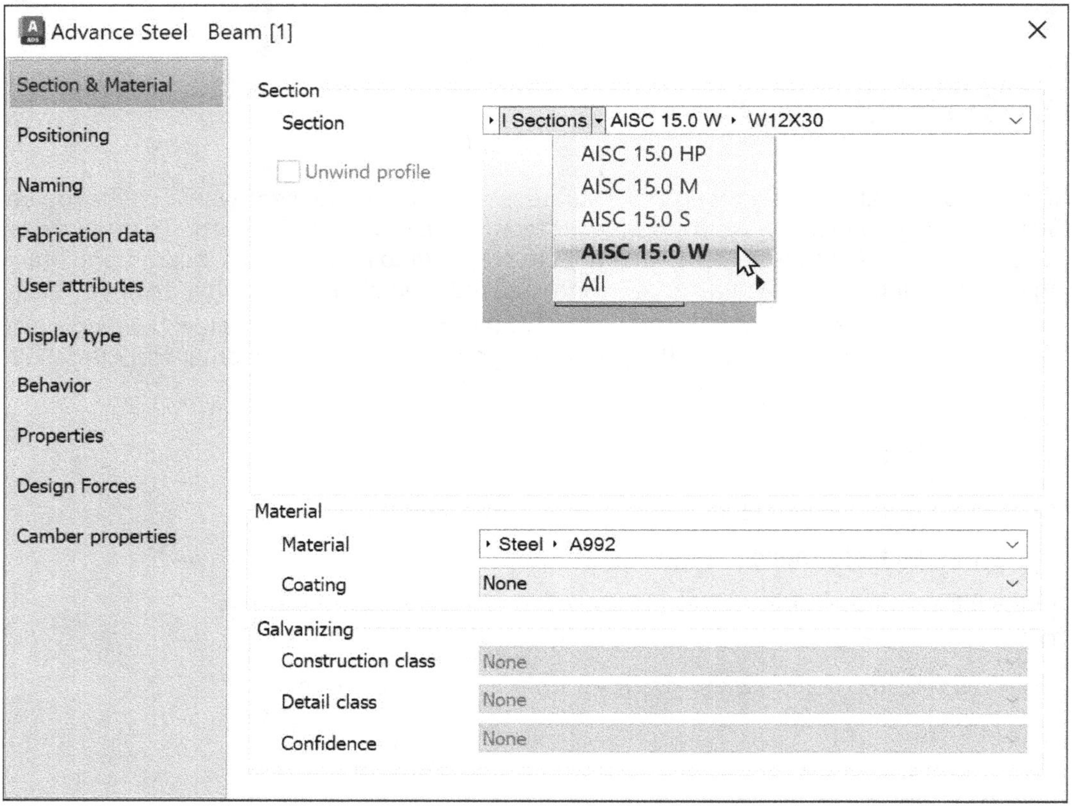

Figure 2–1

- As soon as you start a new project, specify the Project Settings for information such as the name and number of the project.

How To: Start an Autodesk Advance Steel Project

1. In the Quick Access Toolbar or Application Menu, click ⬜ (New).

2. In the *Select template* dialog box, select the appropriate template (as shown in Figure 2–2) and click **Open**.

Figure 2–2

3. The new drawing is opened in the default **SE Isometric** view, ready for you to start modeling in 3D.

4. In the *Home* tab>*Settings* panel, click ⬛ (Project settings).

5. In the *Project data* dialog box, fill in the *Project Info* (as shown in Figure 2–3) and click **OK**.

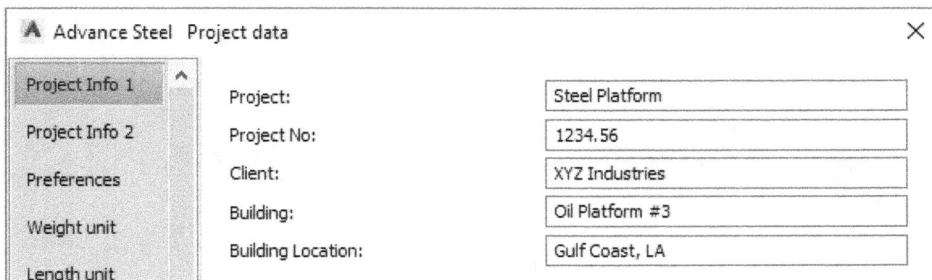

Figure 2–3

6. Save the drawing to the top level folder where you want the rest of the project files to be stored.

2.2 Adding Structural Grids

Structural grids are the underlying basis for most steel structures. Laying out a building grid helps you to accurately locate the columns and other objects used in a model. Grids in the Autodesk Advance Steel software are two separate axis groups on the X- and Y-axis, as shown in Figure 2–4.

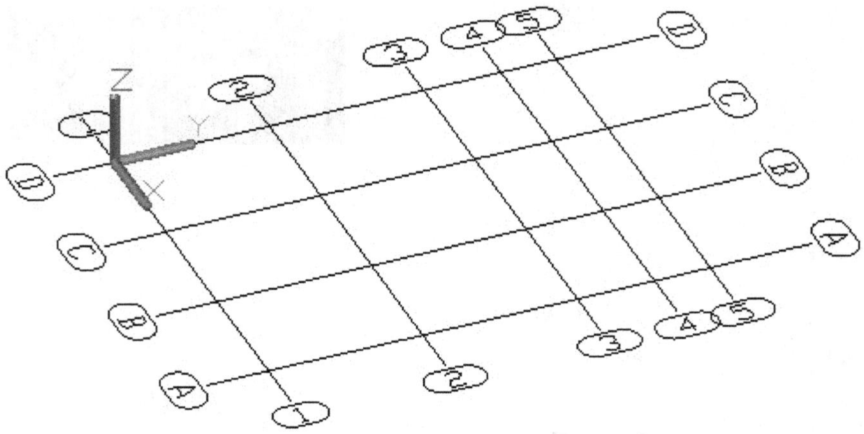

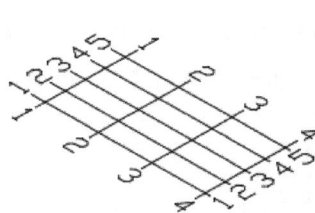

Figure 2–4

• Grids are created in the current UCS, and are automatically placed on the correct layer.

• If you zoom in or out and the grid numbers do not display as expected (as shown in Figure 2–5), type **RE** (Regenerate the model) to update the view, as shown in Figure 2–6.

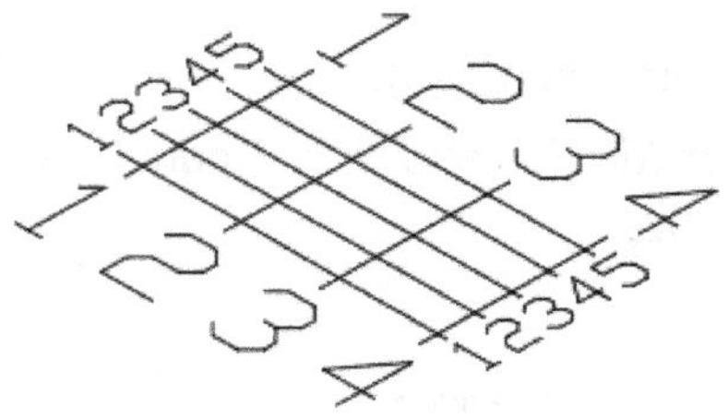

Figure 2–5   **Figure 2–6**

How To: Add a Rectangular Building Grid

1. In the *Home* tab>*Objects* panel or the *Objects* tab>*Grid* panel, click ⊞ (Building Grid).
2. Pick two diagonal points that define the extents of the grid.
    *   Select the origin first and then the second point.
    *   You can type X,Y coordinates separated by a comma.
    *   If you are working in feet and inches, ensure that you include the foot mark (') in the distance.
3. Two grid groups are created with four equal axes in each direction, as shown in Figure 2–7.

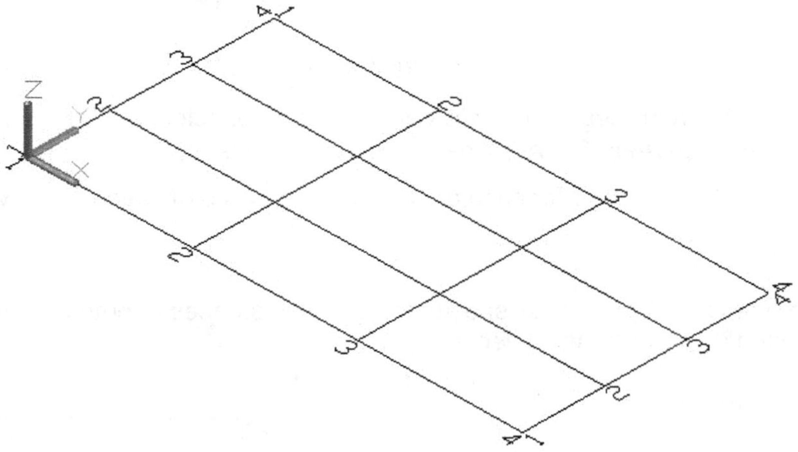

Figure 2–7

4. Modify the grid properties as required.

Modifying Grid Properties

Grid properties enable you to set the overall size and distances of the grid, display labels, add secondary axes, and set the display and grip options of the axes. You can use grips on the ends of the axes to change the length of the grid lines, as shown in Figure 2–8.

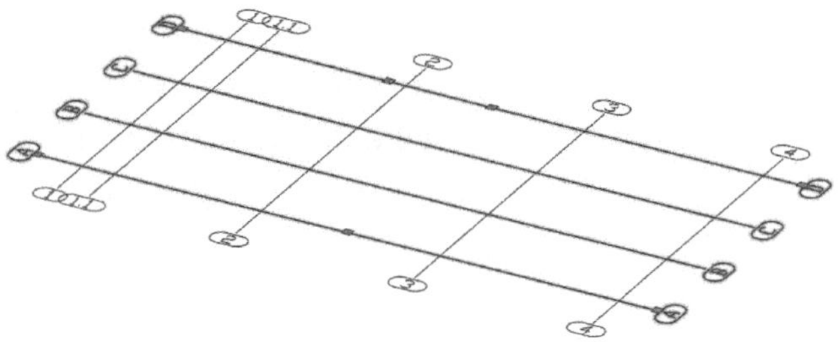

Figure 2–8

- To modify the grid properties, double-click on a grid axis, or select a grid axis, right-click and select **Advance Properties**. The *Axes, parallel* dialog box displays.

- The same dialog box displays for each direction, so be aware of which group you select.

Total Tab

In the *Total* tab (shown in Figure 2–9), specify the overall distances of both groups of axes and how the numbering displays for the selected axes.

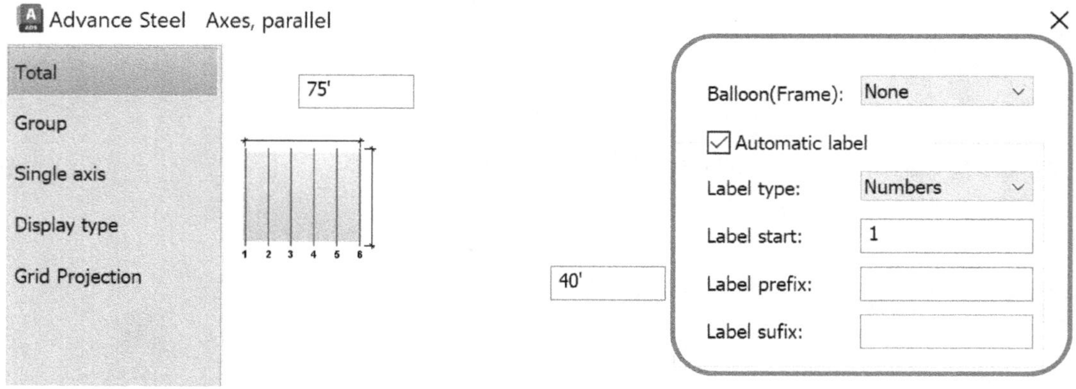

Figure 2–9

- **Balloon (Frame):** Select either **None** or **Edging**, which adds an oval box around the numbers.

- Select **Automatic label** to update the selected group of axes. Then specify the Label type (**Numbers**, **Capital letters**, or **Small letters**), *Label start*, *Label prefix*, and *Label suffix*, as required.

Group Tab

In the *Group* tab (shown in Figure 2–10), specify the *Number* of axes and the *Distance (D)* between axes for the overall group. Changing one automatically changes the other based on the overall size of the grid.

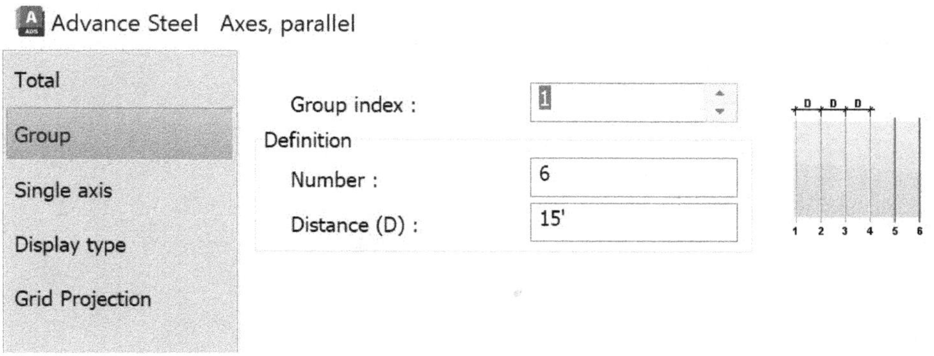

Figure 2–10

- The selected group is highlighted in the drawing.

- If there is more than one group in the selection, you can select the group you want to modify using the *Group index* option.

Single Axis Tab

In the *Single Axis* tab (shown in Figure 2–11), make modifications to the grid axes independently.

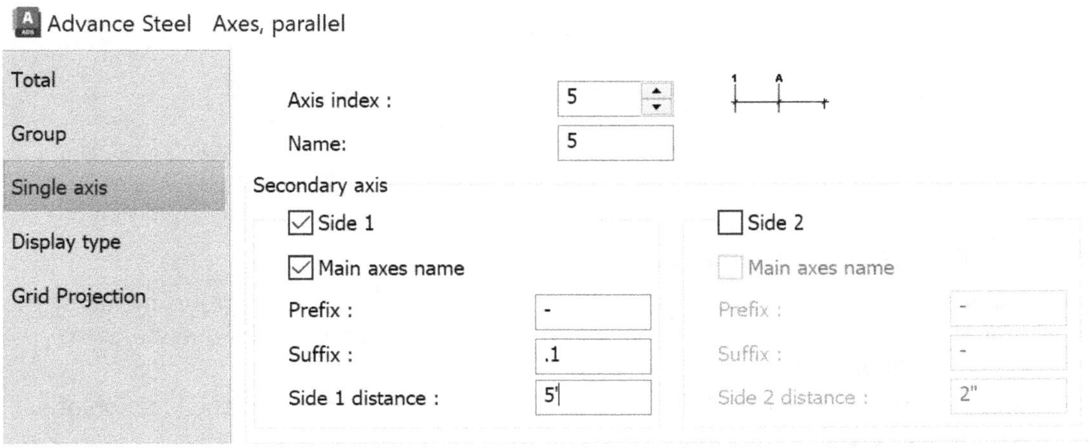

Figure 2–11

- **Axis Index:** Select the axis you want to modify. The selected axis is highlighted in red.
- **Name:** Modify the name of the selected axis if automatic labeling is turned off.
- **Secondary axis:** Add a secondary axis to either side of the selected axis and specify a *Prefix*, *Suffix*, and distance, as shown in Figure 2–11 and Figure 2–12.

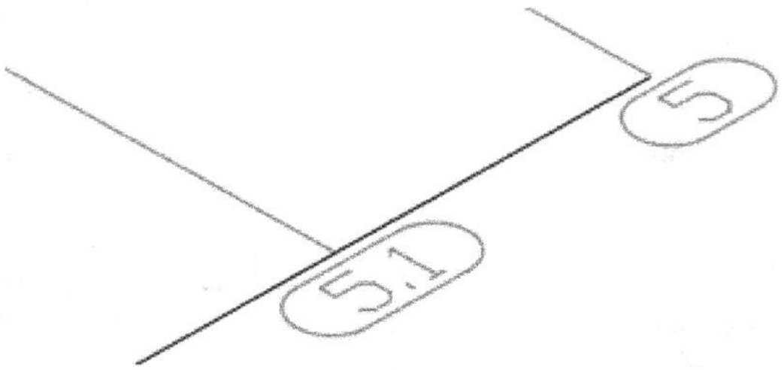

Figure 2–12

Display Type Tab

In the *Display type* tab (shown in Figure 2–13), select how you want the grids and their grips to display.

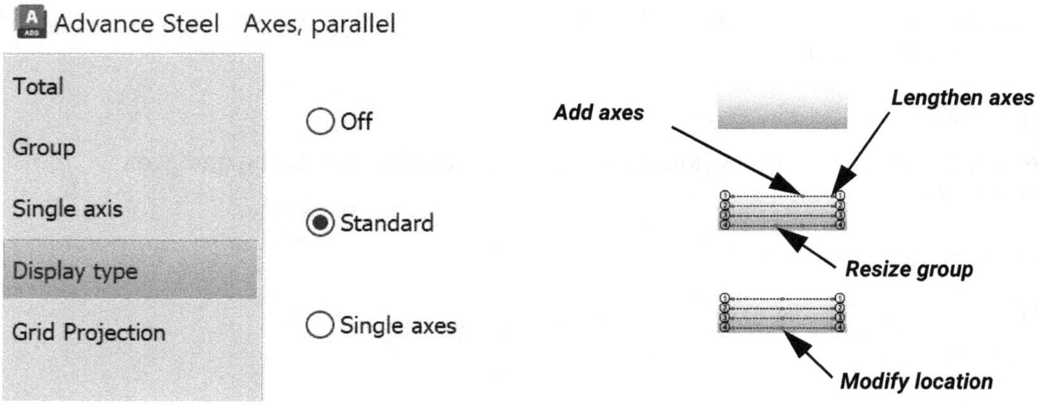

Figure 2–13

- **Off:** Toggles off the selected set of grids.
- **Standard:** Toggles on grips that enable you to modify the selected group, including the length of the axes, the width of the entire group, and add more axes outside of the original group.
- **Single axes:** Toggles on grips that enable you to modify the length and location of each axis separately.

Grid Projection Tab

- In the *Grid Projection* tab, select the Project grid vertically for mode views and drawings check box (as shown in Figure 2–14) to display grids visible in an activated model view and in drawings for multistory buildings.

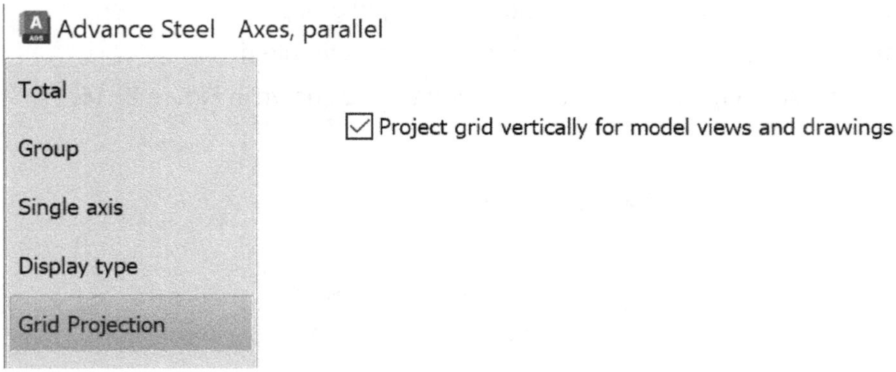

Figure 2–14

Additional Grid Tools

Structural Grids are often created with a diverse set of distances and might include angles and curves, as shown in Figure 2–15. Several other grid tools are available on the *Objects* tab to create and modify grid axes and grid groups.

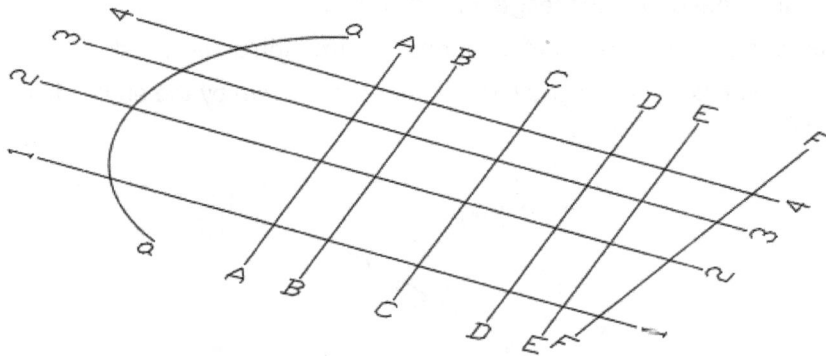

Figure 2–15

- To create an odd sized grid, it is easiest to add individual grid lines without creating a grid group. If you do this, you must update the numbering manually.

How To: Add a Grid with 4 Axes

1. In the *Objects* tab>*Grid* panel, click ⫿⫿ (Grid with 4 axes).
2. Pick two points that define the length of the grid axis.
3. Pick a point that defines the direction and length of the grid group.
4. In the *Axes, parallel* dialog box, set additional information as required.
5. Close the dialog box. A group of four grid lines is created, as shown in Figure 2–16.

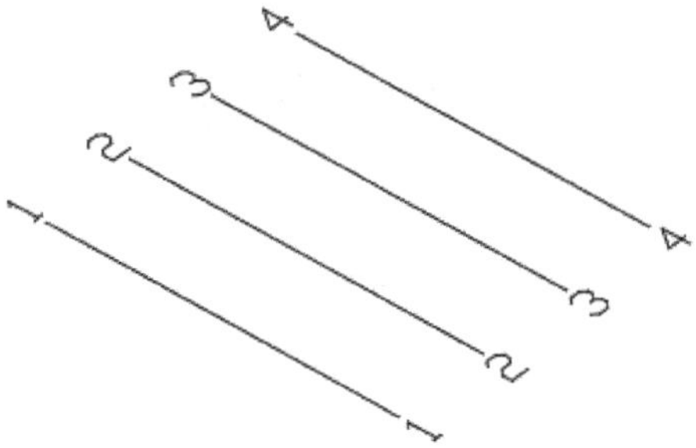

Figure 2–16

How To: Add a Single Axis

1. In the *Objects* tab>*Grid* panel, click ⫿⫿ (Single axis).
2. Pick two points that define the length of the grid line.
3. In the *Axes, parallel* dialog box, set additional information as required.
4. Close the dialog box. A single grid line is created as shown by the angled line in Figure 2–17.

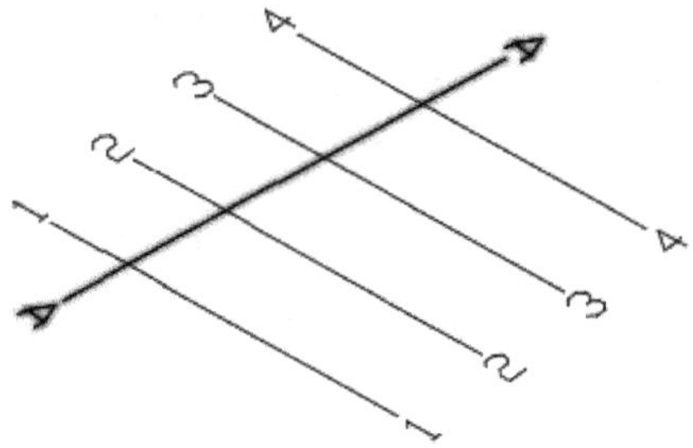

Figure 2–17

• Single grid axes are not connected with any grid group.

How To: Add a Curved Axis

1. In the *Objects* tab>*Grid* panel, click ⌒ (Curved grid with single axis).
2. Pick a start point and end point for the curved grid.
3. Pick a point on the edge of the arc.
4. In the *Axes, parallel* dialog box modify the radius of the arc (as shown in Figure 2–18) and set additional information as required.
5. Close the dialog box. A single curved grid line is created, as shown in Figure 2–19.

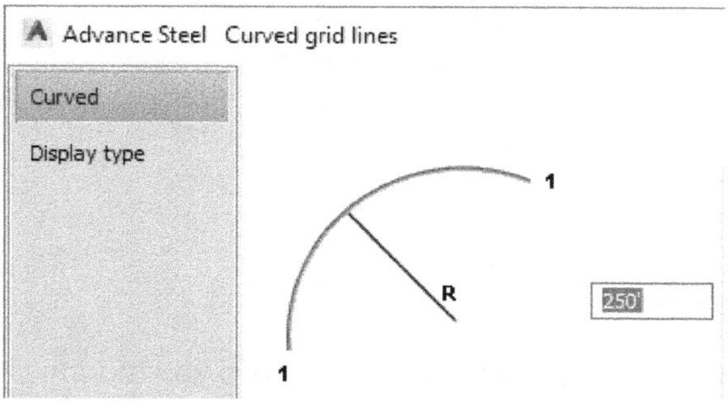

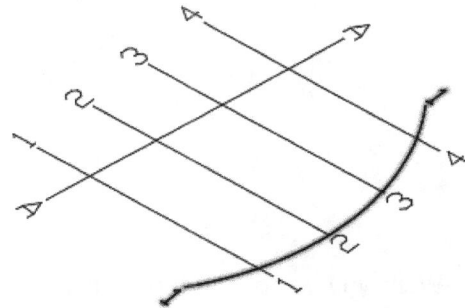

Figure 2–18

Figure 2–19

How To: Add Grid Axes to a Group

1. In the *Objects* tab>*Grid* panel, click ⊞ (Add axes).
2. Select the grid axis after which the group is to be inserted and press <Enter>. Note that you can select more than one axis if you want the added axes in more than one location.
3. Enter the number of axes. You can add one or more evenly spaced axes.
4. Enter or select a point to define the distance between axes.
5. The new axes are added and automatically incremented based on the first axis selected.

- You can create a group of variable distance grid lines by repeating the process several times with different distances. For example, start with a single grid line. Then, use **Add Axes** to add one grid at a time with the different distances. Repeat the process in the perpendicular direction, as shown in Figure 2–20.

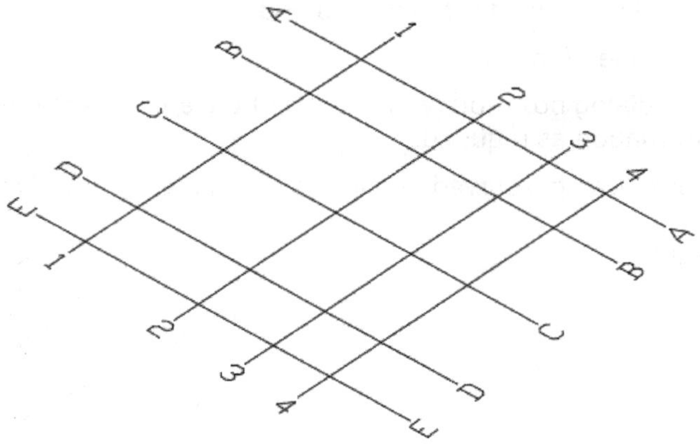

Figure 2–20

- While you do not need to have grids placed in groups, using groups can help you modify numbering and make other changes.

How To: Remove Grid Axes from a Group

1. In the *Objects* tab>*Grid* panel, click ⊞ (Delete axes).
2. Select the grid axes that you want to remove.
3. Press <Enter>. The axes are removed and the remaining axes are renumbered.

How To: Extend or Trim Axes

1. In the *Objects* tab>*Grid* panel, click ⊞ (Extend axes) or ⊠ (Trim axes).

2. Select one or more boundary objects (as shown in Figure 2–21) and press <Enter>.

   - Boundary objects can be other grids or AutoCAD® objects (such as lines, arcs, and polylines).

3. Click on one or more axes that you want to modify, as shown in Figure 2–22.

   - For **Extend**, click on the grid line end closest to the boundary object.

   - For **Trim**, click on the side of the grid line you want to be cut off.

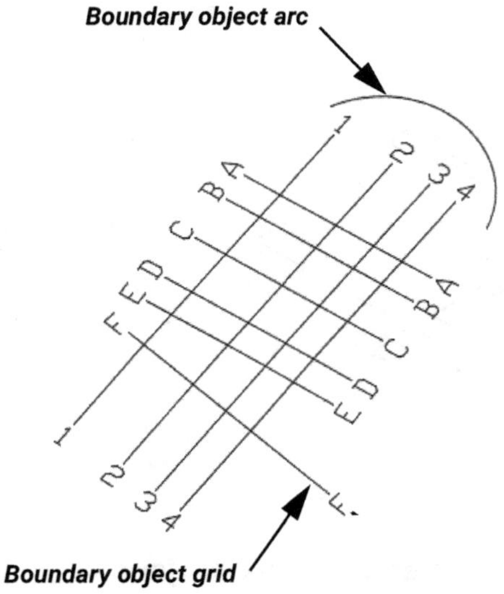

Figure 2–21

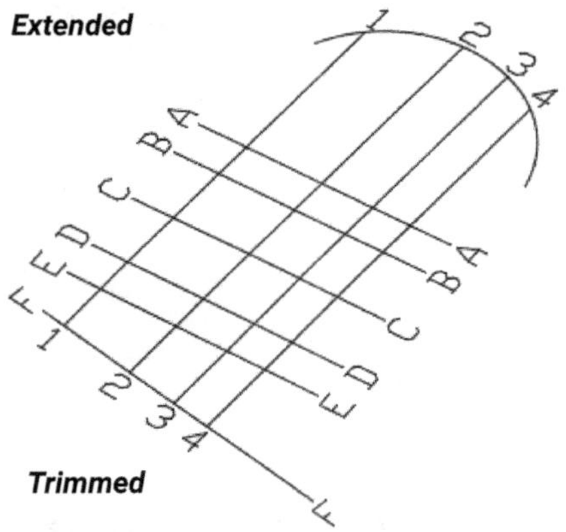

Figure 2–22

Practice 2a
Start a Project and Add Structural Grids

Practice Objectives

- Add a grouped building grid and modify the axes of the group.
- Add an additional grid to the group.

In this practice, you will start a new project based on the Autodesk Advance Steel template. You will set up project information and then use the **Building Grid** and **Add Axes** commands to add a structural grid, as shown in Figure 2–23.

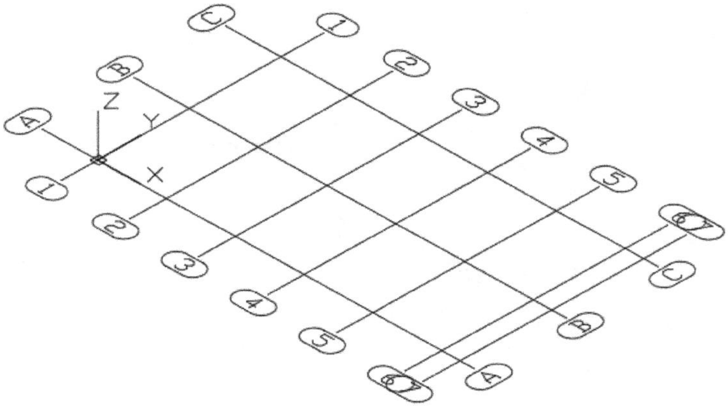

Figure 2–23

Task 1: Start a new Autodesk Advance Steel project and specify project information.

1. In the Quick Access Toolbar click ☐ (New).
2. In the *Select template* dialog box select **ASTemplate.dwt** and click **Open**.
3. Save the drawing to the practice files folder as **Steel Platform**.

4. In the *Home* tab>*Settings* panel, click ▦ (Project Settings).

5. In the *Project data* dialog box, in the *Project Info 1* tab, enter the following information:
   - *Project:* **Steel Platform for XYZ Industries**
   - *Project No:* **1234.56**
   - *Client:* [Your company name]
   - *Building:* **Steel Platform**
   - *Building Location:* [Your home town]
6. In the *Project Info 2* tab, enter any additional information you want to add.
7. Click **OK**.

Task 2: Add a structural grid.

1. In the *Home* tab>*Objects* panel, click ⊞ (Building Grid).
2. For the origin point, enter **0,0**.
3. For the second point, enter **75', 40'**, ensuring that you include the foot marks. The new grid displays as shown in Figure 2–24.

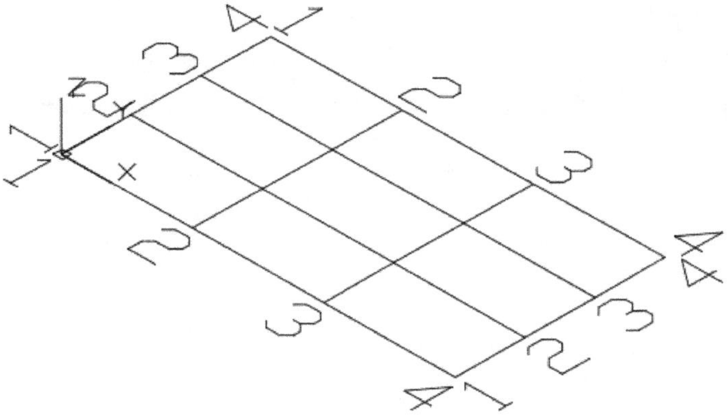

Figure 2–24

4. Double-click on one of the vertical grid lines to open the *Axes, parallel* dialog box.
5. In the *Total* tab, change *Balloon (Frame)* to **Edging**.

6. In the *Group* tab, change *Number* to **6**. Note that the *Distance* automatically updates to **15'**, as shown in Figure 2–25.

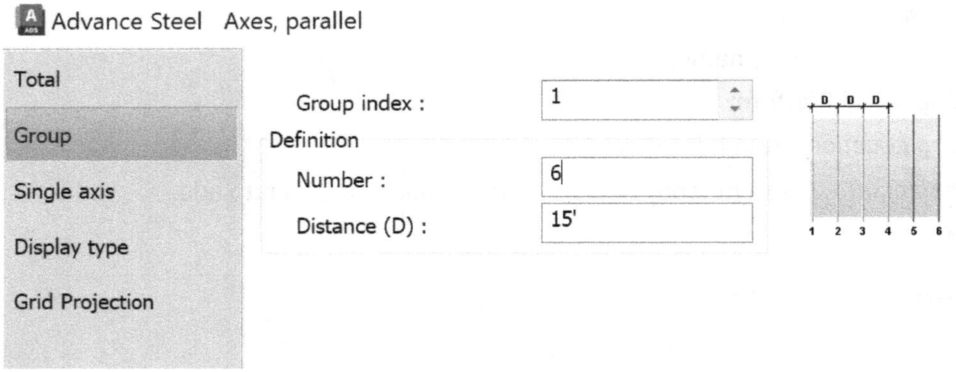

Figure 2–25

7. Close the dialog box.
8. Double-click on a horizontal grid line.
9. In the *Axes, parallel* dialog box, in the *Total* tab, change *Balloon (Frame)* to **Edging** and *Label type* to **Capital letters**.
10. In the *Group* tab, change *Distance* to **20'**. Note that *Number* automatically updates to **3**.
11. Close the dialog box to review the new grid layout, shown in Figure 2–26.

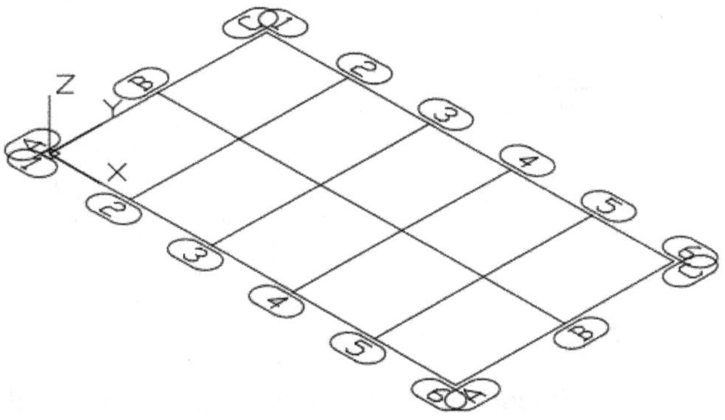

Figure 2–26

12. Select the grid groups and use the endpoint grips to stretch the labels out, as shown in Figure 2-27.

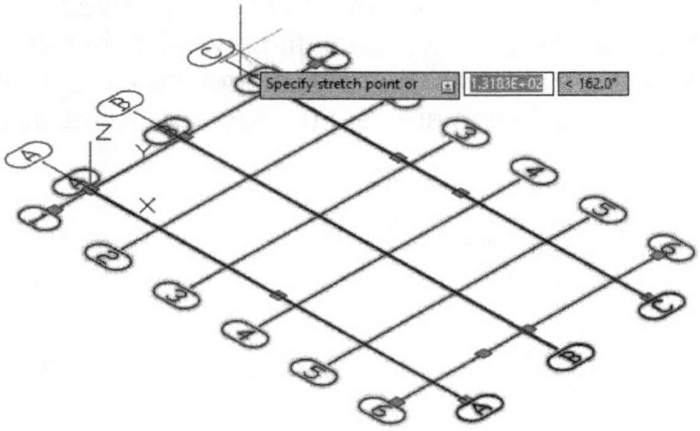

Figure 2-27

13. Save the drawing.

Task 3: Add a single grid line to the grid group.

1. In the *Objects* tab>*Grid* panel, click ⊞ (Add axes).

2. Select grid line 6 and press <Enter>.

3. For *Number of axes*, enter **1** and then press <Enter>.

4. For *Distance between axes*, enter **4'** and then press <Enter>. The modified grid displays with an automatically incremented grid line, as shown in Figure 2-28.

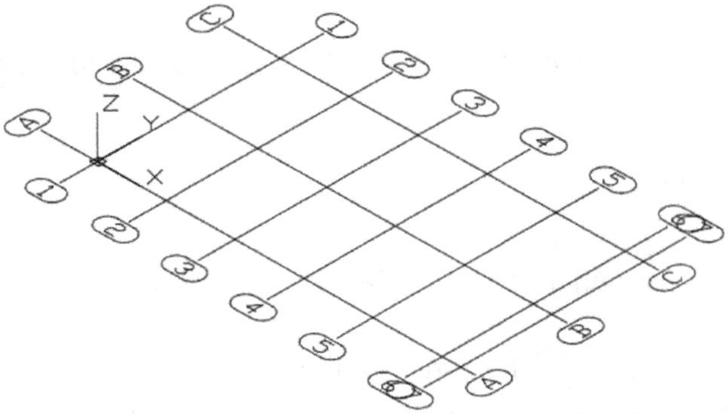

Figure 2-28

5. Save the drawing.

End of practice

2.3 Creating Levels

Levels can be used to specify the height at which objects are created (as shown in Figure 2–29), and to control how much of a model displays at or between levels. If you have a multistory building, you can create a level at each story and then toggle off all of the levels other than the one you are working on. You can also set the Workplane for base and height of columns using Layers.

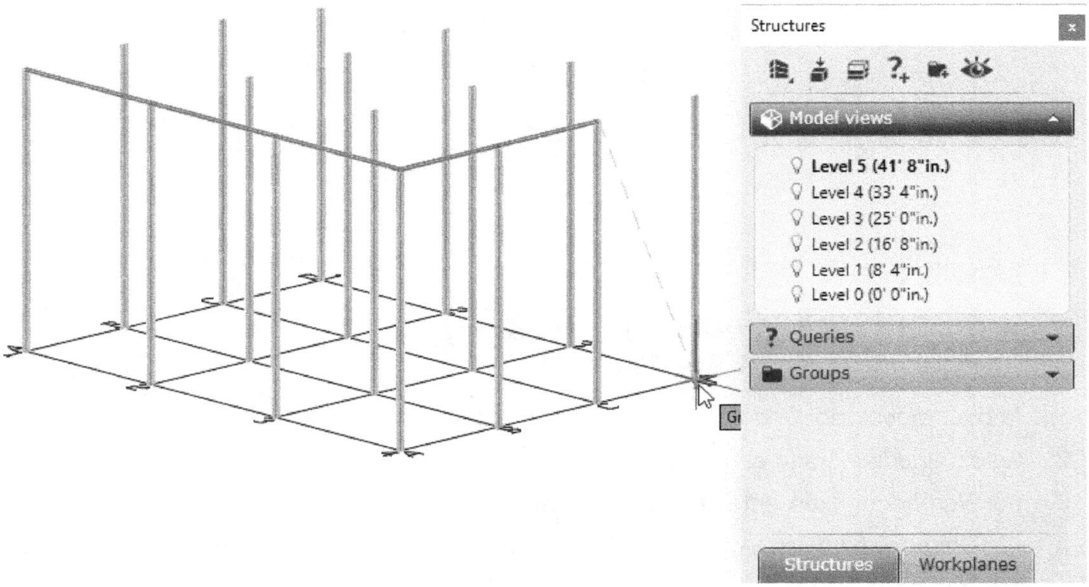

Figure 2–29

* Levels are created in the *Project Explorer*, which can be docked to the side of the screen if required.

How To: Create Levels

1. In the *Home* tab>*Project* panel, click ▭ (Project Explorer).

2. In the *Project Explorer*, click ▭ (Create Level Above).

3. In the *Create level* dialog box, select **Add level above the building** or **Add level below the building**.

4. Type the *Name* for the level. Names automatically increment when you are adding them, so it is best to use a name and number based system.

5. Set the *Altitude* from the bottom or the *Height* from a selected *Base level*, as shown in Figure 2–30. Note that one updates the other automatically.

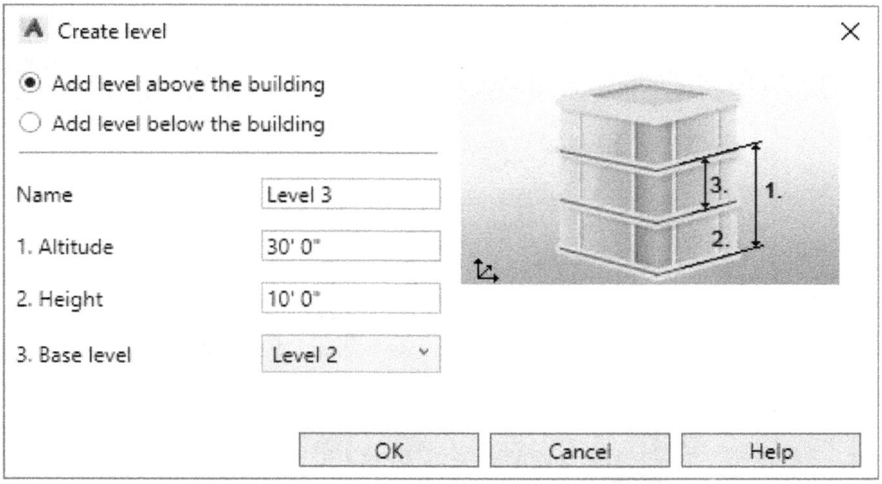

Figure 2–30

- Levels are not included in the template by default, so the first level you create should be the ground level of the building.

6. Repeat the process to add more levels. The new levels display in the *Project Explorer*, and the active level displays in bold, as shown in Figure 2–31.

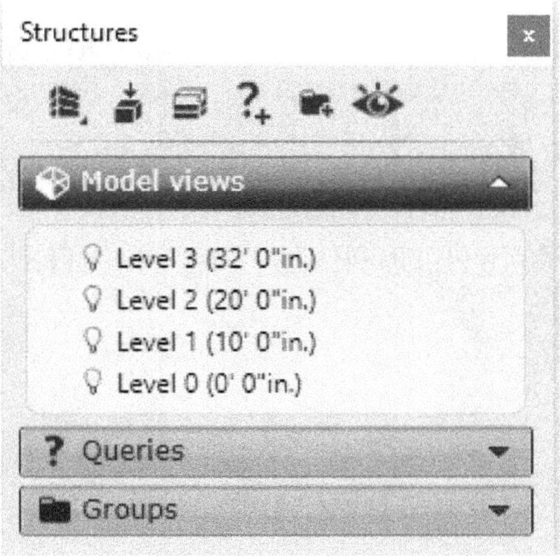

Figure 2–31

How To: Activate and View Levels

1. Open the *Project Explorer* if it is not already open.

2. Right-click on level name and click **Activate**. The view switches to the **Top** view, zooms in on the origin of the UCS, and the level height displays, as shown in Figure 2–32.

   Note: Objects are placed on the active level.

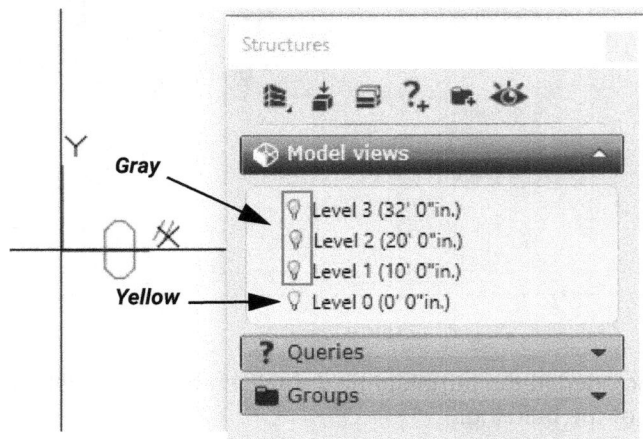

Figure 2–32

3. Return to a 3D view.

4. Click on the light-bulb beside any level names you want to display. In the example shown in Figure 2–32, levels above **Level 0** would not display, as they are not toggled on.

- To deactivate levels, in the *Project Explorer*, right-click on the active level and select **Deactivate**.

- If no level is active, objects are drawn on the current UCS, or at actual snap points (such as the top node of a column).

How To: Set Workplanes

1. In the *Project Explorer*, select the *Workplanes* tab.

2. Click on the bottom button and then click on the one of the levels in the list, as shown in Figure 2-33.

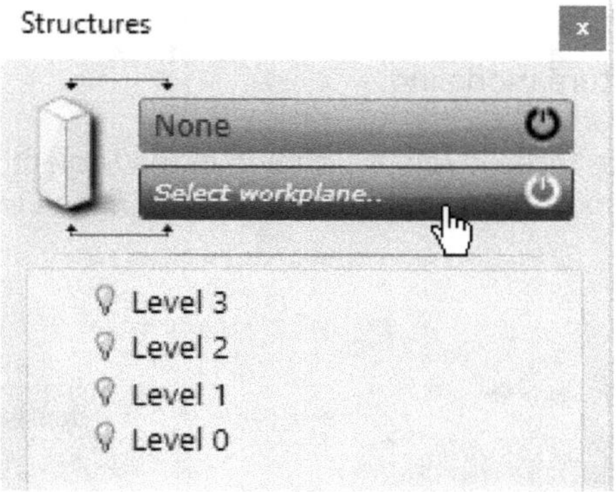

Figure 2-33

3. Click the top button and select the Level for the height of the column.

4. Return to the *Structures* tab to modify any of the Level views, as required.

💡 Hint: Hiding and Isolating Objects

You might need to limit the number of items that display. You can do this by temporarily hiding or isolating objects.

1. Select the objects you want to hide or isolate.

2. In the Status Bar, expand (Isolate Objects) and select one of the following:

3. **Isolate Objects:** Only displays objects you have selected.

4. **Hide Objects:** Toggles off the display of the selected objects.

5. To restore the objects, in the Status Bar, expand (Isolate Objects) and select **End Object Isolation**.

- To select multiple copies of the same object (such as all columns), select one of the objects, and then right-click on it and select **Select Similar**.

Practice 2b
Create Levels

Practice Objectives

- Create levels and toggle them on and off.
- Activate and deactivate levels.
- Set the workplanes for future modeling.

In this practice, you will create three levels, as shown in the *Project Explorer* in Figure 2–34. You will then set up workplanes so that you can draw columns from Level 0 to Level 1, as shown in Figure 2–35.

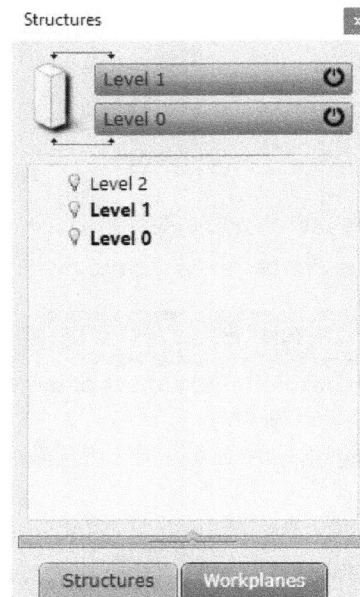

| Figure 2–34 | Figure 2–35 |

1. In the practice files folder, open **Platform-Levels.dwg**.

2. In the *Home* tab>*Project* panel, click ▦ (Project Explorer).

3. In the *Project Explorer*, click ▤ (Create Level Above).

4. In the *Create level* dialog box, select **Add level above the building**.

5. Use the default *Name* of **Level 0** and *Altitude* of **0** and click **OK**. The grid display changes as shown in Figure 2–36.

6. Repeat the process and create two more levels, as shown in Figure 2−36 and as specified below. Note that the name automatically increments.

    *   Level 1: *Height* = **12'-0"**, *Base level* = **Level 0**.
    *   Level 2: *Height* = **18'-0"**, *Base level* = **Level 1**.

Figure 2−36

7. Note that Level symbols are placed as you add levels, as shown in Figure 2−36. You might need to zoom in to see them.

8. Zoom back out to see the full grid by typing **Z** <Enter> **A** <Enter> in the command line.

9. In the *Project Explorer*, right-click on **Level 2** and select **Deactivate**, as shown in Figure 2–37. Note that this also toggles off Level 2, though no elements are on Level 2 at the moment.

10. Click on the lightbulb beside **Level 1**, as shown in Figure 2–38. This toggles off the elements between Level 1 and Level 0, and the full grid symbol displays.

Figure 2–37

Figure 2–38

11. Toggle off **Level 0**. If this project had objects (such as columns or beams) on any level, they would all display.

12. Double-click on **Level 1**. This activates the level and also changes the view to the **Top** view, as shown in Figure 2–39. Remember that the active level is where objects are drawn unless using a specific UCS or snap override.

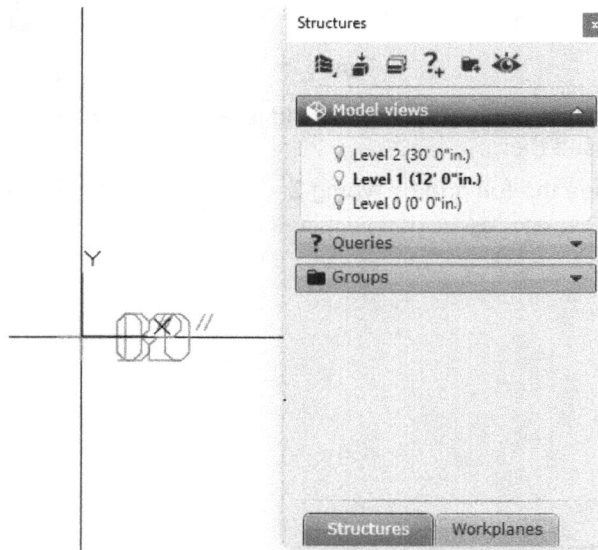

Figure 2–39

13. Return to the **SE Isometric** view.

14. Right-click on **Level 0**, select **Activate**, and then toggle on **Level 1**.

15. In the *Project Explorer*, select the *Workplanes* tab.

16. Set the top Level to **Level 1** and the bottom Level to **Level 0**, as shown in Figure 2–40. You are now ready to start modeling on Level 0 to a height of Level 1.

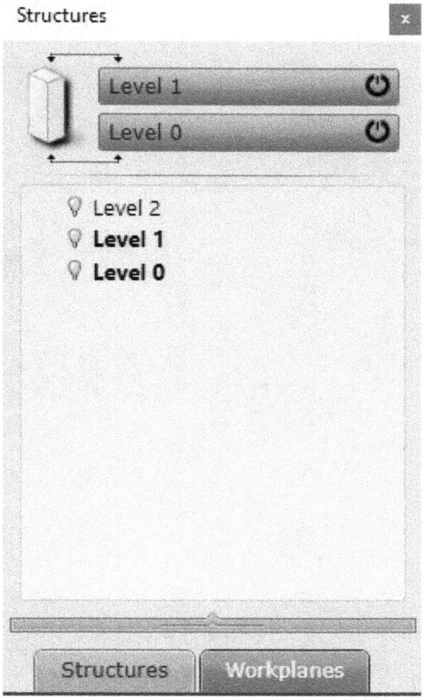

Figure 2–40

17. Save the drawing.

End of practice

2.4 Modeling Columns and Beams

The final step to creating basic steel structures is to add in columns and beams. In the Autodesk Advance Steel software, there are a variety of commands that give you access to different section types, such as rolled I-sections, channels, T-sections, and more. There are also tools for welded and double beams, as well as concrete and timber beams. In each case, you start by drawing the beam or column location and then modifying the Advance Properties, as shown in Figure 2–41.

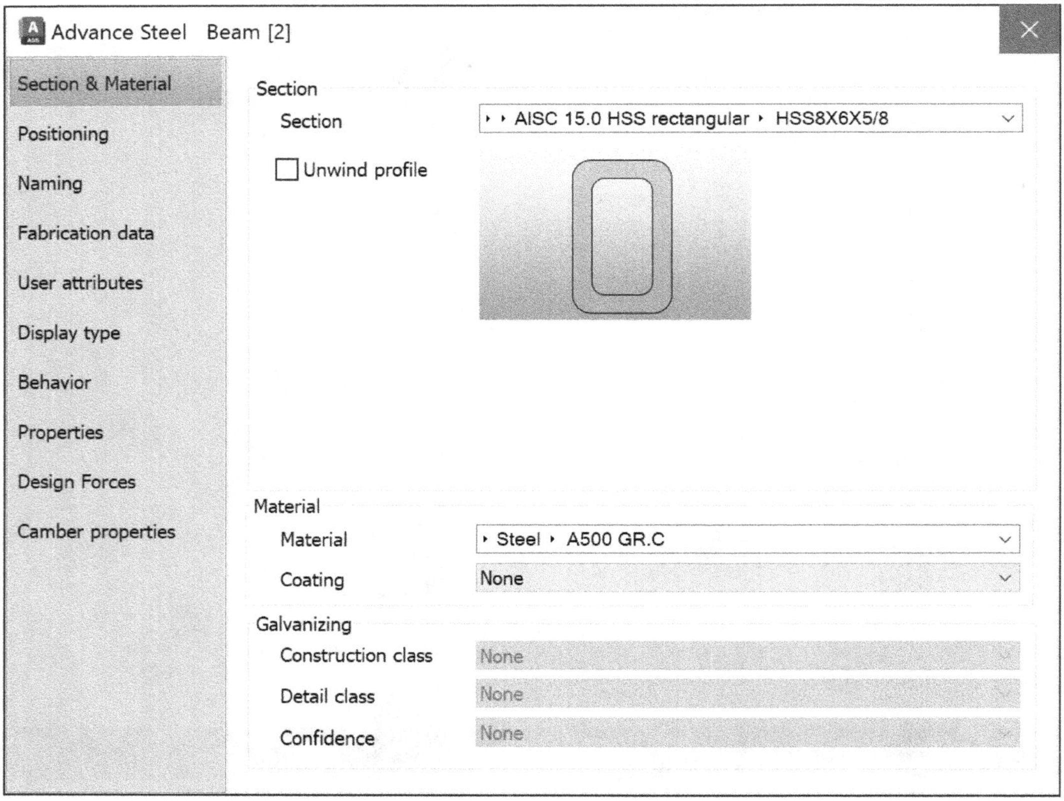

Figure 2–41

- Column objects are identified as beams, but the model role differs. It is helpful to use the **Column** command to assign its role as a column for documentation.

How To: Place Columns

1. In the *Home* tab>*Objects* panel or *Objects* tab>*Beams* panel, click ⌘ (Column).
2. Place the base of the column. The height is determined automatically.
   - You can specify the levels to define the height in the *Project Explorer>Workplane* tab.
   - Toggle on the **Grid Intersection Points** snap for quick and accurate placement.
3. Continue placing columns. Press <Enter> to finish.
4. The *Beam* dialog box displays, enabling you to access tabs for the *Section & Material* (shown in Figure 2–42), *Positions*, *Naming*, and other options.

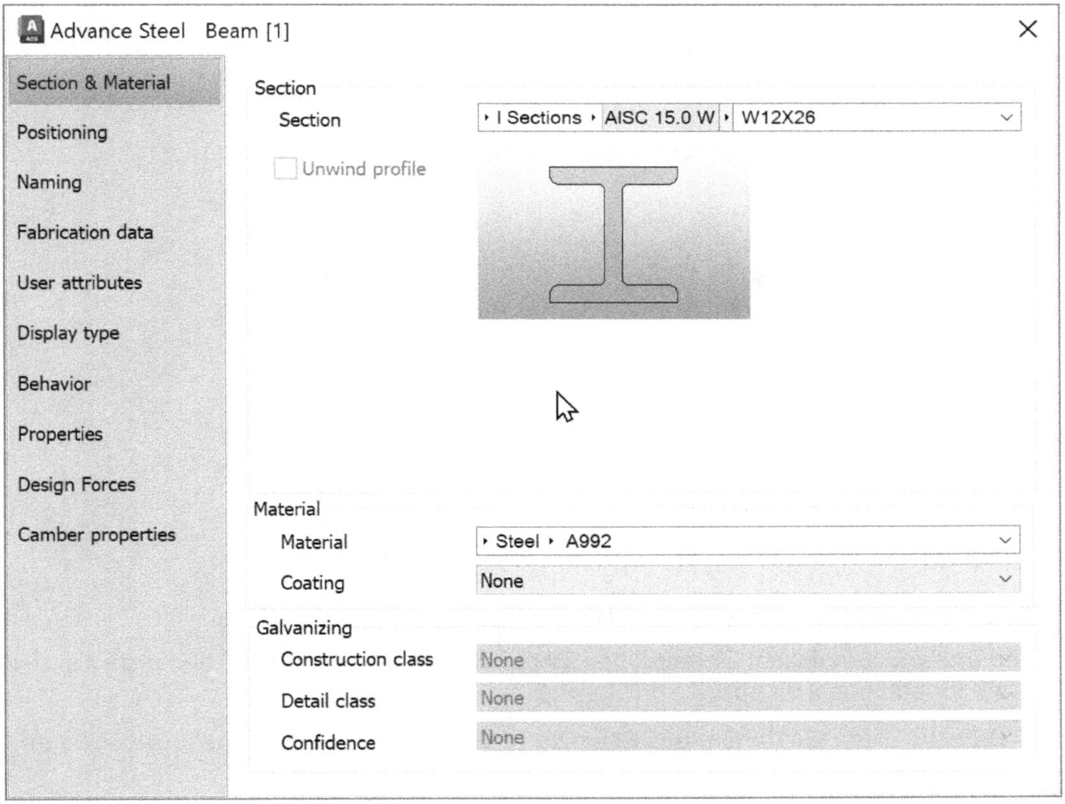

Figure 2–42

- Stand-alone columns can be copied and arrayed using standard AutoCAD commands.

How To: Draw Beams

1. In the *Home* tab>*Objects* panel or *Objects* tab>*Beams* panel, click ⊥ (Rolled I section) or any of the other section types in the drop-down list as shown in Figure 2−43. Alternately, In the *Objects* tab>*Beams* panel, click 🗁 (Continuous Beam).

    *   The *Rolled I* section is the top of the list. If you have most recently used another shape in the list, that name is listed first.

    *   *Welded Beams*, *Double Channel Beams*, and *Concrete Beams* also have similar drop-downs, as shown on the *Objects* tab>*Beams* panel in Figure 2−44.

Figure 2−43

Figure 2−44

2. Click two points for the start and end of the first beam.

3. Note that you remain in the command and can continue selecting additional start and end points for additional beams.

    *   If you are using the **Continuous Beam** command, the end point of the previous beam will be the start point for the next beam.

4. Press <Enter>.

5. The *Beam* dialog box displays, as shown in Figure 2–45.

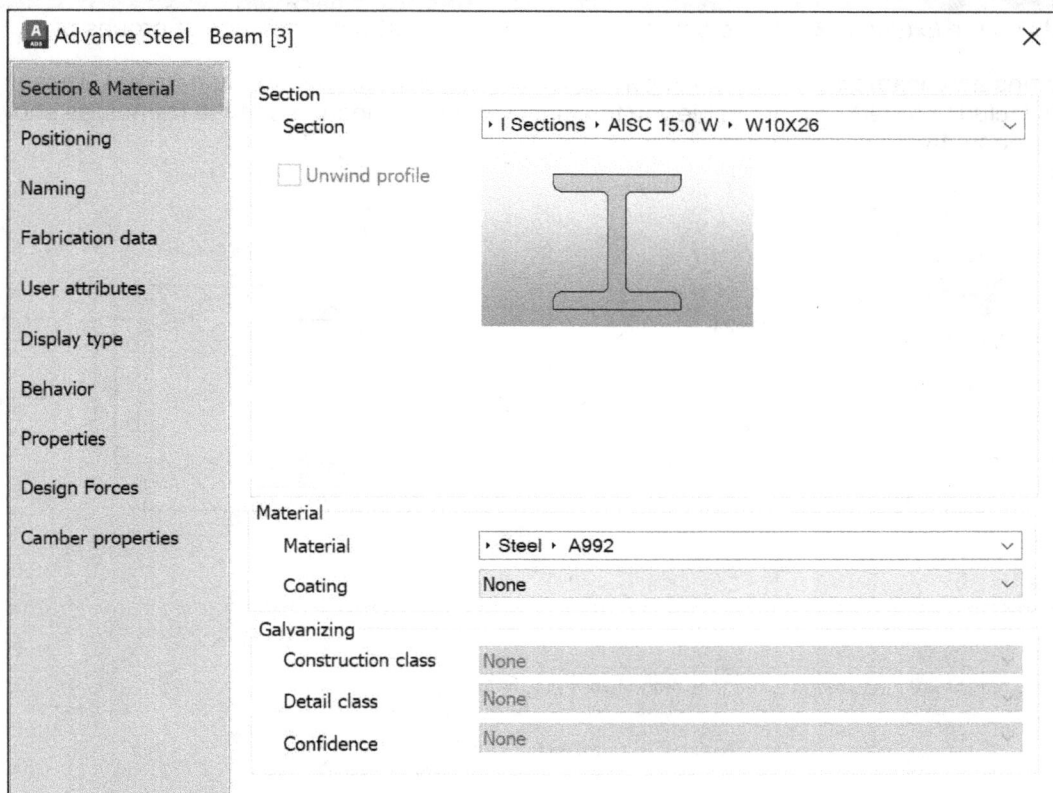

Figure 2–45

6. Adjust the properties in the *Section & Material*, *Positioning*, and other tabs as required.

- You can create beams from AutoCAD polylines, lines, and arcs using the ⚙ (Beam, polyline) and ⌐I (Beam, from line) commands.

- The ⌐ (Curved Beam) command prompts you to select a start point, end point, and circle point that defines the radius.

💡 Hint: Extended Modeling Tools

Tools on the *Extended Modeling* panel can help you create structures made of groups of

columns and beams, such as ⌂ (Portal/Gable Frame), and ⊓ (Mono-pitch frame). These tools include properties that enable you to adjust various components of the frames, as shown in Figure 2–46.

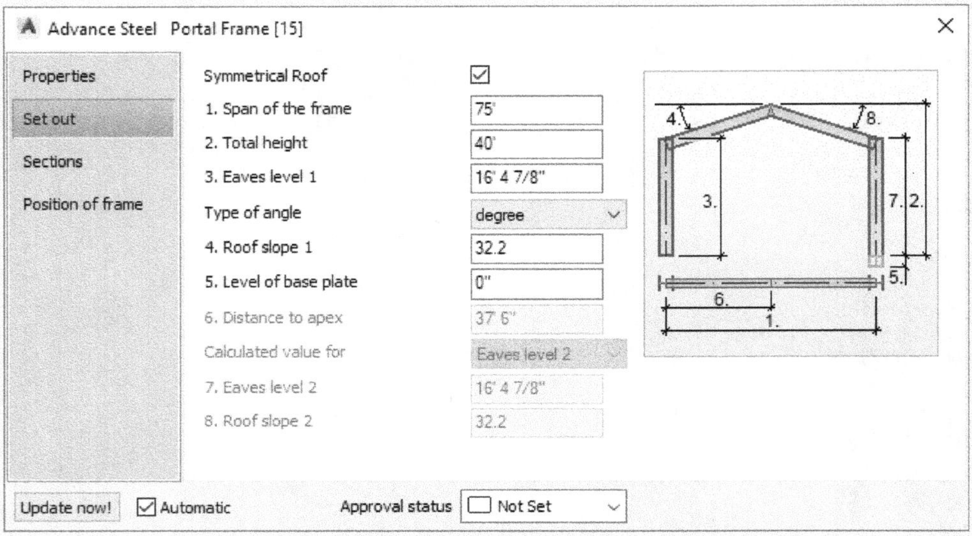

Figure 2–46

Modifying Columns and Beams

Columns and beams can be modified using grips to lengthen or shorten the objects, as shown for a column in Figure 2–47. To modify sizes, positioning, and other data for any Autodesk Advance Steel object, either double-click on the object, or select the objects, right-click, and select **Advance Properties**.

You can also select the beam and hover your cursor over the endpoints grip to show editing options, as shown in Figure 2–48. After selecting an option, you can specify the new length by typing in the dimensions or dragging your cursor in the direction you want and clicking.

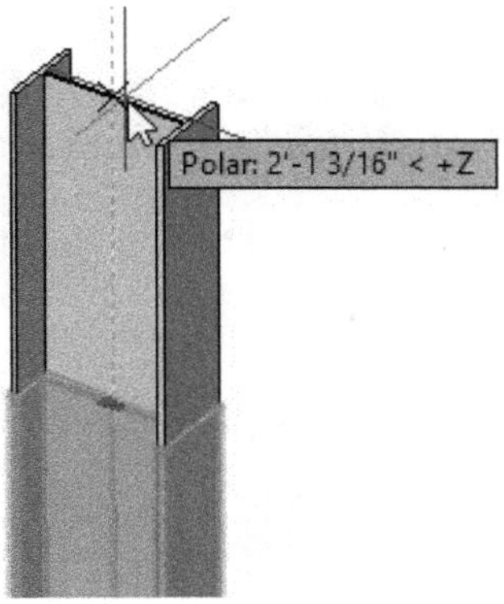

Figure 2–47 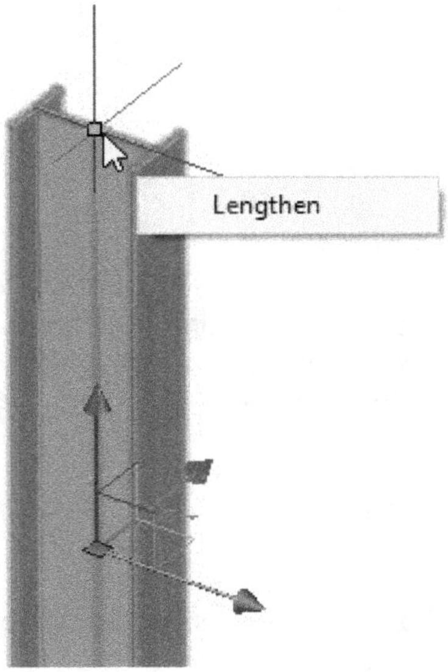   Figure 2–48

- You can also split or merge beams and columns.

- Columns are beam objects. Any tools that can be used on beams can also be used on columns.

How To: Split Beams

1. In the *Home* tab>*Objects* panel or the *Objects* tab>*Beams* panel, click 🖉 (Split Beams).

2. Select the beam you want to split and then press <Enter>.

3. Select one or more split points using object snaps to ensure that you are touching the beam. Press <Enter>. Each split point creates a cut in the beam, as shown in Figure 2–49.

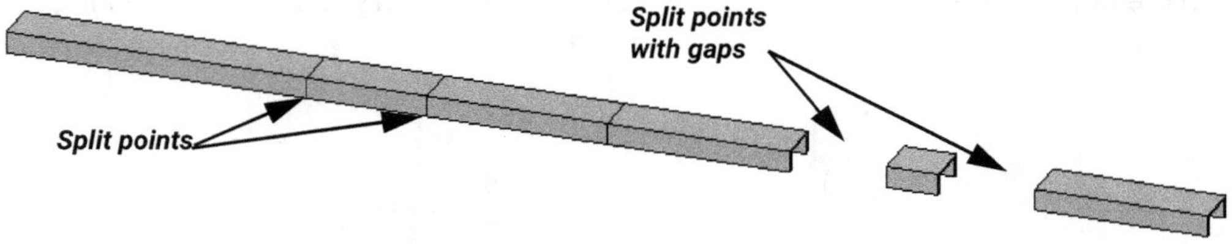

Figure 2–49

- If you want a gap between points, after you select the beam and before you select the split points, type **G** for Gap and then press <Enter>. Enter a distance for the gap and press <Enter>. When you select the split points, a gap is created that is centered on the selected point, as shown in Figure 2–49, above. The gap is set and reused by the command until you change it.

How To: Merge Beams

1. In the *Home* tab>*Objects* panel or *Objects* tab>*Beams* panel, click 🖉 (Merge Beams).

2. Select the beams you want to merge and press <Enter>.

   - Beams must be in the same line.
   - If you select beams with different properties, the first beam that you select controls the merged beam's properties.
   - If the beams are not touching, they are extended and merged.
   - If the beams have any overlapping portions, they are trimmed and merged.

Practice 2c
Model Columns and Beams

Practice Objectives

- Add and modify columns.
- Add and modify beams.

In this practice, you will add a column and modify its Advance Properties. You will then copy the column to other grid locations. You will change some columns to a different profile and specify positioning for several individual columns. You will also add beams using the **Continuous Beam** command and modify the Advance Properties for the entire group of beams. Finally, you will add framing between some of the beams. The final model is shown in Figure 2-50.

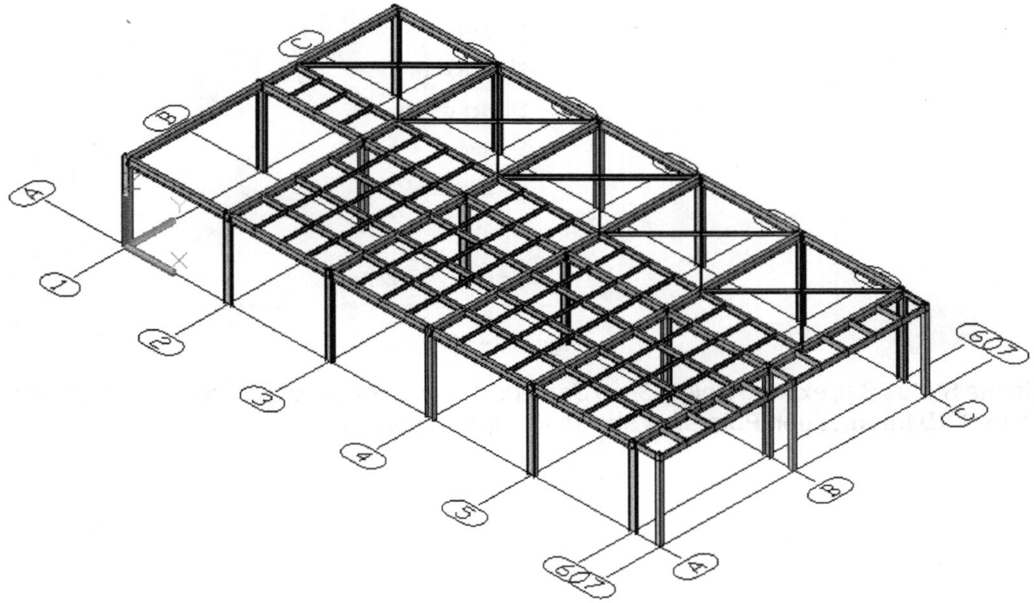

Figure 2-50

Task 1: Add columns.

1. In the practice files folder, open **Platform-Columns.dwg**.

2. In the upper left corner of the view window, set the *View Control* to **SE Isometric** and the *Visual Style* to **Conceptual**, as shown in Figure 2–51.

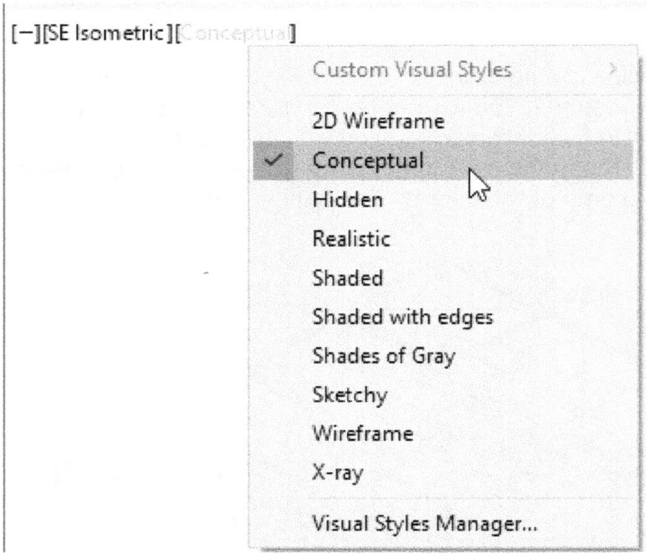

[−][SE Isometric][Conceptual]

Custom Visual Styles	>
2D Wireframe	
✓ Conceptual	
Hidden	
Realistic	
Shaded	
Shaded with edges	
Shades of Gray	
Sketchy	
Wireframe	
X-ray	
Visual Styles Manager...	

Figure 2–51

3. In the Status Bar, expand *Object Snap Settings*. Clear all of the object snaps except for **Node** and **GRID Intersection Points**, as shown in Figure 2–52.

Figure 2–52

4. Open the *Project Explorer* (if it is not already open) and ensure that the workplanes are set up as shown in Figure 2–53.

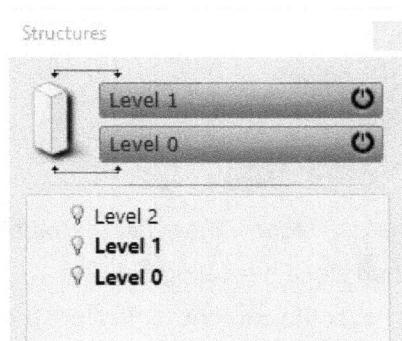

Figure 2–53

5. In the *Home* tab>*Objects* panel, click ▭ (Column).

6. Select the intersection of grid lines **1** and **A**, and then press <Enter>.

7. In the *Beam* dialog box, *Section & Material* tab, expand AISC 15.0 W and select **W12x30,** as shown in Figure 2–54.

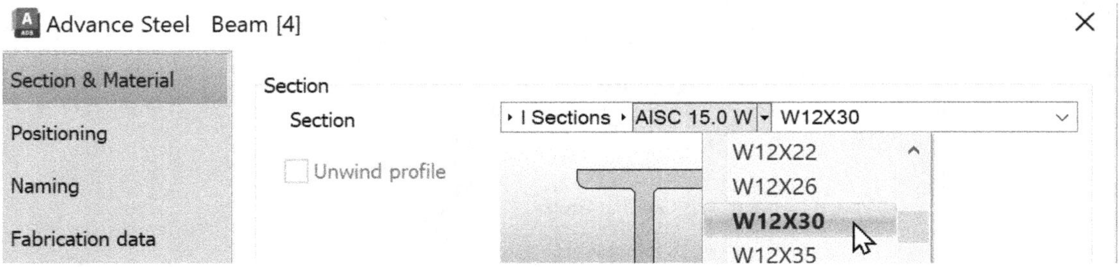

Figure 2–54

8. In the *Positioning* tab, set the *Offset* and *Angle* as shown in Figure 2–55.

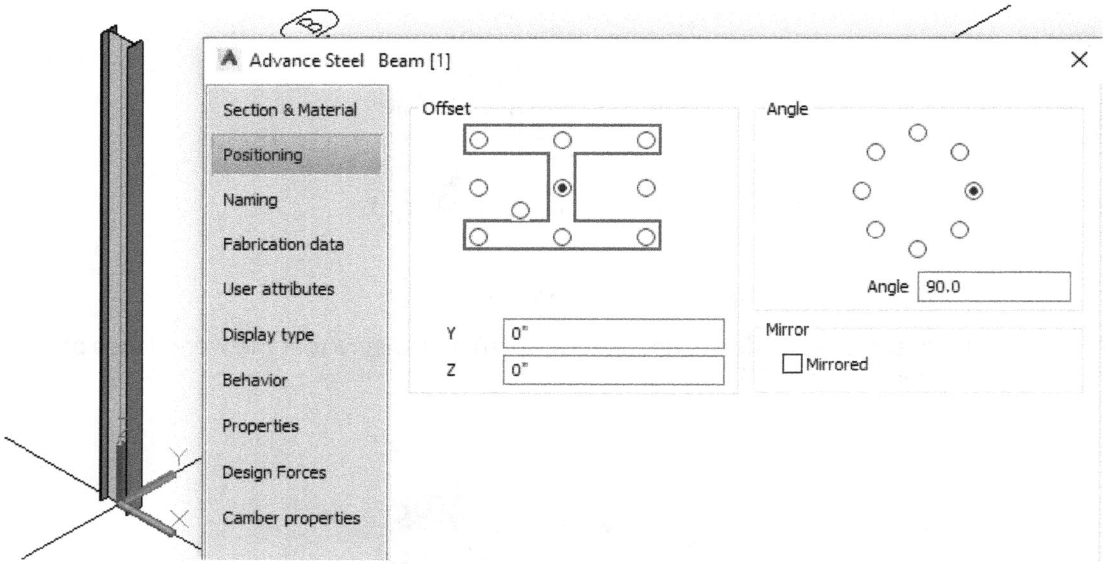

Figure 2–55

9. In the *Naming* tab, ensure that the *Model Role* is set to **Column**.

10. Review the other tabs and then close the dialog box.

11. Zoom in on the column and hover the cursor over it. Note that the column has been placed on the correct layer, as shown in Figure 2–56.

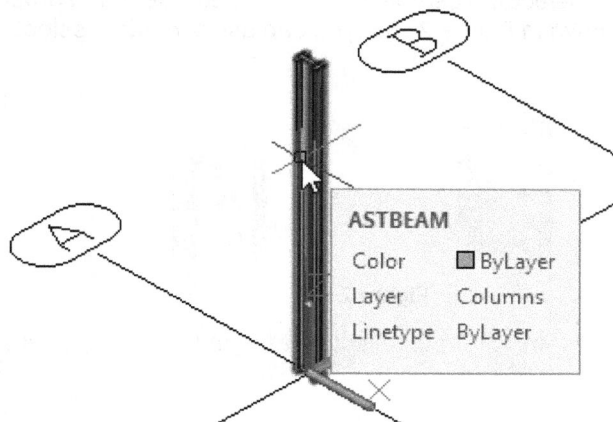

Figure 2–56

12. Type **DI** to start the **Distance** command. Click on the bottom and top of the column to display the default size of **12'-0"**, as defined by Level 0 and Level 1. Note that the Node snap connects to these end points.

13. **Copy** the column to the rest of the grid intersections, as shown in Figure 2–57.

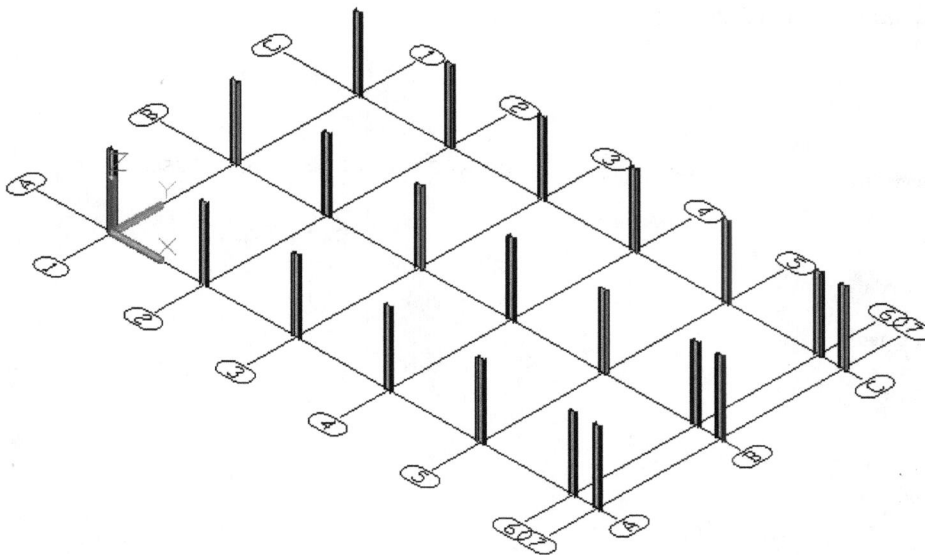

Figure 2–57

14. In the *Project Explorer>Workplanes* tab, right-click on Level 1 and select **Attach elements**.

15. At the command line for *Select the objects to bind to the workplane*, select all of the beams (you can use a window selection) and press <Enter>.

16. At the command line for *Select the geometry points*, select the red geometry points at the top of the beams, as shown in Figure 2–58 (you can use a window selection), and press <Enter>.

Figure 2–58

Note: You can rotate the view in order to select just the red geometry points.

17. Save the drawing.

Task 2: Modify the column type and positioning for selected columns.

1. Select the three columns along grid line 7.

2. Right-click and select **Advance Properties**.

3. In the *Beam* dialog box, on the *Section & Material* tab, change the *Section* to **Rectangular Hollow Section>AISC 15.0 rectangular>HSS 8X6X5/16** and the *Material* to **Steel>A500 GR.C**, as shown in Figure 2–59.

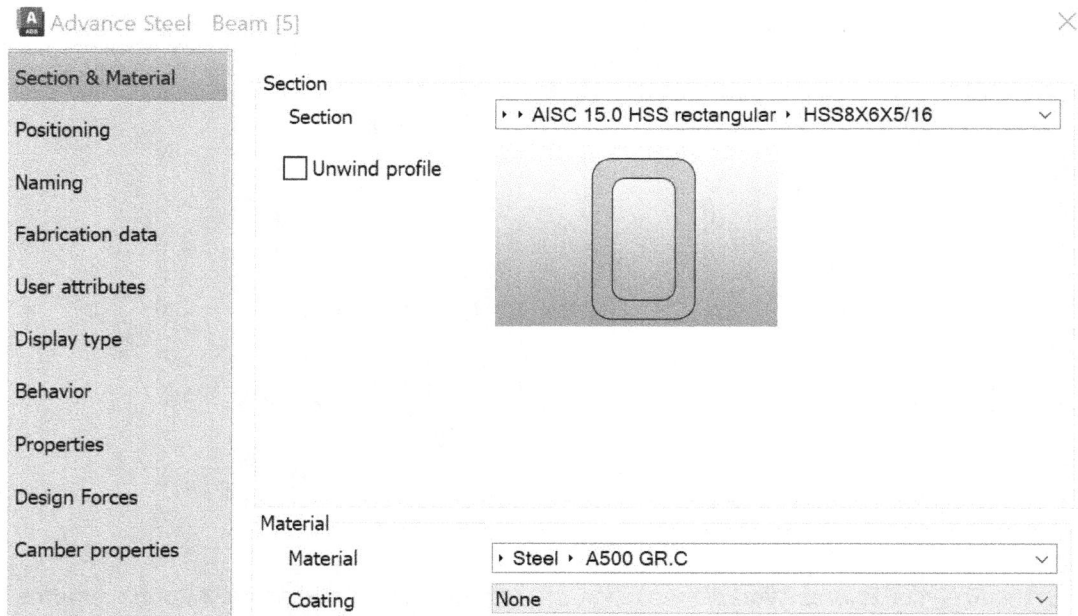

Figure 2–59

4. Close the dialog box.

5. Switch to the **Top** view.

6. Select each of the HSS columns separately and modify the *Positioning* so that they sit inside the grid lines, as shown in Figure 2–60.

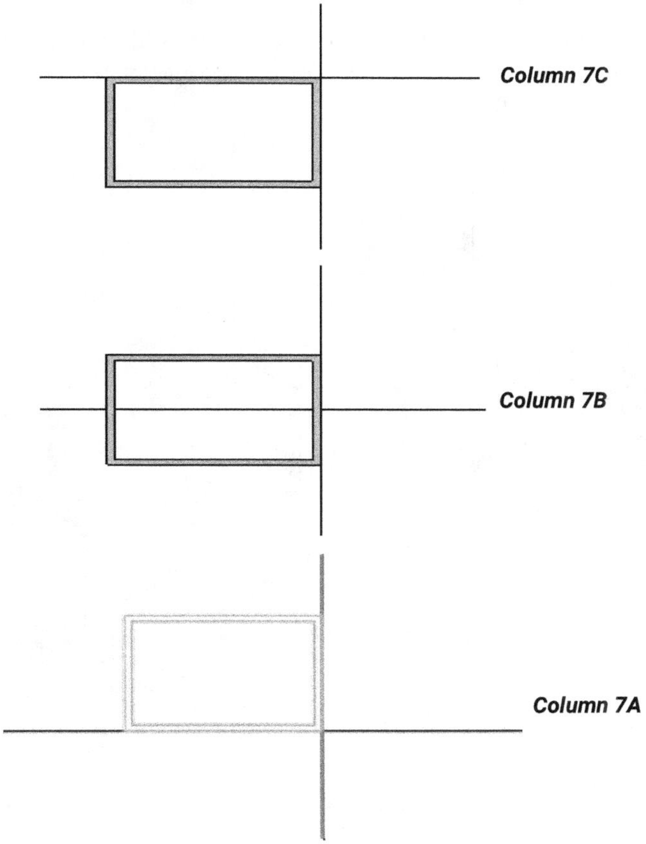

Column 7C

Column 7B

Column 7A

Figure 2–60

7. Return to the **SE Isometric** view and review the changes.

8. Save the drawing.

Task 3: Add beams.

1. In the *Objects* tab>*Beams* panel, click ⏣ (Continuous Beam).

2. Select the start point of the beam on the top node of column A1.

3. Select the end point of the beam on the top node of column A2. Then continue selecting the top node of the outside line of columns (as shown in progress in Figure 2–61) all the way around until you return to column A1.

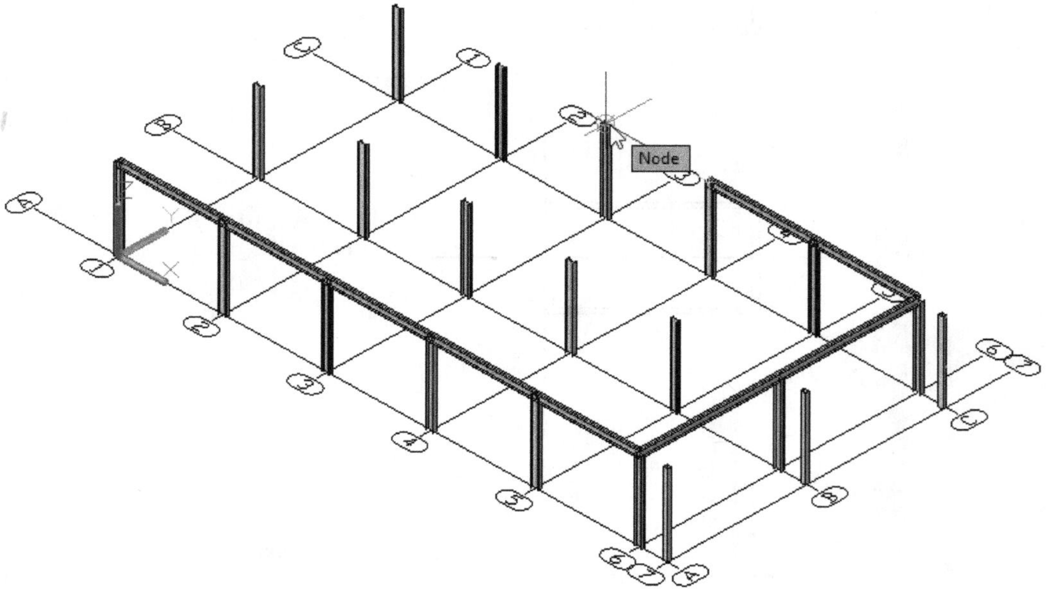

Figure 2–61

4. Press <Enter>.

5. In the *Beam* dialog box, *Section & Material* tab, set the *Section* to **I Sections>AISC 15.0 W>W12x30**.

6. In the *Positioning* tab, set the *Offset* as shown in Figure 2–62. All of the beams created in this process update with the new information.

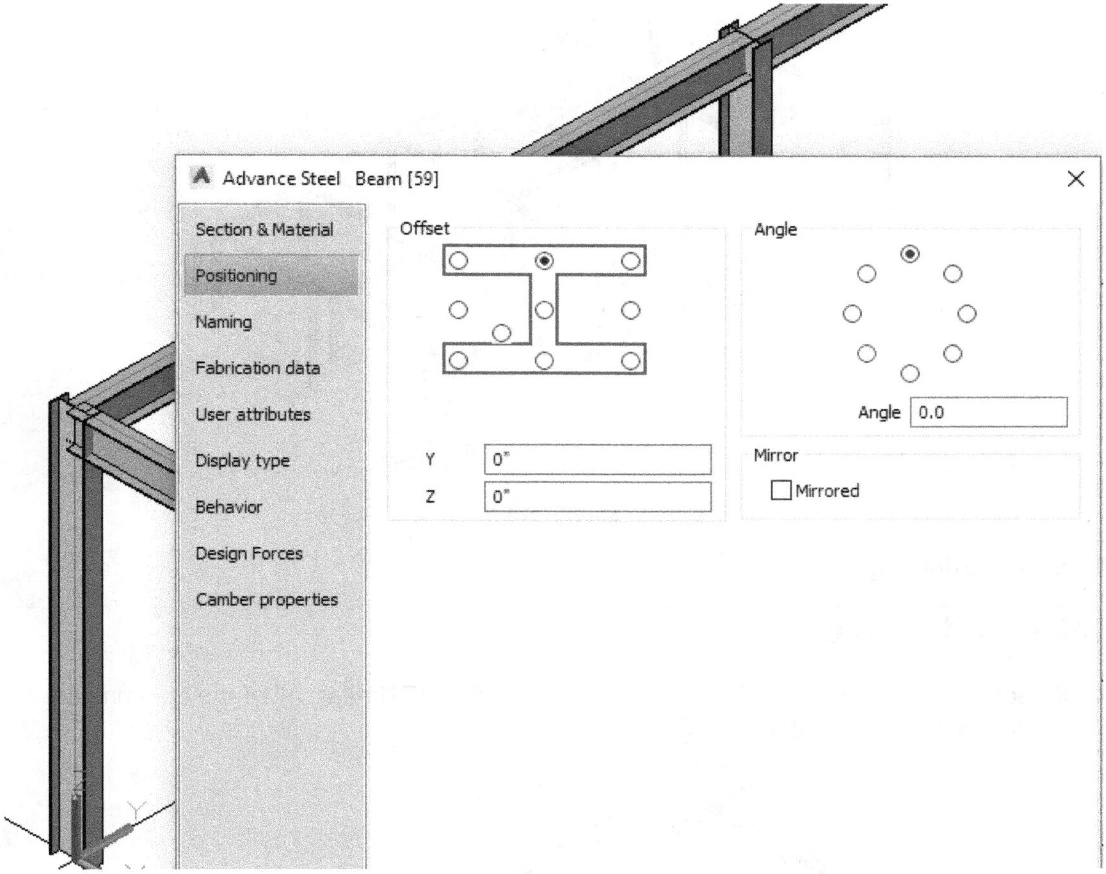

Figure 2–62

7. Add interior beams using the same size and positioning, as shown in Figure 2–63.

Figure 2–63

8. Save the drawing.

Task 4: Add framing.

1. Select one of the columns. Right-click and select **Select Similar**. All of the columns are highlighted, as shown in Figure 2–64.

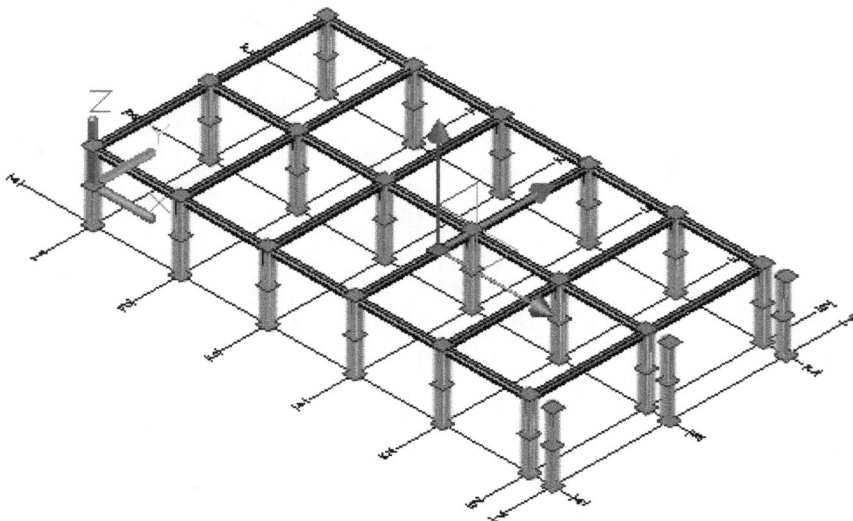

Figure 2–64

2. In the Status Bar, expand (Isolate Objects) and select **Hide Objects**. The columns are hidden, but the rest of the beams remain.

   Note: The columns are hidden to keep you from snapping to their nodes by mistake.

3. In the *Project Explorer*, right-click on **Level 1** and select **Activate**.

4. Return to the **SE Isometric** view.

5. In the *Home* tab>*Objects* panel, click (Rolled I section).

6. Hover over the intersection of Grid 2A so that the Grid Intersection snap displays. Move the cursor to the right along the Y-axis with polar tracking and Dynamic Input on, and then enter **5'**, as shown in Figure 2−65.

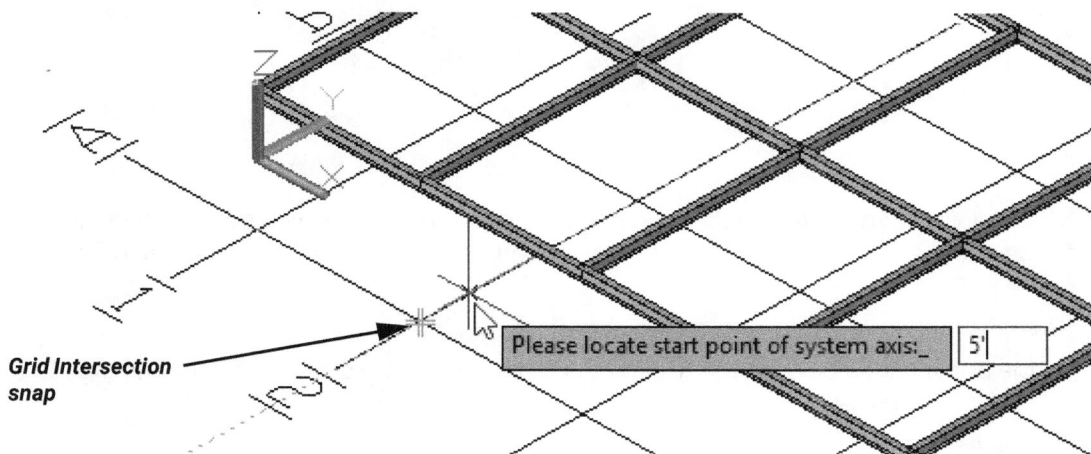

Grid Intersection
snap

Please locate start point of system axis:_   5'

Figure 2−65

7. Hold <Ctrl> and right-click to open the snap overrides list. Select **Perpendicular** and then select Grid 3. The new beam is placed on **Level 1**, as shown in Figure 2–66.

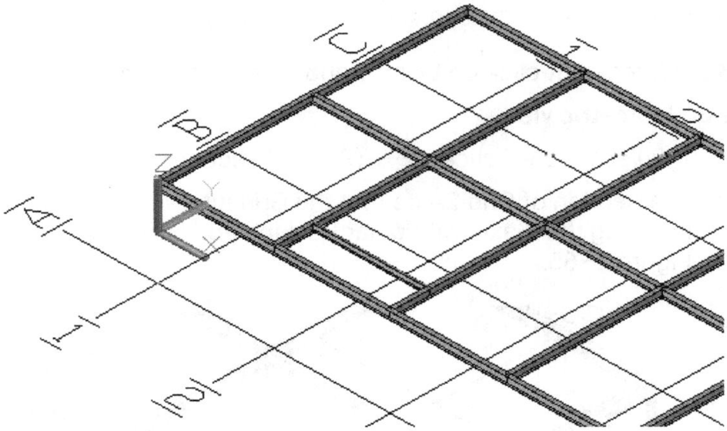

Figure 2–66

- Snapping to the grid with the level set to **Level 1** ensures that you do not accidentally select one of the many nodes of the beam objects, which could give you the wrong location.

8. Press <Enter>.

9. In the *Beam* dialog box, set the *Section* to **I Sections>AISC 15.0 W>W10x19** and the *Material* to **Steel>A992**.

10. Close the dialog box.

11. Use the AutoCAD **Copy** command to add beams **5'-0"** apart in the same bay, and then another beam **5'-0"** on the other side of the main center beam, as shown in Figure 2–67.

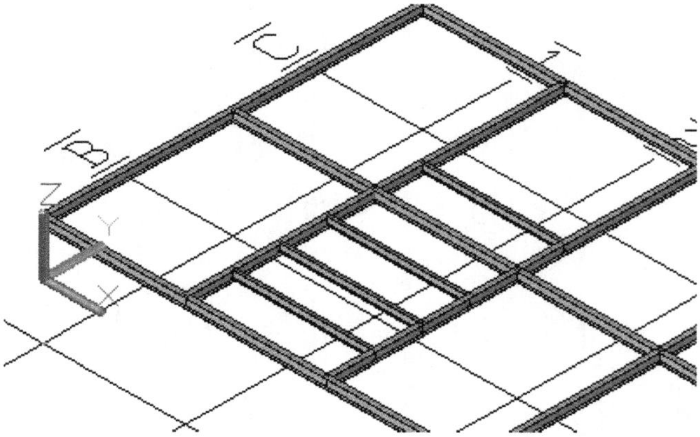

Figure 2–67

12. In the *Home* tab>Objects panel, expand ⊥⊤ (Rolled I section) and select ⊏ (Channel section).

13. Place the beam **3'-9"** (typical) from Grid 2 and perpendicular to the top of the beam, as shown in Figure 2–68. You can use temporary osnaps, then press <Shift> and right-click in the drawing area and select **Perpendicular**. Ensure that you snap to the red insertion axis line (shown in Figure 2–69), and not one of the other parts of the beam.

Figure 2–68

Figure 2–69

14. Press <Enter> and set the *Section* to **AISC 15.0 C Channels>C6x10.5** and the *Material* to **Steel>A36**.

15. In the *Positioning* tab, set *Offset* as shown in Figure 2–70.

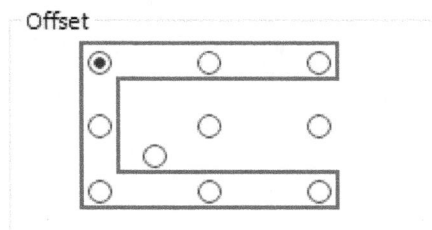

Figure 2–70

16. Zoom in and ensure that the channel is placed as expected.

17. Use the **Copy** commands as required to place the rest of the framing as shown in Figure 2–71. To copy between bays, it is easiest to snap to the endpoint or node of the grid lines away from the steel objects.

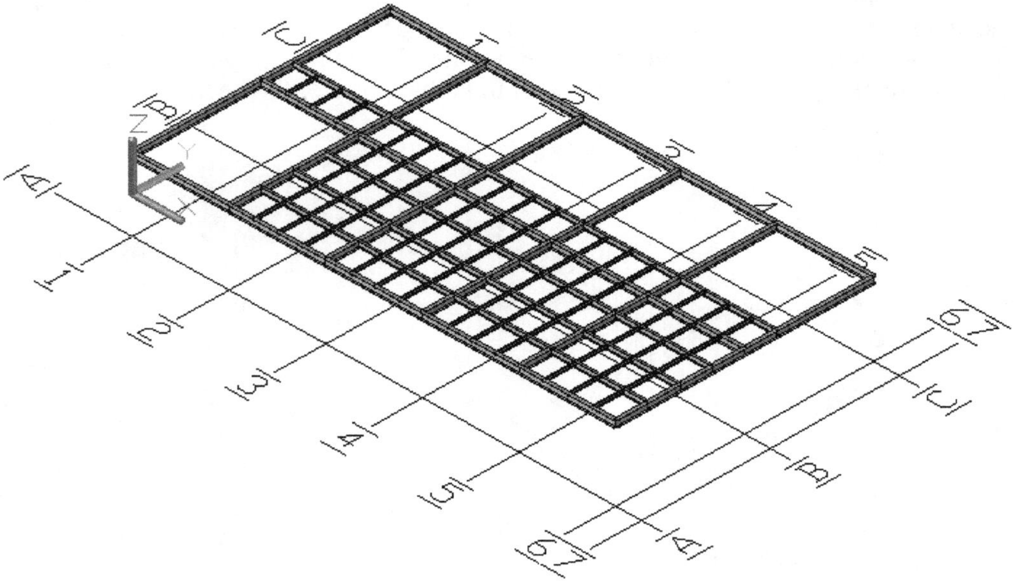

Figure 2–71

18. Save the drawing.

19. If you have time, add the beams along grid 7 and the diagonal beams on the back of the platform, as shown in Figure 2–72.

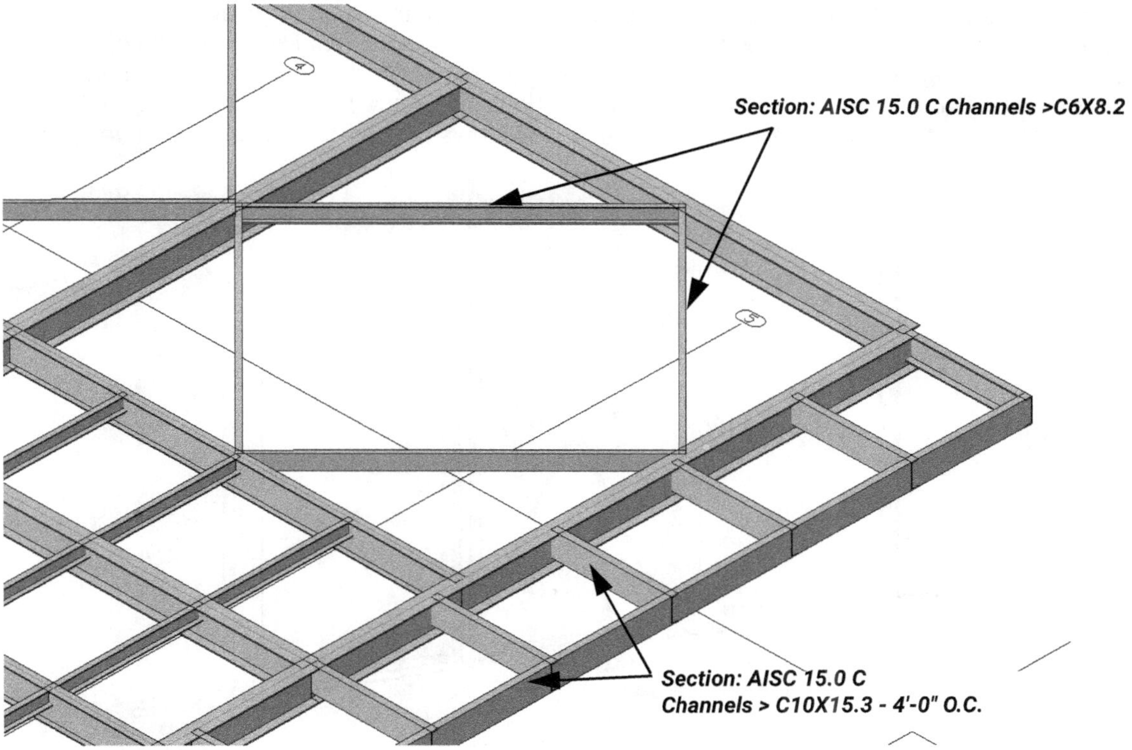

Section: AISC 15.0 C Channels >C6X8.2

Section: AISC 15.0 C Channels > C10X15.3 - 4'-0" O.C.

Figure 2–72

20. When you are finished adding beams (see Figure 2–50 at the beginning of the practice), in the *Project Explorer*, right-click on **Level 1** and select **Deactivate**.

21. Toggle off all of the levels.

22. In the Status Bar, expand (Isolate Objects) and then select **End Object Isolation**.

23. Save the drawing.

End of practice

2.5 Adding Bracing

The Bracing command is a macro that uses two diagonal points to identify an area that needs bracing. Once used, you can then make changes in the *Advance Joint Properties* dialog box to specify the sizing and arrangement of the braces. Before you start the command, you first need to set the UCS icon to the correct orientation, as shown in Figure 2–73. To help you see the part of the model that you are working on, you can set up views based on grid lines that hide the rest of the model.

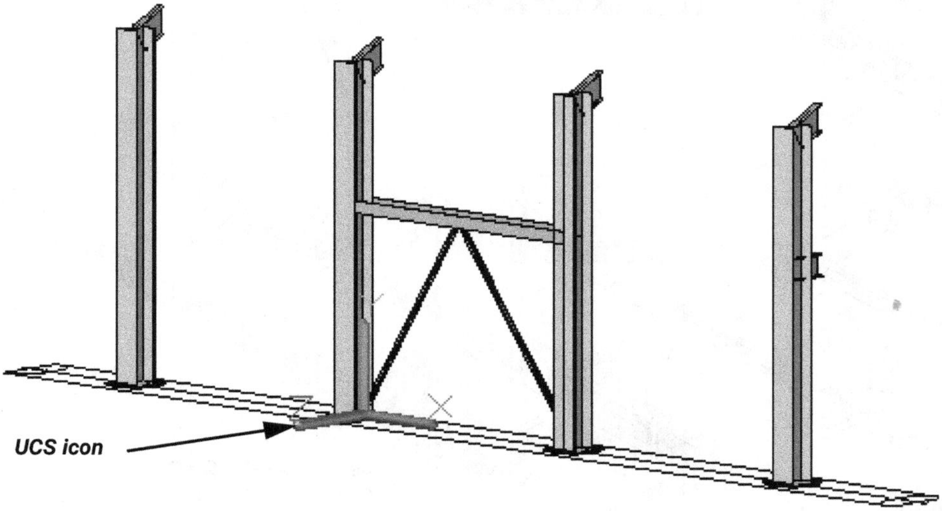

UCS icon

Figure 2–73

- Use the tools found in the *Advance Tool Palette*> [icon] (UCS) category to quickly set the correct location. For example, if you want to place the UCS on a slanted beam, use [icon] (UCS at Object).

How To: Add Bracing

1. Set the UCS so that the XY plane is in the correct orientation for the braces.

2. In the *Home* tab>*Extended Modeling* panel or the *Extended Modeling* tab>*Structural Elements* panel, click [icon] (Bracing).

3. Select two diagonal points that define the bracing elements. The *Bracing* dialog box displays with the *Properties* tab selected, as shown in Figure 2–74.

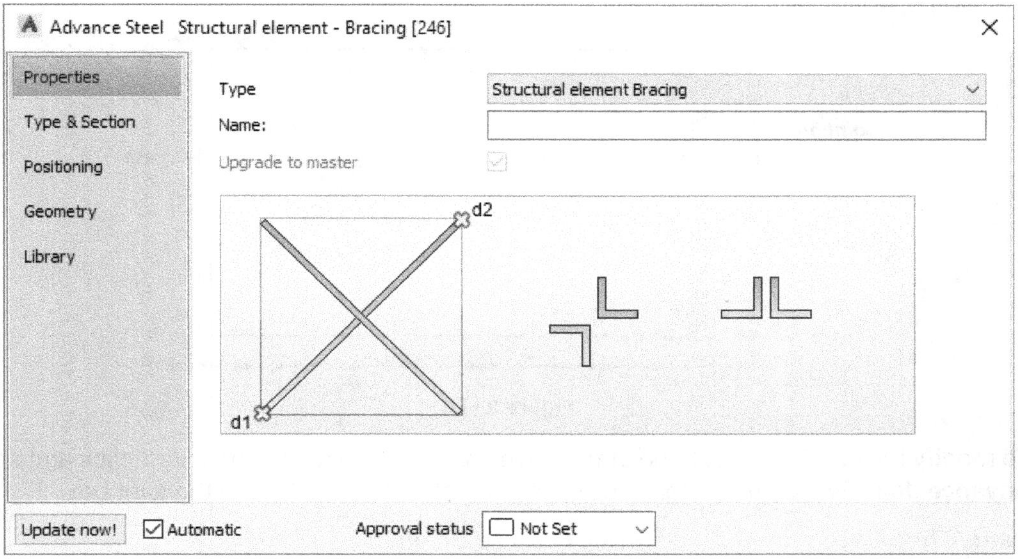

Figure 2–74

4. In the *Type & Selection* tab, specify the *Bracing Type* and other member options. The *Bracing Type* options are shown in Figure 2–75.

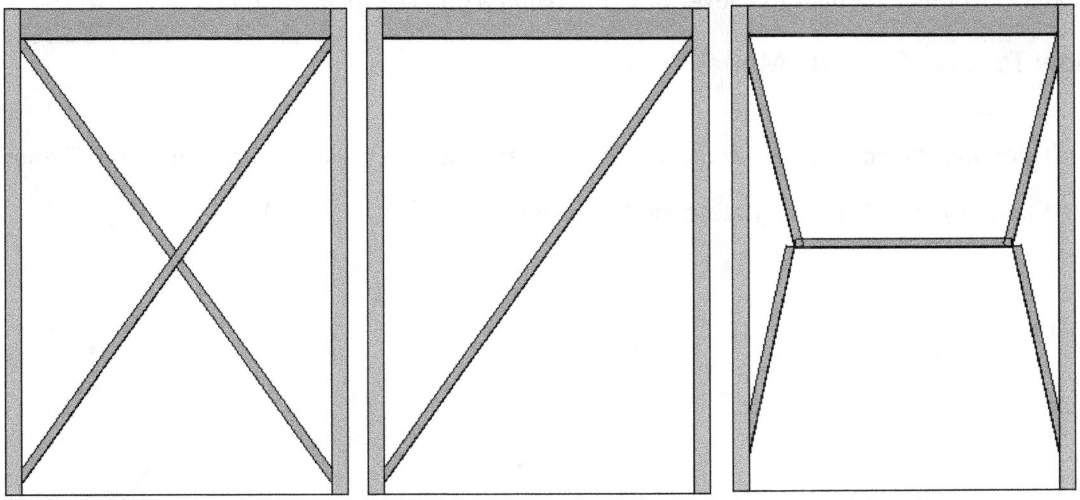

Crossed Bracing Type   *Single Bracing Type*   *Inserted Bracing Type*

Figure 2–75

5. Make modifications in the other tabs as required. For example, to get the bracing design shown in Figure 2–76, set the *Bracing Type* to **Single** and then, in the *Geometry* tab, set *Number of fields* to **2**.

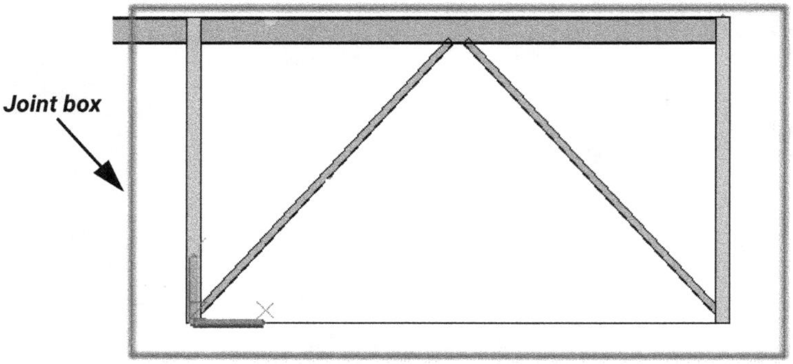

Joint box

Figure 2–76

- To modify the bracing, select one of the elements or the joint box, then right-click and select **Advance Joint Properties**. Alternatively, you can also double-click on the joint box.

    Note: The blue box around the bracing is called a joint box.

Grid Model Views

Bracing is often created along one grid line at a time. You can limit the display by creating a model view that is similar to a level view, but using a grid line as the cutting plane.

How To: Create a Grid Model View

1. Open the *Project Explorer* and click ▲ (Create new model view), or expand ▦ (Project Explorer) and click ⬜ (Create model views).

2. In the *Choose the definition method* dialog box (shown in Figure 2–77), select (At grid line).

Figure 2–77

3. In the model, select a grid line and press <Enter>.

4. Type a name for the view and press <Enter>.

5. Zoom in on the icon (as shown in Figure 2–78) and select the default view direction for the view.

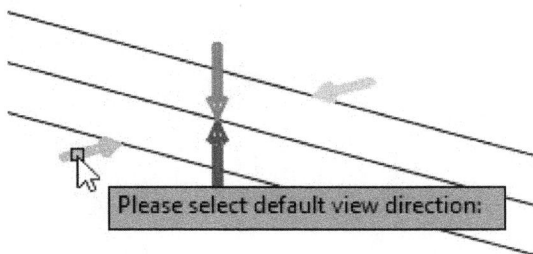

Please select default view direction:

Figure 2–78

6. To display only the elements along the selected grid, toggle off any level views and select the grid view, as shown in Figure 2-79.

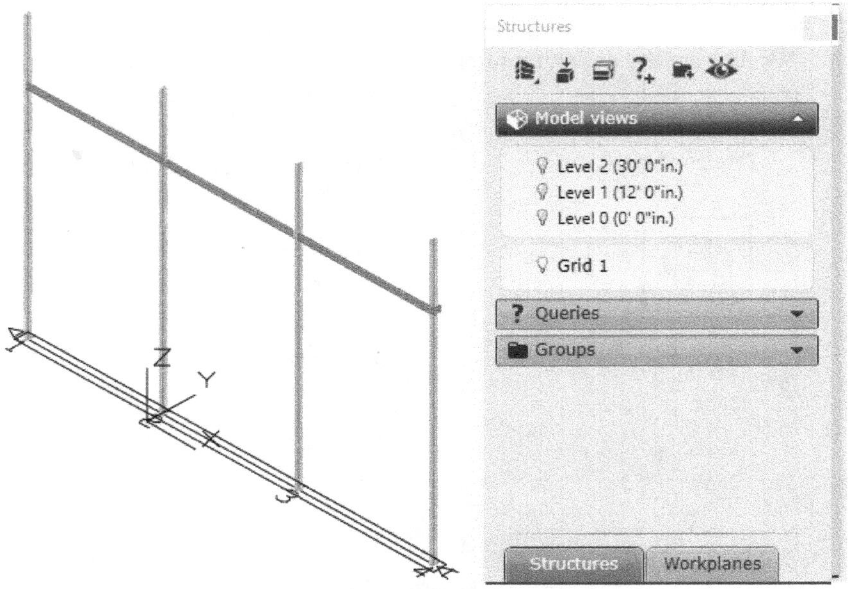

Figure 2-79

*   If more of the view displays than expected, you might need to modify the properties of the view. In the *Project Explorer*, right-click on the grid view name and select **Properties**. In the *Model View* dialog box>*Clipping* tab, ensure that the **Clip front** and **Clip back** options are selected, as shown in Figure 2-80.

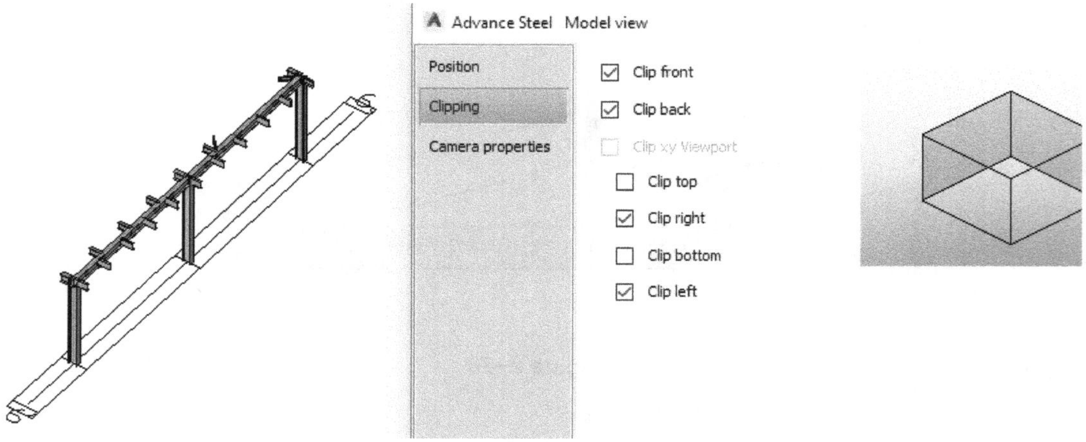

Figure 2-80

💡 Hint: Match Properties

When you have placed objects in a view and want to update other objects to match, you can

use the AutoCAD command 🖽 (Match Properties). This command updates the layer and all of the Advance Properties of the object.

1. In the *Tools* tab>*Properties* panel, click 🖽 (Match Properties), or type **MATCHPROP** in the command line.

2. Select the source object.

3. The cursor changes to a paintbrush, as shown in Figure 2–81. Select select the destination objects.

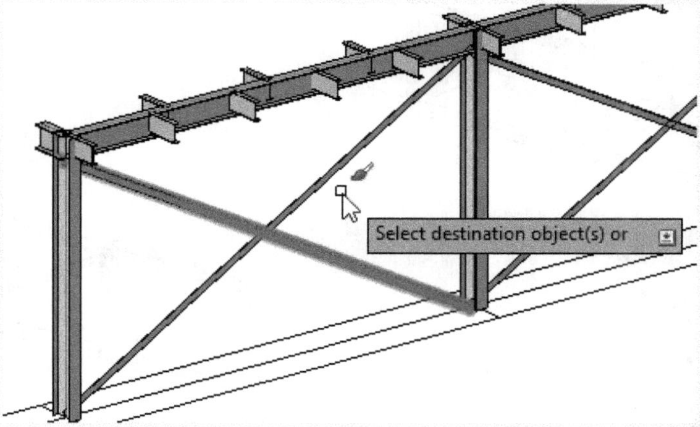

Figure 2–81

Practice 2d
Add Bracing

Practice Objectives

- Create model views based on grid lines.
- Set the UCS to the required orientation to add angled beams.
- Add braces to the structure.

In this practice, you will create grid model views for grid lines 1 and 6. You will then set the UCS to the correct orientation and draw braces, as shown in Figure 2–82.

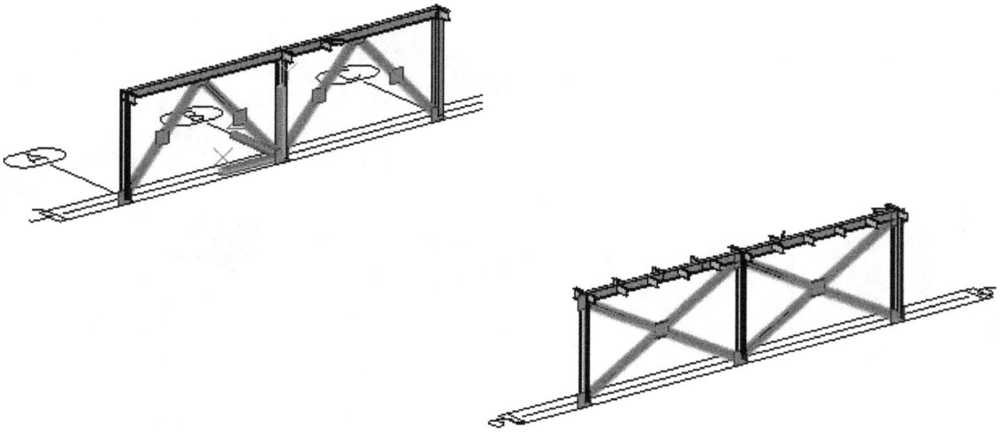

Figure 2–82

Task 1: Create a grid view.

1. In the practice files folder, open **Platform-Bracing.dwg**.
2. Open the *Project Explorer*.
3. In the *Project Explorer*, click ⬛ (Create new model view).
4. In the *Choose the definition method* dialog box, select ⬛ (At grid line).
5. In the model, select **Grid1** and press <Enter>.
6. At the *Give view name* prompt, type **Grid 1** and press <Enter>.

7. Rotate the view and zoom in as required to display the view direction arrows. Select the arrow pointing into the structure (as shown in Figure 2−83) and press <Enter>.

Figure 2−83

8. Return to the **SE Isometric** view. The new model grid view displays, as shown in Figure 2−84.

Figure 2−84

9. Create another model view for Grid 6.

---

10. In the *Project Explorer*, ensure that no level views are on, and then toggle on **Grid 1**. Only the elements on Grid 1 display, as shown in Figure 2−85.

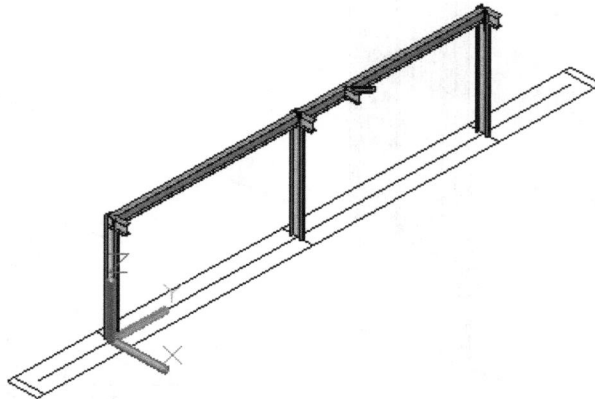

Figure 2−85

11. Save the drawing.

Task 2: Set the UCS and add bracing.

1. Rotate the view so that you can see the middle column from the outside of the structure, as shown in Figure 2−86.
2. In the *Advance Tool Palette>* ▨ (UCS) category, click ▨ (Move UCS).
3. Click the Grid Intersection Point snap at the base of the middle column to place the UCS as shown in Figure 2−86. Note that the UCS is still not facing the right direction.

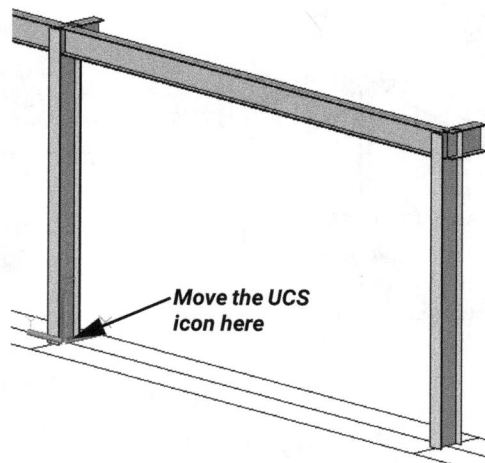

Move the UCS icon here

Figure 2−86

4. In the *Advance Tool Palette>* ▧ (UCS) category, click ▧ (Rotate UCS around X) and ▧ (Rotate UCS around Y) until the XY Axis are aligned as shown in Figure 2–87.

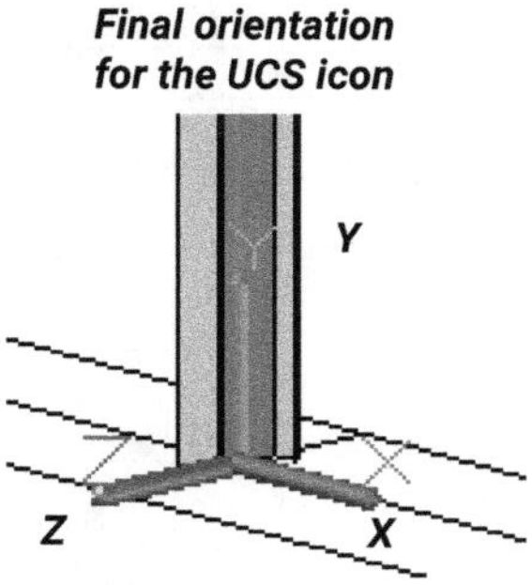

**Final orientation
for the UCS icon**

Figure 2–87

5. In the *Home* tab>*Extended Modeling* panel, click ⌃ (Bracing).

6. For the start point, select the same location as the UCS icon.

7. For the end point, select the top node of the opposite column.

8. In the *Structural element - Bracing* dialog box>*Type & Section* tab, set the following, as shown in Figure 2–88:

- *Bracing type*: **Single**

- *Section size*: **AISC 15.0 Angle identical>L3-1/2X3-1/2X1/4**

- *Model role:* **Vertical Bracing**

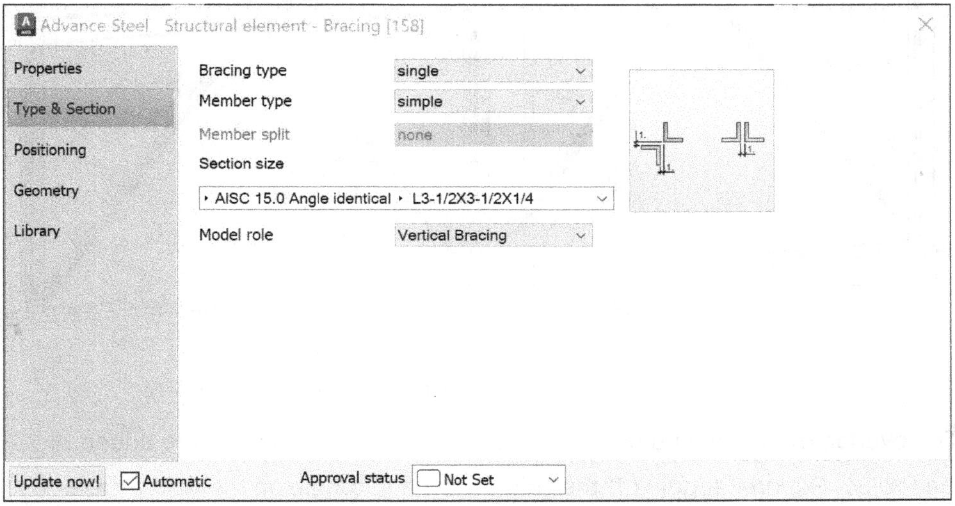

Figure 2–88

9. In the *Geometry* tab, set *Number of fields* to **2** and *Offset from top* to **1'-0"**, as shown in Figure 2–89.

Figure 2–89

10. Close the dialog box.

11. Repeat this process for the opening next to it (as shown in Figure 2–90), then press <Enter>.

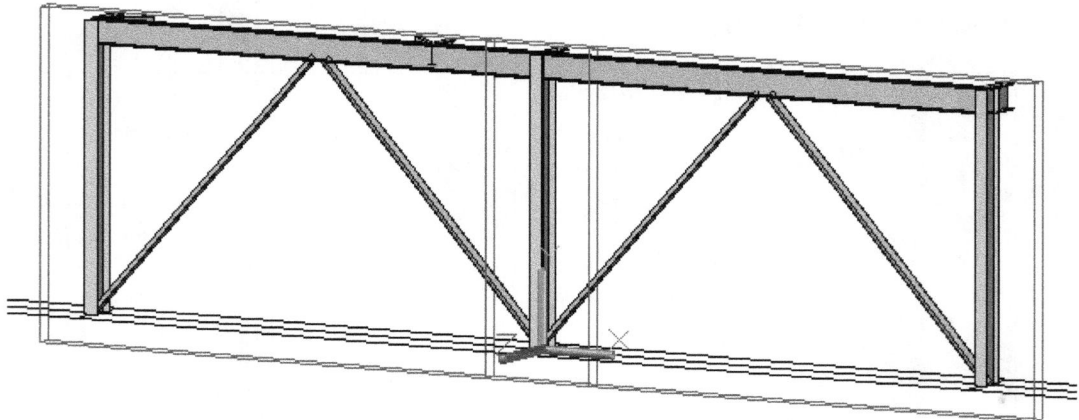

Figure 2–90

- The overlap of beams and braces is resolved when connections are added.

12. In the *Project Explorer*, toggle off the **Grid 1** view and toggle on the **Grid 6** view.

13. Create a new model view for Grid line 6.

14. Rotate the view as required. Note that Grid 7 still displays as shown in Figure 2–91.

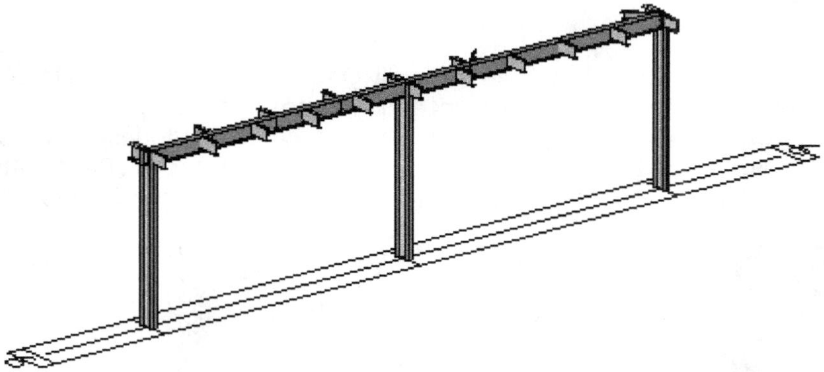

Figure 2–91

15. Move the UCS to Column **A6**.
16. Start the **Bracing** command and add the cross braces shown in Figure 2–92.

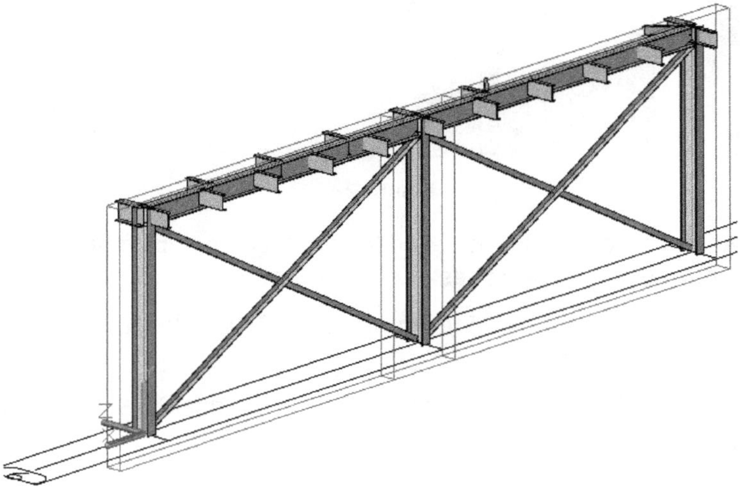

Figure 2–92

17. Toggle off the **Grid 6** view so that the full model displays.
18. Set the UCS to **World**.
19. Save the drawing.

End of practice

2.6 Integrating with the Autodesk Revit Software

Autodesk Revit structural steel models are typically created to a level of detail that provides a good starting point, but needs additional work and information before it can be used for fabrication. To save time recreating models, fabricators that use Autodesk Advance Steel can import basic steel models (as shown in Figure 2–93) from the Autodesk Revit software and then continue detailing the model with required connections and documentation.

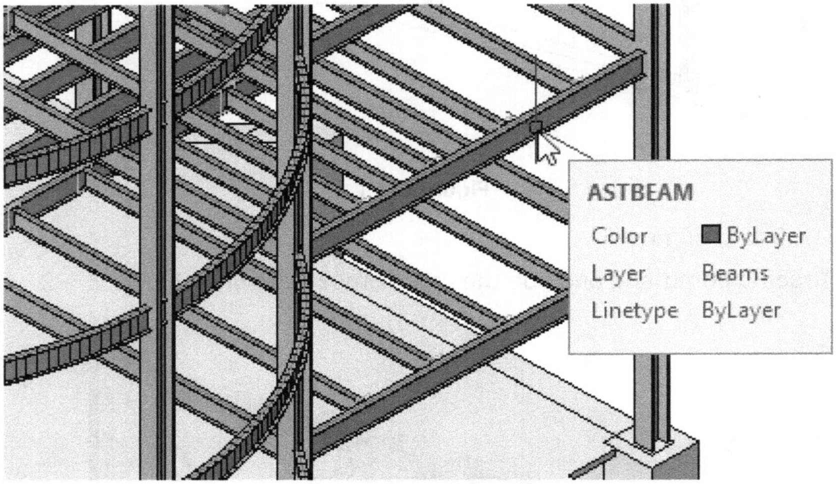

Figure 2–93

- Autodesk Revit models must first be saved to the Steel markup language (.SMLX) file before they can be imported into the Autodesk Advance Steel software.

- You can export Autodesk Advance Steel models as .SMLX files and then import them into the Autodesk Revit software.

- You can also synchronize changes in the Autodesk Advance Steel model with .SMLX files.

How To: Import Autodesk Revit Models

1. In the *Export & Import* tab>*Revit* panel, click 🔲 (Import from .SMLX model file).

2. In the *Open* dialog box, navigate to the folder location, select the .SMLX file, and then click **Open**. The file is opened in the Autodesk Advance Steel drawing, as shown in Figure 2−93.

    *   Steel elements from Autodesk Revit models become Autodesk Advance Steel objects.

    *   If a material used by the Autodesk Revit model is not used in the Autodesk Advance Steel software, you are prompted to select the appropriate information, as shown in Figure 2−94.

Figure 2−94

- To send an Autodesk Advance Steel model to an Autodesk Revit user to review, in the

  Export & Import tab>*Revit* panel, click (Export to .SMLX model file). The main steel objects are automatically converted to Autodesk Revit elements and the connectors retain their information but display as connector elements, as shown in the Autodesk Revit software in Figure 2–95.

Figure 2–95

- You can pass information back and forth with Autodesk Revit users by using

   (Synchronize current model with .SMLX file). This opens the *Synchronisation* dialog box, shown in Figure 2–96.

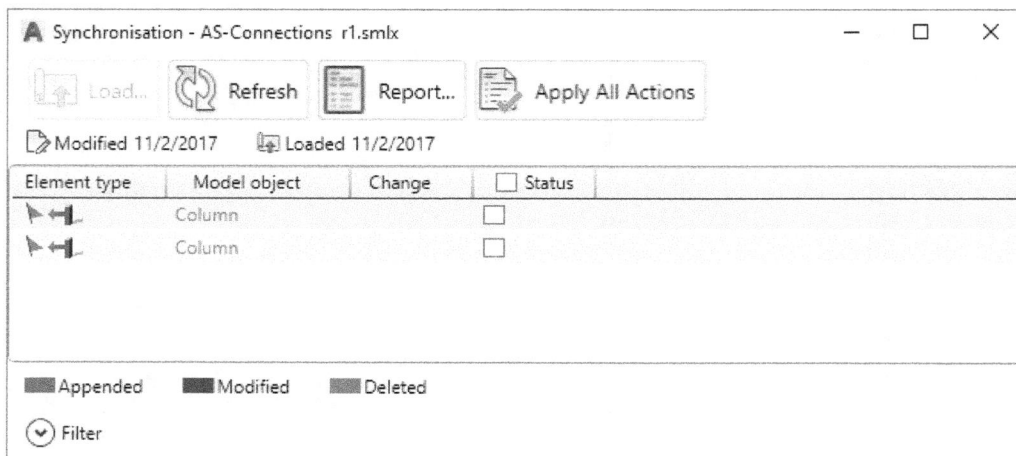

Figure 2–96

Practice 2e
Integrate with the Autodesk Revit Software

Practice Objectives

- Import an Autodesk Revit model that has been saved as an .SMLX file.
- Merge columns.

In this practice, you will import a model of a canopy created in the Autodesk Revit software and saved as a .SMLX file. You will then investigate the properties of the new objects. Finally, you will merge columns that extend from the base platform to the canopy, as shown in Figure 2–97.

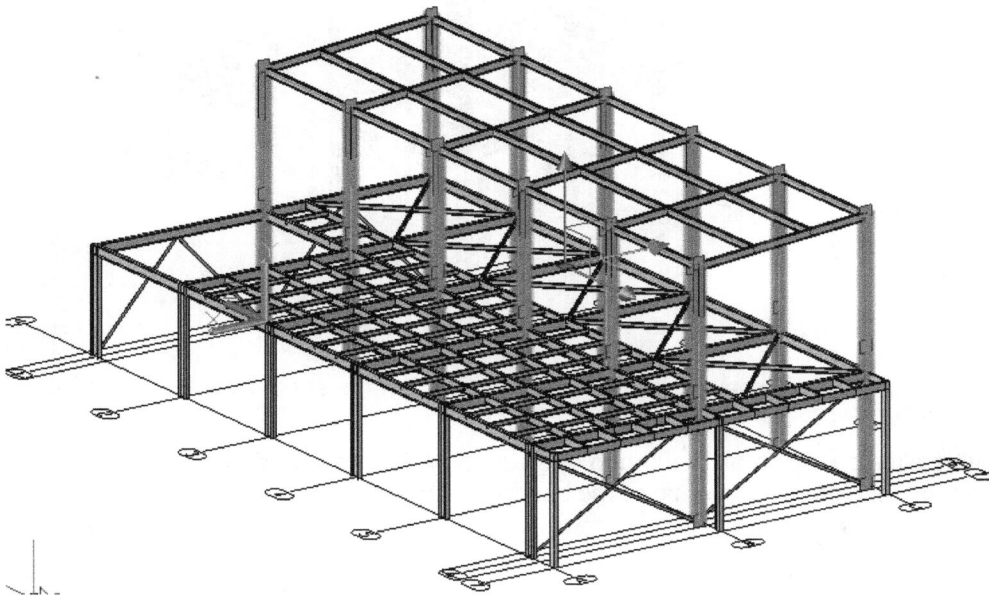

Figure 2–97

Task 1: Import a file created in the Autodesk Revit software.

1. In the practice files folder, open **Platform-Revit.dwg**.

2. In the *Export & Import* tab>*Revit* panel, click 🖾 (Import from .SMLX model file).

3. In the *Open* dialog box, navigate to the practice files folder, select **Platform-Canopy.smlx**, and click **Open**. The model elements are added to the drawing at the correct level, as shown in Figure 2–98.

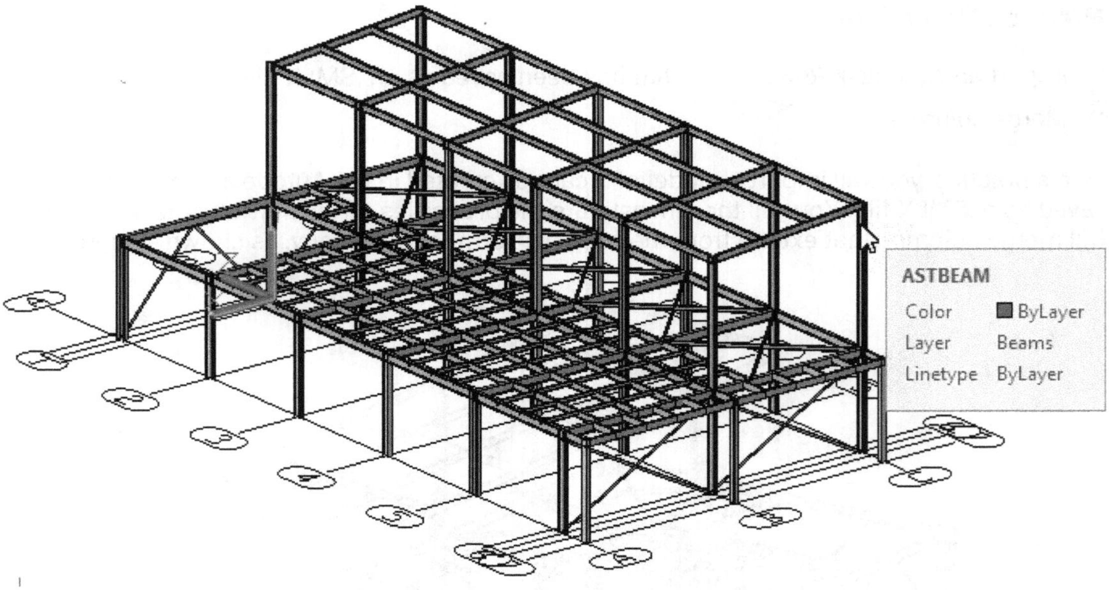

Figure 2–98

4. Hover over one of the new columns. Note that the layer is set to **Beams**, as shown in Figure 2–98.

5. Double-click on the same column to open up the Advance Properties and review the information. Note that the size of the beams is correct.

6. Save the drawing.

Task 2: Merge columns.

1. In the *Home* tab>*Objects* panel, click ![icon] (Merge Beams).

2. Select one of the columns below the canopy on Grid C, then select the canopy column connected to it and press <Enter>.

3. The two columns now become one, as shown in Figure 2–99.

    Note: The model view Grid 6 is toggled on for this figure.

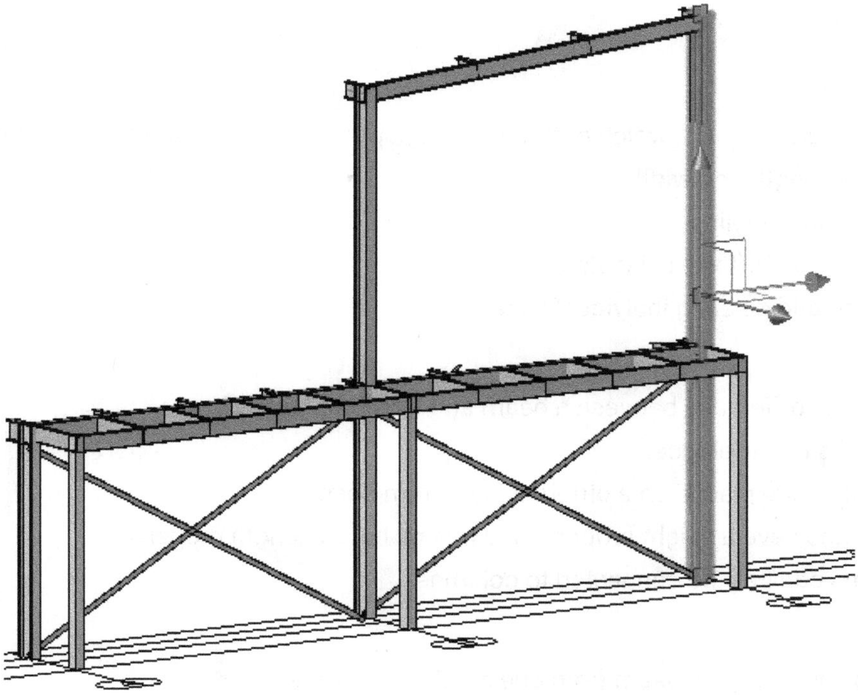

Figure 2–99

4. Press <Enter> to start the **Merge Beams** command again and repeat the process of selecting the base column and then the canopy column for each of the other pairs of beams. Make a Model View for Grid B and repeat Merge Beams on Grid B.

5. Save the drawing.

End of practice

Chapter Review Questions

1. Autodesk Advance Steel projects can be started using any AutoCAD template.
   a. True
   b. False

2. When you create grids, which of the following options are available? (Select all that apply.)
   a. Overall length and width
   b. Automatic labeling
   c. Prefixes or Suffixes in labels
   d. Adding and deleting individual axes

3. What is the difference between a beam and a column object?
   a. There is no difference.
   b. Columns are placed on a different layer from beams.
   c. Columns have a height property and beams have a length property.
   d. Beams can only be connected to columns.

4. How do you change a beam from one section type to another?
   a. Start the process using the tool in the drop-down menu using that specific section shape.
   b. Select the beam and change the section in the *Advance Tool Palette*.
   c. Select the beam and change the section in the *Advance Properties* dialog box.
   d. Use the **Change Beam Section** command.

5. In the *Project Explorer*, which of the following options would you use to limit a view to display beams and columns along only one line of a grid?

   a. ![icon] (Create new query)

   b. ![icon] (Create level above)

   c. ![icon] (Create new model view)

   d. ![icon] (Create new group)

6. What do you first need to do when you are creating a angled beam or bracing?

   a. Set up the bracing properties in the *Advance Properties* dialog box.

   b. Set the UCS so that the X- and Y-axes are in the correct direction.

   c. Add a beam and then rotate it into place.

   d. In the *Project Explorer*, set *Structure* to **Bracing**.

7. What can you do when you are working with a project that includes an Autodesk Revit model?

   a. Import a .RVT file

   b. XREF the model

   c. Import the model as a WBLOCK

   d. Import an .SMLX file

Command Summary

Button	Command	Location
Grids		
	Add axes	• **Ribbon**: *Objects* tab>*Grid* panel
	Building Grid	• **Ribbon**: *Home* tab>*Objects* panel • **Ribbon**: *Objects* tab>*Grid* panel
	Curved grid with single axis	• **Ribbon**: *Objects* tab>*Grid* panel
	Delete axes	• **Ribbon**: *Objects* tab>*Grid* panel
	Extend axes	• **Ribbon**: *Objects* tab>*Grid* panel
	Grid with 4 axes	• **Ribbon**: *Objects* tab>*Grid* panel
	Grid with groups by distance	• **Ribbon**: *Objects* tab>*Grid* panel
	Single Axis	• **Ribbon**: *Objects* tab>*Grid* panel
	Trim axes	• **Ribbon**: *Objects* tab>*Grid* panel
Beams		
	Beam, from line	• **Ribbon**: *Home* tab>*Objects* panel • **Ribbon**: *Objects* tab>*Beams* panel
	Beam, polyline	• **Ribbon**: *Home* tab>*Objects* panel • **Ribbon**: *Objects* tab>*Beams* panel
	Bracing	• **Ribbon**: *Home* tab>*Extended Modeling* panel • **Ribbon**: *Extended Modeling* tab>*Structural Elements* panel
	Channel Section	• **Ribbon:** *Home* tab>*Objects* panel
	Column	• **Ribbon**: *Home* tab>*Objects* panel • **Ribbon**: *Objects* tab>*Beams* panel
	Concrete Beam	• **Ribbon**: *Home* tab>*Objects* panel • **Ribbon**: *Objects* tab>*Other Objects* panel
	Continuous Beam	• **Ribbon**: *Objects* tab>*Beams* panel

Button	Command	Location
	Curved Beam	• **Ribbon**: *Home* tab>*Objects* panel • **Ribbon**: *Objects* tab>*Beams* panel
	Double Channel - back to back (and similar)	• **Ribbon**: *Home* tab>*Objects* panel • **Ribbon**: *Objects* tab>*Beams* panel
	Merge Beams	• **Ribbon**: *Objects* tab>*Beams* panel
	Mono-pitch Frame	• **Ribbon**: *Home* tab>*Extended Modeling* panel • **Ribbon**: *Extended Modeling* tab>*Structural Elements* panel
	Portal/Gable Frame	• **Ribbon**: *Home* tab>*Extended Modeling* panel • **Ribbon**: *Extended Modeling* tab>*Structural Elements* panel
	Rolled I Section (and similar)	• **Ribbon**: *Home* tab>*Objects* panel • **Ribbon**: *Objects* tab>*Beams* panel
	Split Beams	• **Ribbon**: *Objects* tab>*Beams* panel
	Welded Beam, tapered (and similar)	• **Ribbon**: *Home* tab>*Objects* panel • **Ribbon**: *Objects* tab>*Beams* panel
General Tools		
	Create Level Above	• **Project Explorer (Structures palette)**
	Create new model view	• **Project Explorer (Structures palette)** • **Ribbon**: *Home* tab>*Project* panel>expand *Project Explorer*
	Isolate Objects	• **Status Bar**
	Match Properties	• **Ribbon**: *Tools* tab>*Properties* panel • **Command Line:** MATCHPROP
	New	• **Quick Access Toolbar** • **Application Menu**
	Project Explorer	• **Ribbon**: *Home* tab>*Project* panel
	Project Settings...	• **Ribbon**: *Home* tab>*Settings* panel

Button	Command	Location
Integrate with Revit		
	Export to *.SMLX model file	• **Ribbon**: *Export & Import* tab>*Revit* panel
	Import from *.SMLX model file	• **Ribbon**: *Export & Import* tab>*Revit* panel
	Sychronize current model with *.SMLX file	• **Ribbon**: *Export & Import* tab>*Revit* panel

Creating Connections

Connections are the most crucial part of creating steel buildings. Even if you receive existing models that include columns, beams, and braces, they rarely include the level of information that is required for a full fabrication project. The Autodesk® Advance Steel software includes many preconfigured connections that include plates, bolts, and other objects that you can apply using the *Connection Vault*. You can also create custom connections that are based on groups of connections and the individual elements, such as plates, bolts, and welds.

Learning Objectives

- Add connections using tools found in the *Connection Vault*.
- Work with Autodesk Advance Steel *Joint Properties*.
- Create joint groups.
- Edit beam intersections with miters, cutbacks, and copes.
- Create plates, including flat and folded plates.
- Add openings, corner cuts, and similar features to plates and beams.
- Add bolts, anchors, and welds.
- Create custom connections.

3.1 Working with the Connection Vault

The *Connection Vault* contains macros that add connectors, and trim and cope steel elements in the process of applying the connectors. For example, before applying a series of clip angles, the beams touching the column shown in Figure 3–1 overlap. After the connections are applied, the beams are cut back and angles and bolts are applied, as shown in Figure 3–2.

Figure 3–1

Figure 3–2

- Joint boxes display after you apply a connection.

- Modifications to a connection are made in the *Connection Properties* dialog box (also known as Advance Joint Properties), shown for a base plate in Figure 3–3. The information in the dialog box changes depending on the type of connection that is selected.

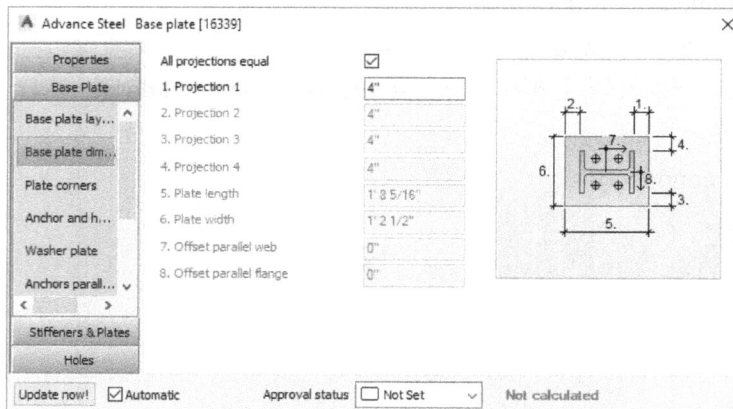

Figure 3–3

- Once a connection is in place, you can double-click on the joint box to open the *Advance Joint Properties* dialog box. Alternatively, select a non-beam object in the connection, right-click and select **Advance Joint Properties**.

The Connection Vault

The *Connection Vault* (shown in Figure 3–4) is divided into categories, with multiple connection types in each category. Each connection tool includes a graphic representation of what it does and a description that includes the selection order of elements.

- In the *Home* tab>*Extended Modeling* panel, click ▐▐▄ (Connection Vault).

  Note: The selection order of the tool is especially important when you are working with three or more objects to select.

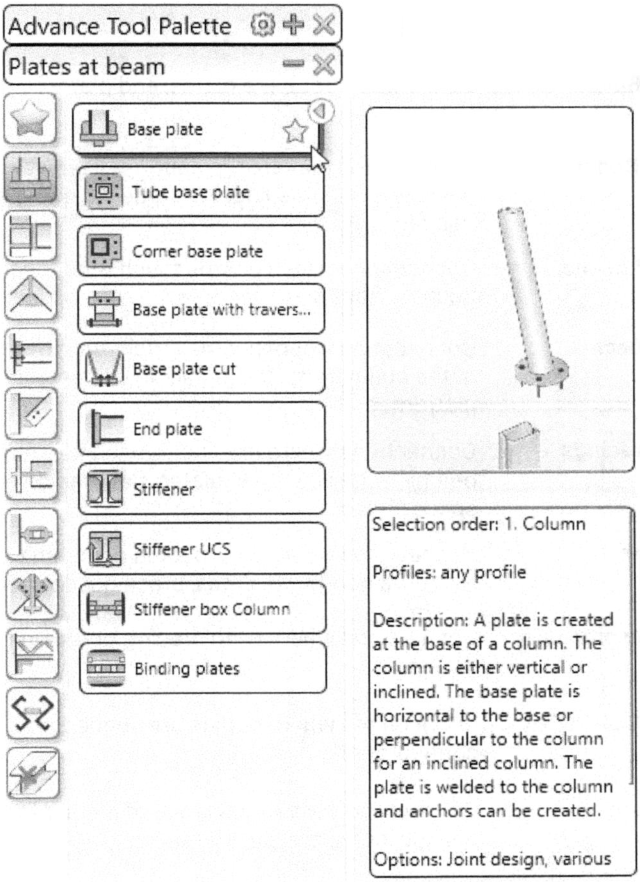

Figure 3–4

- The *Connection Vault* can be docked under the *Advance Tool Palette*, as shown in Figure 3–4, above.

Connection Vault Categories

There are multiple options available, that are divided into categories, as shown in the table below.

	Favorites	Add the connections you use most to this category. Hover the cursor over the tool you want to add to your favorites and click **Add to favorites**, as shown below.
	Plates at Beam	Connections for base plates, end plates, and stiffeners.
	Column - Beam	Connections that are typically used between a column and a beam, such as the knee of frame, seat angles, and moment flanges.
	Beam end to end	Connections where beams touch other beams, including an apex haunch, front plate splice, and purlin splice.
	Platform beams	Connections where beams touch other beams (columns), but not at the beam endpoint, including clip angles, end plates, and platform plates.
	General bracings	Connections where bracing (angled beams) intersect with other beams, such as gusset plates, flat bracing with tension bolts, and splice plates.
	Tube connections	Connections where tube beams are connected to other beams, including sandwich plates, brace gussets, and brace angles.
	Turnbuckle bracings	Connections where wind bracing or tension rods are used.
	Purlins & Cold rolled	Connections where purlins are connected with beams in various configurations.
	Miscellaneous	Connections for various types of elements, such as stairs, railings, bolts, and punch marks.
	Fabricator specific macros	Connections where fabricators requested specific methods.
	Old - not updated anymore	Connections that have been used in the past, but are no longer as frequently used.

Creating Multiple Connections

When you need to apply the same connections to multiple sets of objects, you can create joint groups. For example, a base plate connection was first created using standard anchor bolts. Then, after the connections were applied using joint groups, the original connection was updated to j-bolts, and the rest of the joint groups were updated as well, as shown in Figure 3–5.

> *Note:* Joint groups display as a grey bounding box.

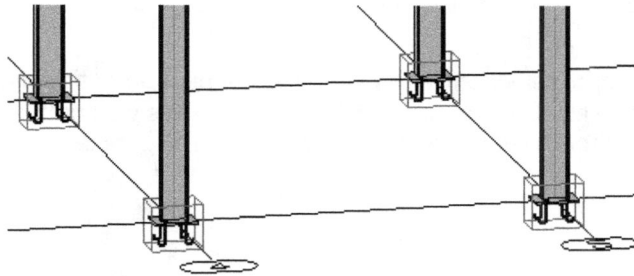

Figure 3–5

Joint groups can be created using any of the following tools:

- **Create Joints in a Joint Group**
- **Create Joints in a Joint Group, Multiple**
- **Propagate Joints**

The difference between these tools is the order in which you select the objects to be connected.

> *Note: Only Connections which have identical situations within the model can be propagated.*

How To: Create Joints in a Joint Group

1. In the *Advance Tool Palette*>🛠 (Tools) category, click 🗒 (Create joint in a joint group).
2. Select an object in a connection (such as a plate or bolt) and press <Enter>.
3. Depending on the type of connection you selected, you are prompted for the same object types as the original process. For example, if you are adding a base plate, you are prompted to select the next column.
4. Follow any other prompts as required for the specific connection type.

How To: Create Joints in a Joint Group, Multiple

1. In the *Advance Tool Palette>* [icon] (Tools) category, click [icon] (Create joint in a joint group, multiple).

2. Select an object in the template connection (such as a plate or bolt) and press <Enter>.

3. Depending on the type of connection you selected, you are prompted for the same object types as the original process. In this case, select all of the first type of object (such as a column), press <Enter>. Then select all of the second type of object (such as a beam), and press <Enter>.

- To update a joint group, select any element in the connection or the joint box (if displayed). Then, right-click and select **Advance Joint Properties**. If this was not the original connection, in the dialog box, select **Upgrade to master**, as shown in Figure 3–6.

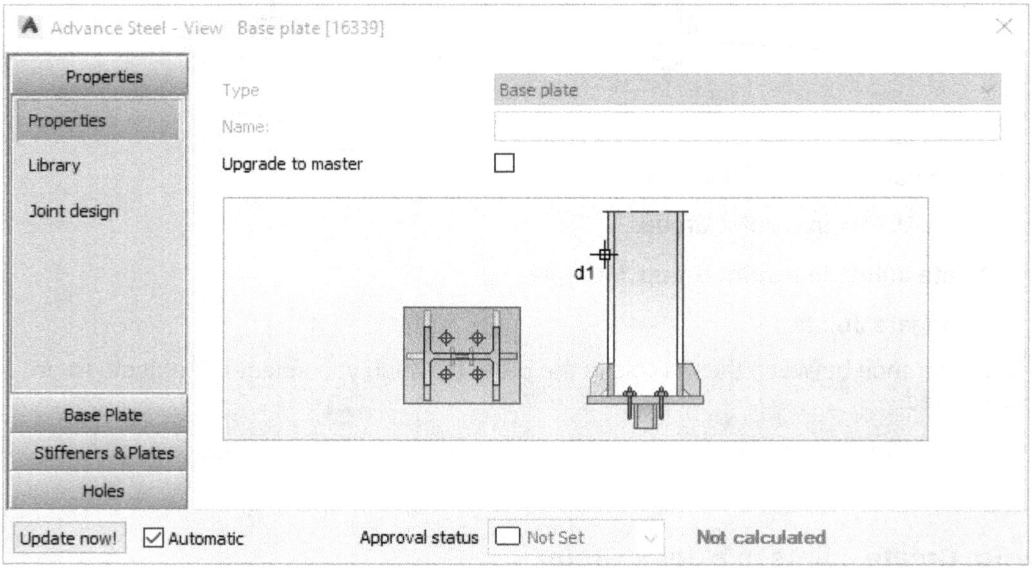

Figure 3–6

- If you delete a joint box, each piece becomes separate (e.g., anchors, beams, plates, etc.). To delete an entire joint group, select one of the objects in the connection and press <Delete>.

- If you want to copy joints but not have them in a joint group, you can use the [icon] (Create by template) or [icon] (Create by template, multiple) tools. These tools work the same as the joint group tools, but do not include the connection with other joints.

How To: Propagate Joints

1. Select the joint box for the connection in the Model.

2. Right-click and select **Propagate Joint** from the menu, as shown in Figure 3-7.

The results of the propagation of original connection are joined in a single group of connections. The original connection is the master connection.

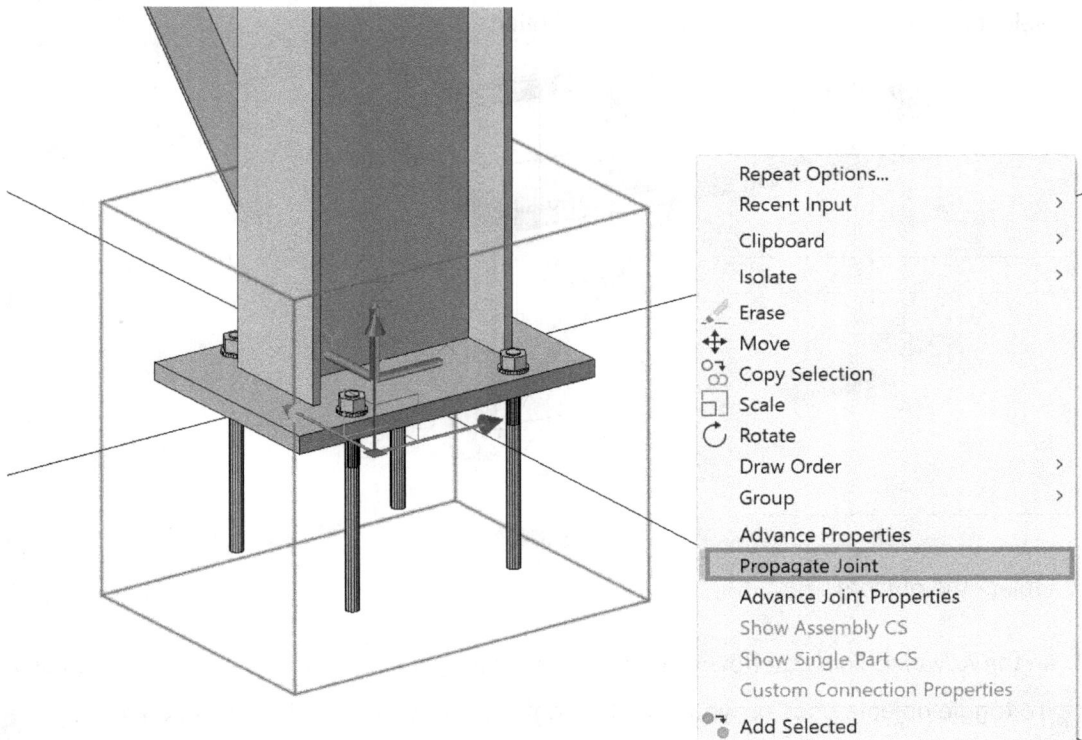

Figure 3-7

🔆 Hint: Toggling Objects Off and On

Joint boxes (shown in Figure 3–8) are one of the many types of objects that you can toggle off if you do not want them to display in a model. You can temporarily toggle them off using

⧉ (Isolate Objects), but this setting resets if you open and close the drawing. Instead, toggle them off using an Autodesk Advance Steel command.

- To select all of the objects in a model that are the same type (e.g., joint boxes or bolts), select one, right-click and select **Select Similar**.

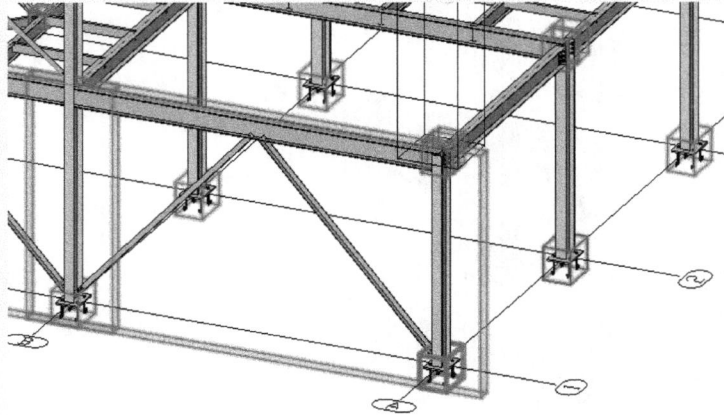

Figure 3–8

1. Select the objects.

2. In the *Advance Tool Palette>* 🔲 (Quick views) category, click 🔲 (Selected objects off).

- To toggle objects back on, you need to toggle on everything. In the *Advance Tool Palette>* 🔲 (Quick views) category, click 🔲 (All Visible).

Advance Joint Properties Libraries

The Advance Joint Properties includes a Library of connections, as shown in Figure 3–9. Note that only some of the connections might be appropriate for a situation. When you create a connection, you can save it to the library where it can be reused for other connections in the current installation of the software.

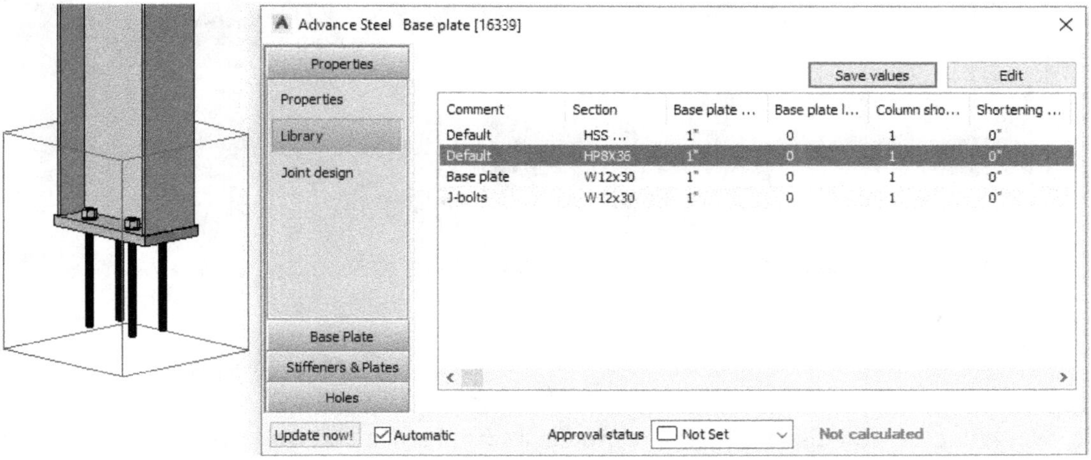

Figure 3–9

- The connection libraries are saved on your computer or server. To use the connections from another user's library, open a project from that user that includes the connections. You can also merge libraries into other installations of the Autodesk Advance Steel software through the Management Tools.

How To: Add a Connection to the Library

1. Set up the connection in the Advance Joint Properties.
2. In the *Library* tab, click **Save Values** and then click **Edit**.

3. In the *Library* dialog box (shown in Figure 3–10), enter a name for the new group.

	Comment	Section	Base plate thickness	Base plate layout	Column shortening type	Shortening extension value
1	Default	HSS 2X2X'	1"	0	1	0"
2	Default	HP8X36	1"	0	1	0"
22002	Base plate	W12x30	1"	0	1	0"
22003	J-bolts	W12x30	1"	0	1	0"

OK   Cancel   New line   Delete line   Help

Figure 3–10

4. Note that you can make adjustments to the information in this dialog box, and even add or delete lines, but it is often easier to do make these changes in the main dialog box tabs, and then save a new set of values.

5. Click **OK**.

> **Hint: Deleting Joint Boxes**
>
> If you delete a joint box, the objects that form the connection remain as separate entities, as shown in Figure 3–11.

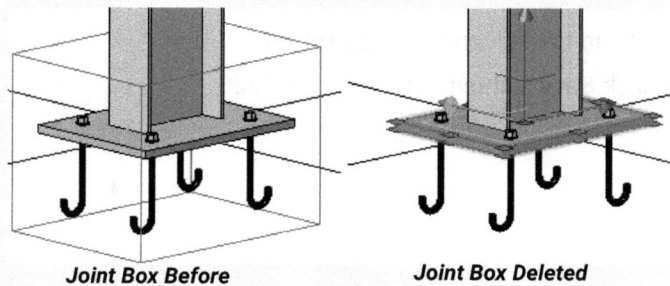

Joint Box Before   *Joint Box Deleted*

Figure 3–11

If you want to completely delete all of the elements in the connection, in the *Extended Modeling* tab>*Joint Utilities* panel, click (Delete All), and then select the joint box. All of the connection objects are removed.

Practice 3a
Work with the Connection Vault

Practice Objectives

- Use the tools found in the *Connection Vault*.
- Create joint groups.

In this practice, you will add a base plate connection to one column, and then create joint groups that you apply to the other columns. You will then update the joint group with a different bolt style and note how it replicates throughout the model. You will then add a clip angle to a column and beam intersection, modify the joint properties, and copy it to other intersections, as shown in Figure 3–12.

- If you have time, you can also add other connections, such as gusset plates for bracing.

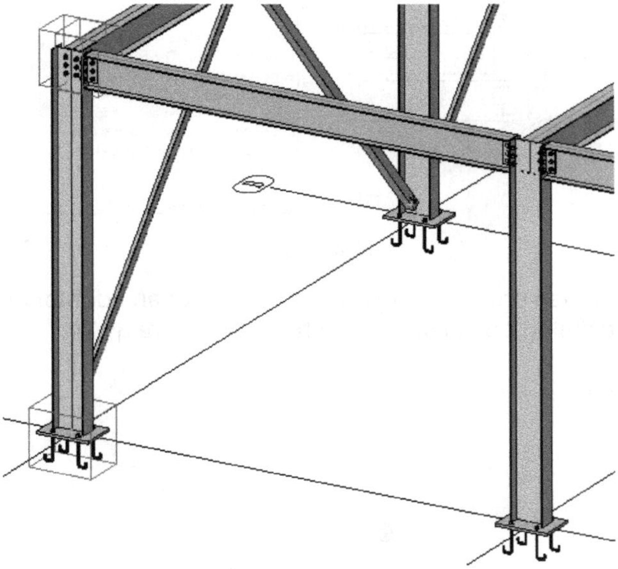

Figure 3–12

Task 1: Add a base plate connection.

1. In the practice files folder, open **Platform-Connections.dwg**.
2. Zoom in on the base of Column A1.
3. In the *Home* tab>*Extended Modeling* panel, click ▨ (Connection Vault).
4. Click and drag the *Connection Vault* so that it is docked under the *Advance Tool Palette*.

5. In the *Connection Vault>* *(Plates at beam)* category, click (Base plate), shown in Figure 3-13.

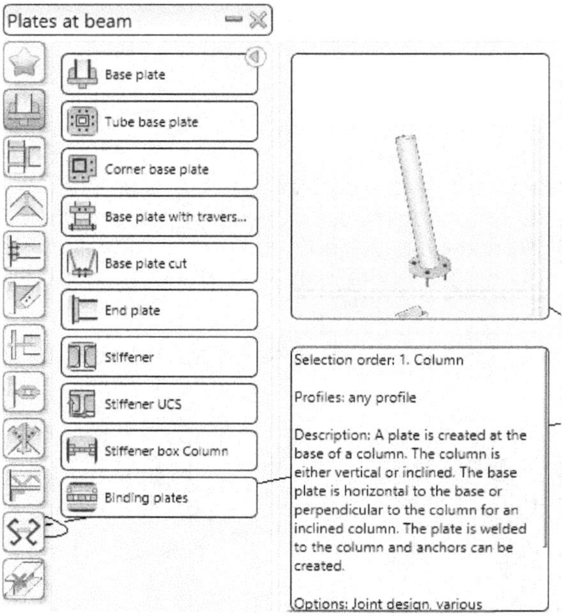

Figure 3-13

6. Select column A1. When prompted if you want to select an additional concrete object, click **No**. The *Base plate* dialog box displays as shown in Figure 3-14.

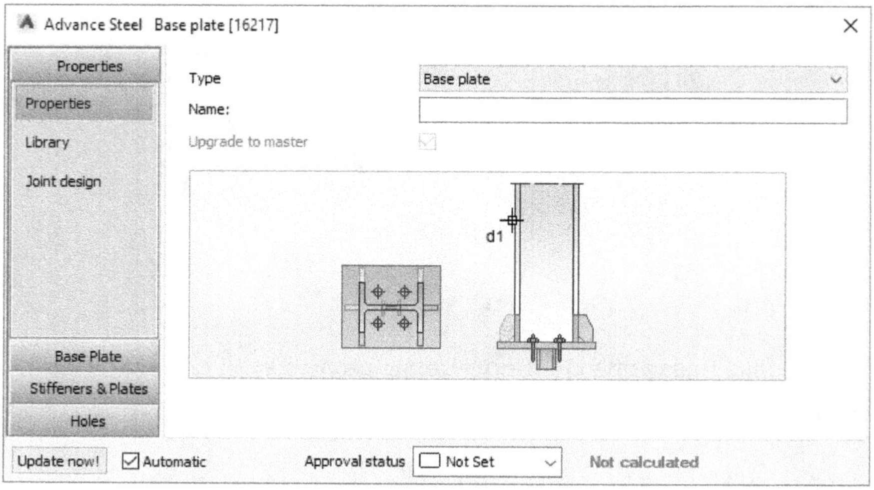

Figure 3-14

7. Move the dialog box so that the new connector displays as shown on the right in Figure 3–15.

8. Click on the *Library* tab, and note that there are at least two **Default** connector options. Click on each to display what they look like, and then select the larger of the two, as shown in Figure 3–15.

    *   Additional options might display if this copy of the software has been used before. Library options are set by the user.

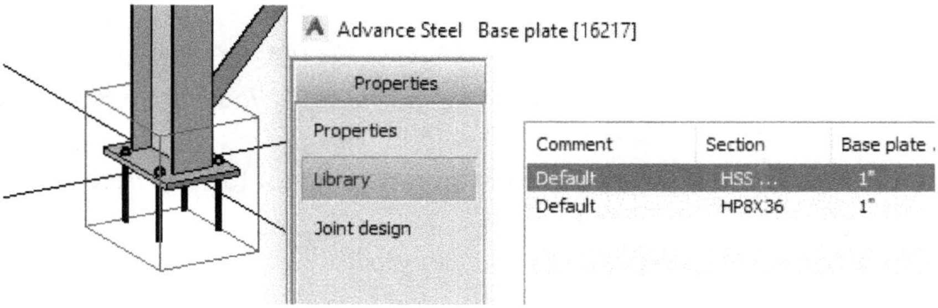

Comment	Section	Base plate .
Default	HSS ...	1"
Default	HP8X36	1"

Figure 3–15

9. Click on the *Base Plate* tab and go through the sub tabs to note the defaults and test different options.

10. To finish, return to the *Properties>Library* tab and select the top default option. Then, in the *Base plate dimensions* tab, change *Projection 1* to **4"** and press <Enter> to apply the change, as shown in Figure 3–16.

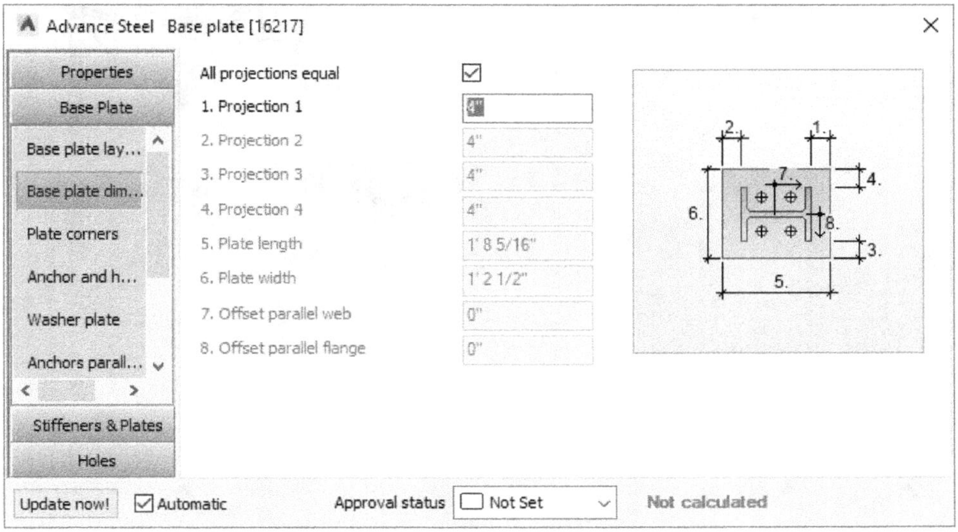

Figure 3–16

11. Close the dialog box and save the drawing.

Task 2: Create and apply joint groups.

1. Rotate the view so that all of the bottom columns display.

2. In the *Advance Tool Palette*>⬚ (Tools) category, click ⬚ (Create joint in a joint group).

   Note: You could continue using this command and add the base plate to the other columns, but it might be faster to use the other joint group tool.

3. Select one of the objects in the original connection and press <Enter>.

4. Select one of the other columns. When you are prompted if you want to select an additional concrete item, press <Enter> to accept the default **No**.

5. Select another column and select **No** when prompted about the concrete column.

6. In the *Advance Tool Palette*>⬚ (Tools) category, click ⬚ (Create joint in a joint group, multiple).

7. Select the original base plate and press <Enter>.

8. Select the rest of the columns that use the same size base plate (as shown in Figure 3–17) and press <Enter>.

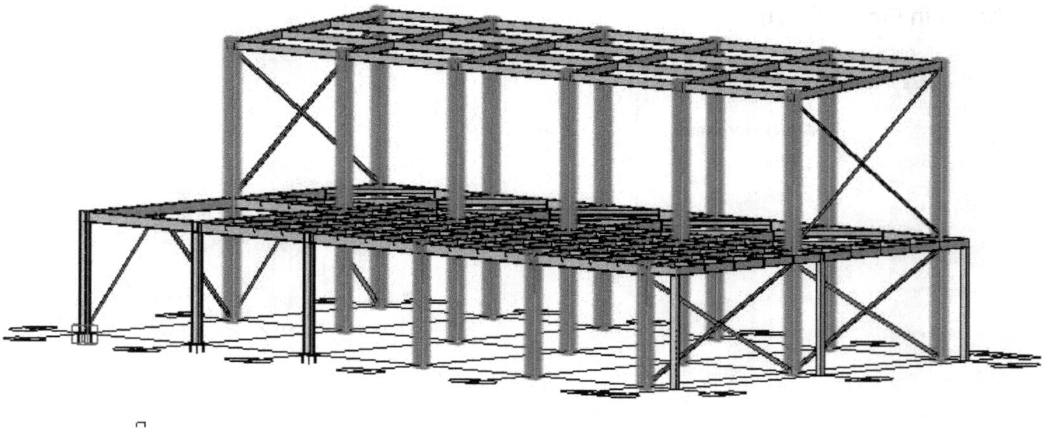

Figure 3–17

9. Connections are added to the base of each column as shown in Figure 3–18.

Figure 3–18

10. Save the drawing.

Task 3: Modify the properties of a joint group.

1. Zoom in on one of the new connections (i.e., not the one you originally created). Select one of the bolts, right-click and select **Advance Joint Properties**.

2. In the *Properties>Properties* tab, select **Upgrade to master**.

3. In the *Base Plate>Anchor and holes* tab, change the *Anchor type* to **US J-Round Anchors**.

4. The rest of the grouped connectors update, as shown in Figure 3–19.

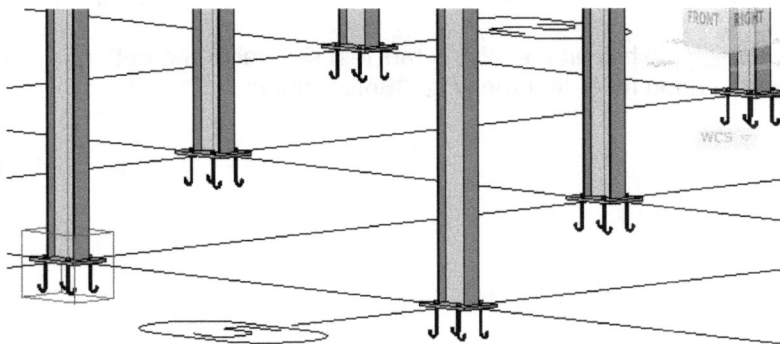

Figure 3–19

5. Save the drawing.

Task 4: Add clip angles and joint groups.

1. Zoom in on the top of Column A1, as shown in Figure 3-20.

2. In the *Connection Vault>* ▣ (Platform beams) category, click ▤ (Clip angle).

3. Read the selection order in the tooltip and then make the appropriate selections between the column and the connecting beam.

    - Remember to press <Enter> between the selections.

    - The connecting beam is automatically trimmed back and the new clip angles and bolts display, but are not the right size, as shown in Figure 3-21.

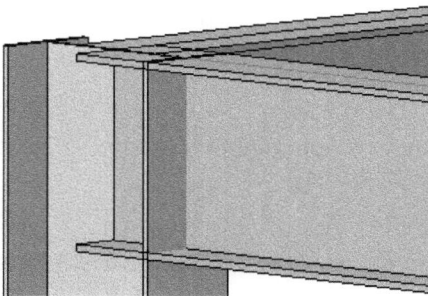

Figure 3-20

Figure 3-21

4. In the *Clip angle* dialog box, in the *Library* tab, test several of the options. Note that none of the options are a good fit. Select the **W12-3Bolts** option.

5. In the *Clip angle* tab, change the *Angle profile size* to **AISC 15.0 Angle not identical> L4X3X3/8**, as shown in Figure 3–22.

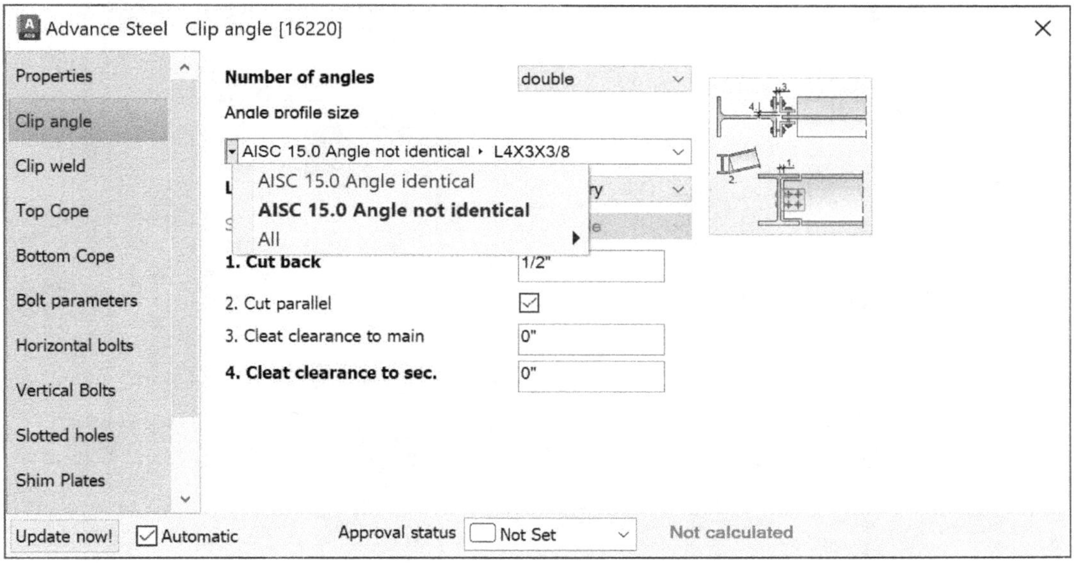

Figure 3–22

6. In the *Horizontal bolts* tab, change the *Back mark* to **2"**. The modified clip angle and bolts display, as shown in Figure 3–23.

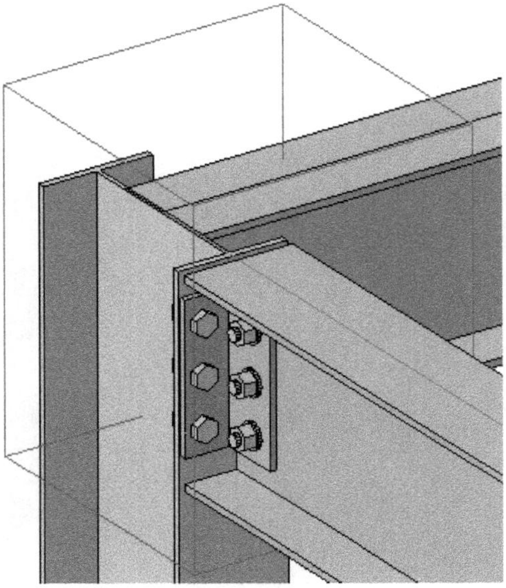

Figure 3–23

7. In the *Library* tab, click **Save values** and then click **Edit**.

8. In the *Library* dialog box, type **W12x30-3Bolts** as a new name (as shown in Figure 3–24) and click **OK**.

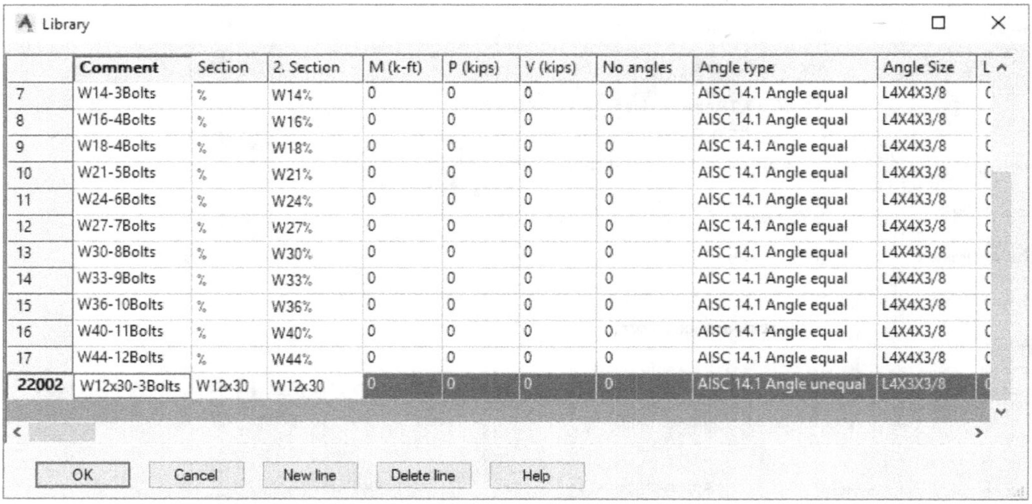

	Comment	Section	2. Section	M (k-ft)	P (kips)	V (kips)	No angles	Angle type	Angle Size	L
7	W14-3Bolts	%	W14%	0	0	0	0	AISC 14.1 Angle equal	L4X4X3/8	(
8	W16-4Bolts	%	W16%	0	0	0	0	AISC 14.1 Angle equal	L4X4X3/8	(
9	W18-4Bolts	%	W18%	0	0	0	0	AISC 14.1 Angle equal	L4X4X3/8	(
10	W21-5Bolts	%	W21%	0	0	0	0	AISC 14.1 Angle equal	L4X4X3/8	(
11	W24-6Bolts	%	W24%	0	0	0	0	AISC 14.1 Angle equal	L4X4X3/8	(
12	W27-7Bolts	%	W27%	0	0	0	0	AISC 14.1 Angle equal	L4X4X3/8	(
13	W30-8Bolts	%	W30%	0	0	0	0	AISC 14.1 Angle equal	L4X4X3/8	(
14	W33-9Bolts	%	W33%	0	0	0	0	AISC 14.1 Angle equal	L4X4X3/8	(
15	W36-10Bolts	%	W36%	0	0	0	0	AISC 14.1 Angle equal	L4X4X3/8	(
16	W40-11Bolts	%	W40%	0	0	0	0	AISC 14.1 Angle equal	L4X4X3/8	(
17	W44-12Bolts	%	W44%	0	0	0	0	AISC 14.1 Angle equal	L4X4X3/8	(
22002	W12x30-3Bolts	W12x30	W12x30	0	0	0	0	AISC 14.1 Angle unequal	L4X3X3/8	(

OK   Cancel   New line   Delete line   Help

Figure 3–24

9. Close the *Clip angles* dialog box.

10. Repeat the process and add clip angles to the other beam, as shown in Figure 3–25.

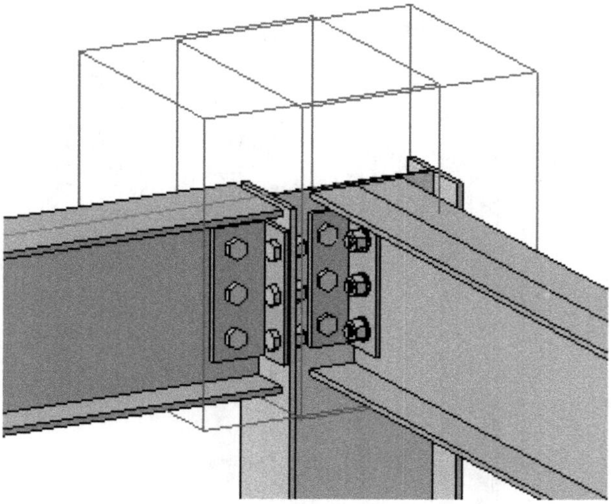

Figure 3–25

11. Save the drawing.

Task 5: Propagate a Connection.

1. Rotate the view so that the outside columns display.

2. Select the joint box connection, right-click and select **Propagate Joint** from the contextual menu.

3. Repeat step 2 for the other clip angle connection.

4. Add a double-sided clip angle on Column 1B and the two beams on each side, as shown in Figure 3–26.

5. Propagate the double-sided clip angle, as shown in Figure 3–27.

6. Save the drawing.

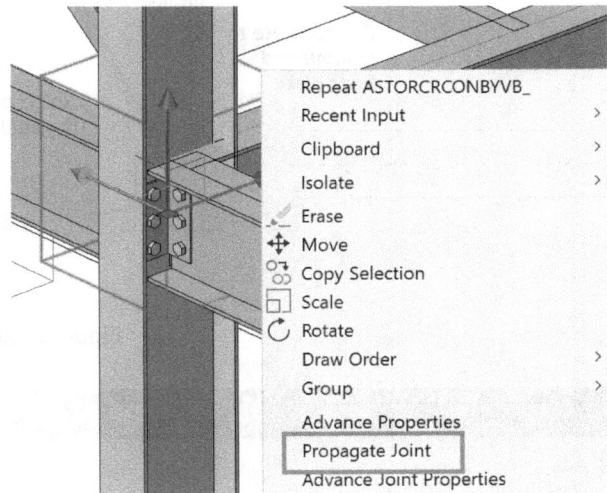

Figure 3–26   **Figure 3–27**

7. Save the drawing.

Task 6: Additional connections.

1. Add other connections, such as the gusset plate for 2 diagonals and gusset plate to column and base plate, as shown in Figure 3–28.

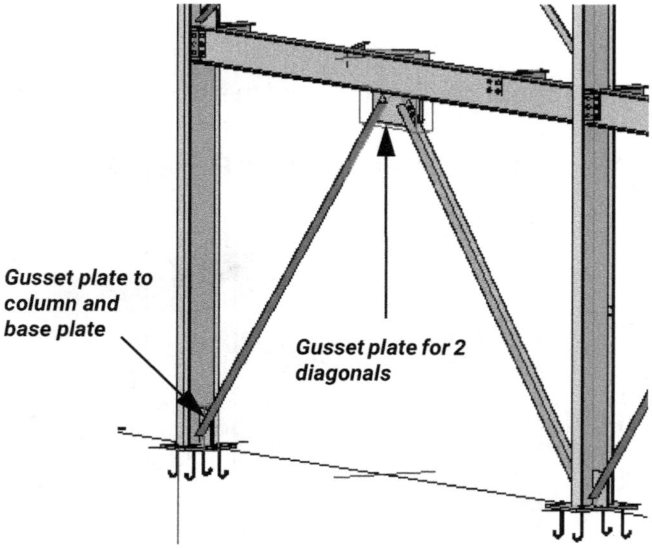

Figure 3–28

End of practice

3.2 Editing Beam Intersections

If you are going to create a connection from scratch, you first need to modify the overlapping beams (as shown in Figure 3–29) or otherwise modify the structural elements. You can do this through a variety of custom beam features, including miters (as shown in Figure 3–30), saw cuts, shortening, and coping tools.

Figure 3–29

Figure 3–30

Note: Shortening is where you adjust the length of a column or beam to accommodate a joint plate.

How To: Shorten Beams

1. In the *Advance Tool Palette>* (Features) category, click (Shorten).
2. Select the column or beam you want to shorten near the end that you want to modify.
3. In the *Shorten beam* dialog box (shown in Figure 3–31), specify the *Reference* (i.e., the amount of the beam to remove). You can also specify a value in degrees to rotate the cut about the Y or Z axis.

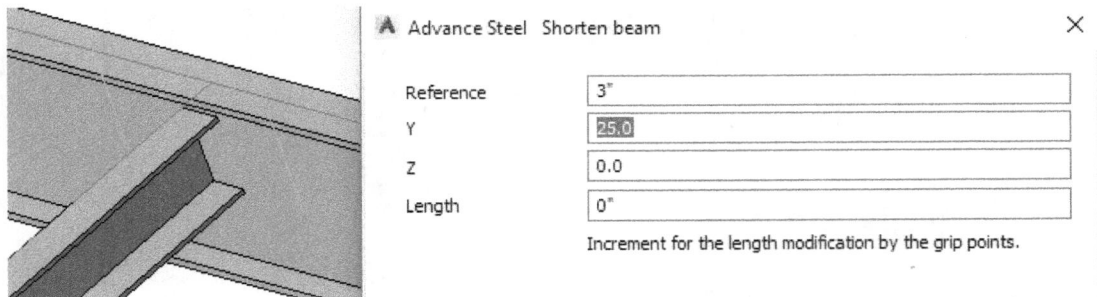

Advance Steel   Shorten beam   ✕

Reference	3"
Y	25.0
Z	0.0
Length	0"

Increment for the length modification by the grip points.

Figure 3–31

- Feature contours displays as a green box. You can double-click on the box to open the related *Advance Properties* dialog box.

- If you are working with angled beams, or do not know the angle you want to cut off, set the UCS to the required position and orientation where you want the cut to be made and then use the [icon] (Shorten at UCS) command.

How To: Cut and Miter Structural Objects

1. If you want to cut one object against another, in the *Advance Tool Palette*>[icon] (Features) category, click [icon] (Cut an object). Alternatively, if you want both objects to be cut at the corner, click [icon] (Miter).

2. Select the section to cut against and press <Enter>.

3. Select the section to cut and press <Enter>.

   Note: Only one selection can be made at a time.

4. In the *Properties* tab of the dialog box (shown in Figure 3–32), specify the *Type of cut* and then make any other required modifications.

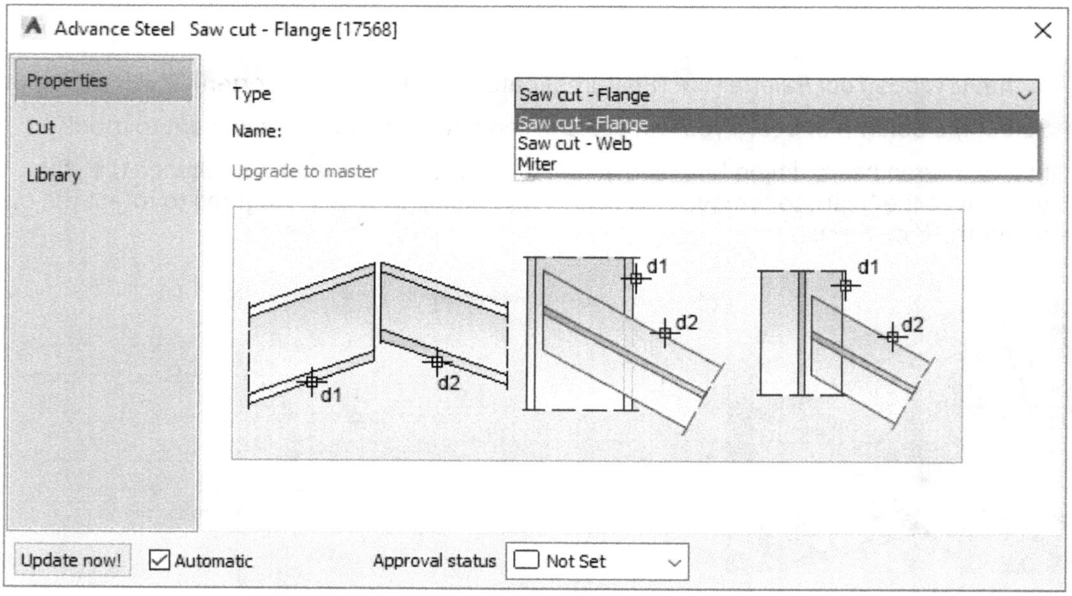

Figure 3–32

How To: Cope Beams

1. In the *Advance Tool Palette>* (Features) category, click (Cope).

2. Select the beam at the end you want to modify.

3. In the *Cope* dialog box>*Shape* tab (shown in Figure 3–33), make changes to the shape.

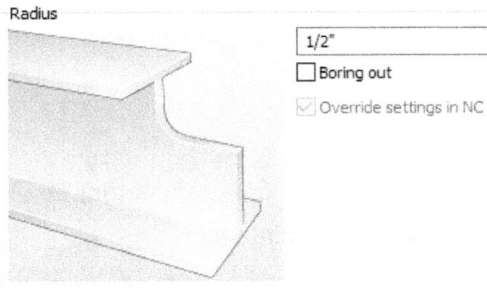

Advance Steel Cope		✕
Shape	Width X	3 15/16"
Corner finish	Depth	13/16"
	Length	3/8"
Increment for the length modification by the grip points.		
	Depth	1/16"
Increment for the depth modification via the grips.		

Figure 3–33

4. In the *Corner finish* tab, specify if you want a radius (as shown in Figure 3–34) or to show **Boring out** (as shown in Figure 3–35).

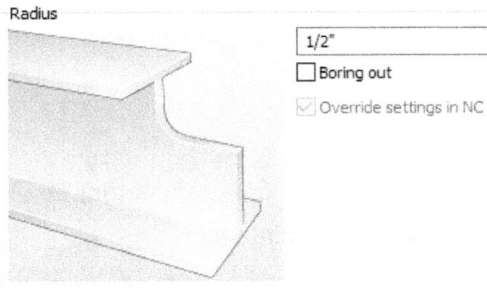
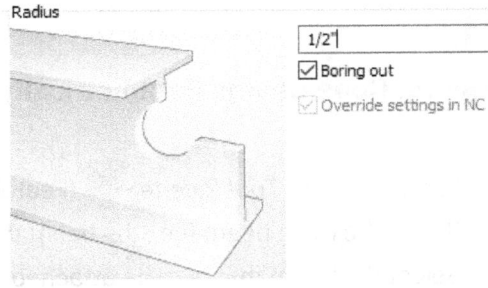

Radius — 1/2" — ☐ Boring out — ☑ Override settings in NC

Radius — 1/2" — ☑ Boring out — ☑ Override settings in NC

Figure 3–34   **Figure 3–35**

- If you want the cope at an angle, use the ⌐ (Cope, skewed) command. This command includes additional X- and Z-axis angles, as shown in Figure 3–36.

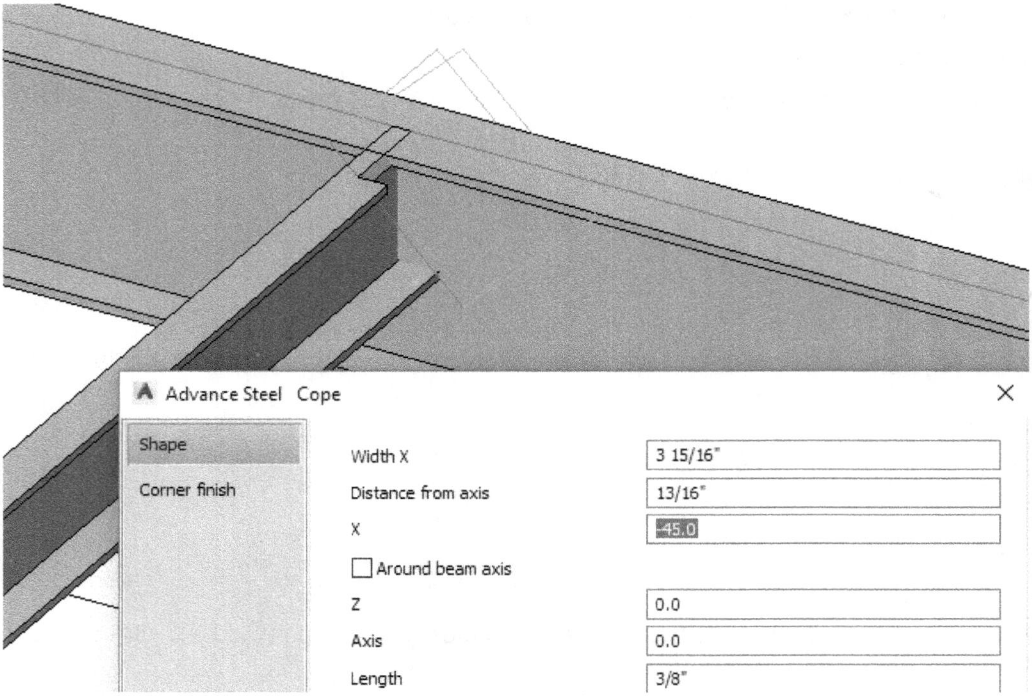

Figure 3–36

How To: Cope Beams Parametrically

1. In the *Advance Tool Palette>* ▨ (Features) category, click ⌐ (Cope, parametric).
2. Select the main beam (i.e., the beam that is not cut) and press <Enter>.
3. Select the beam that is to be attached and press <Enter>.

4. In the *Properties* tab>*Type* field, select **Parametric cope** or **Parametric contour**, as shown in Figure 3–37.

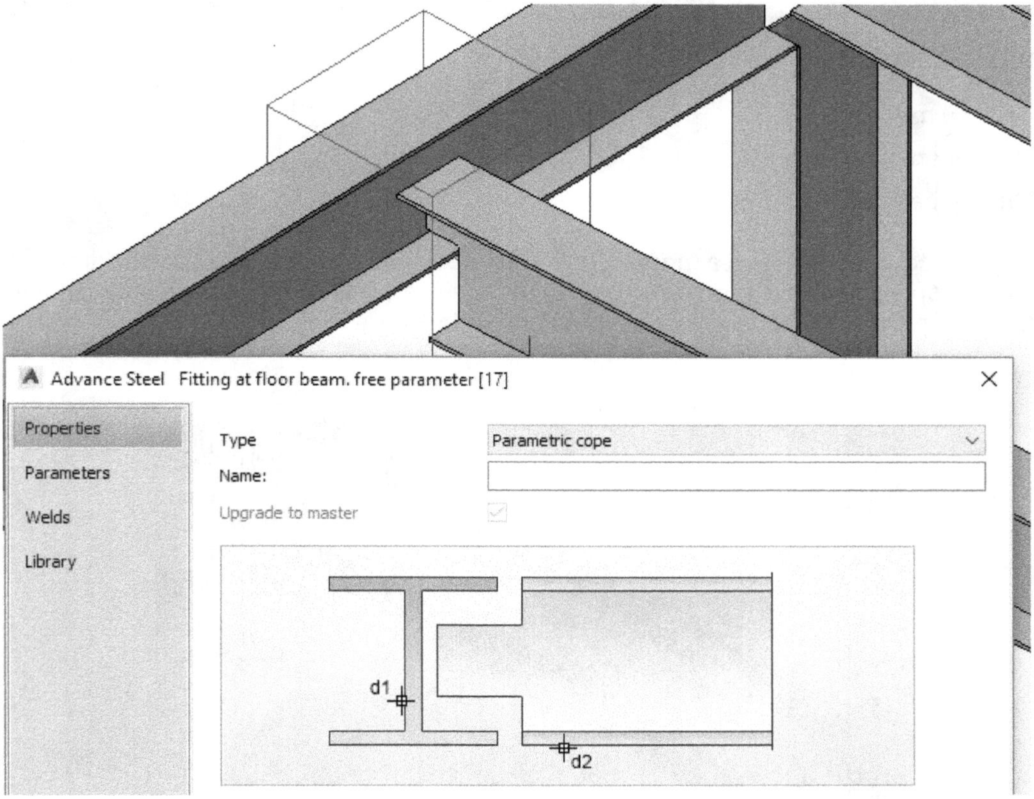

Figure 3–37

5. Go through the *Parameters* and *Welds* tabs to set the size and weld information. You can also access or save options from the *Library* tab.

*   For more control and when you need to have the cut go through beams, use the

    (Element contour - rule) command.

Practice 3b
Edit Beam Intersections

Practice Objectives

- Set up a model view.
- Miter beam intersections.
- Shorten beams.

In this practice, you will set up a model group that enables you to work in a specific area without having to display the entire model. You will then modify a small platform for a stair landing by mitering beams and shortening a column, as shown in Figure 3-38.

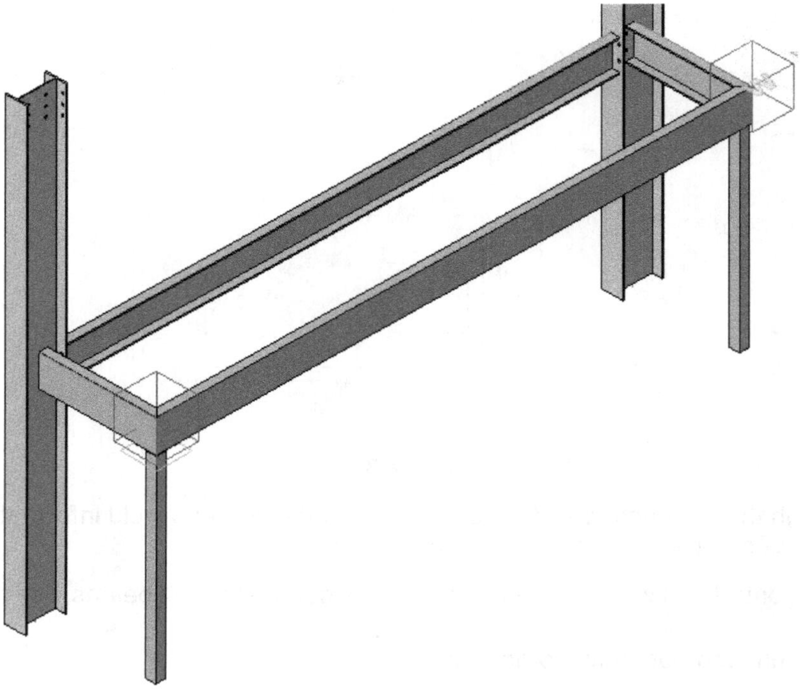

Figure 3-38

Task 1: Set up a model group for viewing.

1. In the practice files folder, open **Platform-Landing.dwg.**

2. Change the view to **SW Isometric** and zoom in on the small landing shown in Figure 3−39.

Landing

Figure 3−39

3. In the *Project Explorer>Structures* tab, toggle on **Model views>Level 1**.

   - If the top platform does not toggle off, right-click on **Level 1** and select **Properties**. In the *Level properties* dialog box, click **OK**. Then, toggle **Level 1** off and on again.

4. In the *Project Explorer>Structures* tab click ▇ (Create new group).

5. In the *Group properties* dialog box, for the *Name*, type **Landing** and click **OK**.

6. Right-click on the new *Landing* group and select **Add elements**, as shown in Figure 3–40.

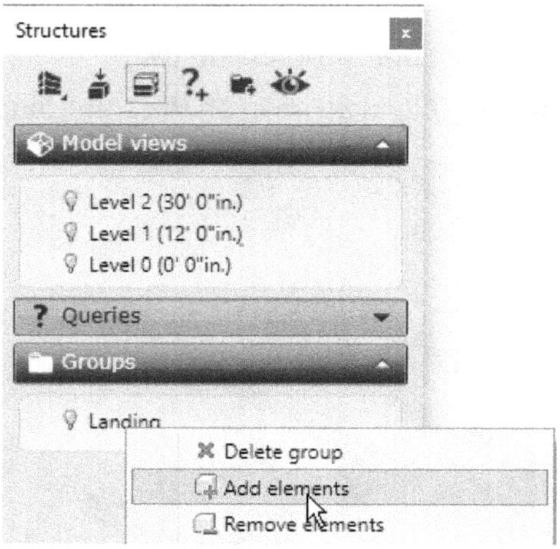

Figure 3–40

7. Select the columns and beams (but not the base plates) that form the landing, as well as the two supporting larger columns, and then press <Enter>.

8. Toggle on the *Landing* group. Note that only the selected elements display, as shown in Figure 3–41.

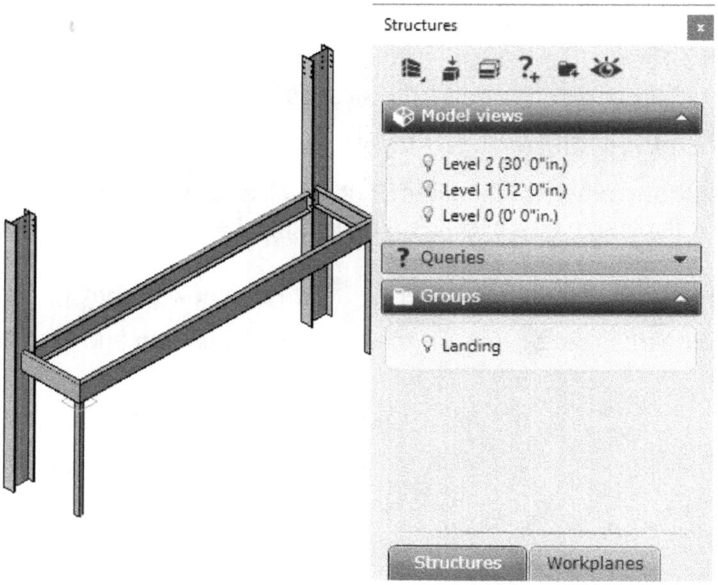

Figure 3–41

9. Save the drawing.

Task 2: Miter corners and shorten a column.

1. In the *Advance Tool Palette>* (Features) category, click (Miter).

2. Select one of the channel beams, press <Enter>, then select the other channel beam, and press <Enter>. The corner is mitered and the dialog box displays as shown in Figure 3–42.

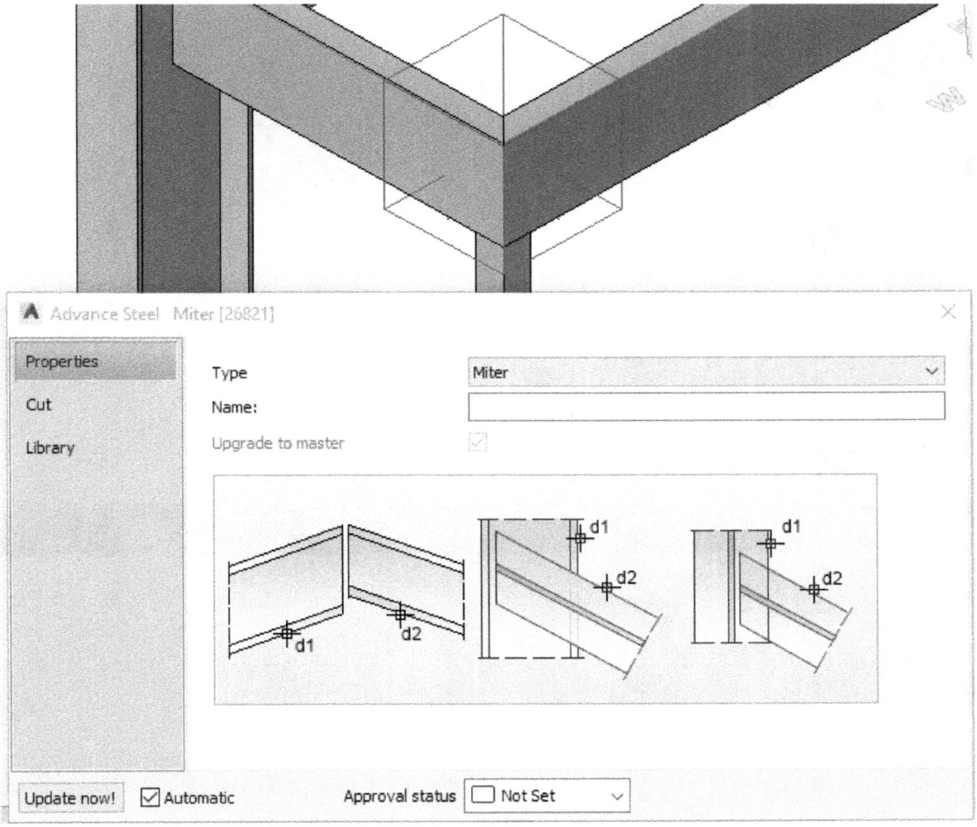

Figure 3–42

3. Close the dialog box.

4. Repeat the process on the other corner.

5. In the *Advance Tool Palette>* (Features) category, click (Shorten).

6. Click on the small column on Grid line 1 near the top of the column.

7. In the *Shorten beam* dialog box, set *Reference* to **1/4"** (as shown in Figure 3−43) and close the dialog box.

Figure 3−43

8. Save the drawing.

End of practice

3.3 Creating Plates

Many plates are automatically added when you use the connection vault, but you might need to create custom plates for specific connections or for large elements, such as pipes and ducts. There are a variety of plate tools that are included in the Autodesk Advance Steel software, including flat plates, conical and twisted plates, and a variety of modification tools. These tools are found on the *Objects* tab>*Plates* panel, as shown in Figure 3–44.

Figure 3–44

- When adding plates (or the objects that will become plates), the UCS impacts the orientation.

Creating Flat Plates

Flat plates can be created in many shapes and sizes. If required, you can use Advance Steel Features tools to shape the plates and add cutouts, as shown in Figure 3–45.

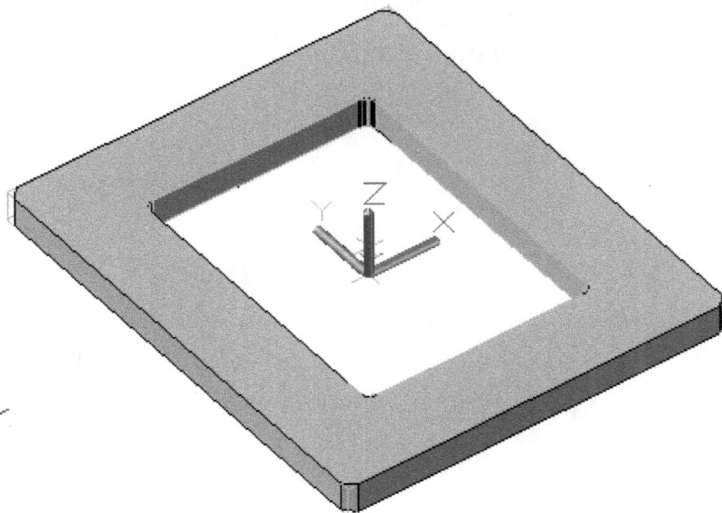

Figure 3–45

How To: Create a Flat Plate

1. Set the UCS to the orientation required for the plate.

2. In the *Home* tab>*Objects* panel or the *Objects* tab>*Plates* panel, select the method you want to use and follow the prompts, as outlined below and as shown in Figure 3−46.

☐	**Rectangular plate, 2 points**	Select two diagonal points that define the width and length of the plate.
☐	**Rectangular plate, 3 points**	Select three points that define the corner point, the X-direction and length, and the Y-direction and angle of the plate.
◺	**Plate at polyline**	Select an existing closed polyline that extrudes to the thickness specified in the *Plate* dialog box.

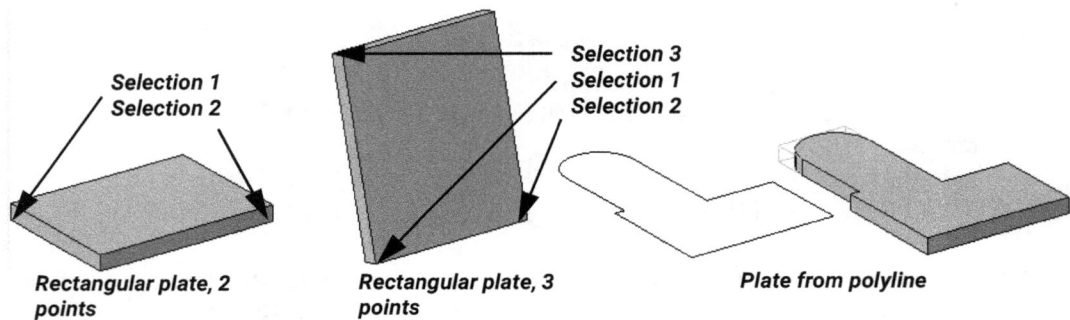

Figure 3−46

Note: *No matter how you create the plate, you can always change it in the Plate dialog box.*

3. After you have selected the points or polyline, in the *Plate* dialog box>*Shape & Material* tab (shown in Figure 3–47), specify the *Width X, Length Y, Thickness,* and other information. Go through the other tabs to continue specifying the required information.

    Note: Plate at polyline only has the **Thickness** *option.*

A Advance Steel Plate [16263]		X
Shape & Material	Width X	7"
Positioning	Length Y	4"
Naming	☐ Convert to polygon	
	Thickness	1/2"
Fabrication data	Material	▸ Steel ▸ A36
User attributes	Coating	None
Display type	Length increment	0"
Behavior	Galvanizing	
	Construction class	None
	Detail class	None
	Confidence	None

Figure 3–47

Additional Flat Plate Creation Tools

Additional flat plate tools are available on the *Objects* tab>*Plates* panel:

⊡	**Rectangular plate, center**	Select a center point for the plate, which is placed at a default size. In the *Plate* dialog box, specify the *Width X, Length Y,* and thickness.
△	**Polygon plate**	Select at least three points that define a multi-sided outline. In the *Plate* dialog box, specify the thickness.
◯	**Circular plate**	This is a macro that creates a square plate and automatically processes the edges to create a circular plate at the origin of the UCS.

Modifying Plates

You can modify plates by splitting and merging them. To make complex modifications to an already complex plate, you might want to use the **Plate to polyline** command, then modify the polyline and remake the plates, as shown in Figure 3–48. If required, plates can also be scaled.

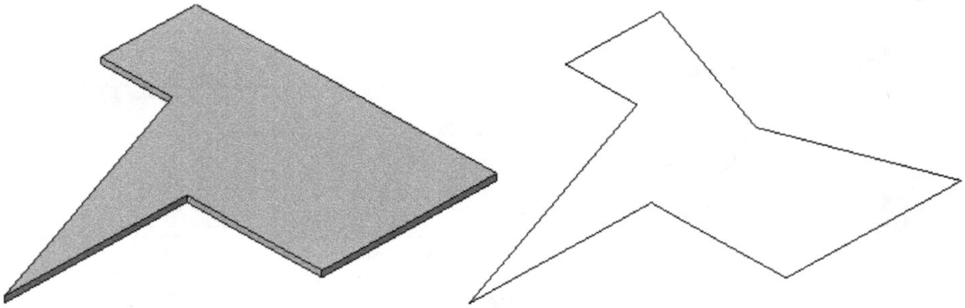

Figure 3–48

	Split plates at line	Draw a line or polyline. Select the plate and then select the line or polyline. You can select multiple plates and multiple lines. The split lines must cross the selected plates.
	Split plates by 2 points	Select the plate and then select two points on the plate to create the dividing line.
	Merge plates	Select two or more plates to merge. The plates must be touching or overlapping and in the same plane.
	Plate to polyline	Select a plate. A polyline is created that follows the outline of the plate. You can either keep or delete the plate.
	Shrink/ expand poly plate	Select the plate to shrink or expand. Enter a negative number to shrink the plate and a positive number to expand it.

Creating Folded Plates

Folded plates are used for many different situations, such as stairs, large pipes and ducts, as well as smaller connections. You can create folded plates by using existing flat plates or by using objects such as circles, polylines, and splines as templates to create plates that are conical (from closed objects) or twisted (from open objects).

How To: Create a Folded Plate without Position Adjustment

1. Create the plates, ensuring that they are touching.

2. In the *Home* tab>*Objects* panel or the *Objects* tab>*Plates* panel, click ⏣ (Create folded plate - without position adjustment).

3. Select the plate to connect to.

4. Select the plate you want to be connected to the previous plate.

5. In the *Folded Plate Relation* dialog box (shown in Figure 3–49), specify the *Angle*, *Justification*, and *Radius*.

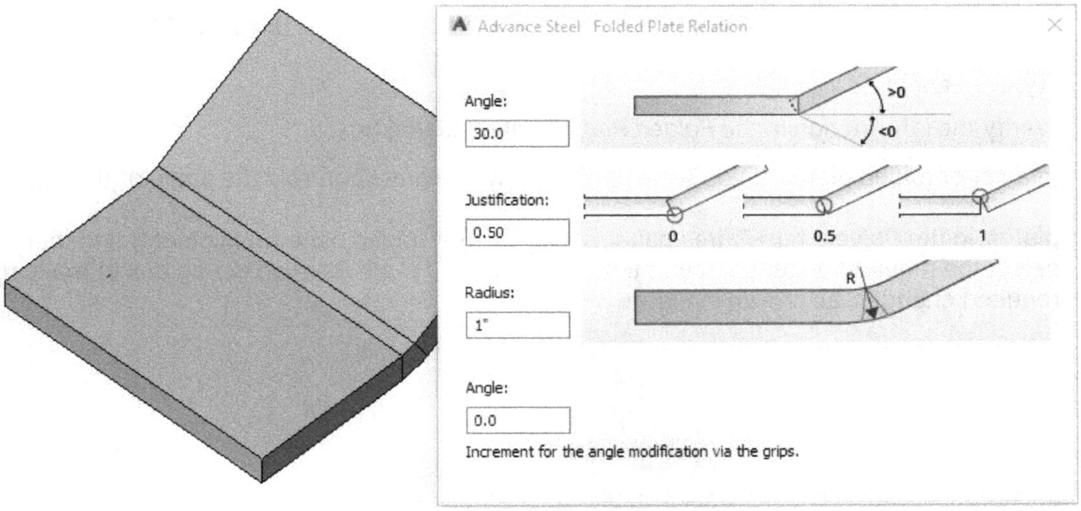

Figure 3–49

- To reopen the *Folded Plate Relation* dialog box, double-click on the radius object. Double-clicking on the plate opens the *Folded Plate* dialog box, which enables you to adjust the *Thickness*, *Material*, and other standard properties.

How To: Create a Folded Plate with Position Adjustment

1. Create the plates. Note that the plates do not need to be touching, as shown in Figure 3–50.

2. In the *Objects* tab>*Plates* panel, click 🔩 (Create folded plate - with position adjustment).

3. Select the main plate near the edge where you want the other plate to connect.

4. Select the plate to be connected on the edge you want to make the connection. The plate is moved, as shown in Figure 3–51.

Figure 3–50

Figure 3–51

5. Type an angle.

6. Verify the information in the *Folded Plate Relation* dialog box.

*   In a set of folded plates, if you want to change which plate controls the angle of the other

    plates, in the *Objects* tab>*Plates* panel, click 🔩 (Set folded plate main object) and then select the required plate. You can then use the *Folded Plate Relation* dialog box to make the required changes, as shown in Figure 3–52.

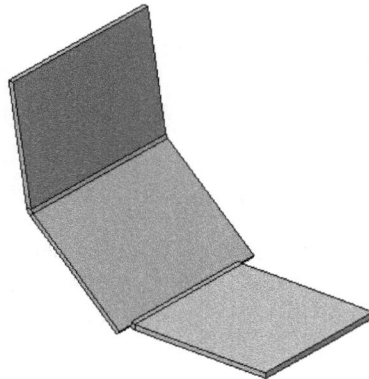

Figure 3–52

How To: Create a Conical Folded Plate Using Contours

1. Draw two elements at different heights, such as the circles shown in Figure 3–53. These define the ends of the folded plate, as shown in Figure 3–54. You can also use other closed elements such as polygons.

Figure 3–53

Figure 3–54

2. In the *Home* tab>*Object* panel or *Objects* tab>*Plates* panel, click ⬦ (Create conical folded plate).

3. At the *Select start shape type* prompt, select **Contour**.

4. Select the first object and press <Enter>.

5. At the *Select end shape type* prompt, select **Contour**.

6. Select the second object and press <Enter>.

7. In the *Conical Folded Plate* dialog box (shown in Figure 3–55), complete the required information.

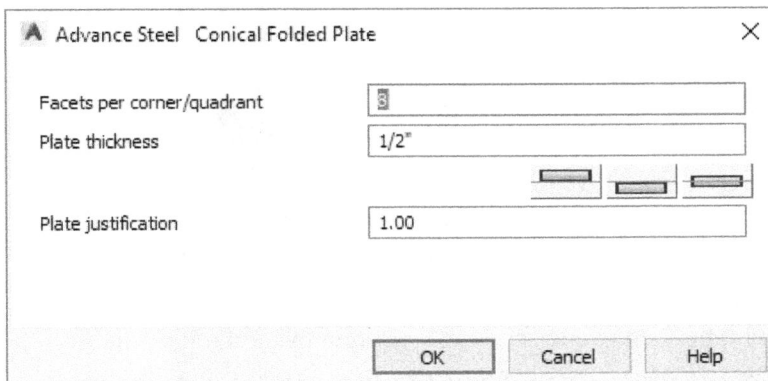

Figure 3–55

- Note that while you cannot modify the number of faces once the plates have been created,

  you can delete the plate or use the ⌒⊓ (Plate to polyline) tool to extract the original contours.

- The *Beam* option enables you to select the end of a beam as the contour line, as shown with two different sizes of circular hollow sections in Figure 3–56.

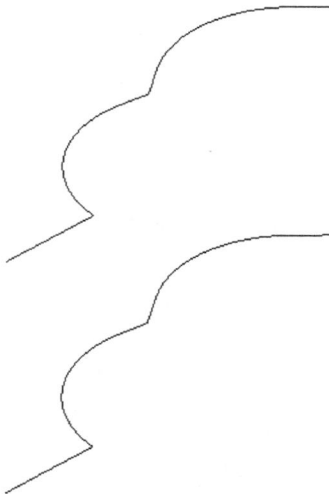

Figure 3–56

How To: Create a Twisted Folded Plate

1. Draw two open elements such as arc, lines, splines or the polylines shown in Figure 3–57. These define the ends of the folded plate as shown finished in Figure 3–58.

Figure 3–57	Figure 3–58

2. In the *Objects* tab>*Plates* panel, click 𝓐 (Create twisted folded plate).
3. Select the first entity close to the end where you want the plates to start.
4. Select the second entity on the same end.

5. In the *Twisted Folded Plate* dialog box, specify the *Number of division points* and other information (as shown in Figure 3–59) and click **OK**.

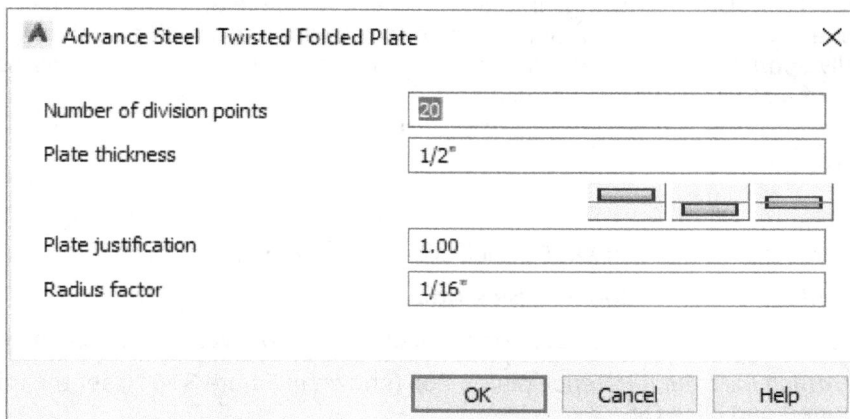

Figure 3–59

- When you create documentation using the Autodesk Advance Steel software, folded plates are automatically unfolded. To preview how the plates would display earlier in the process

  (as shown in Figure 3–60), in the *Objects* tab>*Plates* panel, click ⊕ (Check unfolding), select the plate, and at the *Display unfolded representation* prompt, click **Yes**. Note that this creates a separate object that you can delete when you are finished reviewing it.

Figure 3–60

💡 Hint: Using External References (XREFs)

You can import AutoCAD® drawings into an Autodesk Advance Steel drawing using the XREF Manager. This creates and maintains a link to the original drawing so that it automatically updates if any changes are made. You can use this for large drawings, such as another part of a steel structure or smaller drawings such as a plate.

How To: Attach an XREF

1. In the *Export & Import* tab>*XREF* panel, click 🖼️ (XREF Manager).

2. In the *XREF Manager* dialog box, click **Attach**.

3. In the *Select Reference File* dialog box, select the file you want to attach and click **Open**.

4. In the *Attach External Reference* dialog box (shown in Figure 3–61), set the information as required and click **OK**.

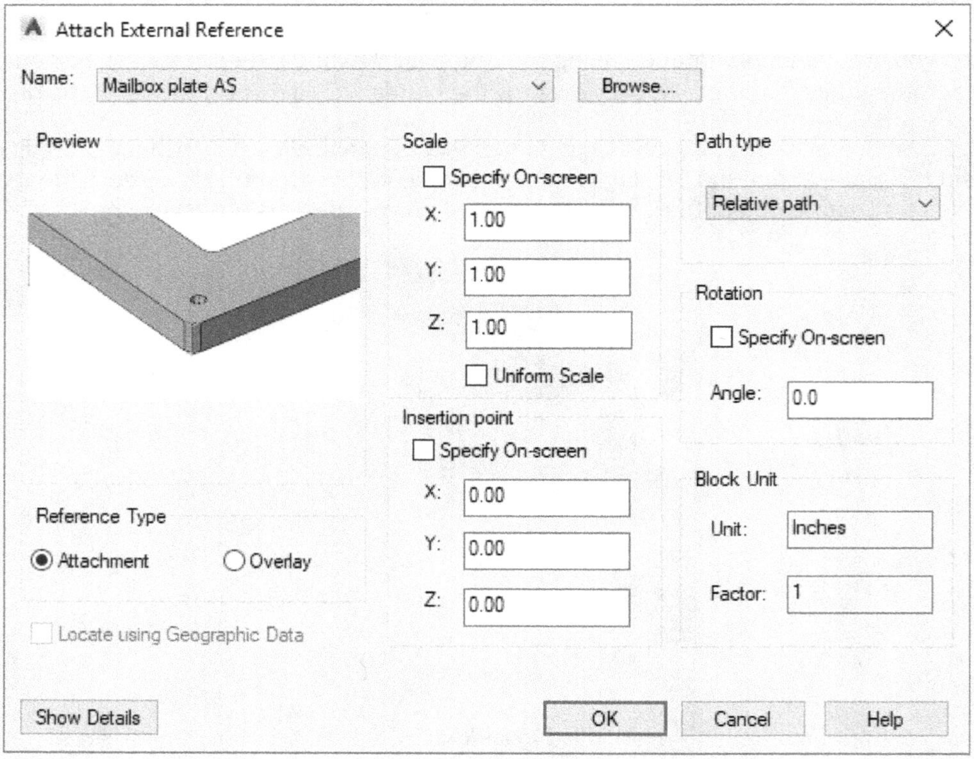

Figure 3–61

3.4 Adding Features to Plates and Beams

You might need to create cutouts in plates or beams so that wires or other objects can pass through or create a fillet for plate corners, as shown in Figure 3-62. To do this, you can use the feature tools in the *Advance Tool Palette>* (Features) category, shown in Figure 3-62.

> **Note:** *Features display as green outlines.*

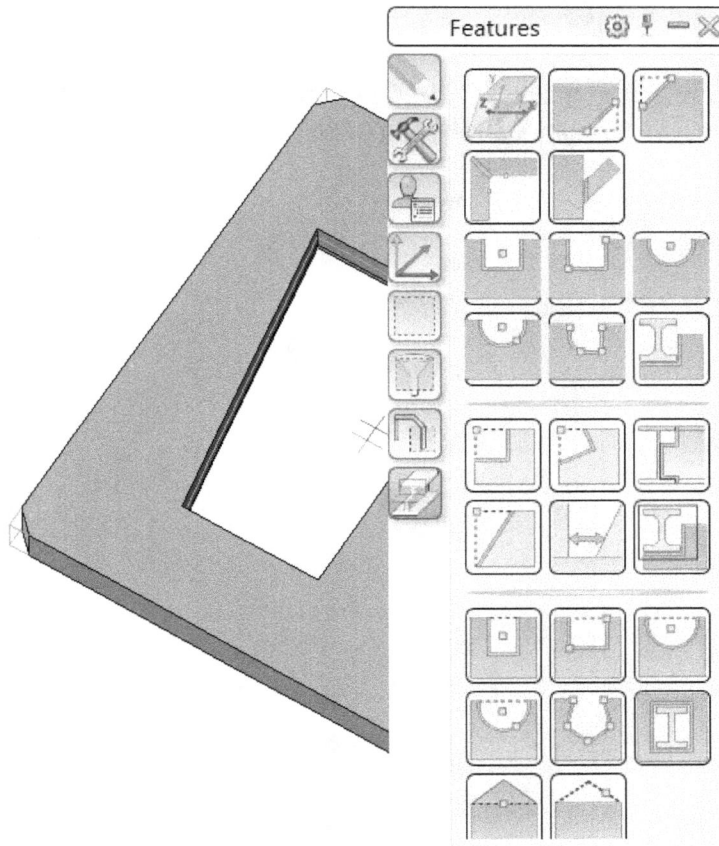

Figure 3-62

- The green icons (UCS related) can be used on both plates and beams. You must set the UCS to the correct orientation before you can use these tools.

- The orange icons can only be used with plates. The direction of the cut is always perpendicular to the plate.

How To: Add a Bevel Cut or Corner Cut

1. Set the UCS to the orientation you need for the object you are working on. These tools work on both beams and plates.

2. In the *Advance Tool Palette>* (Features) category, click (Bevel cut) to cut along an edge, or (Corner cut) to cut off a corner.

3. Select object near the edge or corner.

4. In the dialog box, specify the *Type* and sizes, as shown for a Bevel cut in Figure 3–63.

   • Cut *Types* include **Straight**, **Convex**, and **Concave**.

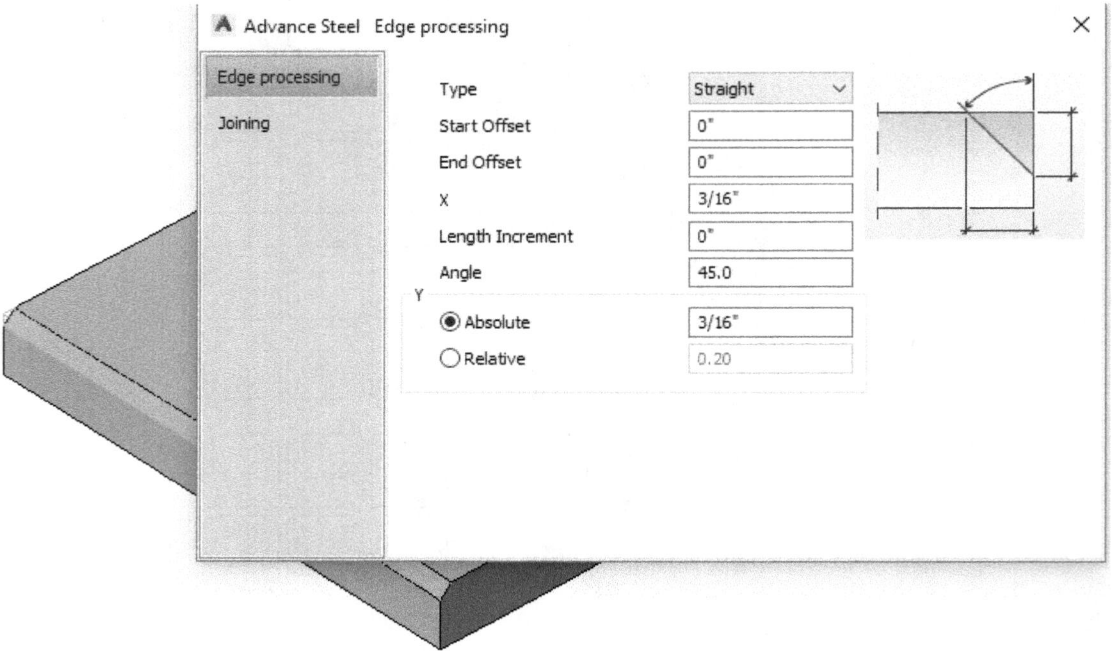

Figure 3–63

How To: Add Features to Plates and Beams with UCS Location

1. Place the UCS on the plate or beam that you want to cut.

2. In the *Advance Tool Palette>* (Features) category, select the method you want to use from the list below. The green icons are UCS related, and the orange icons are for plates only.

UCS/Plate Only

	Rectangular contour, center	Creates a rectangular hole based on a center point.
	Rectangular contour, 2 pointspoly	Creates a rectangular hole based on two diagonal points.
	Circular contour, center	Creates a circular hole based on a center point.
	Circular contour, 2 points	Creates a circular hole based on a center point and another point on the radius.
	Polygon contour	Creates a polygonal hole based on a premade polyline or by selecting multiple points.
	Element contour	Creates a hole based on a selected object, such as a beam.

3. Select the object you want to modify and then follow the prompts for the specific method.

4. In the *Contour processing* dialog box (shown for a rectangular contour in Figure 3–64), modify the information as required for the feature you are creating.

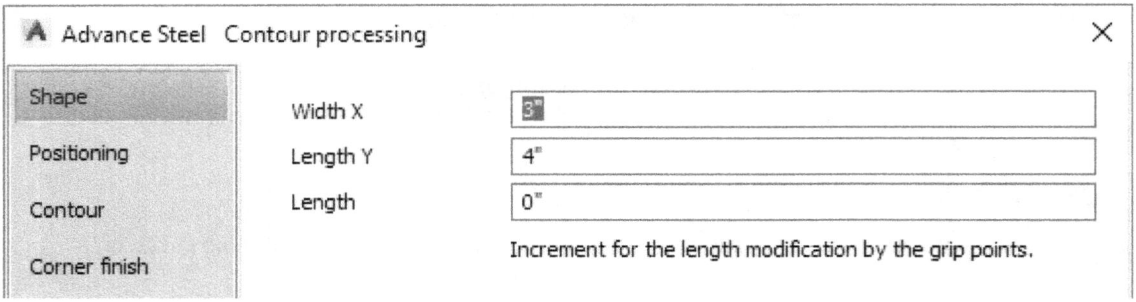

A Advance Steel Contour processing		✕
Shape	Width X	3"
Positioning	Length Y	4"
Contour	Length	0"
Corner finish		Increment for the length modification by the grip points.

Figure 3–64

- If you have created a plate using the **Polygon Plate** or **Plate at polyline** commands, you can use the  (Insert corner) and  (Remove corner) tools to modify the shape.

Practice 3c
Create Plates

Practice Objectives

- Create plates in place.
- Create plates in a separate drawing, and then create an XREF to them in a project.

In this practice, you will add a plate at the top of a column, as shown in Figure 3–65. You will also create a plate in a separate drawing and place it into the model as an AutoCAD external reference.

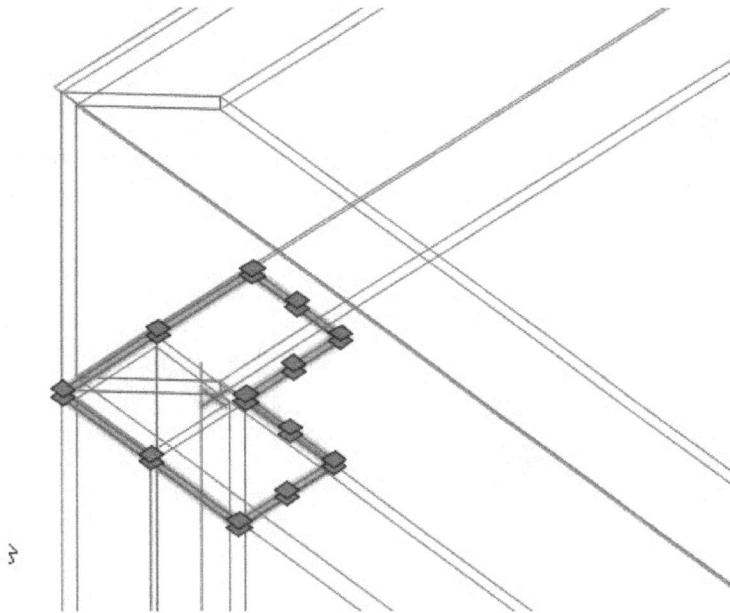

Figure 3–65

Task 1: Draw a polyline and use it to create a plate.

1. In the practice files folder, open **Platform-Plates.dwg**.
2. Select the small column closest to the edge that has been shortened. In the Status Bar,

    click (Isolate Objects) and then select **Isolate Objects**.
3. Set the UCS to the top of the shortened column.
4. Toggle on **ORTHO**.

5. Zoom in on the top of the column and draw the polyline shown in Figure 3–66.

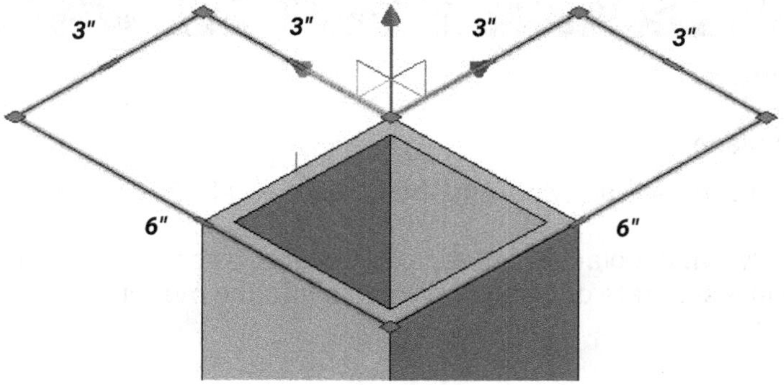

Figure 3–66

6. In the *Home* tab>*Objects* panel, click ⌐ (Plate at polyline).

7. Select the polyline and press <Enter>. In the *Plate* dialog box, set the *Thickness* to **1/4"**, as shown in Figure 3–67. In the *Positioning* tab, ensure that the plate is sitting above the top of the column and close the dialog box.

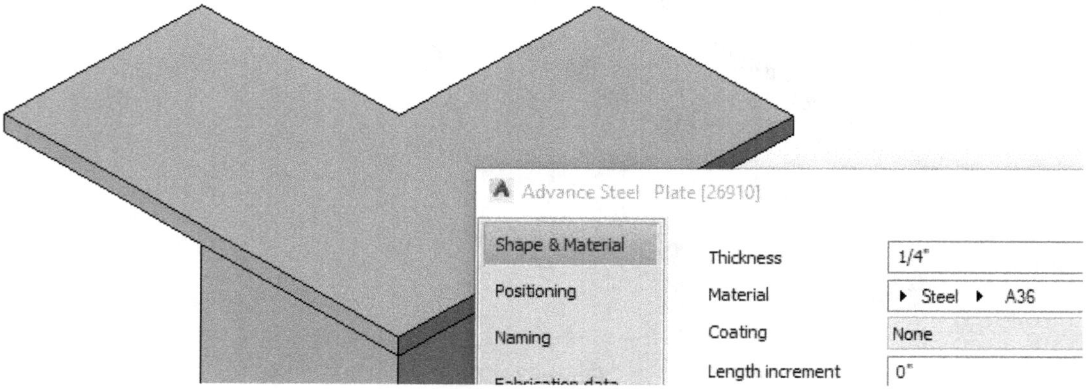

Figure 3–67

8. In the *Project Explorer*, right-click on the *Landing* group and select **Add elements**.

9. Select the new plate and press <Enter>. Note that this will help for when you want to use this group again.

10. End the object isolation.

11. In the *Project Explorer*, toggle off **Level 1** and the *Landing* group.

12. Save the drawing.

    • You will mirror the plate to the other side once it is connected to the beam and column.

Task 2: Create a holder plate.

1. Start a new drawing based on **ASTemplate.dwt**.

2. Save the drawing in the practice files folder as **Holder.dwg**.

3. In the *Home* tab>*Objects* panel or *Objects* tab>*Plates* panel, click ⬚ (Rectangular Plate, center).

4. For the start point, enter **0,0**.

5. In the dialog box, for the Width x enter **9-3/4"** and for the Length y enter **11-5/8"**.

6. In the *Plate* dialog box, ensure that the *Width* and *Length* are correct, and then set the *Thickness* to **3/4"**.

7. Close the dialog box. The plate is created as shown in Figure 3–68.

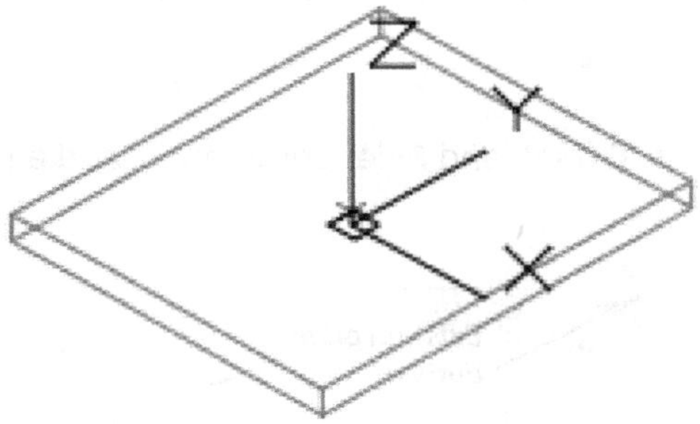

Figure 3–68

8. Save the drawing.

Task 3: Add features and a cutout.

1. In the *Advance Tool Palette*> ✎ (Features) category, click ▦ (Corner cut).

2. Select the plate near one of the corners.

3. In the *Edge processing* dialog box, set the *Type* to **Straight** and the *X* and *Y* to **1/4"**, as shown in Figure 3−69. Note that the *Angle* automatically becomes **45.0**.

Figure 3−69

4. Repeat the process for each of the 4 corners. The size is saved so you do not have to modify it each time.

5. Go to the **Top** view.

6. Using the **Plate to Polyline, Offset**, and **Fillet** commands, add a polyline inside the plate, as shown in Figure 3−70.

Figure 3−70

7. In the *Advance Tool Palette>* ▱ (Features) category, click ⬡ (Polygon, contour).

8. Select the edge of the outer plate.

9. Type **P** to start the Polyline option. Select the inner polyline and press <Enter>.

10. Change the *Visual Style* to **Conceptual** and rotate the view to display the cutout, as shown in Figure 3−71.

Figure 3−71

11. Save the drawing.

Task 4: Add the holder as an XREF to the model.

1. Return to **Platform-Plates.dwg**.

2. Set the UCS to the midpoint of the front flange of column B1, **5'-0"** above the base plate, as shown in Figure 3−72.

   • **Hint:** Use **UCS 3-point** to place the UCS at the base, and then select the UCS icon and drag it up **5'-0"**.

   Note: Watch out for object snaps when typing in distances. Sometimes it is safer to toggle object snaps off for the process.

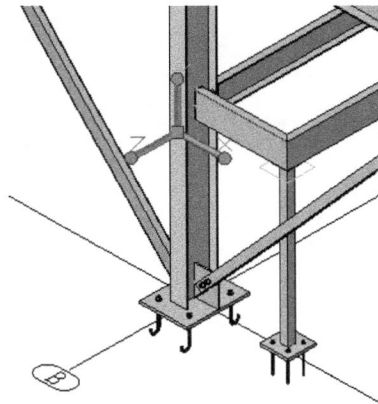

Figure 3−72

3. In the *Export & Import* tab>*XREF* panel, click 🖼️ (XREF Manager).

4. In the *XREF Manager* dialog box, click **Attach**.

5. Navigate to the practice files folder, select **Holder.dwg**, and then click **Open**.

6. In the *Attach External Reference* dialog box, ensure that the *Scale* is set to **1**, the *Reference Type* is set to **Attachment**, and the *Insertion point* is set to **0,0,0**, as shown in Figure 3–73.

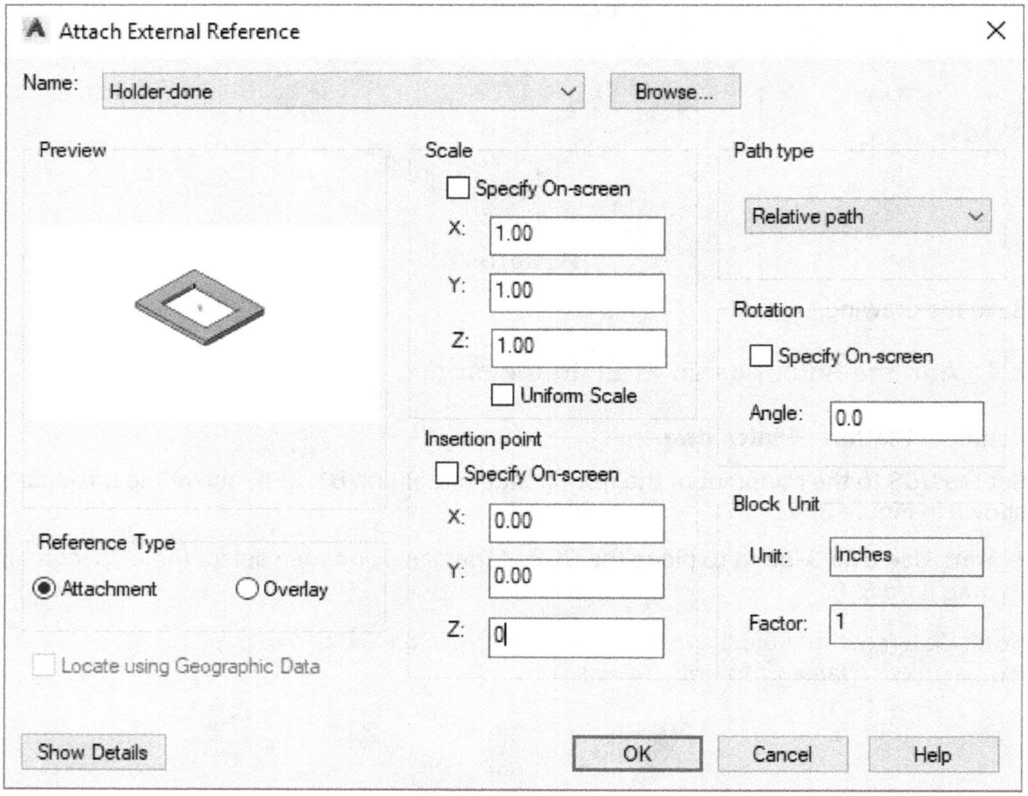

Figure 3–73

7. Note that the holder plate is placed at the UCS, as shown in Figure 3–74.

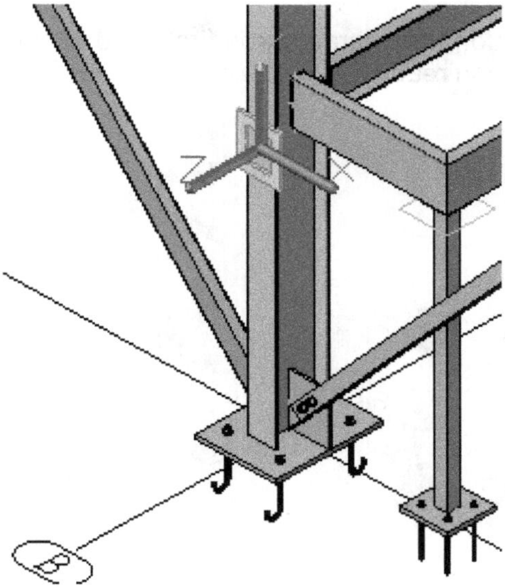

Figure 3–74

8. Return the UCS to the **World** location.
9. Save the drawing.

End of practice

3.5 Adding Bolts and Welds

If you are creating a connection from scratch, you first need to cut back the intersections and create plates. Only then are you ready to add the bolts, anchors, and welds, as shown in Figure 3–75.

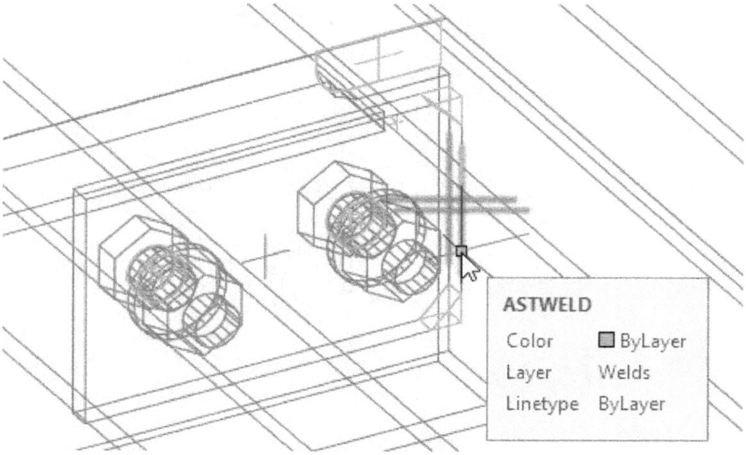

Figure 3–75

- The UCS impacts the placement of bolts and similar objects.

- To help you place bolts and other objects, you might want to draw lines for reference. This can especially help you when the locations of the patterns are not equal distances from the edges, as shown in Figure 3–76.

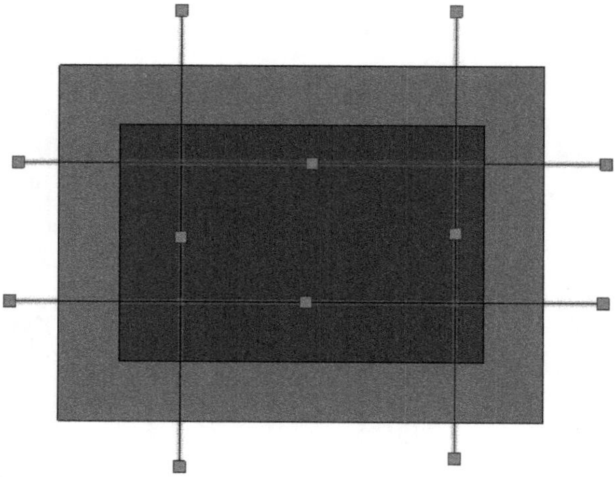

Figure 3–76

How To: Add Bolts, Anchors, Holes, and Shear Studs in a Pattern

1. In the *Home* tab>*Objects* panel or *Object* tab>*Switch* panel (shown in Figure 3–77), toggle through the options in the **bolts/anchors/holes/shear studs** tool to select the type of objects you want to create.

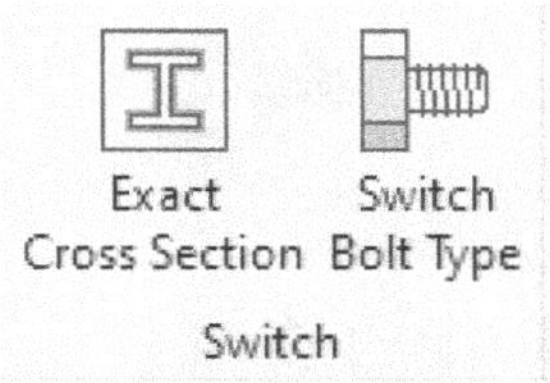

Exact Switch
Cross Section Bolt Type

Switch

Figure 3–77

⊫	Bolts		🔲	Holes
∪	Anchors		⊤	Shear Studs

2. In the *Home* tab>*Objects* panel or *Objects* tab>*Connection objects* panel, click

   ⠿ (Rectangular, 2 points).

3. For bolts and anchors, select the objects you want to connect. For holes and shear studs, select the plate or beam.

4. Select two diagonal points that define the default location of the objects.

5. In the *Definition* tab, specify the *Type* and other information. The required information varies depending on the type of object that you are creating.

    • For example, select **Inverted** to switch the direction of bolts (as shown in Figure 3–78) if the original direction is not reachable.

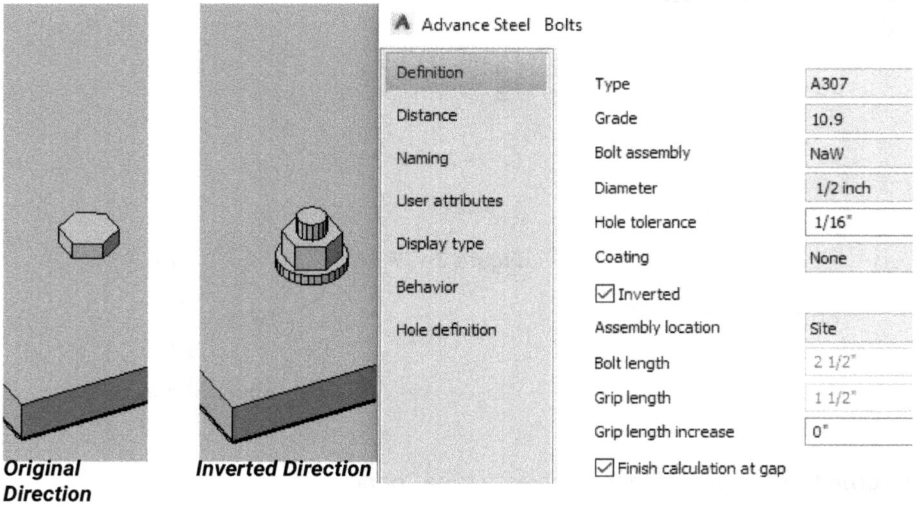

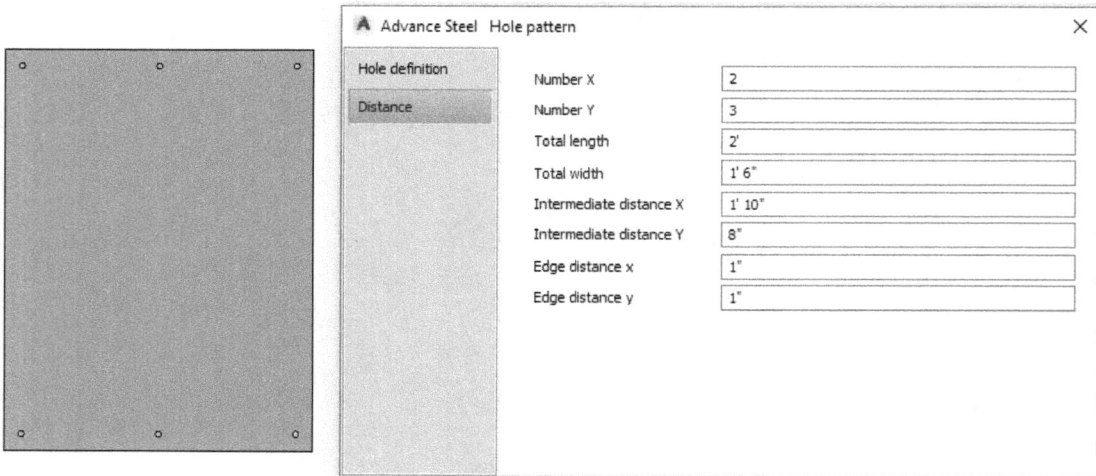

Figure 3–78

6. In the *Distance* tab, specify the *Number* in the *X* and *Y* axis and any of the other options, as shown for a hole pattern in Figure 3–79. All connector types include this tab.

    • To create one bolt, set the Number X and Number Y to 1.

Figure 3–79

7. Go through the rest of the dialog box tabs and apply the required information.

- You can break up a bolt group into individual bolts. In the *Objects* tab>*Connection objects* panel, click ⊞ (Split bolt group).

- You can change bolts and anchors into holes. Select the bolts or anchors, right-click, and select **Change into holes**, as shown in Figure 3–80.

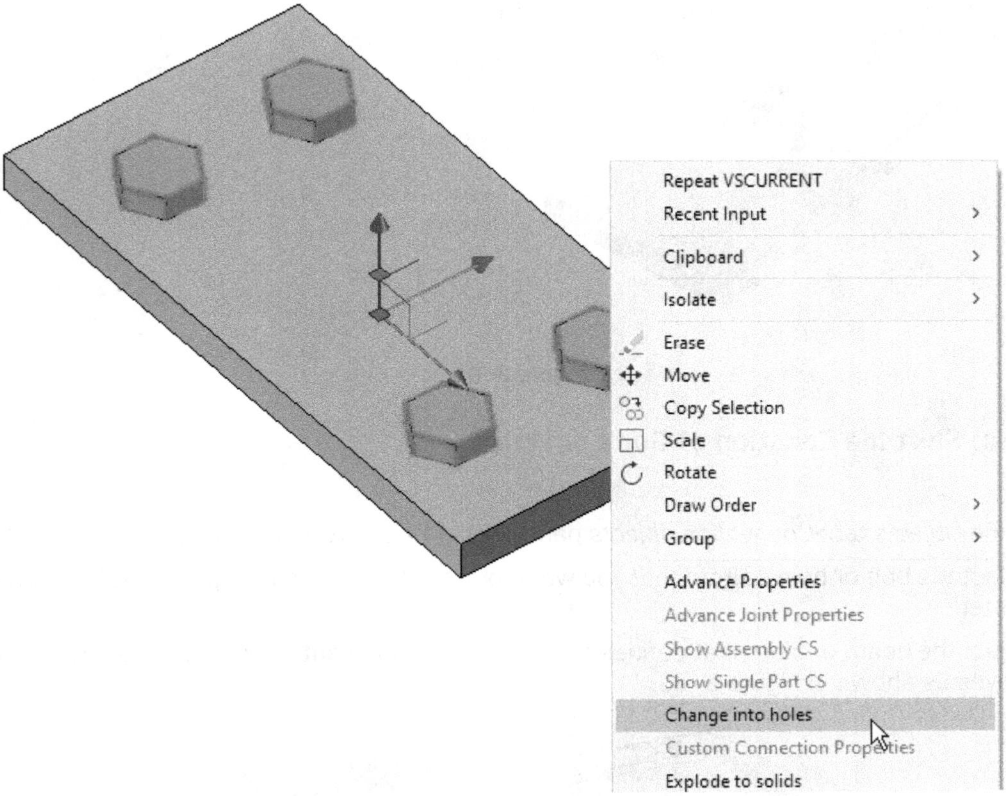

Figure 3–80

- You can change holes into bolts or anchors. Select the holes, right-click, and select **Change into>Bolts** or **Anchors**, as shown in Figure 3-81.

Repeat ASTORCREATEHOLESFROMCONNECTORS
Recent Input   >
Clipboard   >
Isolate   >
Erase
Move
Copy Selection
Scale
Rotate
Draw Order   >
Group   >
Advance Properties
Advance Joint Properties
Show Assembly CS
Show Single Part CS
Change into   >   Bolts
Custom Connection Properties   Anchors

Figure 3-81

How To: Shift the Location of Bolts or Holes

1. In the *Objects* tab>*Connection objects* panel, click ⬓ (Shift bolts/holes).
2. Select the bolt or hole pattern that you want to move (as shown in Figure 3-82) and press <Enter>.
3. Select the beam or plate that defines the area where you want the bolt or hole pattern. It moves as shown in Figure 3-83.

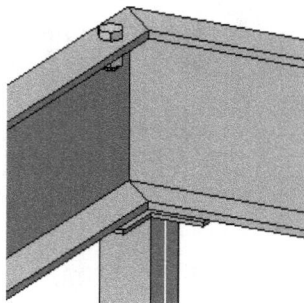

Figure 3-82

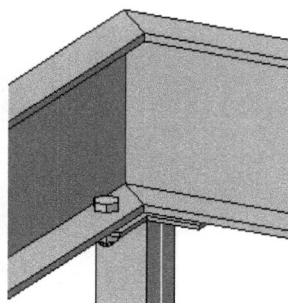

Figure 3-83

How To: Place a Single Weld

1. In the *Home* tab>*Objects* panel or *Objects* tab>*Connection objects* panel, click ✐ (Weld point).
2. Select the objects you want to weld together and press <Enter>.
3. Specify the weld point.
4. In the *Weld* dialog box (shown in Figure 3–84), specify the information about the weld.
    *   The weld displays in the model as a pink cross. The information about the weld is applied when you create the fabrication documents.

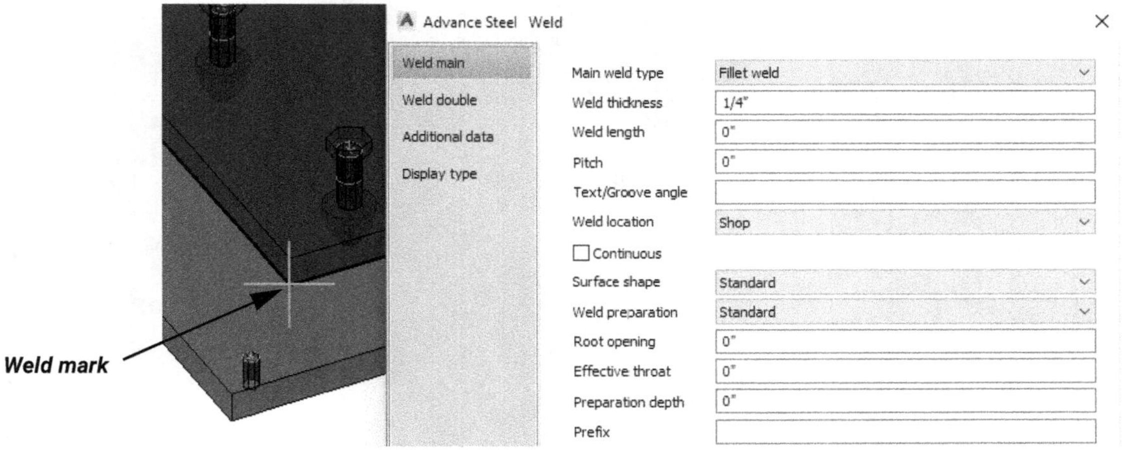

Figure 3–84

How To: Place a Continuous Weld

1. In the *Objects* tab>*Connection objects* panel, click ✐ (Line of Weld).
2. Select the objects to be connected and press <Enter>.
3. Pick points that define the line of welds and press <Enter> to finish.
4. In the *Weld* dialog box, specify the information about the weld.

5. The weld displays as shown in Figure 3–85.

Figure 3–85

Practice 3d
Add Bolts and Welds

Practice Objectives

- Add bolts.
- Add holes and countersinks.
- Add welds.

In this practice, you will add bolts to a plate, as shown in Figure 3–86. You will then mirror the plate, shortening objects and bolts to another column/beam connection. You will add countersinks and other holes to another plate and update the XREF in the primary model. You will then bind the XREF to the drawing and weld it into place, as shown in Figure 3–87.

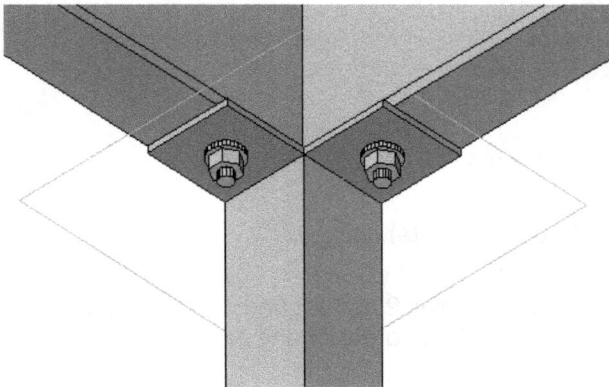

Figure 3–86

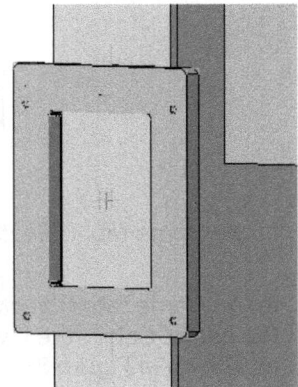

Figure 3–87

Task 1: Add bolts to a plate.

1. In the practice files folder, open **Platform-Bolts.dwg.**

2. Modify the view as required to display only the landing area. Change the *Visual Style* to **X-Ray** or **2D Wireframe** so that the new plate displays through the beam, as shown in Figure 3–88.

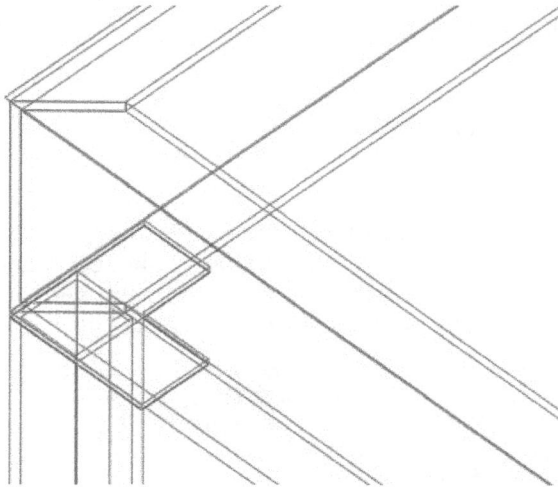

Figure 3–88

3. In the *Home* tab>*Objects* panel, ensure that ⬛ (Bolts) display.

4. In the *Home* tab>*Objects* panel, click ⬛ (Rectangular, 2 points) and select the plate and the beam. Select the lower left corner and upper right corner of one of the legs of the plate, as shown in Figure 3–89.

Upper right corner

Lower left corner

Figure 3–89

5. In the *Bolts* dialog box>*Distance* tab, change *Number X and Number Y* to **1**.

6. In the *Definition* tab, change *Diameter* to **1/2 inch.**

   • If the bolt is at the top of the beam (instead of at the bottom), in the *Objects* tab>

     Connection objects panel, click ⬛ (Shift bolts/holes) and follow the prompts to select the correct part of the beam.

7. Repeat the process to add a bolt to the other leg of the plate, as shown in Figure 3–90.

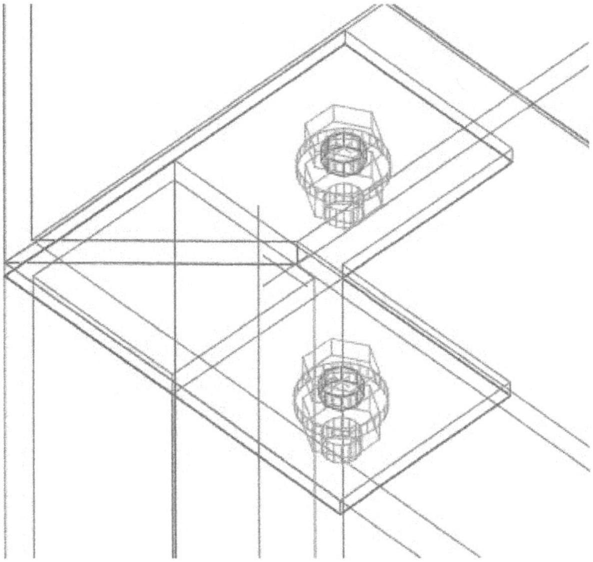

Figure 3–90

8. Save the drawing.

Task 2: Mirror the plate and bolts to the other column.

1. Zoom out so that the full landing displays.

2. In the *Advance Tool Palette*>🛠 (Tools) category, click 🔘 (Advance Copy).

3. In the *Transform elements* dialog box, click **Select objects**.

4. Select the plate, bolts, and the green Shorten beam joint box and press <Enter>.

5. In the *Transform elements* dialog box, select **Mirror** and **2D**, then click the *Select mirror points* selection button, as shown in Figure 3–91.

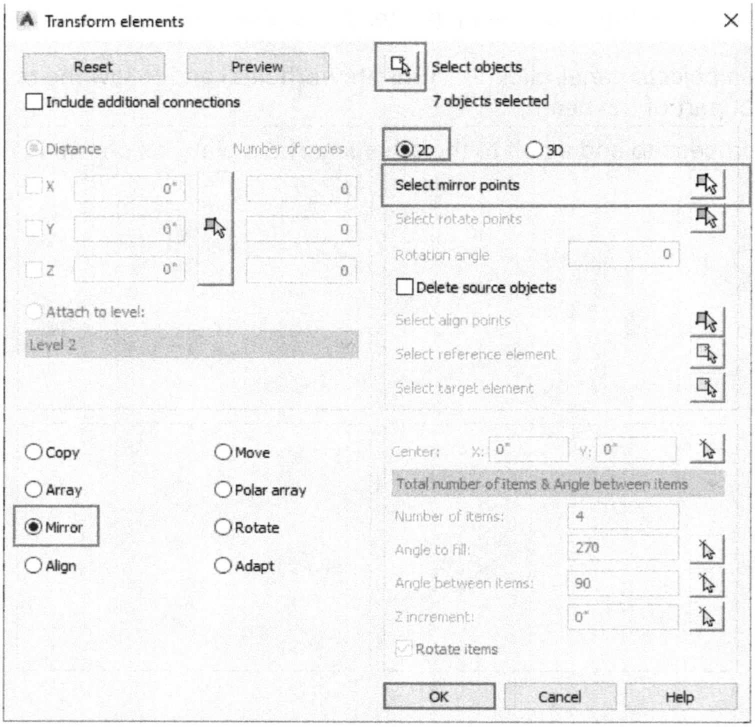

Figure 3–91

6. Select the midpoints of the landing beams, as shown in Figure 3–92.

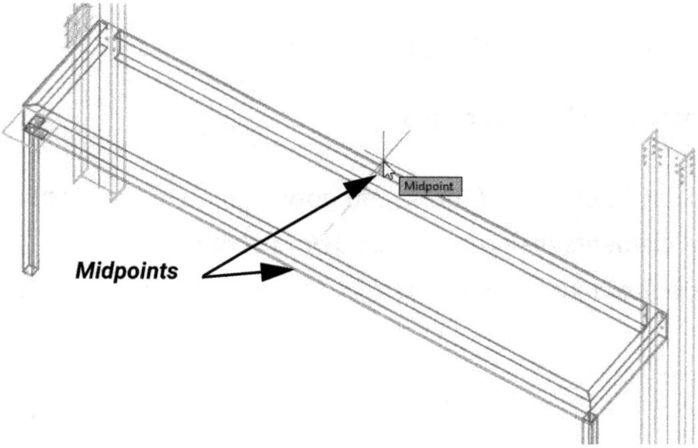

Figure 3–92

7. In the dialog box, click **OK**. The connection is mirrored and copied.

8. Save the drawing.

Task 3: Add holes to the holder plate.

1. In the practice files folder, open **Holder-Plate.dwg**.

2. Go to the **Top** view and set the *Visual Style* to **2D Wireframe**.

3. Add lines as construction lines for bolt placement, as shown dashed in Figure 3−93.

Figure 3−93

4. In the *Home* tab>*Objects* panel, click [icon] (Switch bolts, anchors, holes, shear studs) until the holes icon displays.

5. In the *Home* tab>*Objects* panel, click [icon] (Rectangular, 2 points).

6. Select the plate.

7. For the lower left corner and upper right corner, use the Intersection object snap to select the two locations indicated in Figure 3–94.

8. In the *Hole pattern* dialog box>*Hole definition* tab, set the information shown in Figure 3–94.

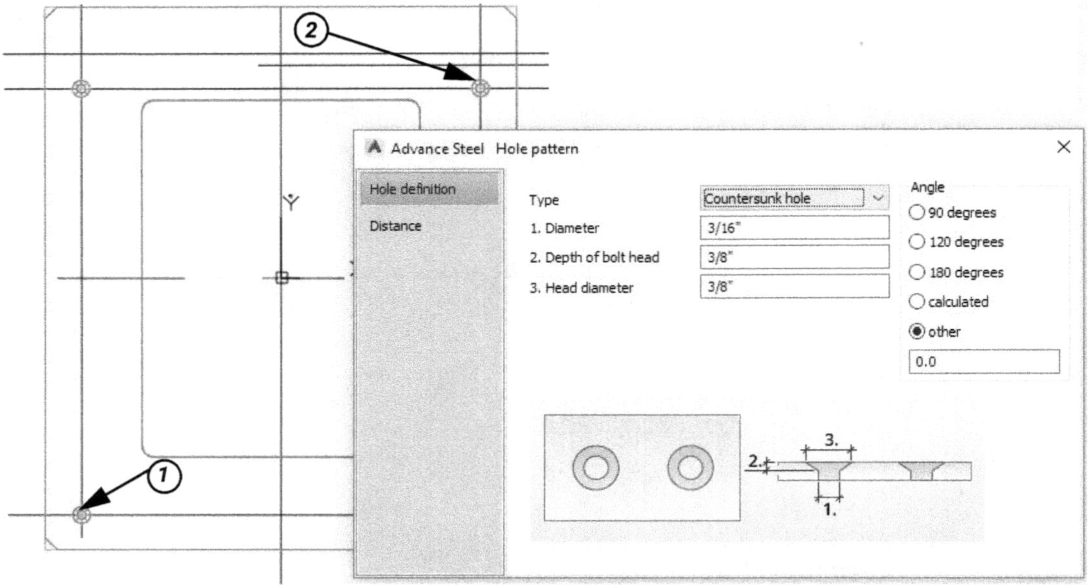

Figure 3–94

9. Review the *Distance* tab. Note that the default number is acceptable, and that the locations were specified by the points you picked.

10. Close the dialog box.

11. In the *Objects* tab>*Connection objects* panel, click ⸬ (Rectangular, center point).

12. Select the plate and then intersection # 3, shown in Figure 3–95.

13. In the *Hole pattern* dialog box>*Hole definition* tab, set the *Type* to **Round hole** and the *Diameter* to **1/16"**.

14. In the *Distance* tab, set *Number X* and *Number Y* to **1**, as shown in Figure 3–95.

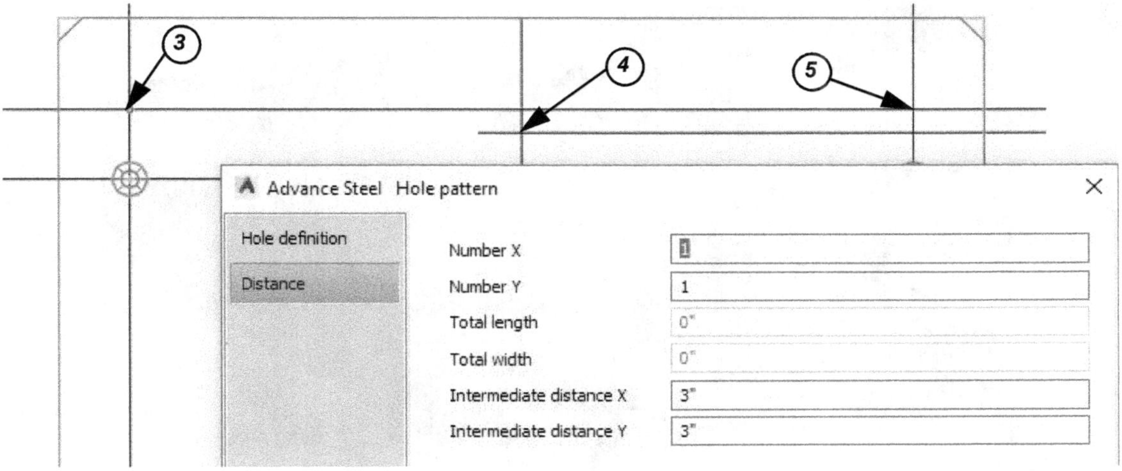

Figure 3–95

15. Repeat the process for holes at intersections #4 and #5, shown in Figure 3–95, above.

16. Rotate the model so you the side of the plate displays. Change the *Visual Style* to **X-Ray**.

17. Modify the UCS so that it parallels the side of the plate.

18. Draw a **1/4"** long line from the intersection of the plate and horizontal line, as shown in Figure 3–96.

19. Start the **Rectangular, center point** command again and place a single **Blind hole** of **1/16"** *diameter*, **1"** *deep* on the side of the plate, as shown in Figure 3–96.

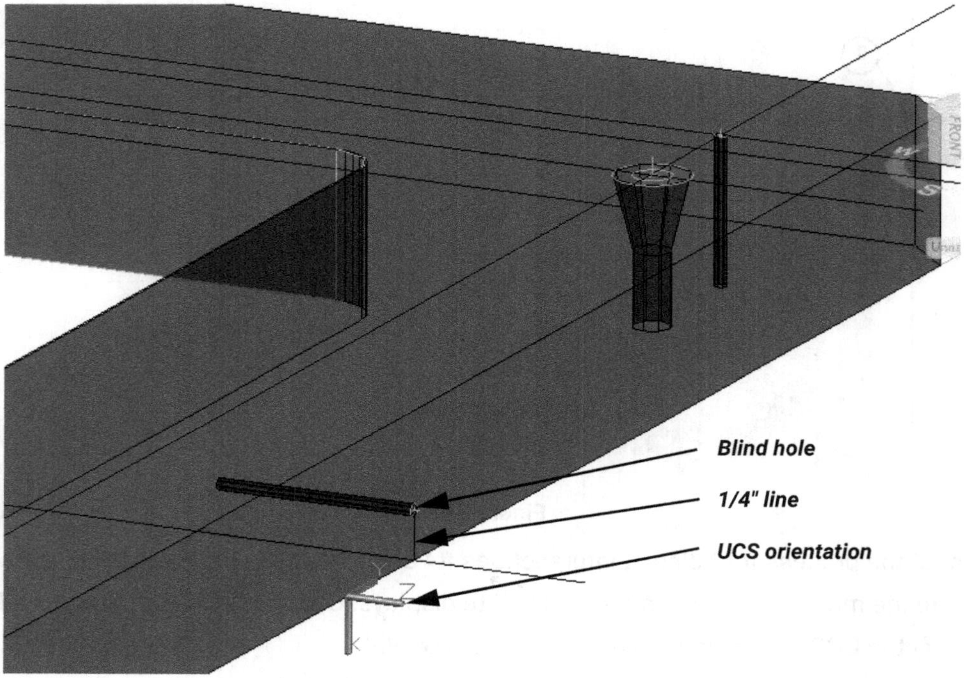

Figure 3–96

20. Return the UCS to **World** orientation.

21. Use **Advance Copy** to mirror the hole to the other side of the plate.

22. Delete all of the lines used to place the holes.

23. Save the plate drawing.

24. Return to **Platform-Bolts.dwg**.

25. Open the *XREF Manager* dialog box. Note that the *Holder-Plate* reference needs to be updated, as shown in Figure 3–97. Select the *Reference Name*, click **Reload**, and then click **OK**.

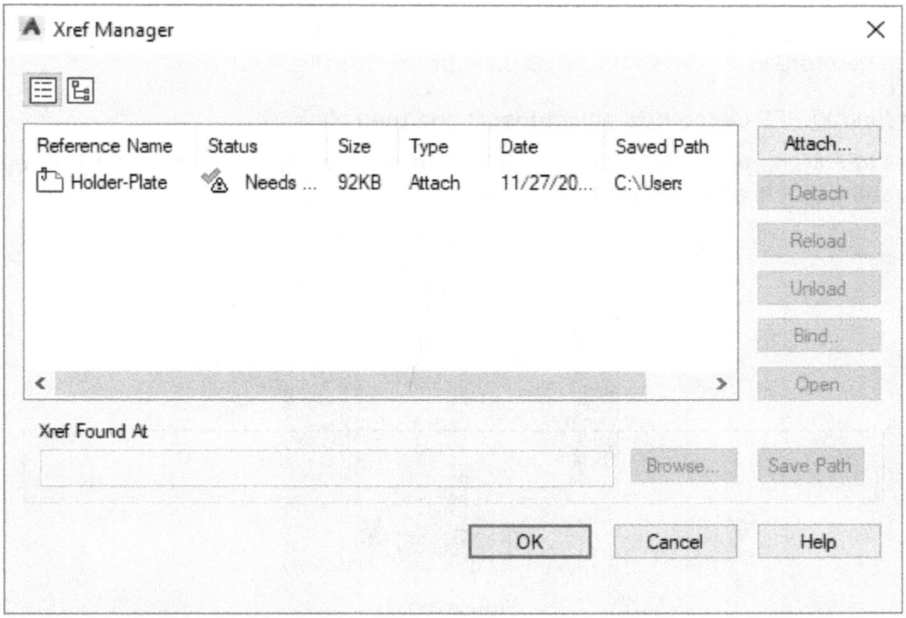

Figure 3–97

26. The holder plate is updated in the main platform, as shown in Figure 3–98.

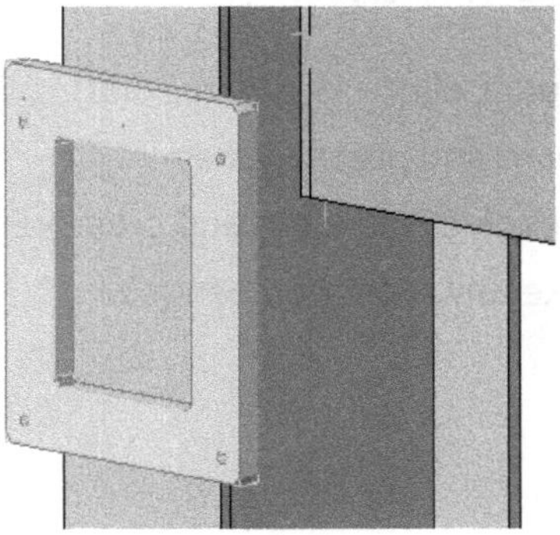

Figure 3–98

27. Save the drawing.

Task 4: Attach the holder plate to the column using welds.

1. Open the *XREF Manager* dialog box. Select **Holder-Plate.dwg** and click **Bind**.

   Note: The holder plate needs to be welded to the beam, but you cannot weld an XREF or a block. Therefore you need to bind it to the current drawing.

2. In the *Bind XREFS* dialog box, select **Insert** and then click **OK**.

3. In the *XREF Manager* dialog box, click **OK**. Note that the plate becomes part of the drawing, but is still a block, as shown in Figure 3–99.

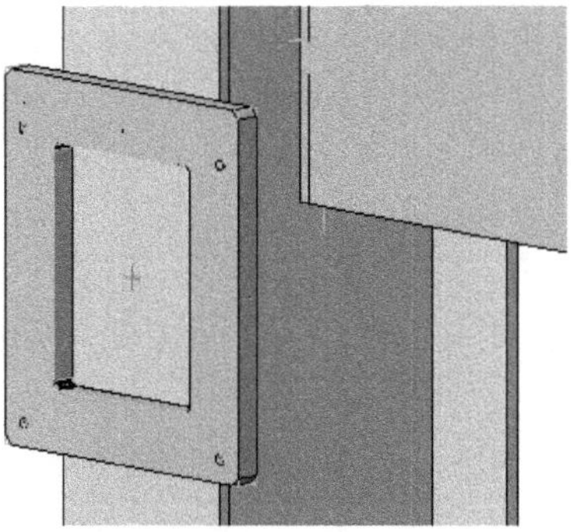

Figure 3–99

4. Select the block and type **X** for Explode. The block becomes a plate that can now be welded to the column.

5. Change the *Visual Style* to **2D Wireframe**.

6. In the *Home* tab>*Objects* panel, click (Weld Point).

7. Select the plate and the beam and press <Enter>.

8. Select the midpoint of the back of the plate and press <Enter>.

9. In the *Weld* dialog box, make any required changes.

10. Repeat the process to add a weld to the lower midpoint of the plate, as shown in Figure 3-100.

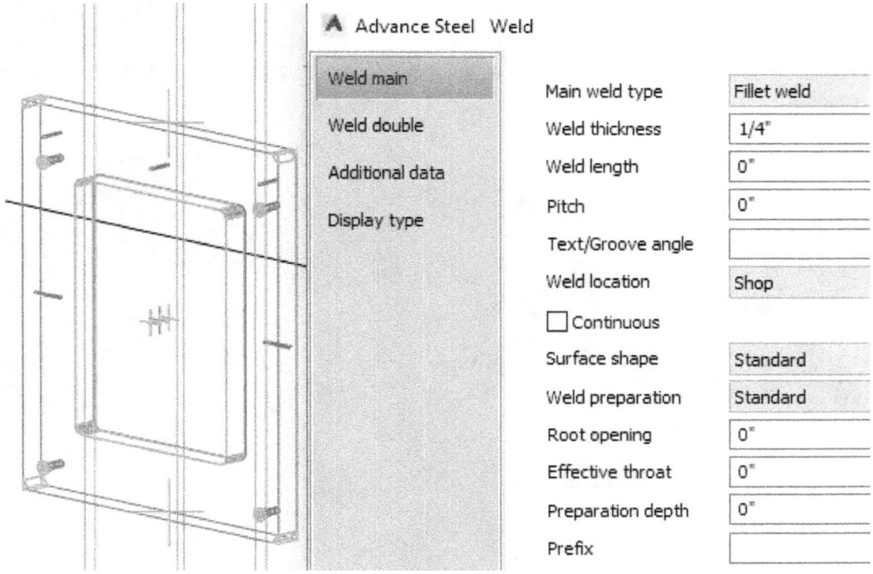

Advance Steel Weld		
Weld main	Main weld type	Fillet weld
Weld double	Weld thickness	1/4"
Additional data	Weld length	0"
Display type	Pitch	0"
	Text/Groove angle	
	Weld location	Shop
	☐ Continuous	
	Surface shape	Standard
	Weld preparation	Standard
	Root opening	0"
	Effective throat	0"
	Preparation depth	0"
	Prefix	

Figure 3-100

11. Modify the view so that the entire platform displays.

12. Save the drawing.

End of practice

3.6 Creating Custom Connections

Custom connections can be created using existing connection objects (as shown in Figure 3–101) and then grouping them together in a template (as shown in Figure 3–102). This is done by creating a custom connection template. You can then insert the template at similar connection points.

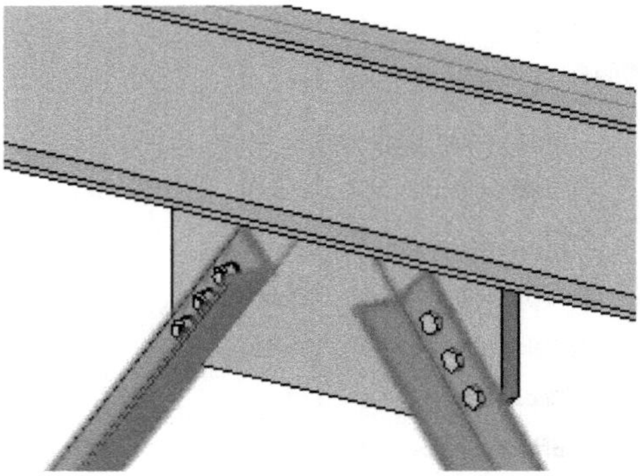

Figure 3–101

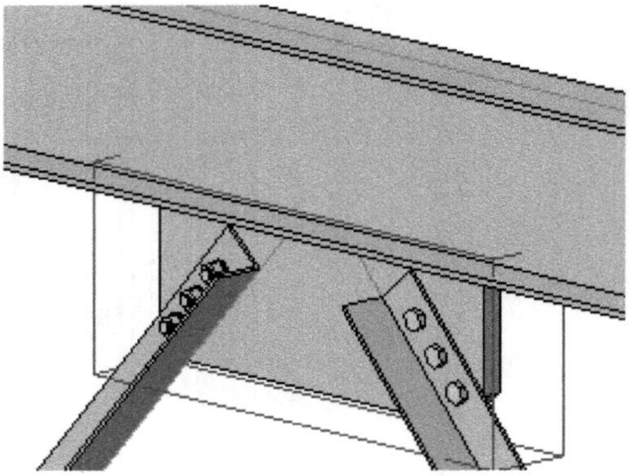

Figure 3–102

How To: Create Custom Connection Templates

1. Set up the connections you want to include in the custom connection. Include objects such as plates and bolts, and features such as cuts and miters.

2. In the *Advance Tool Palette>* (Custom connections) category, click  (Create connection template).

3. In the *Choose the definition method* dialog box, select a method that matches the type of connection you are trying to make, as shown in Figure 3–103.

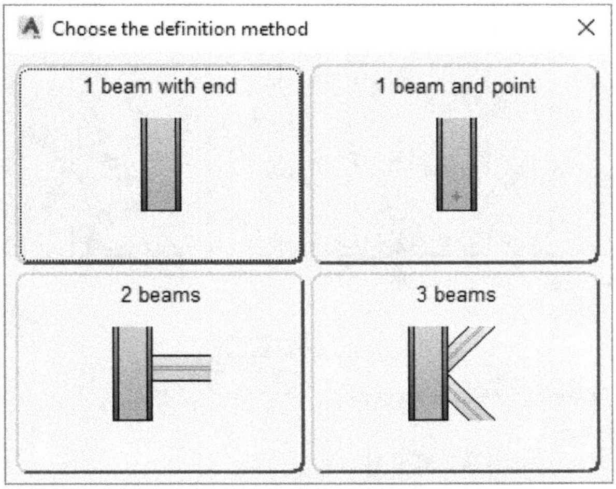

Figure 3–103

- **1 beam with end:** Use when adding objects such as base plates or end plates.
- **1 beam and point:** Use when adding connections at a specific point, such as stiffeners and reinforcing plates.
- **2 beams:** Use when adding connections between two beams.
- **3 beams:** Use when adding connections between three beams.

Note: If you have more than three beams that intersect, you need to create separate custom connections.

4. At the *Select input beam* prompt, select the main beam and press <Enter>.

- Note that the prompts are different for each method. Select the required options and press <Enter> after each, as required.

5. In the *User template* dialog box, type a *Name.* Then, in the *Drivers selection prompts,* type the text to display at the prompts, as shown in Figure 3–104.

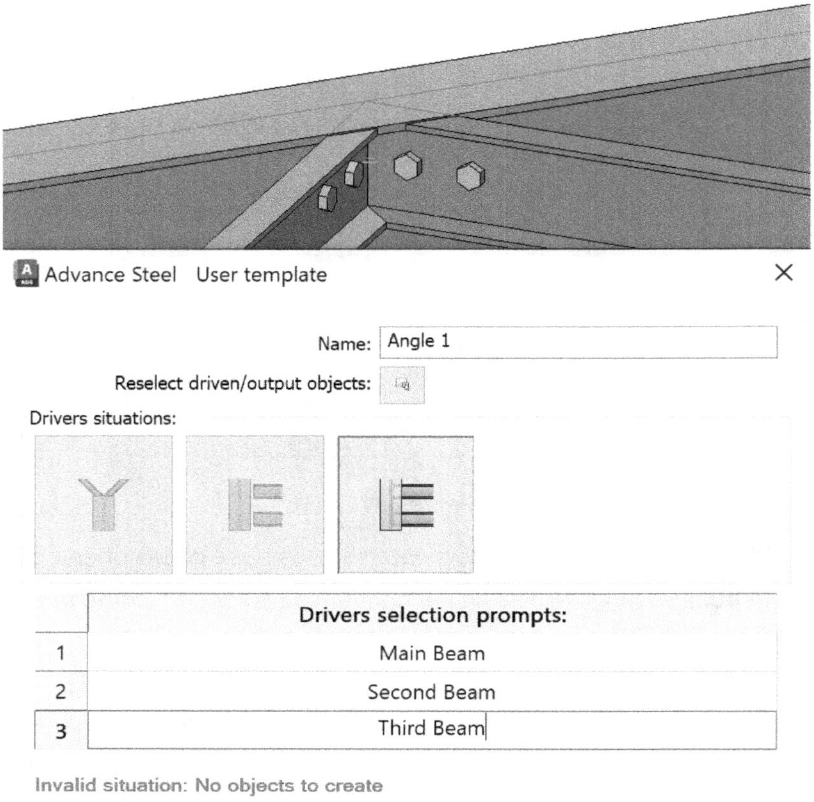

Figure 3–104

6. Click (Reselect driven/output objects), draw a window (not a crossing) around all of the objects that form the custom connection, and press <Enter>.

- Ensure that you select any feature elements, such as miters as well.

- Custom connection objects display as a blue bounding box (as shown in Figure 3–105). You can select the box to copy, modify, and delete the connection.

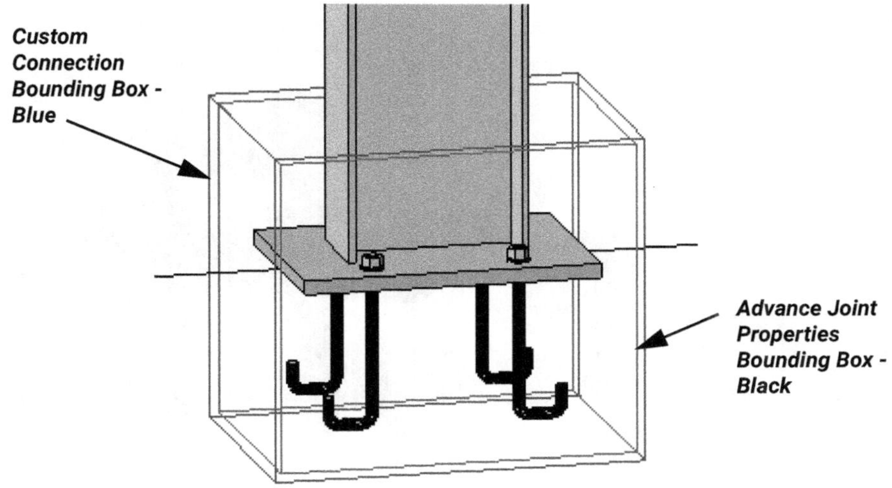

Custom Connection Bounding Box - Blue

Advance Joint Properties Bounding Box - Black

Figure 3–105

- Many of the joints in the *Connection Vault* also cope or otherwise modify beams to fit the connections. If you are creating a custom connection from scratch and including features such as coping, ensure that you select the feature joint box when you select the other connection objects.

How To: Use Custom Connection Templates

1. In the *Advance Tool Palette>* (Custom connections) category, click (Insert connection template).

   Note: In order to use custom connections in another drawing, save the custom connection templates drawing to C:\ProgramData\Autodesk\Advance Steel 2024\USA\Shared\ConnectionTemplates.

2. In the *Connection template explorer* dialog box, select the connection template you want to apply (as shown in Figure 3–106) and click **OK**.

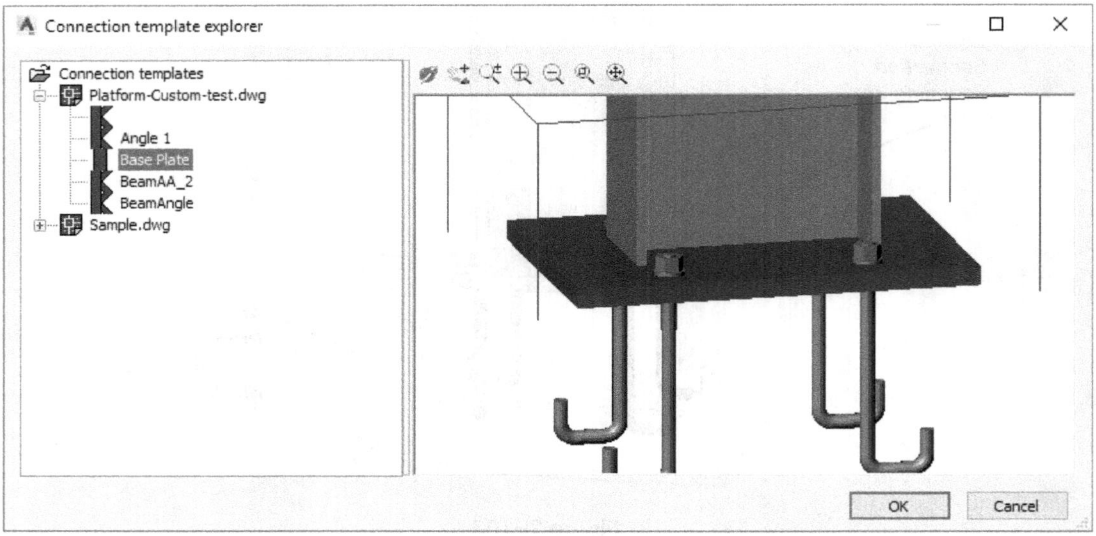

Figure 3–106

3. Follow the prompts to select the elements.

4. In the *User template* dialog box (shown in Figure 3–107), you can specify a *Name* and select **Allow object modification**, if required.

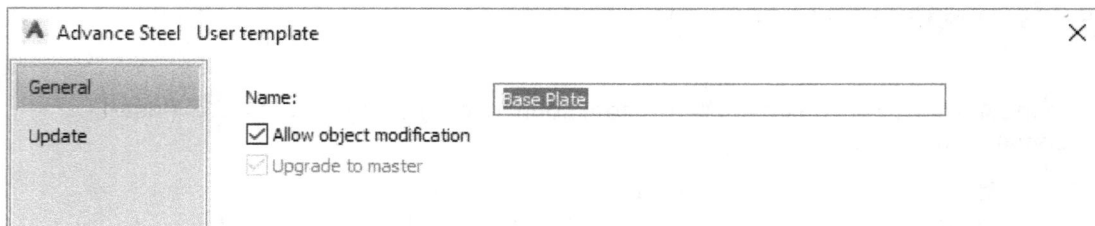

Figure 3–107

- You can open the *User template* dialog box using any of the following methods:
  - Double-click on the blue bounding box.
  - Right-click and select **Advance Properties**.
  - Right-click and select **Advance Joint Properties**.
  - Right-click and select **Custom Connection Properties**.

Practice 3e
Create Custom Connections

Practice Objectives

- Use joint properties stored in a library.
- Create and apply custom connections.

In this practice, you will open a completed model and view a connection and its corresponding library. You will miter beams, add two connections using the *Connection Vault* and the saved connection type from the library. You will then create a custom connection using the two connections (as shown in Figure 3–108) and apply them to similar connections in the model.

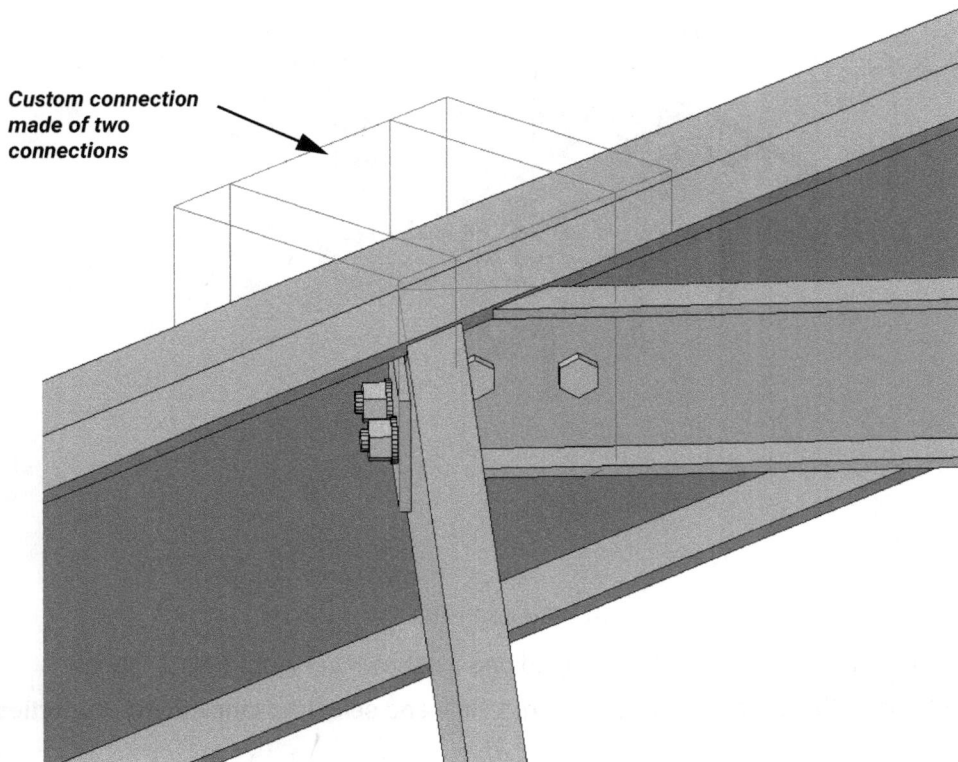

Custom connection made of two connections

Figure 3–108

Task 1: Create connections using Advance Joint Properties stored in a library.

1. In the practice files folder, open **Platform-Introduction.dwg**. This is a completed version of the Platform project that includes Advance Joint Properties Libraries.

2. Set the view to **SE Isometric**.

3. Open the *Project Explorer* and toggle on Level 1 so that the Level 2 structure does not display, as shown in Figure 3–109.

   - If Level 2 still displays, right-click on **Level 1** and select **Properties**. Click **OK** to close the dialog box.

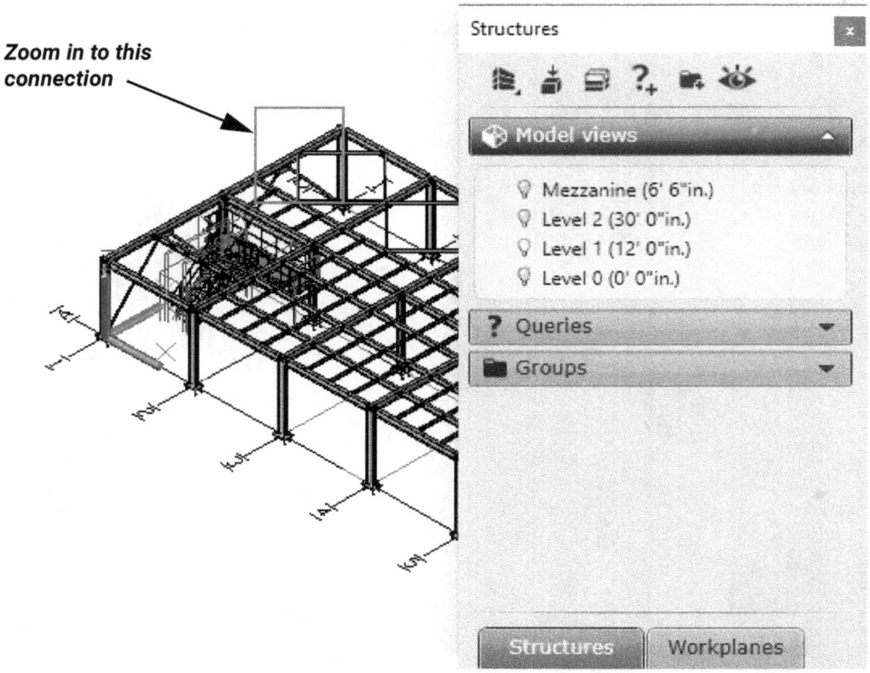

Figure 3–109

4. Zoom in on the connection at the angled beams shown in Figure 3–109, above.

5. Select one of the plates or bolts, then right-click and select **Advance Joint Properties**.

6. In the dialog box, select **Library**, click **Save values**, then click **Edit** and name it **Angled connection** as shown in Figure 3–110. Click **OK**.

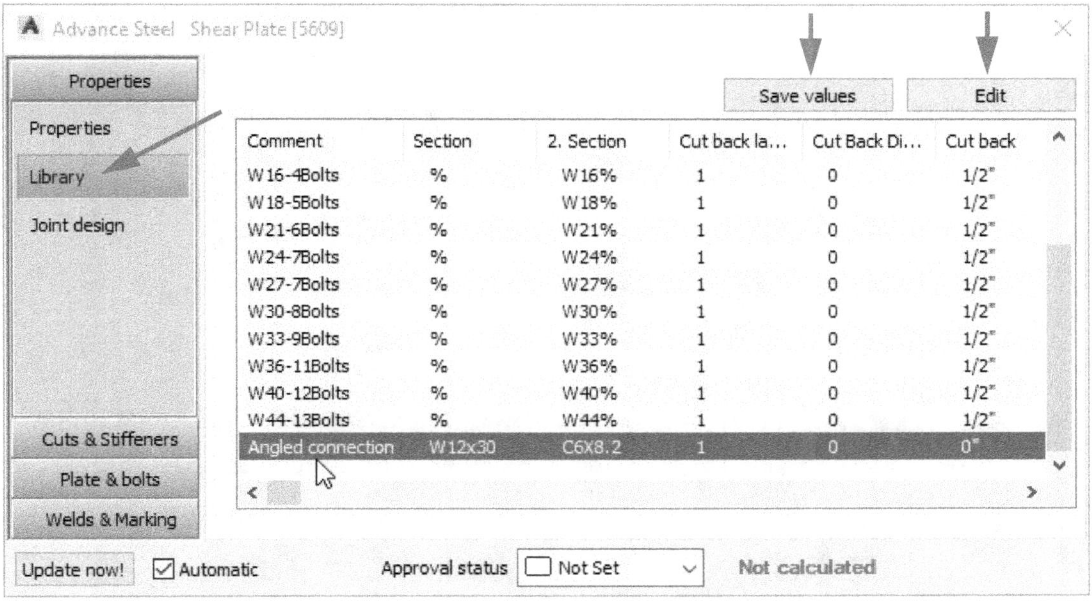

Figure 3–110

7. Close the dialog box. By opening this drawing, you have added the information to your computer and made this the default value for shear plates.

8. Open **Platform-Custom.dwg**.

9. In the *Project Explorer*, toggle on **Level 1** so that the objects on Level 2 do not display.

10. Zoom in on the same intersection that you did in the **Platform-Introduction.dwg**.

11. In the *Connection Vault>* (Platform beams) category, click (Shear Plate).

    Note: Even though these prompts seem to let you select more than one object, they do not actually select multiple objects. You must press <Enter> after each selection.

12. At the *Select Passing main beam* prompt, select the large beam and press <Enter>.

13. At the *Select Secondary beam that is to be attached* prompt, select one of the angled beams that is touching the main beam and press <Enter>.

14. The *Shear Plate* dialog box displays and the default preset connection is applied. Note that the angled beam is automatically coped to the primary beam, as shown in Figure 3–111.

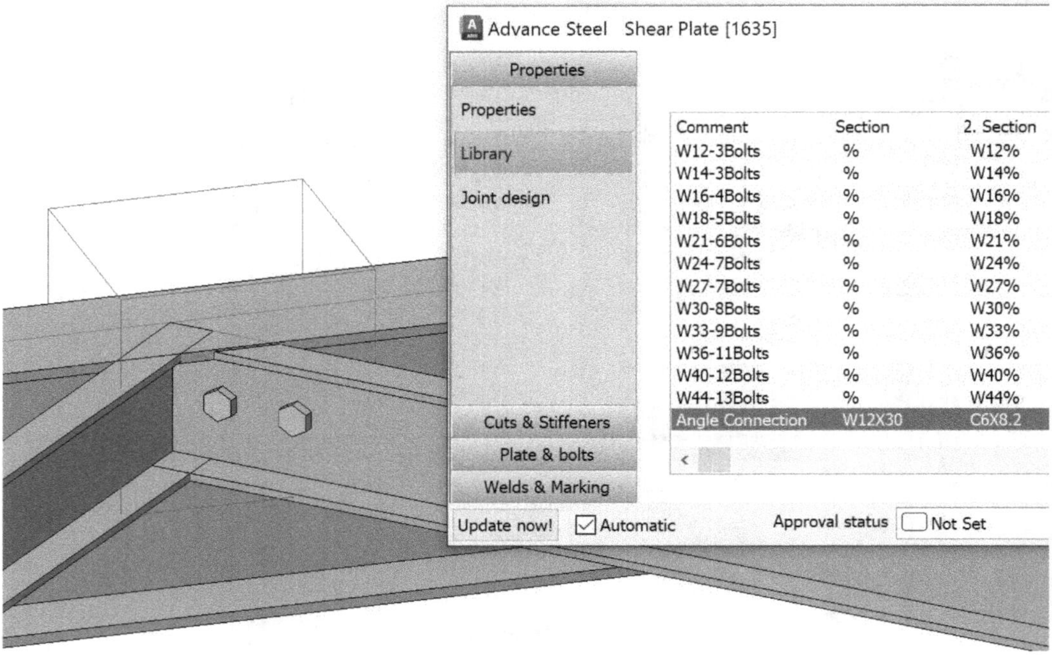

Figure 3–111

15. Repeat the **Shear Plate** command and add a plate to the other angled beam. If needed, in the Shear Plate macro, from the *Plate & bolts* tab>*Plate* layout, change side to **Right**, and from the *Plate & bolts* tab>*Bolts and holes*, select the **Inverted** checkbox.

16. In the *Advance Tool Palette*> (Features) category, click (Miter) and miter the two angled beams together.

17. Save the drawing.

Task 2: Create a custom connection and apply it to other locations.

1. In the *Advance Tool Palette*> (Custom connections) category, click (Create connection template).

---

2. In the *Choose the definition method* dialog box, select **3 beams**, as shown in Figure 3–112.

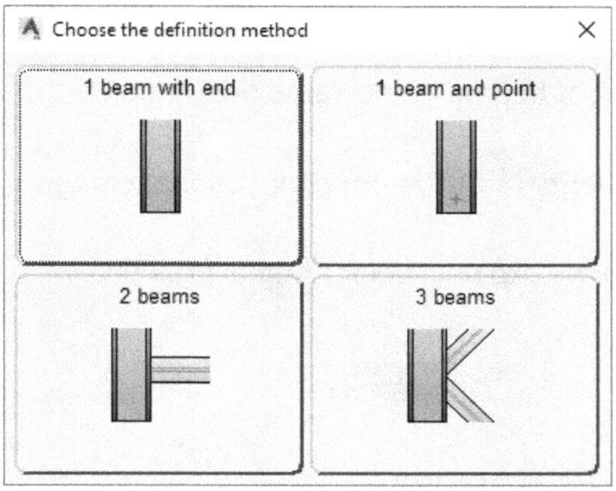

Figure 3–112

3. At the *Select input beam* prompt, select the main beam and press <Enter>.

4. Select one of the angled beams and press <Enter>.

5. Select the other angled beam and press <Enter>.

6. In the *User template* dialog box, in the *Name* field, type **Angle 1**. Then, in the *Drivers selection prompts*, enter the information shown in Figure 3–113.

   • Note how the *Drivers selection prompts* highlight the related object when you click in the name.

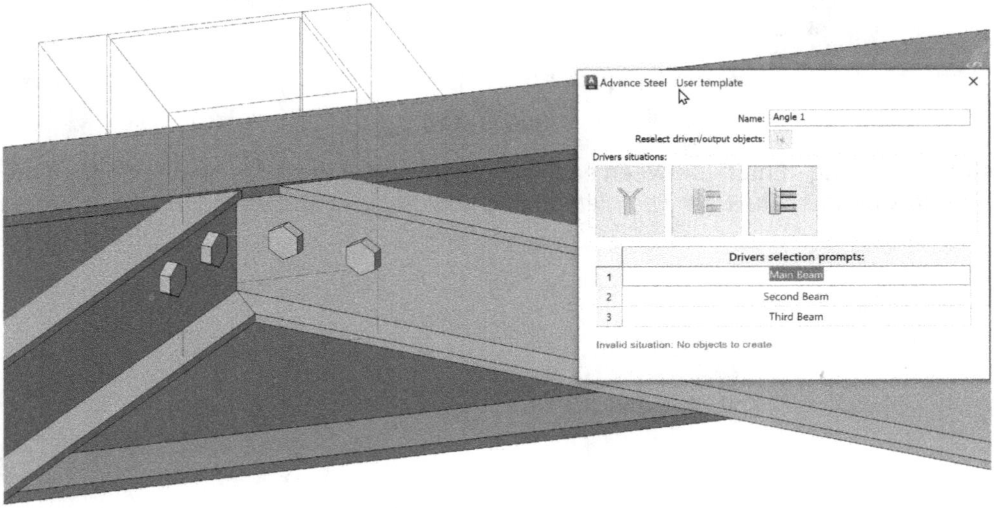

Figure 3–113

7. Select **Reselect driven/output objects** and then draw a window (not a crossing) around both shear plate connections. Press <Enter>.

8. Close the dialog box.

9. Pan over to the next set of beams at the same orientation.

10. In the *Advance Tool Palette>* (Custom connections) category, click (Insert connection template).

11. In the *Connection template explorer,* select **Angle 1** (as shown in Figure 3–114) and click **OK**.

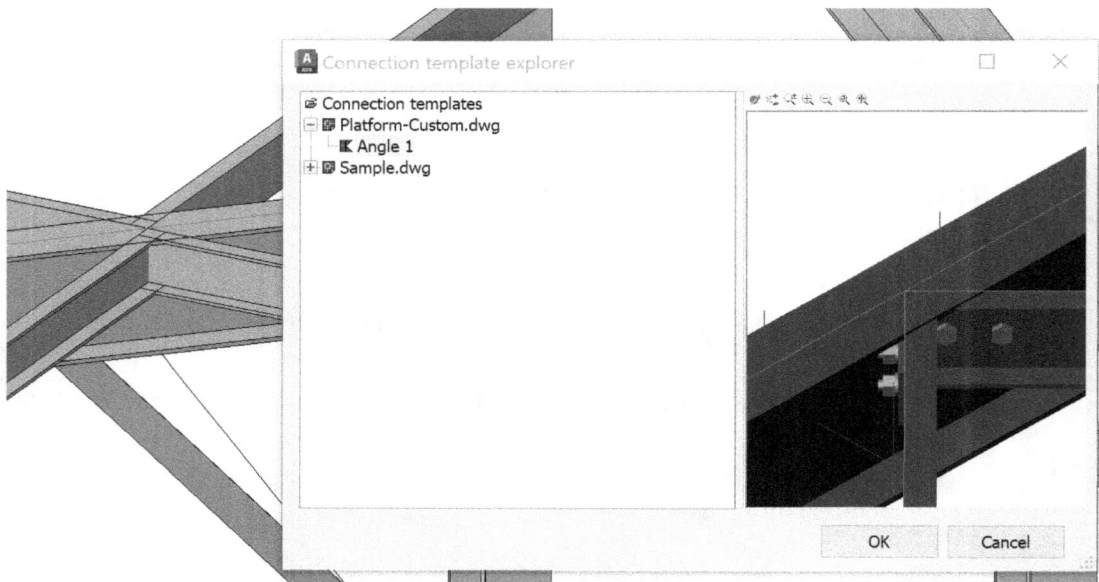

Figure 3–114

12. Without pressing <Enter> between each selection, select the main beam and then the two secondary beams. The *User Template* dialog box displays and the connections are added.

13. Repeat the process on the other intersections that are similar. This custom connection works on both sides of the beams (as shown in Figure 3–115), but not on the beams that connect to the outside beams, as those are at a different angle.

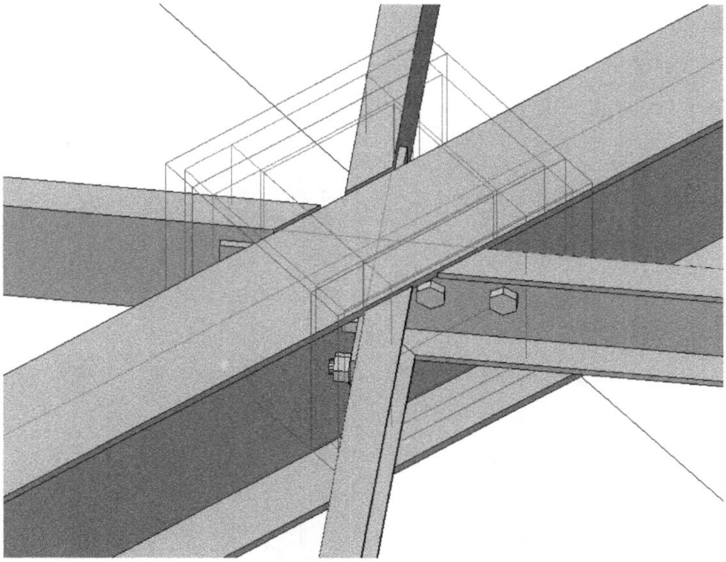

Figure 3–115

14. Save the drawing.

End of practice

Chapter Review Questions

1. Which of the following tools can you use to apply full connections, including plates and bolts to a column and beam intersection?

   a. Project Explorer

   b. Advance Tool Palette

   c. Connection Vault

   d. Connection Objects

2. Which of the elements in Figure 3–116 can you double-click on to open the Advance Steel *Joint Properties* dialog box?

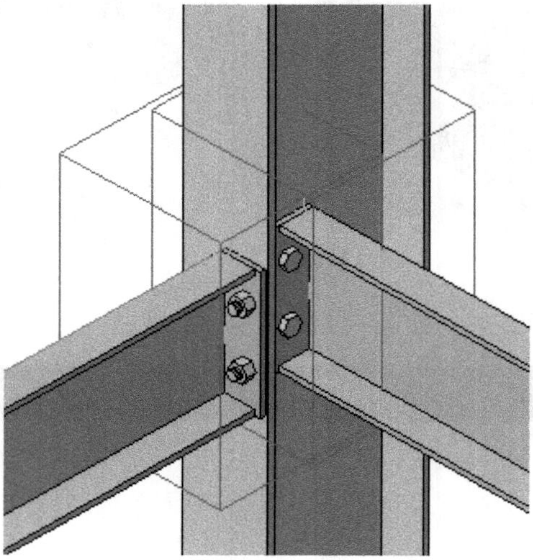

Figure 3–116

   a. Bolts

   b. Beam

   c. Plates

   d. Box

3. Which of the following commands creates a linked copy of all of the connection pieces in a joint so that you can update one connection and have the other update as well?

a. (Create joint in a joint group)

b. (Create connection template)

c. (Create by template)

d. (Add joint to a joint group)

4. When you are creating a custom connection (such as an Access, as shown in Figure 3–117), what kind of objects should you select for the driven/output objects? (Select all that apply.)

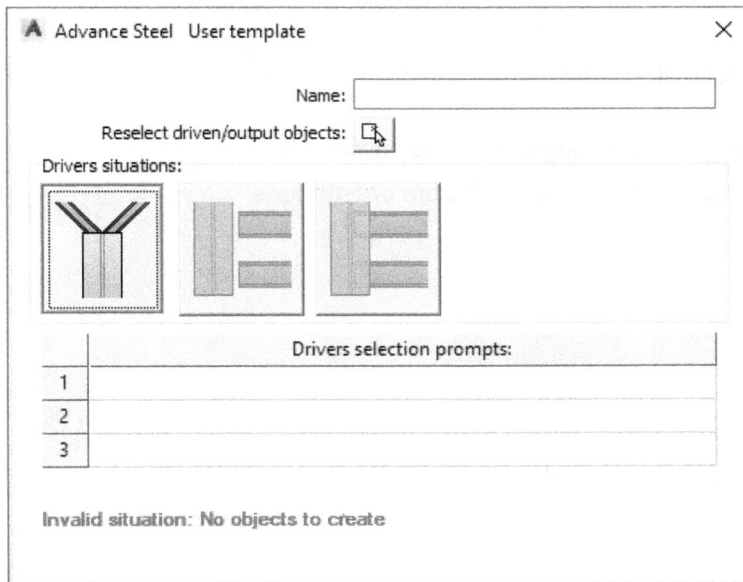

Figure 3–117

a. Beams

b. Columns

c. Bolts

d. Plates

5. Which command modifies a corner where two beams intersect (as shown in Figure 3–118) to create a welded connection (as shown in Figure 3–119)?

Figure 3–118   **Figure 3–119**

a. AutoCAD Fillet

b. Corner cut

c. Cut at object

d. Miter

6. Which of the following must be true when you are creating a folded plate without a position adjustment?

a. The unfolded plates must be touching each other.

b. The unfolded plates can be in different orientations.

7. The holes cut in the plate shown in Figure 3–120 are both square, however they have been cut in different ways. What is the difference between the two holes?

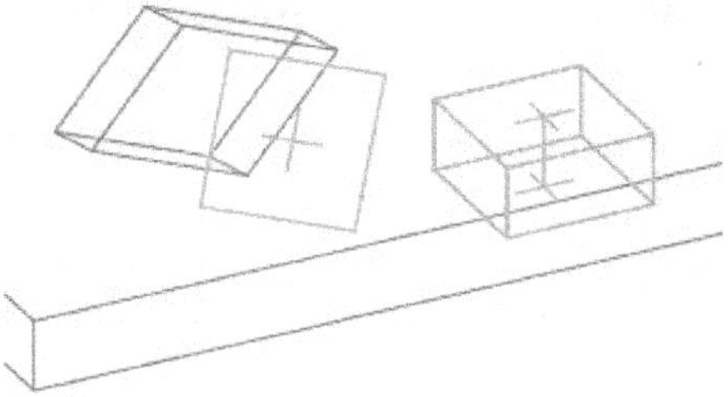

Figure 3–120

a. The angled hole was created around an object, while the straight hole was created using a Feature tool.

b. The angled hole was created with a UCS tool, while the straight hole was created with a plate only tool.

c. Both holes were created by polylines, but at different UCS positions.

d. They were both created using the Element contour tool, but at different UCS positions.

Command Summary

Button	Command	Location
Connections		
	Configure Quick Connect	• **Ribbon:** *Extended Modeling* tab>*Joint Utilities* panel
	Connection Vault	• **Ribbon:** *Home* tab>*Extended Modeling* panel • **Ribbon:** *Extend Modeling* tab>*Joints* panel
	Create connection template	• **Advance Tool Palette:** *Custom connections* category
	Create joint in a joint group	• **Advance Tool Palette:** *Tools* category
	Create joint in a joint group, multiple	• **Advance Tool Palette:** *Tools* category
	Delete All	• **Ribbon:** *Extended Modeling* tab>*Joint Utilities* panel
	Insert connection template	• **Advance Tool Palette:** *Custom connections* category
	Quick connect all	• **Ribbon:** *Extended Modeling* tab>*Joint Utilities* panel
Beam Modification		
	Bevel cut	• **Advance Tool Palette:** *Features* category
	Cope	• **Advance Tool Palette:** *Features* category
	Cope, parametric	• **Advance Tool Palette:** *Features* category
	Cope, skewed	• **Advance Tool Palette:** *Features* category
	Corner cut	• **Advance Tool Palette:** *Features* category
	Cut at object	• **Advance Tool Palette:** *Features* category
	Element contour - rule	• **Advance Tool Palette:** *Features* category

Button	Command	Location
	Miter	• **Advance Tool Palette:** *Features* category
	Shorten	• **Advance Tool Palette:** *Features* category
	Shorten at UCS	• **Advance Tool Palette:** *Features* category
Plates: Creating		
	Check unfolding	• **Ribbon:** *Objects* tab>*Plates* panel
	Circular plate	• **Ribbon:** *Objects* tab>*Plates* panel
	Create conical folded plate	• **Ribbon:** *Objects* tab>*Plates* panel • **Ribbon:** *Home* tab>*Objects* panel
	Create folded plate - with position adjustment	• **Ribbon:** *Objects* tab>*Plates* panel
	Create folded plate - without position adjustment	• **Ribbon:** *Objects* tab>*Plates* panel • **Ribbon:** *Home* tab>*Objects* panel
	Create twisted folded plate	• **Ribbon:** *Objects* tab>*Plates* panel
	Plate at poly line	• **Ribbon:** *Objects* tab>*Plates* panel • **Ribbon:** *Home* tab>*Objects* panel
	Plate to poly line	• **Ribbon:** *Objects* tab>*Plates* panel • **Ribbon:** *Home* tab>*Objects* panel
	Polygonal plate	• **Ribbon:** *Objects* tab>*Plates* panel
	Rectangular plate, 2 points	• **Ribbon:** *Objects* tab>*Plates* panel • **Ribbon:** *Home* tab>*Objects* panel
	Rectangular plate, 3 points	• **Ribbon:** *Objects* tab>*Plates* panel • **Ribbon:** *Home* tab>*Objects* panel
	Rectangular plate, center	• **Ribbon:** *Objects* tab>*Plates* panel
	Set folded plate main objects	• **Ribbon:** *Objects* tab>*Plates* panel

Button	Command	Location
Plates: Modifying		
	Explode plate with features to polygonal plate	• **Ribbon:** *Objects* tab>*Plates* panel
	Merge plates	• **Ribbon:** *Objects* tab>*Plates* panel • **Ribbon:** *Home* tab>*Objects* panel
	Shrink/expand poly plate	• **Ribbon:** *Objects* tab>*Plates* panel
	Split plates at lines	• **Ribbon:** *Objects* tab>*Plates* panel • **Ribbon:** *Home* tab>*Objects* panel
	Split plates by 2 points	• **Ribbon:** *Objects* tab>*Plates* panel • **Ribbon:** *Home* tab>*Objects* panel
Bolts, etc.		
	bolts/anchors/holes/shear studs (toggle)	• **Ribbon:** *Objects* tab>*Switch* panel • **Ribbon:** *Home* tab>*Objects* panel
	Circular, center point	• **Ribbon:** *Objects* tab>*Connection objects* panel
	Line of Weld	• **Ribbon:** *Objects* tab>*Connection objects* panel
	Rectangular, 2 points	• **Ribbon:** *Objects* tab>*Connection objects* panel • **Ribbon:** *Home* tab>*Objects* panel
	Rectangular, center point	• **Ribbon:** *Objects* tab>*Connection objects* panel
	Rectangular, corner point	• **Ribbon:** *Objects* tab>*Connection objects* panel
	Shift bolts/holes	• **Ribbon:** *Objects* tab>*Connection objects* panel
	Split bolt group	• **Ribbon:** *Objects* tab>*Connection objects* panel
	Weld point	• **Ribbon:** *Objects* tab>*Connection objects* panel • **Ribbon:** *Home* tab>*Objects* panel
Features		
	Circular contour, 2 points	• **Advance Tool Palette:** *Features* category

Button	Command	Location
	Circular contour, 2 points, UCS	• **Advance Tool Palette:** *Features* category
	Circular contour, center	• **Advance Tool Palette:** *Features* category
	Circular contour, center, UCS	• **Advance Tool Palette:** *Features* category
	Element contour	• **Advance Tool Palette:** *Features* category
	Element contour, UCS	• **Advance Tool Palette:** *Features* category
	Insert corner	• **Advance Tool Palette:** *Features* category
	Polygon contour	• **Advance Tool Palette:** *Features* category
	Polygon contour, UCS	• **Advance Tool Palette:** *Features* category
	Rectangular contour, 2 points	• **Advance Tool Palette:** *Features* category
	Rectangular contour, 2 points, UCS	• **Advance Tool Palette:** *Features* category
	Rectangular Contour, center	• **Advance Tool Palette:** *Features* category
	Rectangular contour, center, UCS	• **Advance Tool Palette:** *Features* category
	Remove corner	• **Advance Tool Palette:** *Features* category

Additional Tools

Button	Command	Location
	All visible	• **Advance Tool Palette:** *Quick views* category
	Isolate Objects	• **Status Bar**
	Selected objects off	• **Advance Tool Palette:** *Quick views* category
	XREF Manager	• **Ribbon:** *Export & Import* tab>*XREF* panel

Additional Model Objects

Autodesk® Advance Steel model objects include specialty items like grating for flat surfaces and cladding for vertical and slanted surfaces. You can create ladders, stairs, and railings using macros and Advance Joint Properties without having to draw each part separately. To fully complete a model, you might want to add concrete objects, such as columns, footings, and beams. These tools are also available using Autodesk Advance Steel commands.

Learning Objectives

- Add grating and cladding objects to a model.
- Create stairs, railings, and ladders.
- Create concrete footings to indicate where anchors are attached.
- Insert a special part.

4.1 Adding Grating and Cladding

Grating and cladding might be produced and ordered separately from the rest of a steel structure, but you can still include them in a model, as shown in Figure 4–1. Grating is used for walkways, covers, and other flat surfaces, while cladding is typically used for roofs and walls.

ASTBEAM

Color	ByLayer
Layer	Claddings
Linetype	ByLayer

ASTGRATING

Color	ByLayer
Layer	Gratings
Linetype	ByLayer

Figure 4–1

Note: There are three types of grating: standard, variable, and bar grating.

- Grating is considered a type of plate with a hatch pattern, as shown in Figure 4–2. The hatch pattern is for information only and might not match the actual design of the grating.

- Cladding is made from beam objects with a decking-like profile, as shown in Figure 4–3.

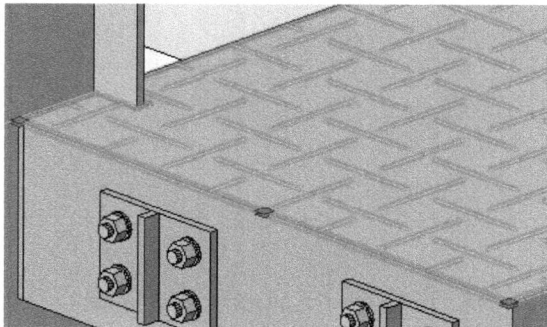

Figure 4–2

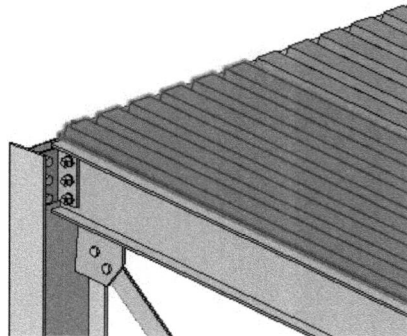

Figure 4–3

- The UCS controls the location and direction of both grating and cladding so remember to set that first.

How To: Add Standard Grating

1. Set the UCS to the orientation you want to add the grating.

2. In the *Home* tab>*Objects* panel or *Objects* tab>*Grating* panel, click ▨ (Standard Grating).

3. Select a center point for the grating.

4. In the *Standard Grating* dialog box>*Shape & Connector* tab (shown in Figure 4–4), set the *Grating class*, *Grating size*, and other details.

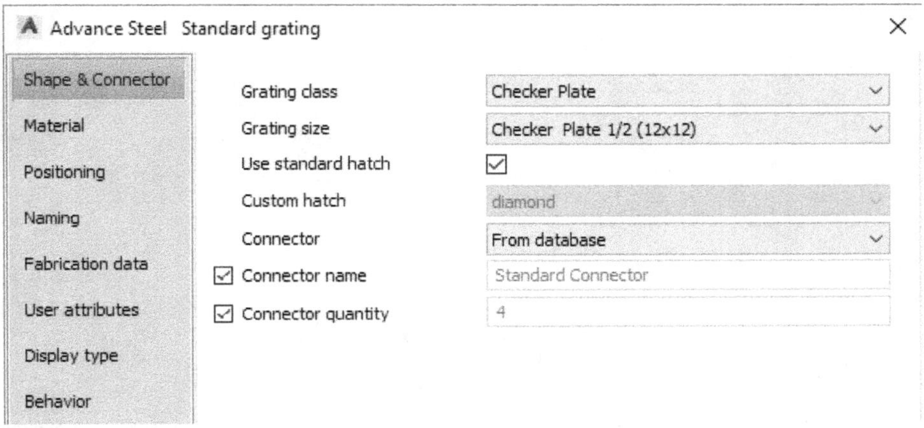

Figure 4–4

5. Select the other tabs and enter the required information. Move, copy, or array the grating, as required.

How To: Add Variable Grating

1. Set the UCS to the orientation you want to add the grating.

2. In the *Home* tab>*Objects* panel or *Objects* tab>*Grating* panel, click ▨ (Variable Grating, rectangular).

3. Select two diagonal points to define the grating area.

4. In the *Variable grating* dialog box>*Shape & Connector* tab, specify the grating type and size, as shown in Figure 4–5.

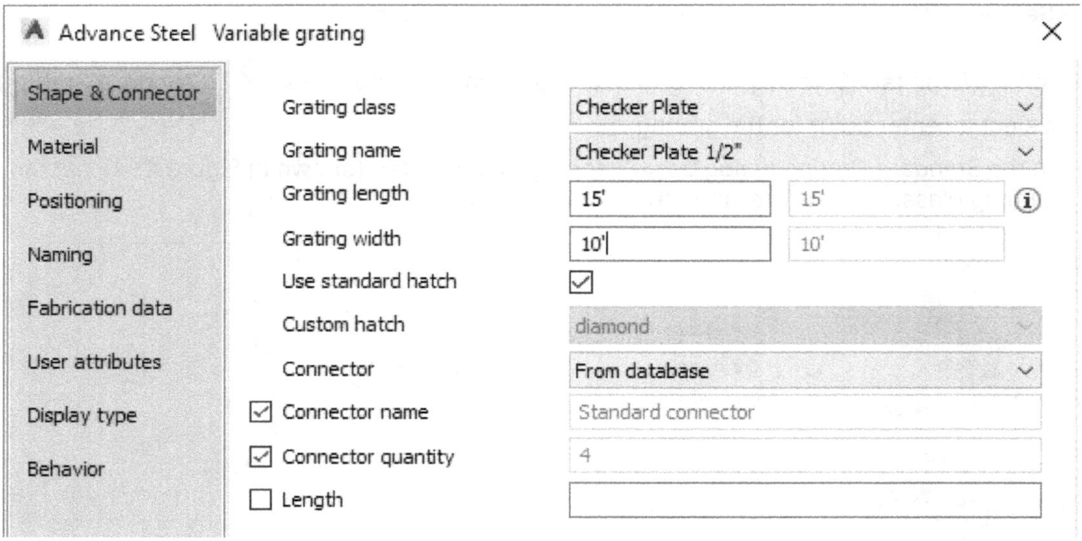

Advance Steel Variable grating	✕

Figure 4–5

5. Select the other tabs and enter the required information.

• To create an odd shaped grating (as shown in Figure 4–6), you can use two different tools. In the *Home* tab>expanded *Objects* panel or *Objects* tab>expanded *Grating* panel, click

  (Variable Grating, polygonal) or (Grating at Polyline).

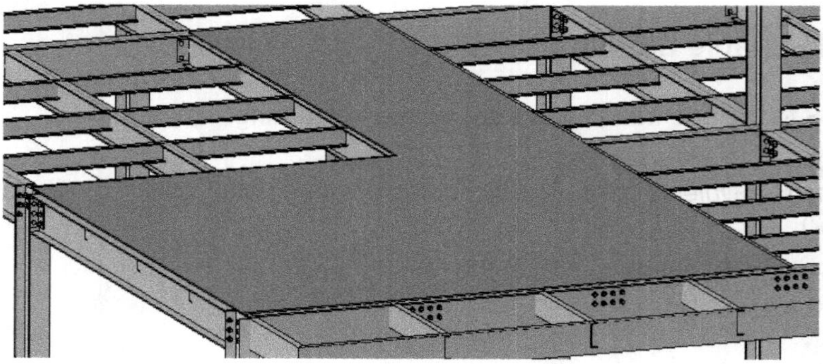

Figure 4–6

How To: Add Bar Grating

1. In the *Home* tab>expanded *Objects* panel or *Objects* tab>expanded *Grating* panel, click

    ▨ (Bar Grating).

2. Select a start point and end point that defines the length of the grating.

3. In the *Bar grating* dialog box>*Shape & Connector* tab (shown in Figure 4–7), specify the grating type and size. To add to the default width of the grating, enter a value in the *Width extension left* or *Width extension right* fields.

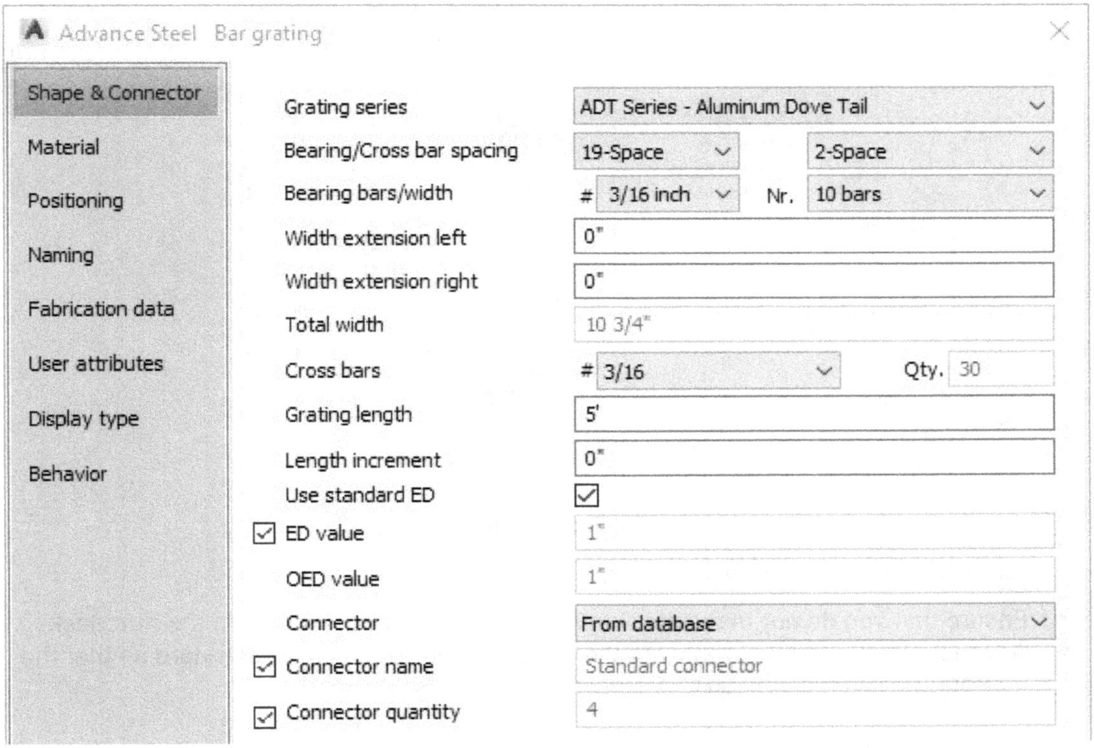

A Advance Steel Bar grating	✕

Shape & Connector	Grating series	ADT Series - Aluminum Dove Tail	⌄
Material	Bearing/Cross bar spacing	19-Space ⌄	2-Space ⌄
Positioning	Bearing bars/width	# 3/16 inch ⌄ Nr.	10 bars ⌄
	Width extension left	0"	
Naming	Width extension right	0"	
Fabrication data	Total width	10 3/4"	
User attributes	Cross bars	# 3/16 ⌄	Qty. 30
Display type	Grating length	5'	
	Length increment	0"	
Behavior	Use standard ED	☑	
☑ ED value		1"	
OED value		1"	
Connector		From database	⌄
☑ Connector name		Standard connector	
☑ Connector quantity		4	

Figure 4–7

4. Select the other tabs and enter the required information.

💡 **Hint: Changing the Color of Layers**

In the default Autodesk Advance Steel templates, grating is set to a bright magenta pink. You can change this default color using the *Layer Properties Manager*.

1. In the *Home* tab>*Layers* panel, click 📑 (Layer Properties).

2. In the *Layer Properties Manager*, beside *Gratings*, click in the *Color* column.

3. In the *Select Color* dialog box, select a color and click **OK**. The layer color changes, as shown in Figure 4-8.

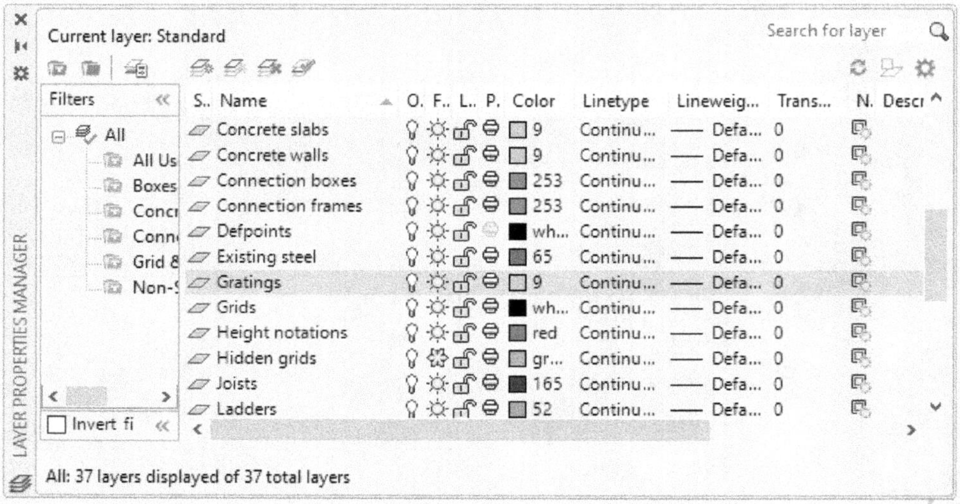

Figure 4-8

4. Close the *Layer Properties Manager*.

• Ensure that you do not make the *Gratings* layer current by mistake. In the Autodesk Advance Steel software, you want the *Current Layer* to remain at **Standard** so that the appropriate layer is applied to the objects when they are created.

How To: Add Cladding

1. Set the UCS in the orientation of the cladding.

2. (Optional) If the shape of the cladding is not rectangular, draw a closed polyline for the boundary.

3. In the *Home* tab>*Extended Modeling* panel, click ⌂ (Define Cladding Area).

4. At the *Do you want to select supporting beams?* prompt, select **Yes** or **No**. (The default is **No**.)

    - If you select **Yes**, select the supporting beams.

5. At the *Create area* prompt, select either **Rectangular** or **From Polyline**.

    - Select two diagonal points for the rectangle, or select an existing polyline. At this point, all that displays is a polyline, as shown in Figure 4–9.

6. In the *Home* tab>*Extended Modeling* panel, click 🏠 (Define Cladding Opening).

7. At the *Create opening* prompt, select either **Rectangular** or **from Polyline**.

8. Select the area object the opening is connected too.

    - Pick two diagonal points for the rectangle, or select an existing polyline. At this point, all that displays is a polyline, as shown in Figure 4–9.

Cladding Area

Cladding Opening

Joint Box

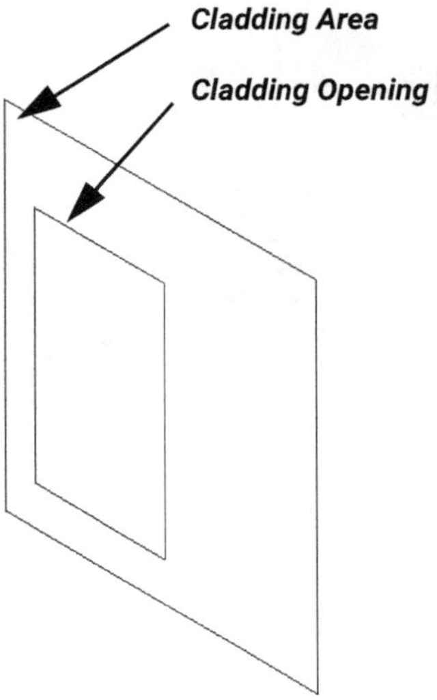

Figure 4–9 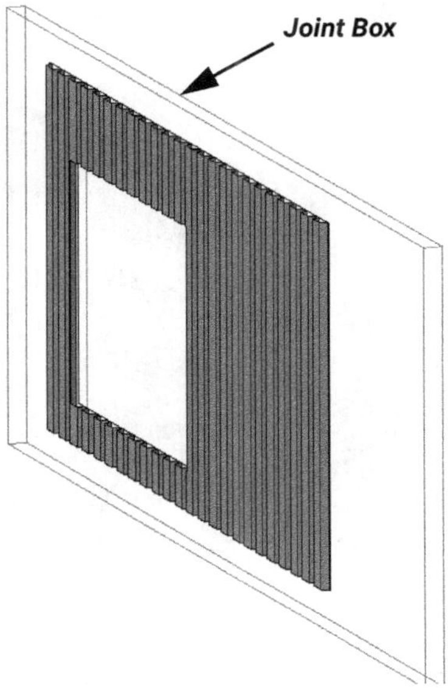   Figure 4–10

9. In the *Home* tab>*Extended Modeling* panel, click 🧊 (Create claddings).

10. Select any cladding areas you created. You do not need to select cladding openings, because they are connected to an area.

11. The cladding is created and displays a joint box, as shown in Figure 4–10, above.

12. In the *Cladding for Roofs and Walls* dialog box, make any required changes.

Practice 4a
Add Grating and Cladding

Practice Objectives

- Place the UCS at the correct angle and location.
- Add grating to a platform.
- Create a cladding area and apply cladding.

In this practice, you will move the UCS to locations where you are creating grating and cladding. You will add grating to at least two platform areas and then add cladding to the roof, as shown in Figure 4–11.

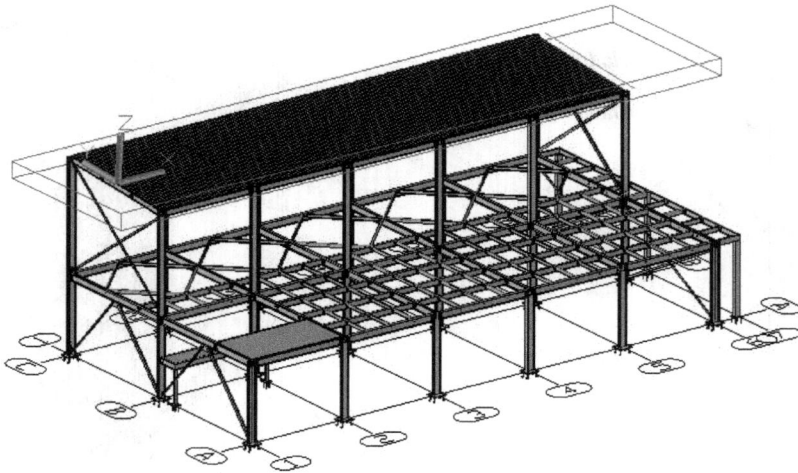

Figure 4–11

Task 1: Add grating to two platform areas.

1. In the practice files folder, open **Platform-Grating.dwg**.
2. Zoom in on the small platform near Column B1.
3. Move the UCS to the top of the small platform.

4. In the *Home* tab>*Objects* panel or *Objects* tab>*Grating* panel, click ▨ (Variable Grating, rectangular).

5. Select two diagonal points to define the grating area.

6. In the *Variable grating* dialog box>*Shape & Connector* tab, ensure that the *Grating class* is set to **Checker Plate**, and the *Grating name* is set to **Checker Plate 1/2"**. Modify the *Grating length* and *Grating width* so that the grating fits on top of the beams, as shown in Figure 4–12. In the *Positioning* tab, ensure that the *Justification* is set to **1.00**.

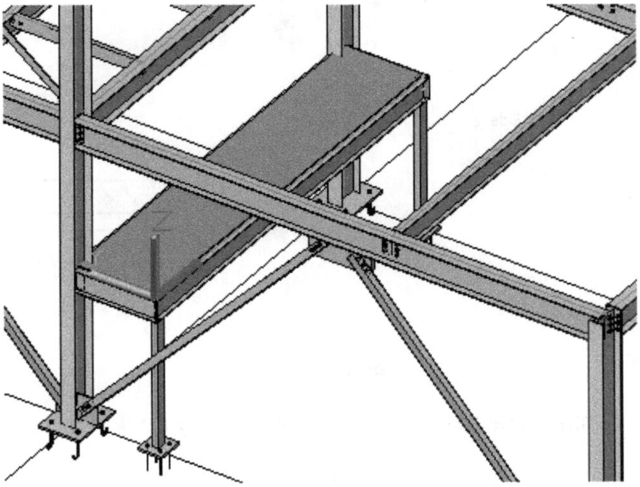

Figure 4–12

7. In the *Objects* tab>*Switch* panel, make sure **Exact Cross Section** is chosen, as shown in Figure 4–13.

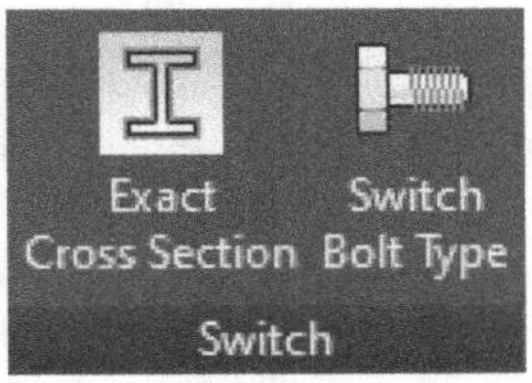

Figure 4–13

8. In the *Advance Tool Palette*>  (Features) category, click  (Element Contour UCS).

    *   Pick the grating, then the column to cut the grating against the column.

9. In the *Contour processing* dialog box, set the Gap width to **1/8"** as shown in Figure 4–14.

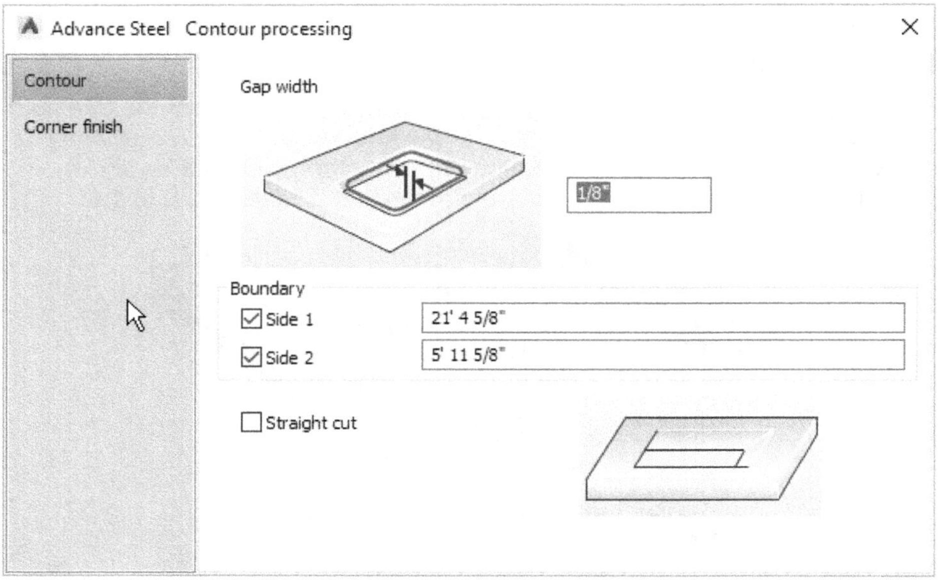

Figure 4–14

10. Repeat on the other side. The result should look like Figure 4–15.

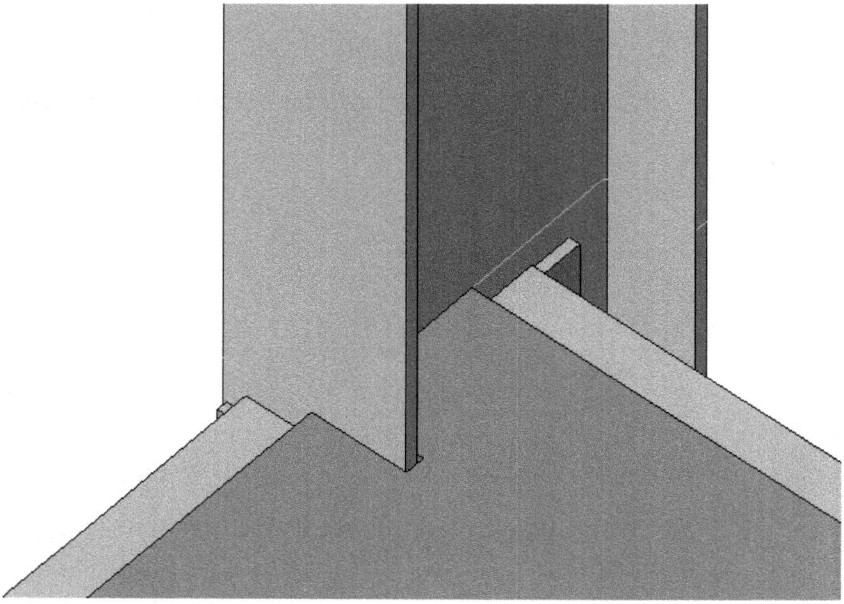

Figure 4–15

11. Move the UCS and add grating on Level 1, as shown in Figure 4–16. You can add grating to other parts of Level 1 (other than the diagonal opening areas) if you have time.

- If required, change the color of the **Grating** layer to a light gray.

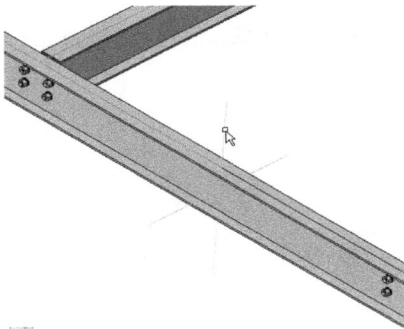

Figure 4–16

12. Save the drawing.

Task 2: Modify the UCS location to a sloping beam.

1. In the *Advance Tool Palette>* (UCS) category, click (UCS at Object).
2. Select on the beam as shown in Figure 4–17 and select **CS** as shown in Figure 4–18.

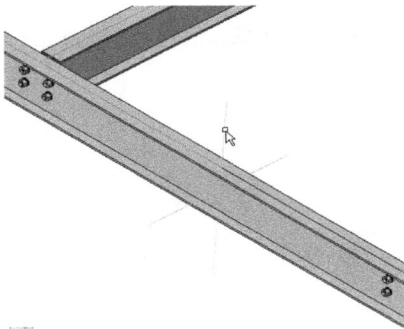

Figure 4–17

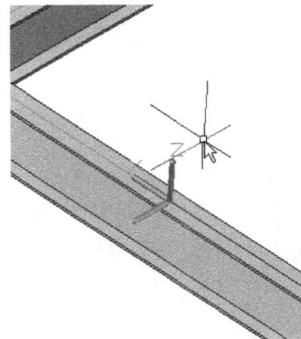

Figure 4–18

3. Save the drawing.

Task 3: Create the cladding.

1. In the *Home* tab>*Extended Modeling* panel, click ⌂ (Define Cladding Area). Use the **Rectangular** option and do not select the supporting beams.

2. For the two points of the rectangle, select the top node of columns B1 and C6 to create the cladding area, as shown in Figure 4–19.

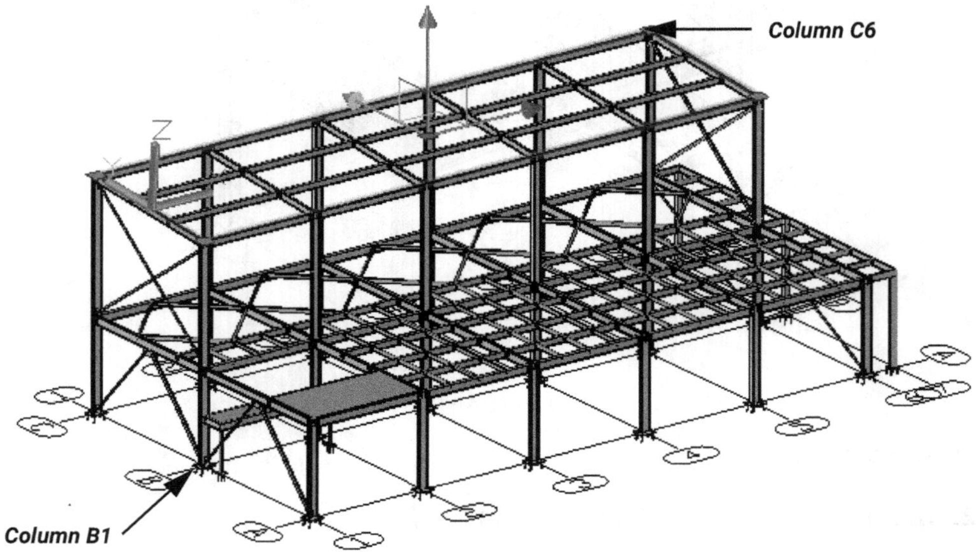

Column C6

Column B1

Figure 4–19

Note: Make sure to turn on Selection Cycling to pick the Cladding Area.

3. In the *Home* tab>extended *Modeling* panel, click 🏠 (Create claddings). Pick the Cladding area.

    Note:

4. In the C*ladding for Roofs and Walls* dialog box>*Cladding general properties* tab, change the *Cladding direction* to **Horizontal cladding**.

5. Continue working through the tabs to verify or change the properties to suit your type of design.

6. Close the dialog box.

7. Reset the UCS to **World**, and then save the drawing.

End of practice

4.2 Modeling Ladders, Stairs, and Railings

Ladders, stairs, and railings are formed using steel sections and connections, as shown in Figure 4–20. You can model each item separately, but the Autodesk Advance Steel software includes macros that make it easy to create, adjust, and connect the objects using specific tools and connections.

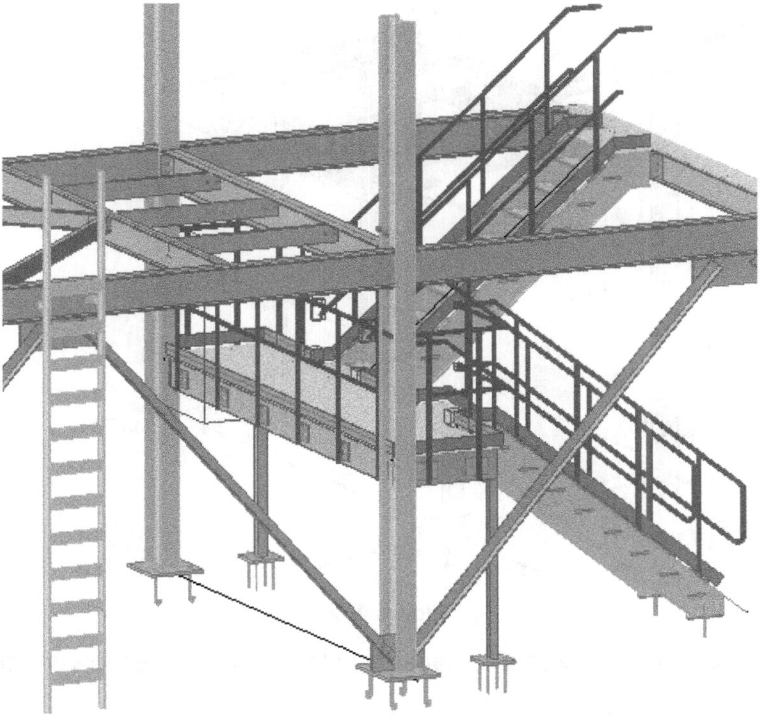

Figure 4–20

- The most important part of modeling ladders, stairs, and railings is to set up the Advance Joint Properties correctly. This takes time to process, as there are many different options.

- When you set up a ladder, stair, or railing style, you can save it to the library. In the related *Advance Joint Properties* dialog box>*Library* tab, click **Save values**. Once saved, click **Edit** and, in the *Library* dialog box, give the style a name.

How To: Create a Ladder

1. Set the UCS so that the Y-axis follows the approach to the ladder (i.e., facing the person climbing the ladder), as shown in Figure 4–21.

    Note: Ladders are placed in the Y-direction.

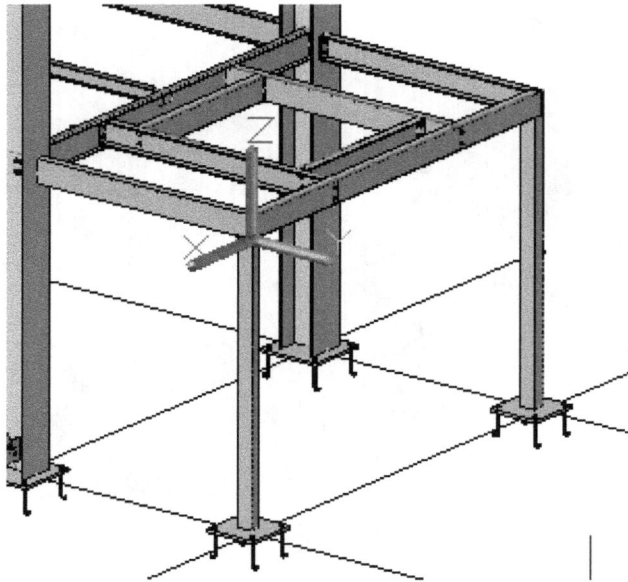

Figure 4–21

2. In the *Home* tab>*Extended Modeling* panel or *Extended Modeling* tab>*Structural Elements* panel, click ○ (Cage Ladder).

    Note: The command is called Cage Ladder, but you can toggle off the cage in Advance Properties.

3. Select the start point of the ladder where you want it to end at the top. This is where you would exit from the ladder.

4. Select a point to define the ladder height down from the first point. You can enter a distance or select an object on that level.

5. Go through the different tabs and sub-tabs in the dialog box. For example, the *Ladder* tab contains several sub-tabs that define parts of the ladder, such as the *Ladder exit* shown in Figure 4–22. The *Cage* and *Exit* tabs also contain sub-tabs.

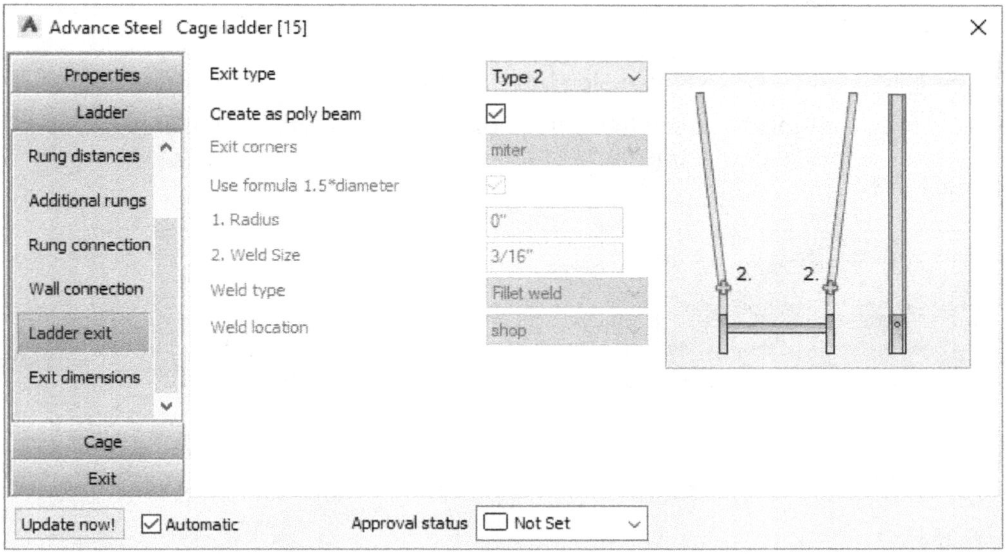

Figure 4–22

- If you want to create a cage with a widened exit, you can start using the parametric options and then modify it manually. A work around is to copy the ladder, create a wider ladder with the required size of exit type, and then delete the joint box so each piece is separate. You can then copy the individual pieces to the main ladder.

How To: Create a Stair

1. Add a sketch line that defines the run of each stair. Typically, this is drawn from the midpoint of the stair location. Include space for landings.

2. In the *Home* tab>*Extended Modeling* panel or *Extended Model* tab>*Structural Elements* panel, click ⚙ (Straight stair).

   Note: Commands for saddled and spiral stairs are also available.

3. Specify the method you want to use for the stair:

   - Start and end point: **0** (default)
   - Length and angle: **1**
   - Height and angle: **2**

4. If using the default method, select the first (bottom) and second points (top).

5. Specify the alignment for the stair:

   - Left: **0**
   - Middle: **1** (default)
   - Right: **2**

6. The stair is placed and the *Stair* dialog box displays.

7. Go through the different tabs in the dialog box. For example, under the *Step - General* tab are several sub-tabs that define parts of the stairs, such as the *Tread type* shown in Figure 4–23.

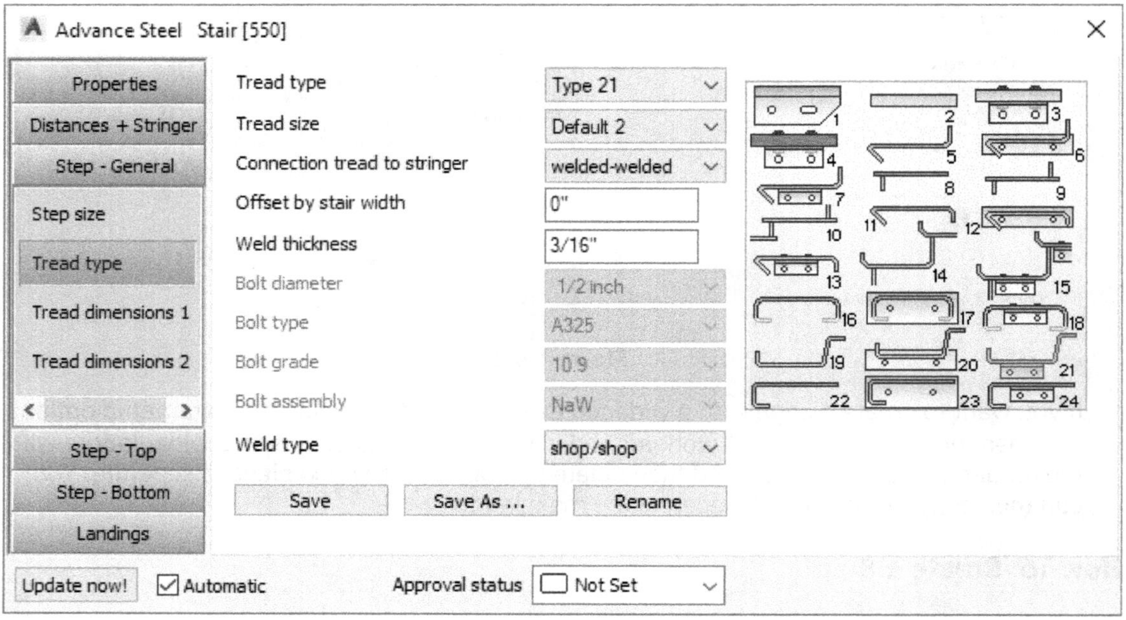

Figure 4–23

- Note that top and bottom landings have different controls.

💡 Hint: Creating Model Views

It can help you to limit the amount of a model that displays as you go through more detailed processes. To do this, you can create a model view. There are several different ways to create model views, including using the UCS, a grid line, or a joint box. In the example shown in Figure 4–24, the joint box of the stair was selected and then the model view box was expanded using grips.

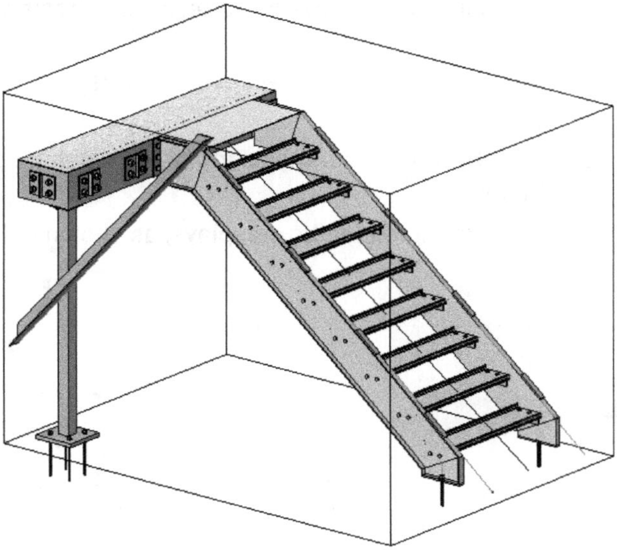

Figure 4–24

1. In the *Project Explorer (Structures)* dialog box, click 🔨 (Create new model view). Alternatively, in the *Home* tab>*Project* panel, expand *Project Explorer*, and then click

   ⬚ (Create Model View).

2. In the *Choose the definition method* dialog box, specify the method you want to use: **One point in UCS, Two points in UCS + front and rear depth, At grid line,** or **At joint box**.

3. Depending on the method you use, select the related points or objects.

4. Type a name for the view and then press <Enter>.

5. Select one of the arrows to specify the default view direction and press <Enter>.

6. In the *Project Explorer*, click on the new model view name to toggle it on.

How To: Create Railings

1. In the *Home* tab>*Extended Modeling* panel, click ⛩ (Hand-railing).

   Note: Railings must be placed on existing objects in the model.

2. Select the base of the beams in the order you want the railing to follow.

3. Select the start point of the railing, which can be anywhere on the beam.

4. Select the end point of the railing. If you have selected several beams, this should be on the last beam.

5. At the *Do you want to select a nosing point relative to the start point* prompt, select **Yes** or **No**.

   * On a set of stairs you might want to set the railing to the noising point of the stair, rather than the top of the stringer.

6. The railing is added and the *Railing* dialog box displays, as shown in Figure 4–25.

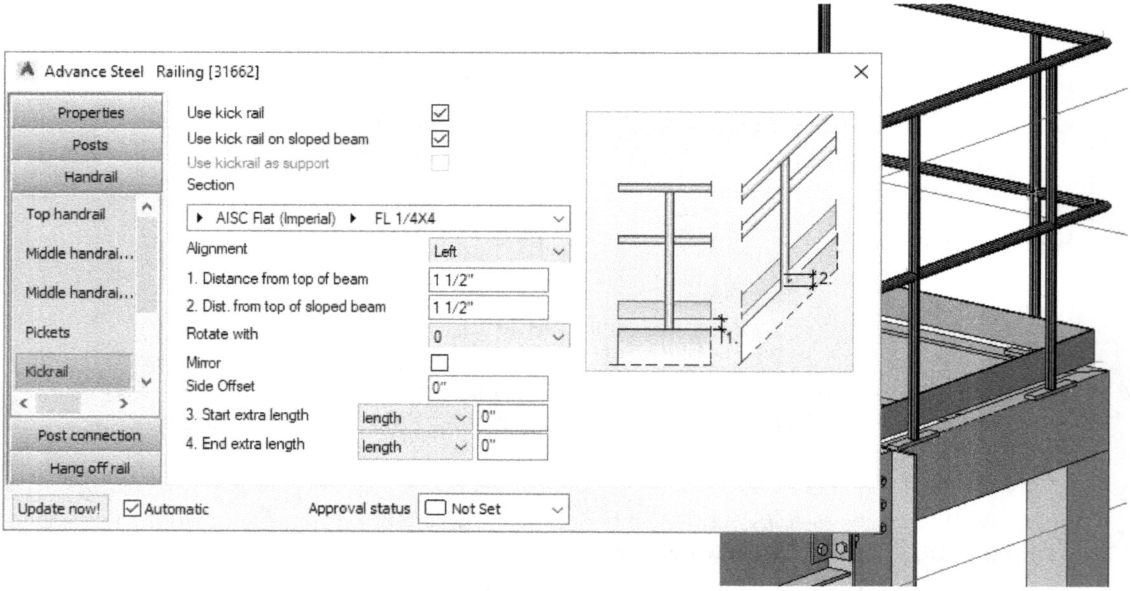

Figure 4–25

7. In the *Railings* dialog box, specify the options on each of the tabs. These options include *Post, Handrail, Post connections,* and more.

💡 Hint: Connecting Stairs and Railings

In the *Connection Vault>* 🖼 (Miscellaneous) category, there are a variety of connectors specifically for stairs and railings, as shown in Figure 4–26 with an example of the Stair Anchor Base Plate connection.

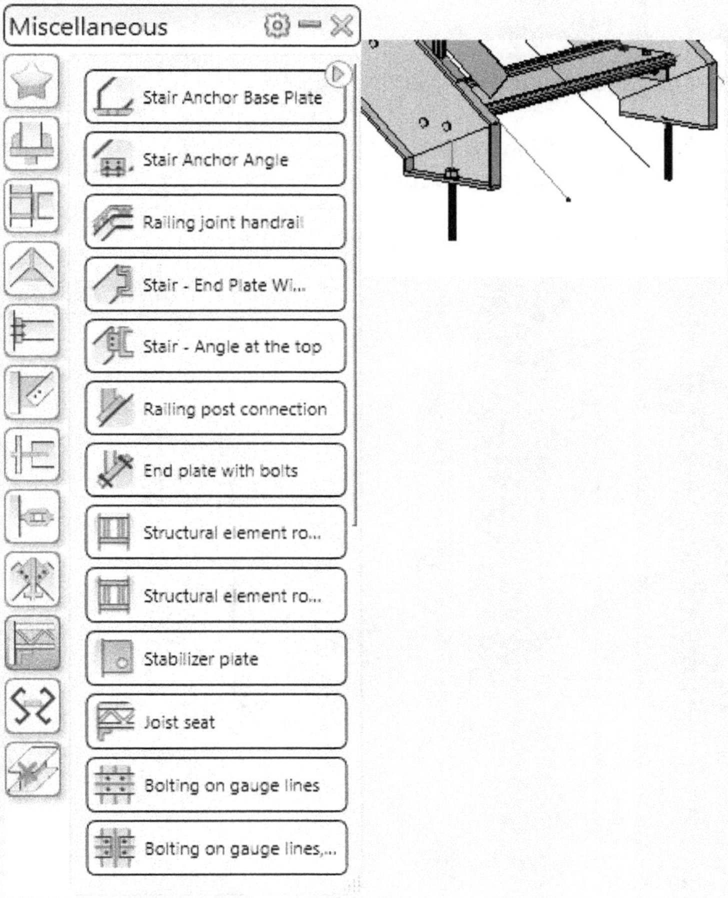

Figure 4–26

Practice 4b
Model Ladders and Stairs

Practice Objectives

- Model ladders with and without cages.
- Model stairs.
- Attach stairs using options in the *Connection Vault*.

In this practice, you will model a ladder without a cage, modifying the steel sections and height (as shown in Figure 4–27), as well as a ladder with a cage. You will also create two stair runs based on existing lines and connect them to the landings, as shown in Figure 4–28. You will use an existing model to assist you in working through the dialog boxes.

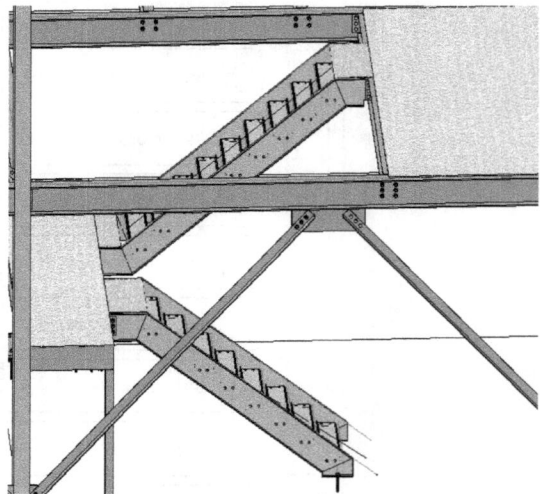

Figure 4–27

Figure 4–28

Task 1: Model a ladder.

1. In the practice files folder, open **Platform-Introduction.dwg**. You will use this drawing as an example of what you need to create.

2. Open **Platform-Stairs.dwg**. Continue working in this drawing.

3. Change the view to **SE Isometric** and then rotate and zoom so that the platform area along grid lines B, C, and 7 displays.

4. Set the UCS so that the Y-axis is pointing away from the platform, as shown in Figure 4–29 at the base of column B7.

5. In the *Home* tab>*Extended Modeling* panel or *Extended Modeling* tab>*Structural Elements* panel, click ⭕ (Cage Ladder)

6. Select the start point of the ladder at the top front midpoint of beam BC6.

7. With ORTHO on, pull the cursor down and enter **12'** for the ladder height. The ladder is added using default values, including a much taller height, as shown in Figure 4–29.

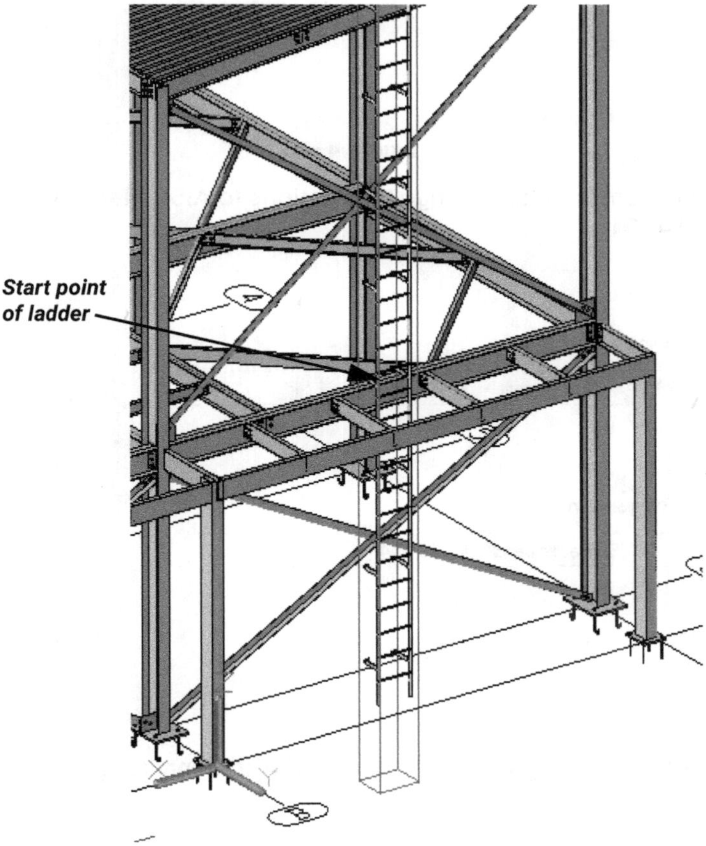

Figure 4–29

8. In the *Cage ladder* dialog box>*Ladder* tab>*Ladder* sub-tab, change the *Height*, *Height extension*, and *Width*, as shown in Figure 4–30.

Figure 4–30

9. In the *Ladder>Sections* sub-tab, change the sections to **AISC Flat (Imperial) > FL 1x4**, as shown in Figure 4–31.

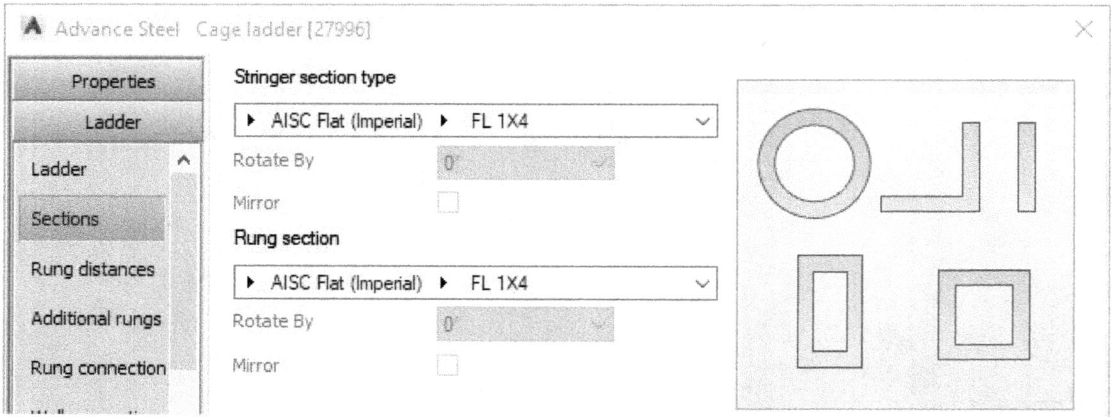

Figure 4–31

10. In the *Ladder>Additional rungs* sub-tab, toggle off **Additional rungs**.

11. Rotate the view so that the top of the ladder displays as shown in Figure 4−32.

12. In the *Ladder>Wall connection* sub-tab, set the *Wall distance, Start distance,* and *Spacing,* as shown in Figure 4−33.

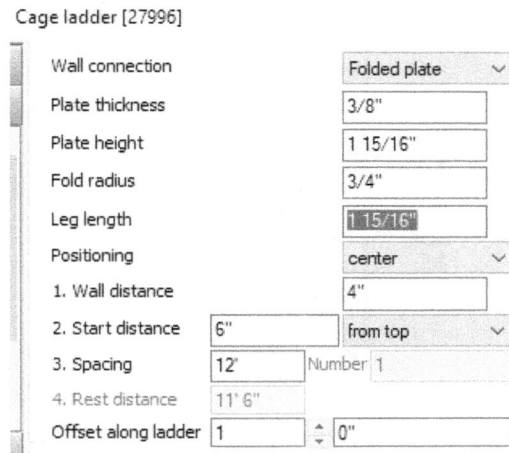

Figure 4−32

Cage ladder [27996]

Wall connection	Folded plate
Plate thickness	3/8"
Plate height	1 15/16"
Fold radius	3/4"
Leg length	1 15/16"
Positioning	center
1. Wall distance	4"
2. Start distance	6" from top
3. Spacing	12' Number 1
4. Rest distance	11' 6"
Offset along ladder	1 0"

Figure 4−33

13. Close the dialog box.

14. Save the drawing.

Task 2: Create a cage ladder.

1. Start the **Advance Copy** command.

2. Select the joint box of the ladder and press <Enter>.

   • If the existing ladder joint box is not displaying, in the *Extended Modeling* tab>*Joint Utilities* panel, click 🄴 (Display) and select one part of the ladder. Then, restart the **Advance Copy** command.

3. In the *Transform elements* dialog box select **Include additional connections**. Then, in the *Distance* area, set the distance to copy the existing ladder by setting *X* to **5'-0"** and *Z* to **12'-0"**, as shown in Figure 4–34.

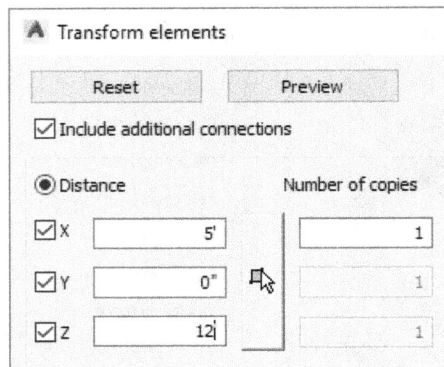

Figure 4–34

4. The copied ladder is added as shown in Figure 4–35.

Figure 4–35

5. Open the Advance Steel Properties for the new ladder. In the *Cage ladder* dialog box>*Ladder* tab>*Ladder* sub-tab, match the settings shown in Figure 4–36.

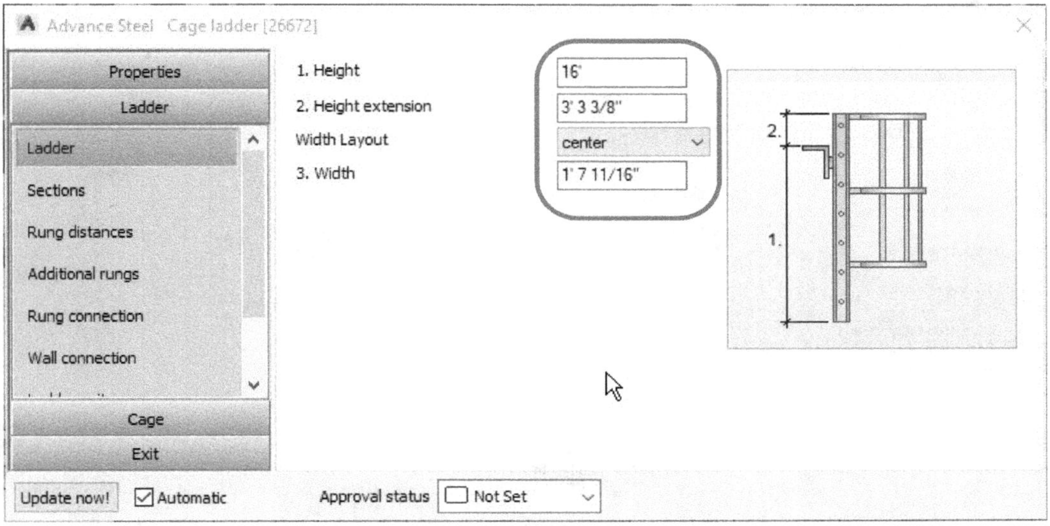

Figure 4–36

6. In the *Wall connection* sub-tab, set *3. Spacing* to 20' as shown in Figure 4–37.

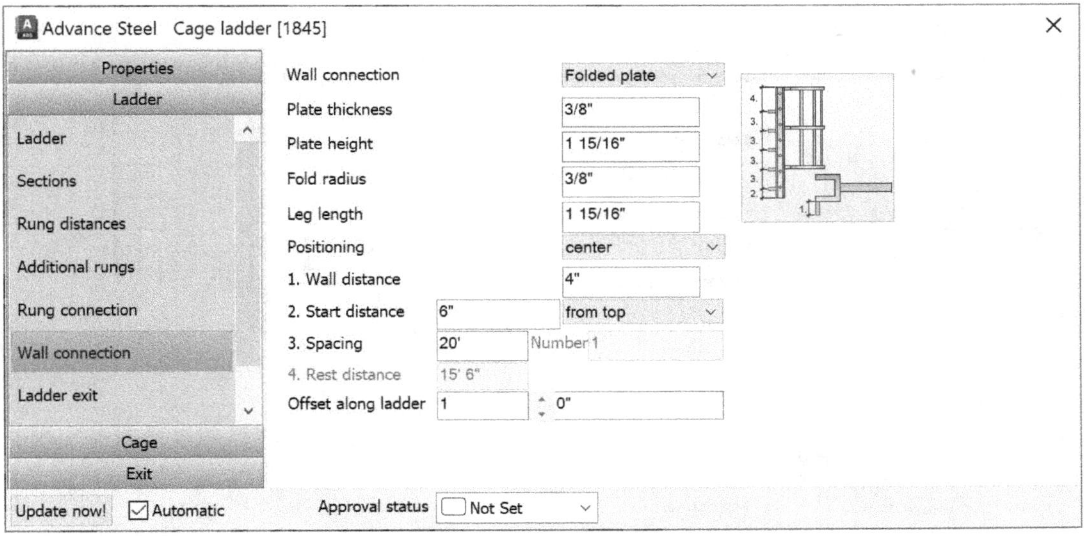

Figure 4–37

7. In the *Cage* tab>*General* sub-tab, change both *Cage* and *First brace* to **Type 2**. For *Height to first brace* and *Brace section*, match the settings shown in Figure 4–38.

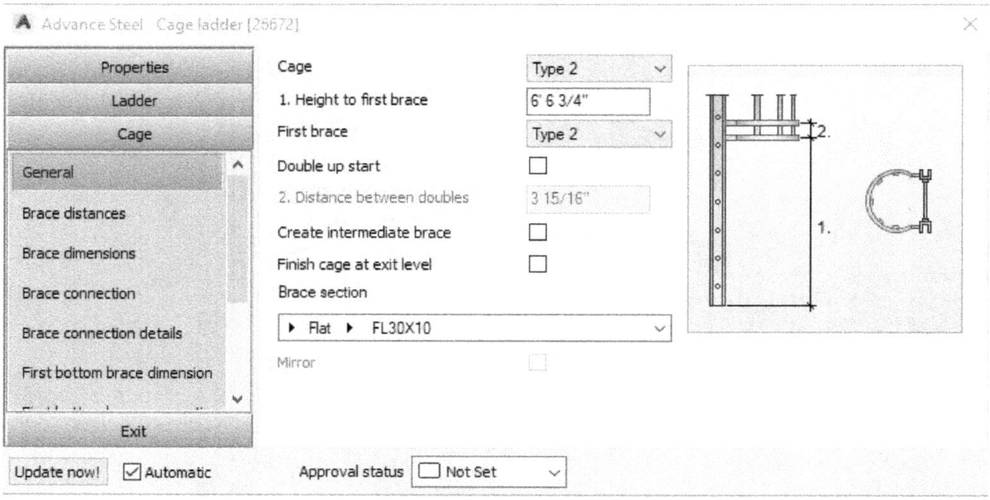

Figure 4–38

8. In the *Cage* tab>*Brace distances* sub-tab, change the *Distance between braces* and *Offset at top* values to match those shown in Figure 4–39.

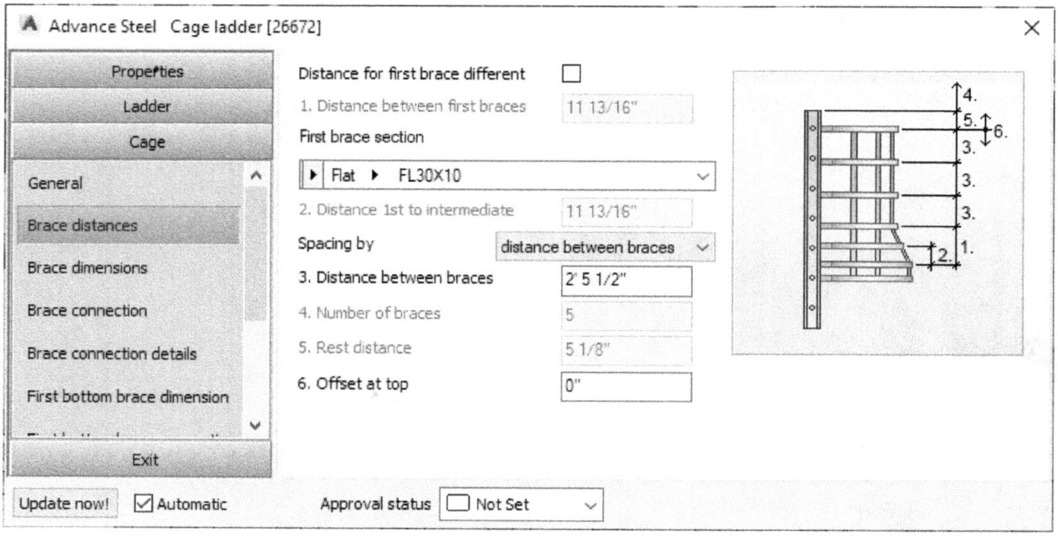

Figure 4–39

9. In the *Cage* tab>*Brace dimensions* sub-tab, match the values shown in Figure 4–40.

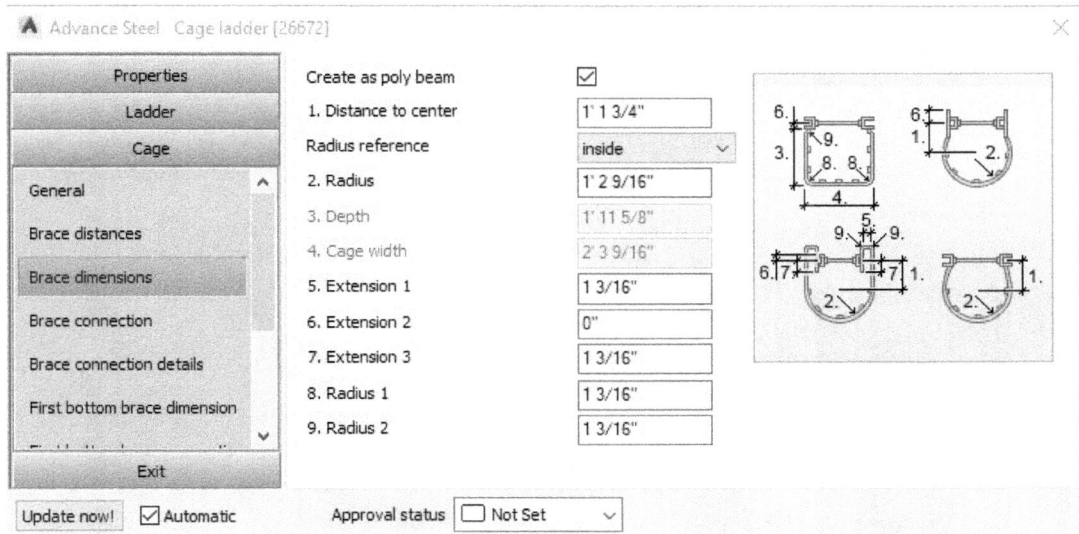

Figure 4–40

10. In the *Cage* tab>*First bottom brace dimension* sub-tab, match the values shown in Figure 4–41.

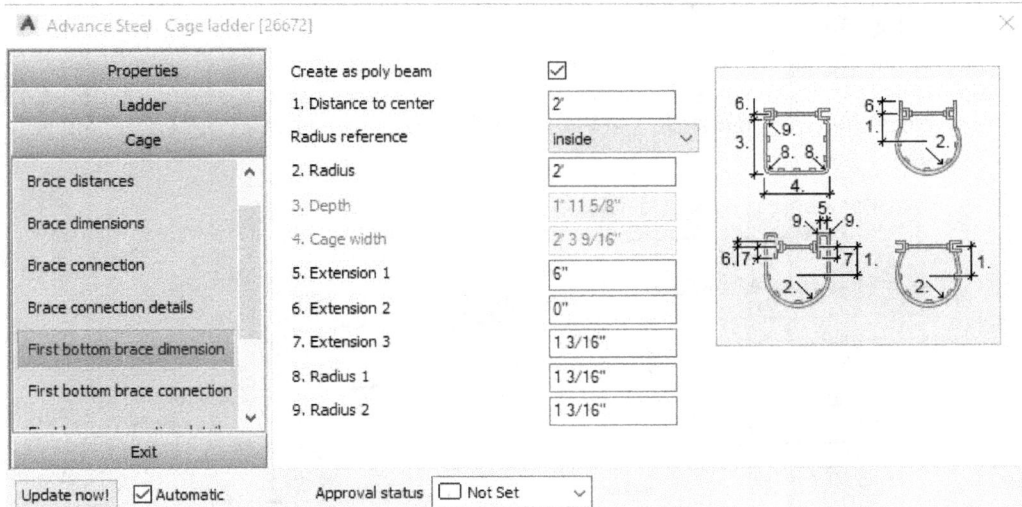

Figure 4–41

11. In the *Cage* tab>*Bands* sub-tab, match the values shown in Figure 4–42.

Figure 4–42

12. Review the options in the other tabs that can be applied to the cage ladder.

13. Set the UCS to **World**.

14. Save the drawing.

Task 3: Model stairs.

1. Rotate the model so that the platforms that include gratings on the other side of the model display. Note the two lines shown in Figure 4–43, which have been created to help you add the stairs.

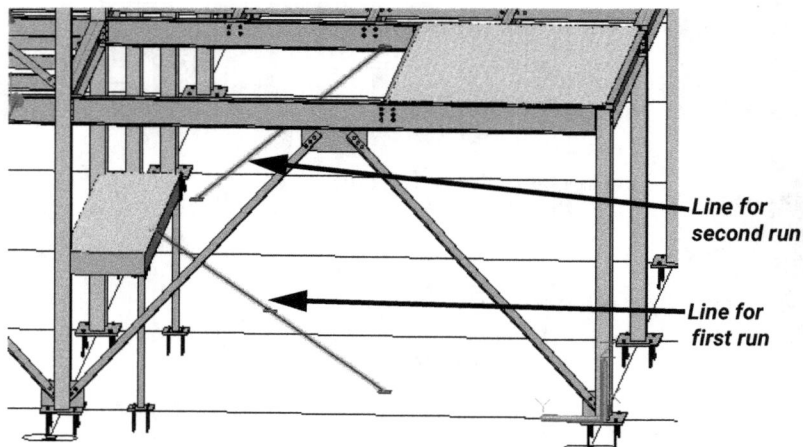

Figure 4–43

2. In the *Home* tab>*Extended Modeling* panel, click ⬚ (Straight stair).

3. Press <Enter> to accept the default **Start and end point** option.

4. Use the Endpoint object snap (override is best) and select the lower end of the line for the first run and then the upper end of the line.

5. Press <Enter> to accept the default **Middle** alignment of the stair. The stair is placed and the *Stair* dialog box displays, as shown in Figure 4–44.

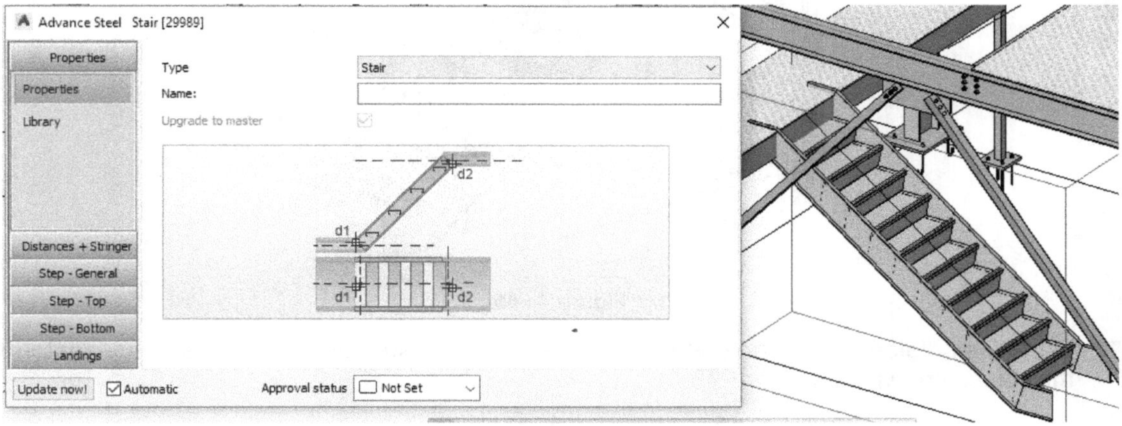

Figure 4–44

6. Switch to the **Platform-Introduction.dwg**.

7. Select one part of the stair, right-click, and select **Advance Joint Properties**. This opens the dialog box for this stair and displays its parameters.

8. Select the *Distance + Stinger* tab. Note that the *Length* is set to **7'-6"**.

9. Select the *Library* tab and ensure that the **Platform Stairs** type exists. If it does not, click **Save Values**, click **Edit**, and then rename the new type as **Platform Stair**.

10. Repeat this process for the second level stairs and save as Platform Stairs 2.

11. Switch to the **Platform-Stairs.dwg** and open the same tab. Note that the *Length* is **8'-6"**. Change it to **7'-6"**. The stair moves back, as shown in Figure 4–45.

Figure 4–45

12. In the *Stair* dialog box>*Distances + Stringer* tab>*Stringer* sub-tab, verify the settings match the values shown in Figure 4–46.

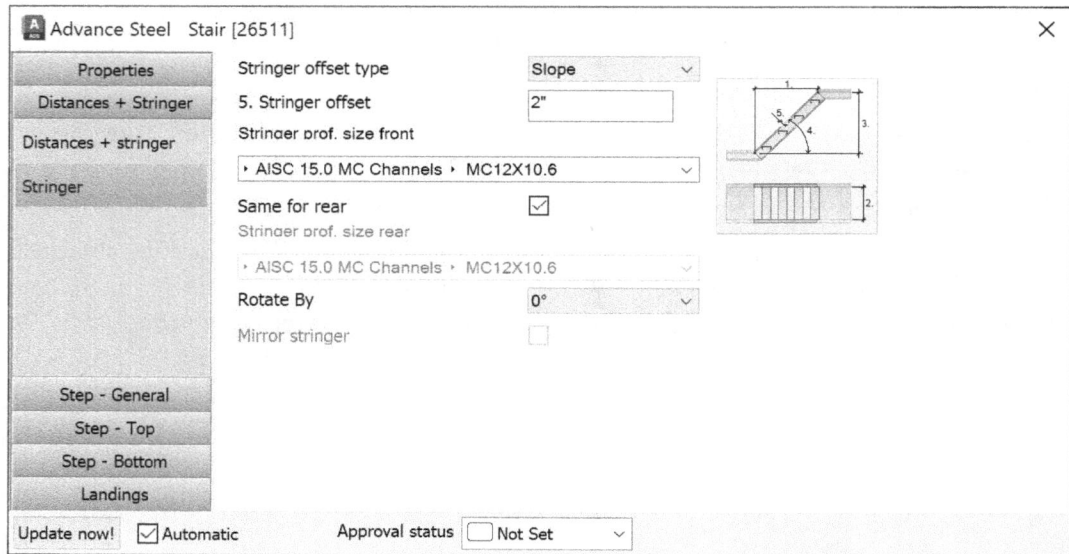

Figure 4–46

13. In the *Step - General* tab>*Tread type* sub-tab, match the settings shown in Figure 4−47.

Figure 4−47

14. In the *Step - General* tab>*Tread dimensions 1* sub-tab, match the settings shown in Figure 4−48.

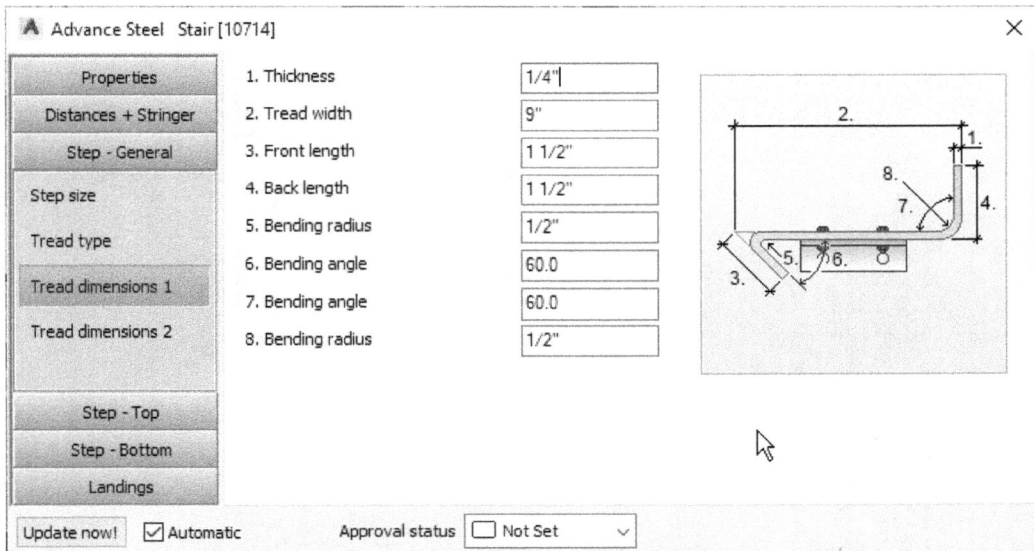

Figure 4−48

15. In the *Step - General* tab>*Tread dimensions 2* sub-tab, match the settings shown in Figure 4–49.

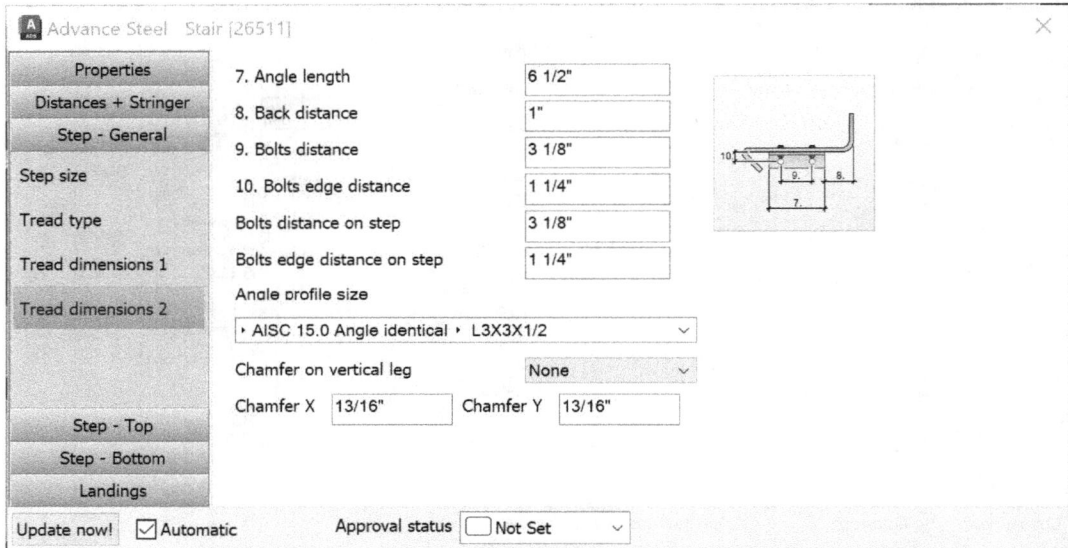

Figure 4–49

16. In the *Step - Top* tab>*Step size* sub-tab, select the **Same as other steps** checkbox, as shown in Figure 4–50.

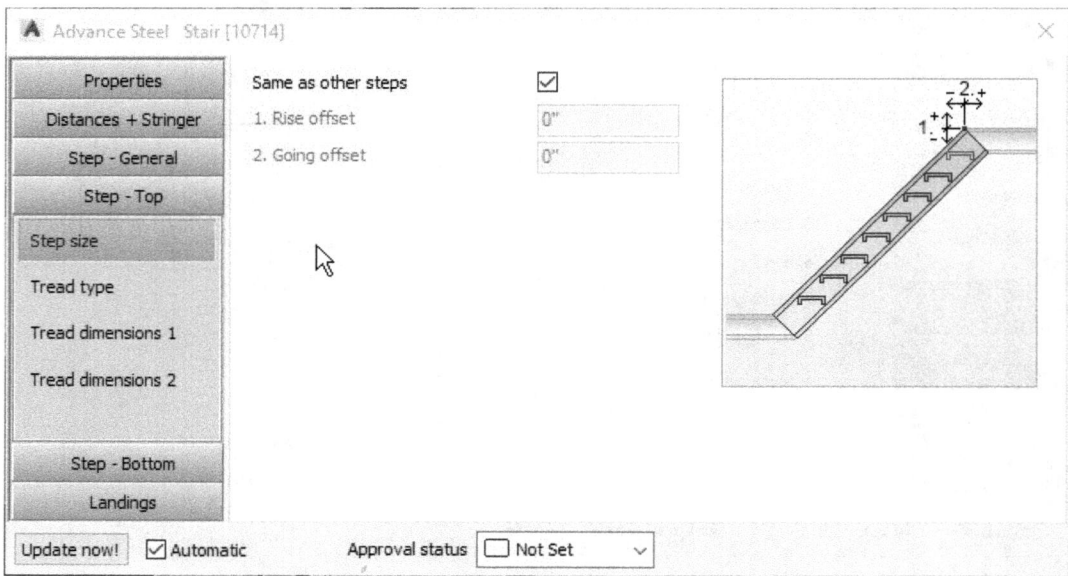

Figure 4–50

17. In the *Step - Bottom* tab>*Step size* sub-tab, select the **Same as other steps** checkbox, as shown in Figure 4–51.

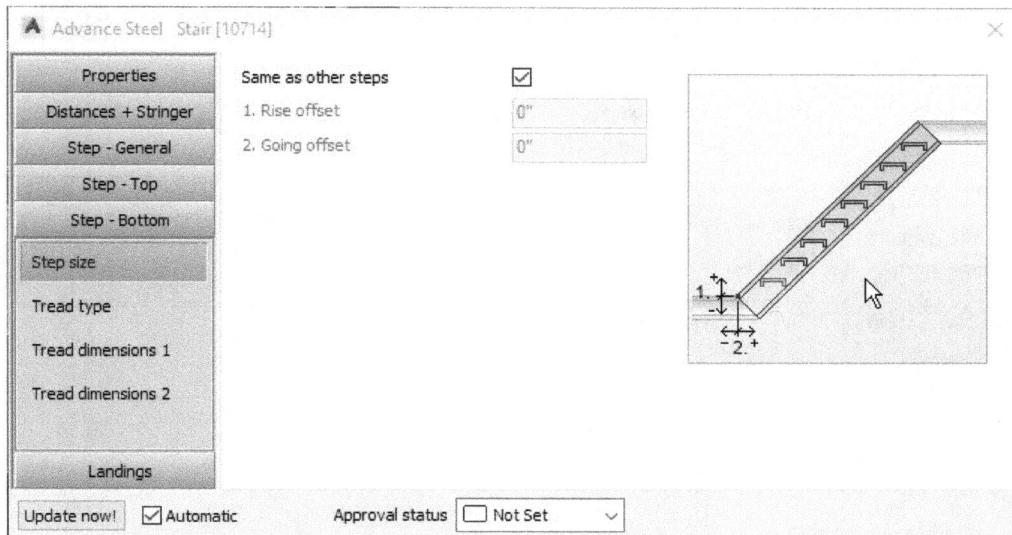

Figure 4–51

18. In the *Landings* tab>*Top landing prof.* sub-tab, clear the **Same for rear** checkbox, as shown in Figure 4–52.

Figure 4–52

19. In the *Landings* tab>*Top landing* sub-tab, change *Landing length (front)* and *Landing length (rear)* to **1' 3 3/16"**. Clear the **Create last tread** checkbox and set *Landing offset* to **0"**, as shown in Figure 4–53.

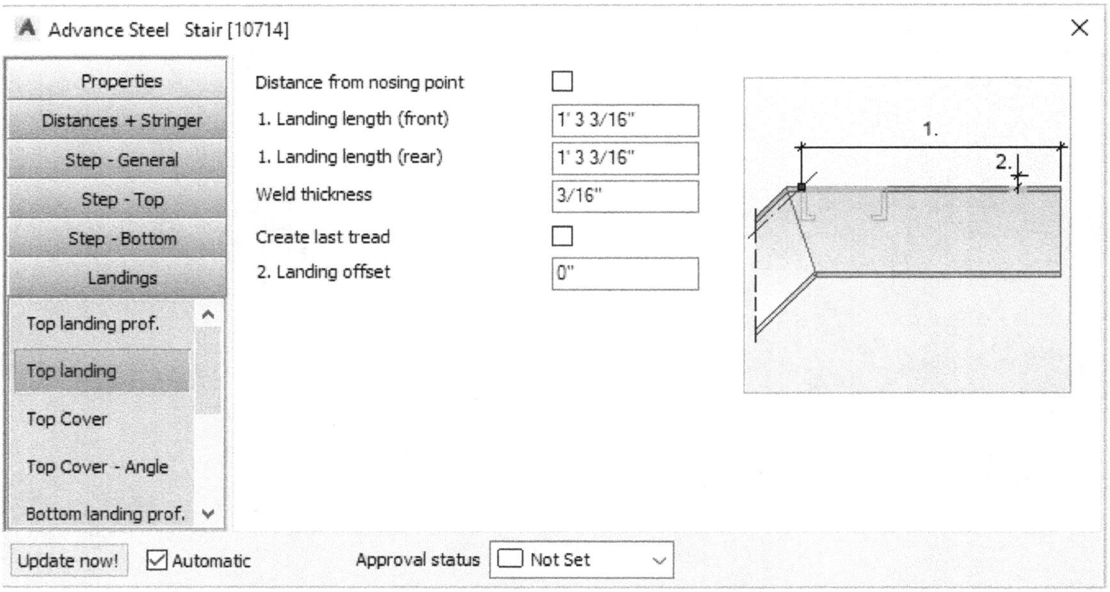

Figure 4–53

20. In the *Landing* tab>*Top Cover* sub-tab, complete the following, as shown in Figure 4–54:

- Set *Cover made from* to **Grate.**
- Set *Grating class* to **Checker Plate**.
- Set *Grating size* to **Checker Plate 3/16"**.
- Set *Offset from Stringer* to **0"**.
- Set *Cover length* to **1' 3 3/16"**.
- Select the check boxes for **Cover on top of stringer** and **Stay on top of landing**.

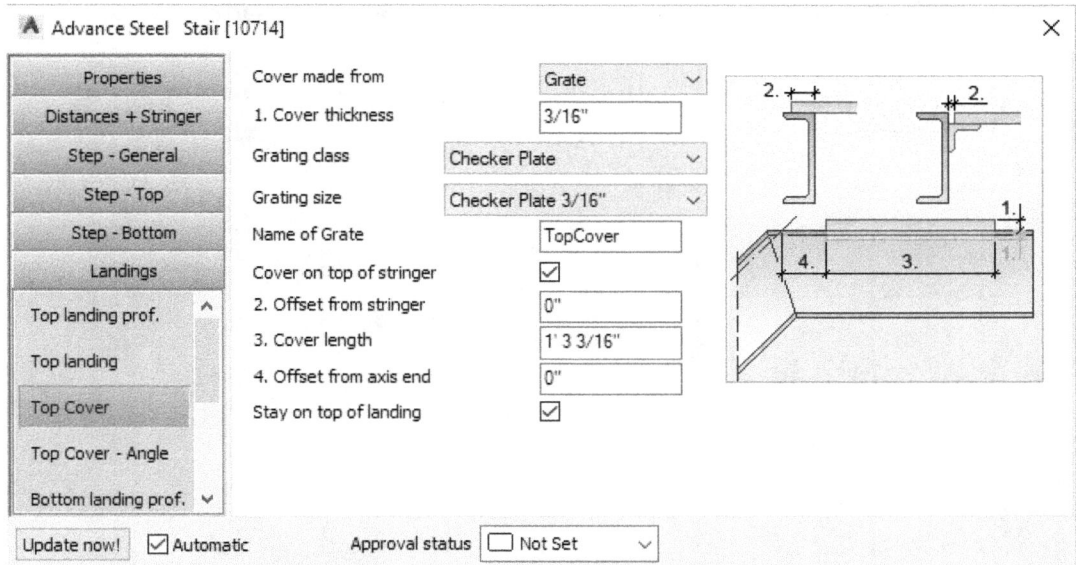

Figure 4–54

21. In the *Landings* tab>*Bottom landing prof.* sub-tab, clear the **Create front** and **Create rear** checkboxes as shown in Figure 4–55.

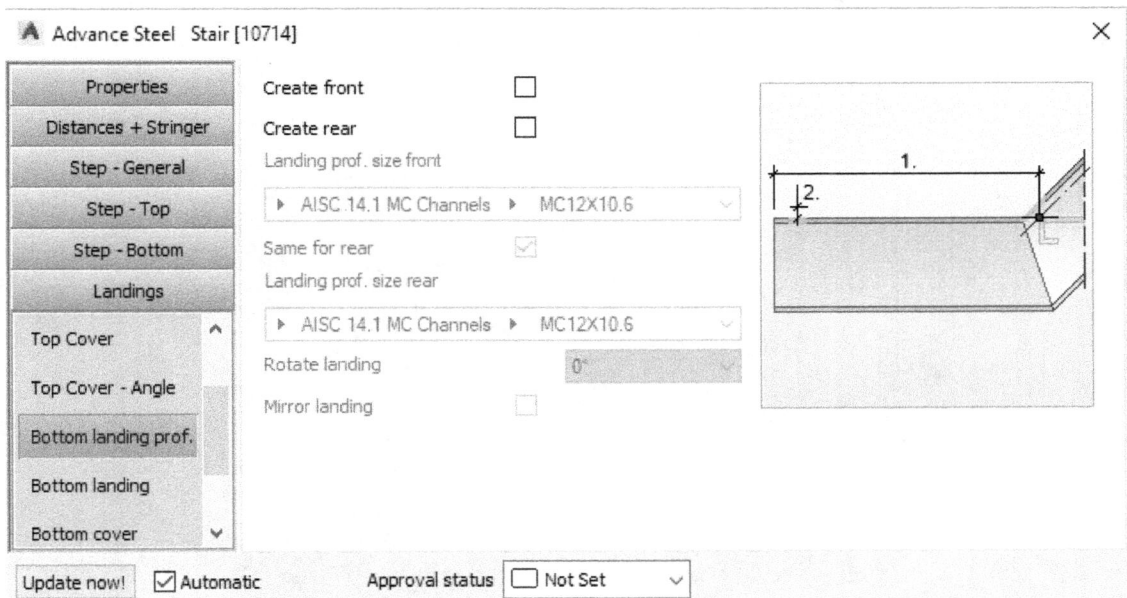

Figure 4–55

22. In the *Landings* tab>*Bottom landing* sub-tab, clear the **Create first tread** checkbox, as shown in Figure 4–56.

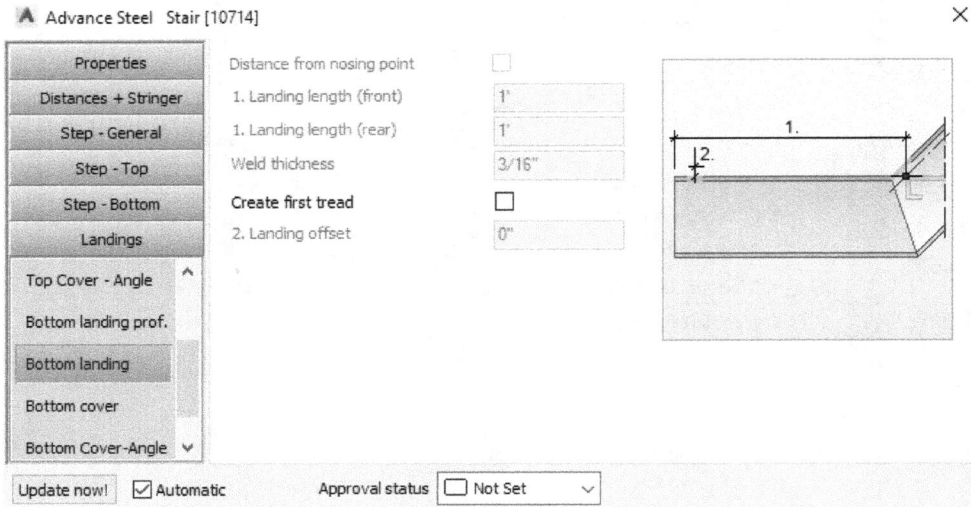

Figure 4–56

23. In the *Library* tab, apply the **Platform Stair** value to update the stairs to match the information in the **Platform-Introduction.dwg**.

24. Repeat the process for the second run of the stair. Apply the Library style **Platform Stair 2** value to the run, as shown in Figure 4–57.

Figure 4–57

25. Save the drawing.

Task 4: Add anchors and clip angles to the stair.

1. In the *Connection Vault>* ▨ (Miscellaneous) category, click ◿ (Stair Anchor Base Plate).

2. Select the end of the stringer and press <Enter>.

3. When prompted to select a reference point, click **No**.

4. In the *Stair anchor base plate* dialog box, set the following on the *General* tab>*Vertical extension* sub-tab, as shown in Figure 4–58:

   - Select *Create vertical profile.*
   - *Height Layout:* **Vertical**
   - *Layout value:* **7 1/2"**
   - *Adjust base level:* **0"**

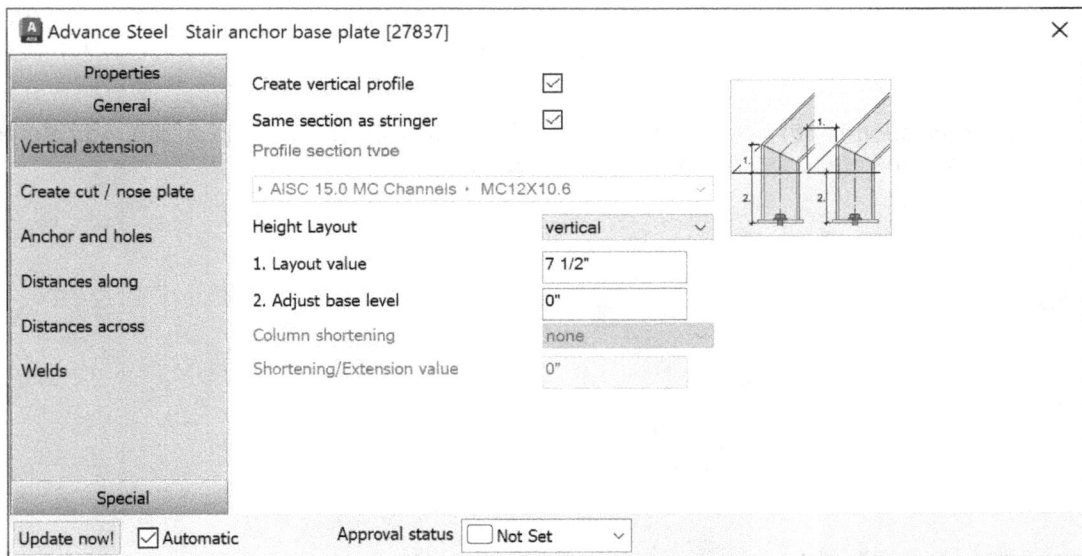

Figure 4–58

5. In the *Properties* tab>*Library* sub-tab, click **Save values**. Then, click **Edit**.

6. In the *Library* dialog box, in the *Comment* column, type **Platform Stair Base Plate** and click **OK**.

7. Repeat the process on the other side of the stairs, as shown in Figure 4–59.

Figure 4–59

8. In the *Connection Vault>* (Platform beams) category, click (Clip angle). Add plates to the top of the stairs as shown in Figure 4–60.

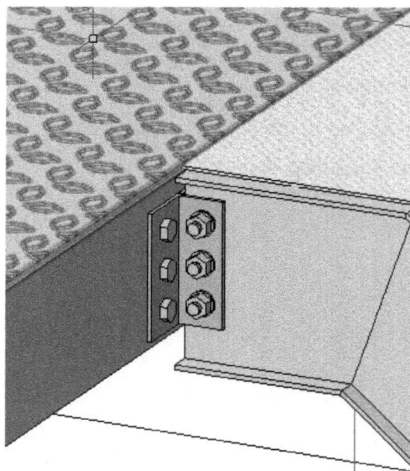

Figure 4–60

9. Save the drawing.

End of practice

Practice 4c
Model Railings

Practice Objectives

- Create a model view using a joint box.
- Model railings on stair stringers.
- Model railings on flat beams.
- Connect railings using options in the *Connection Vault*.

In this practice, you will isolate the stair area of the model by creating a model view using the joint box of the stair. You will then add railings to the stringers of the stairs, making changes to the Properties based on an existing set of stairs in another model. Finally, you will add railings around the landing of the stair and use the *Connection Vault* options to clean up the railing intersections, as shown in Figure 4–61. If you have time, you can also add railings around the first level platform.

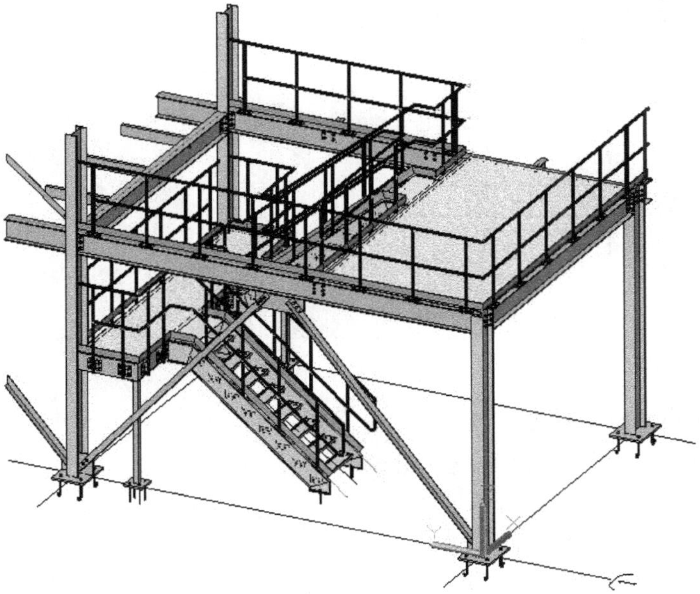

Figure 4–61

Task 1: Set up a model view of the stairs.

1. In the practice files folder, open **Platform-Railings.dwg**.

2. Open **Platform-Introduction.dwg**. Use this drawing as an example of what you need to create.

3. Select one of the stair stringers, open the *Advance Joint Properties* dialog box and then close the dialog box. This toggles on the stair joint box which you can use to create a model view.

4. In the *Home* tab>*Project* panel, click (Project Explorer).

5. In the *Project Explorer (Structures)* dialog box, click (Create new model view).

6. In the *Choose the definition method* dialog box (shown in Figure 4–62), click (At joint box).

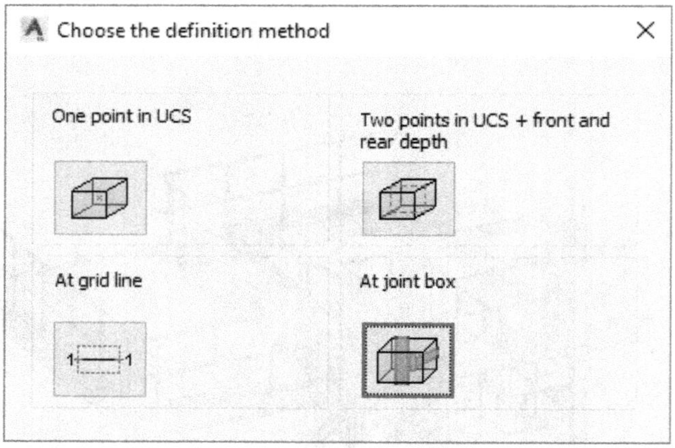

Figure 4–62

7. Select the joint box for the stair.

8. For the view name, enter **Stairs** and press <Enter>.

9. Specify the default view direction by selecting one of the arrows pointing toward the side of the stair and press <Enter>.

10. In the *Project Explorer*, click on the new **Stairs** Model view to switch to it.

11. Use grips to modify the model view box so that the full set of stairs displays, as shown in Figure 4–63.

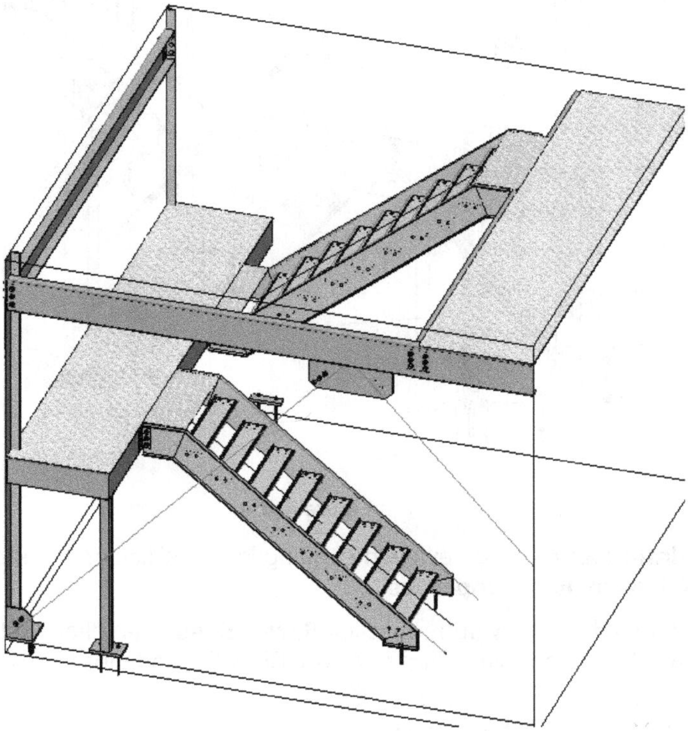

Figure 4–63

12. Save the drawing.

Task 2: Model railings on stairs.

1. In the *Home* tab>*Extended Modeling* panel, click ▥ (Hand-railing).
2. Select the left stringer of the first stair run and the beam that connects to the landing and press <Enter>.
3. Select the start point of the railing at the end of the stringer.
4. Select the end point of the railing as the end of the beam that connects to the landing.

5. At the *Do you want to select a nosing point relative to the start point* prompt, click **No**. The new railing is added using the default properties (as shown in Figure 4–64) and the *Railings* dialog box opens.

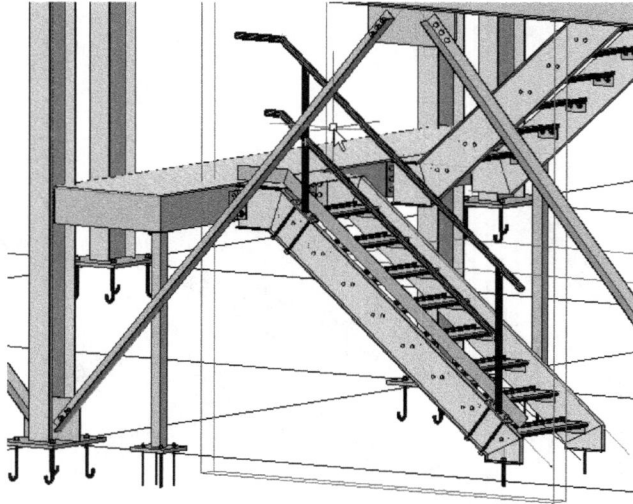

Figure 4–64

6. In **Platform-Introduction.dwg**, review the railing layout. Click on the stair rail, then right-click and select **Advance Joint Properties**.

7. In the *Properties* tab>*Library* sub-tab, click **Save values**, then click **Edit** and name it as **Platform Stair Rail** as shown in Figure 4–65. Click **OK** and close the *Stair* dialog box.

Figure 4–65

8. Click on the stair rail, then right-click and select **Advance Joint Properties**.

9. In the *Properties* tab>*Library* sub-tab, click on **Platform Stair Rail**.

10. In **Platform Railings.dwg**, apply the **Platform Stair Rail** style from the library to the railing.

11. Add a railing to the other side of the stairs, as shown in Figure 4–66. Change the *Handrail>Kickrail>Alignment* for this side of the stair.

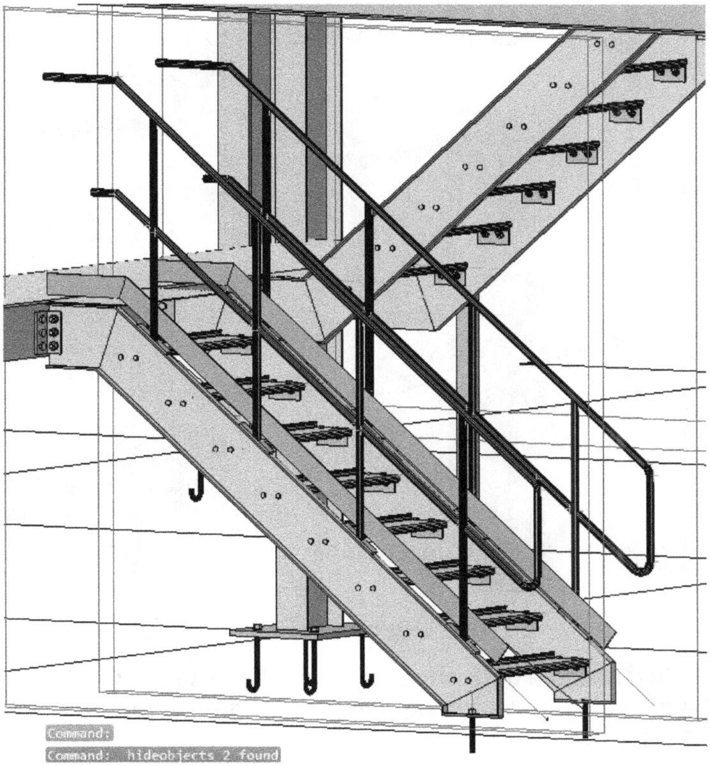

Figure 4–66

12. Add railings to the second run of stairs.

13. Modify the kick plates as required.

14. In the *Handrail>End of handrail* tab, set the *Connection type* to **Nothing**.

15. Select the joint boxes of the railings and toggle them off.

16. Save the drawing.

Task 3: Add railings around the landing.

1. Start the **Railing** command and select the front and outside beams of the landing, as shown in Figure 4–67.

2. For the start point, select the end of the stair. For the end point, select the end of the landing where it intersects the column, as shown in Figure 4–67.

3. Press <Enter> to accept the default of not setting the nosing. The new railing is added, as shown in Figure 4–67.

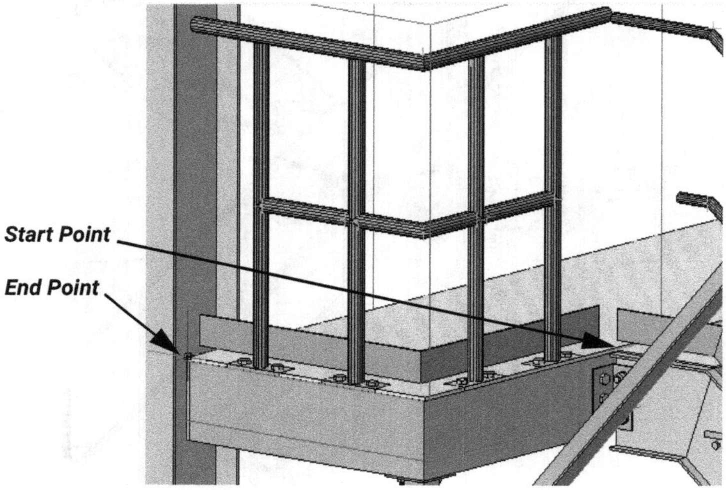

Figure 4–67

4. Note that this railing needs a different connection that attaches to the side of the beam. In the **Platform-Introduction.dwg** drawing, click on the landing rail, then right-click and select **Advance Joint Properties**.

5. In the *Properties* tab>*Library* sub-tab, click **Save values**. Click **Edit** and name it **Platform Landing Rail**.

6. Investigate the railing properties in the finished version of the model. Return to the **Platform Railings.dwg** drawing and then apply the **Platform Landing Rail** style from the Library.

7. Continue adding railings using this style around the lower landing, as shown in Figure 4–68.

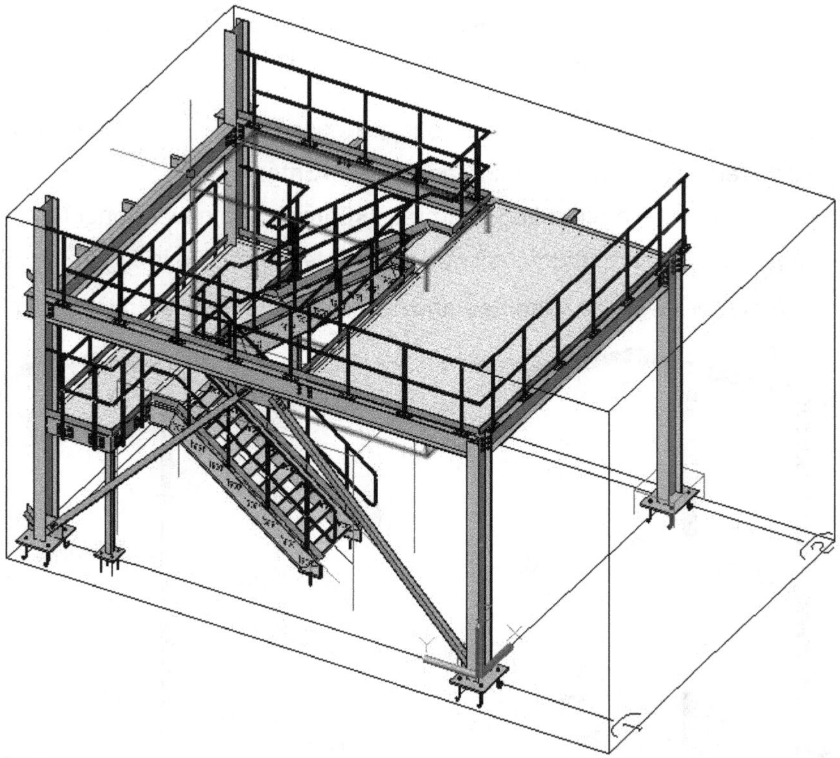

Figure 4–68

8. Save the drawing.

Task 4: Clean up the connections of the railings.

1. Zoom in on the intersection of the stair railings and landing railings. Note that the connections are not yet complete, as seen in Figure 4–69.

2. In the *Connection Vault>* 🖼 (Miscellaneous) category, click 🖋 (Railing joint handrail).

3. Select the top rail of the landing railing and press <Enter>.

4. Select the top rail of the stair railing and press <Enter>. The new connection is made.

5. Repeat the process for the middle rail and kick plate, as shown in Figure 4–70.

    *   *Note:* The model view was modified and some objects temporarily hidden for this view.

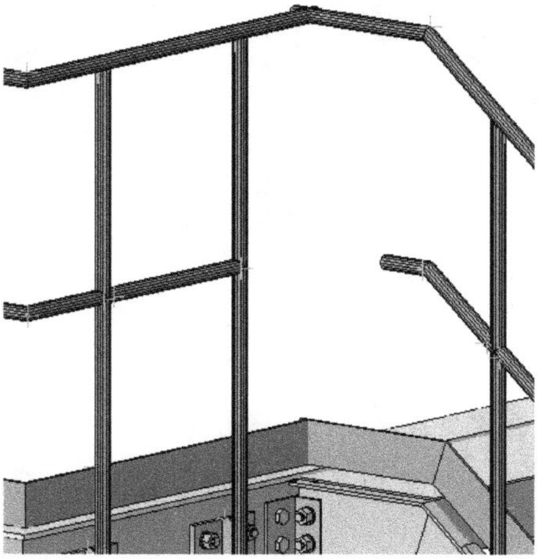

Figure 4–69

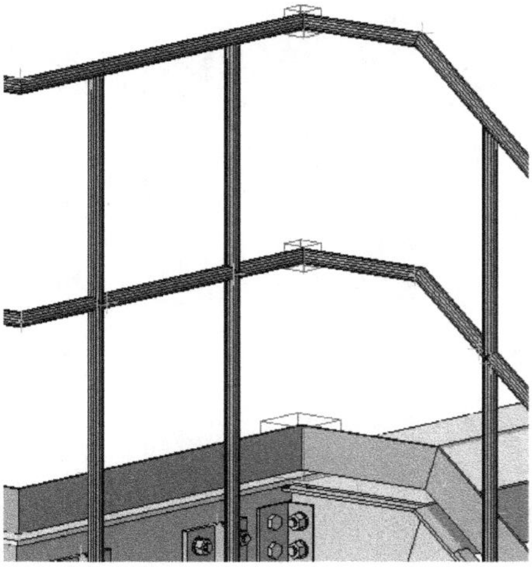

Figure 4–70

6. Repeat the process for the other intersections of landing railings and stair railings.

7. If required, you can add welds at the end of the landing railings where they intersect with the columns.

8. Toggle off the **Stairs** model view.

9. If you have time, add railings along the platform on Level 1.

10. Save the drawing.

End of practice

4.3 Creating Concrete Objects

Steel fabrication structures are often supported by concrete footings. Therefore, to have a complete model, you can create concrete walls, slabs, beams, and columns, as well as isolated and continuous footings, as shown in Figure 4−71.

Figure 4−71

- The UCS impacts the location and orientation of most concrete objects.

- The concrete tools are found in the *Objects* tab>*Other objects* panel, and are created as outlined below with modifications in the related *Advance Properties* dialog box.

	Wall	Pick points where you want the wall to be added (similar to the Line or Polyline commands). Set the *Height, Thickness,* and other information in the *Wall* dialog box. Each wall is considered a separate object.
	Polygonal wall	Set the UCS to the origin and direction of the wall that you draw on the XY plane. Pick points that describe the vertices of the wall. It can help to draw a polyline first and then use object snaps to trace the polyline.
	Slab	Pick two diagonal points that define the extents of the slab.
	Polygonal slab	Pick multiple vertices that define the extents of the slab.
	Concrete beam	Pick two points for the length of the beam.
	Concrete curved beam	Pick a start point, endpoint, and circle point (radius) for the beam.
	Concrete column	Pick the start point of system axis. You can use Workplanes to define the height, just as you would with steel columns.

⬭	**Isolated footing**	Pick a point that defines the position of the footing.
🪨	**Continuous footing**	Pick a start point and end point for the footing.

Concrete Objects Advance Properties

After you pick the points that define the concrete objects, the *Advance Properties* dialog box displays, enabling you to access the various options for the selected objects. For example, when you place an Isolated Footing, you can set or modify the *Shape*, *Length*, *Width*, *Height*, and *Material*, as shown in Figure 4–72.

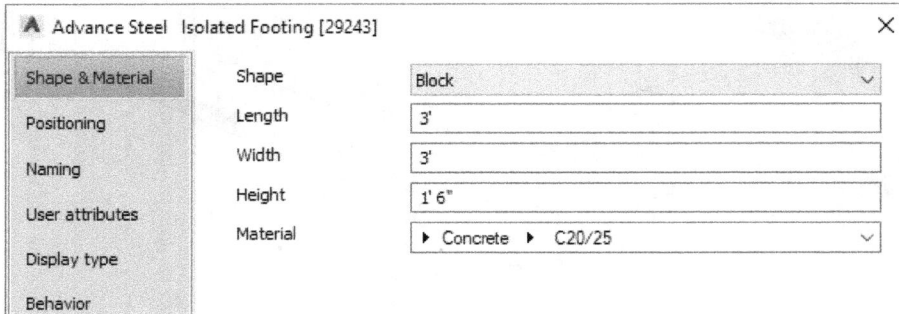

Figure 4–72

- Isolated footings are a specific object type, while continuous footings are beams that are set up using a section and material, as shown in Figure 4–73. Continuous footings are assigned the model role of Concrete Foundation.

Figure 4–73

Practice 4d
Create Concrete Objects

Practice Objectives

- Add isolated footings.
- Add continuous footings.

In this practice, you will add isolated footings at the base of each column. You will then add continuous footings between each column, as shown in Figure 4–74.

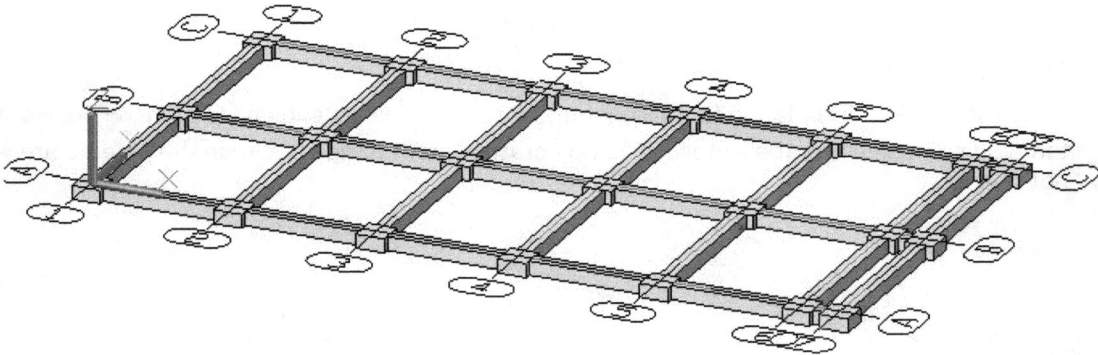

Figure 4–74

Task 1: Add isolated footings at the base of each column.

1. In the practice files folder, open **Platform-Footings.dwg**.
2. Select both sets of grid lines.

3. In the Status Bar, expand ⬚ (Isolate Objects) and select **Isolate Objects**. Note that only the grid lines display.
4. Set the Object Snaps so that only the **GRID Intersection Points** object snap is toggled on.

5. In the *Objects* tab>*Other objects* panel, click 🗀 (Isolated Footing).
6. Pick the grid intersection point A1.

7. In the *Isolate Footing* dialog box, set the *Shape* and sizes as shown in Figure 4–75.

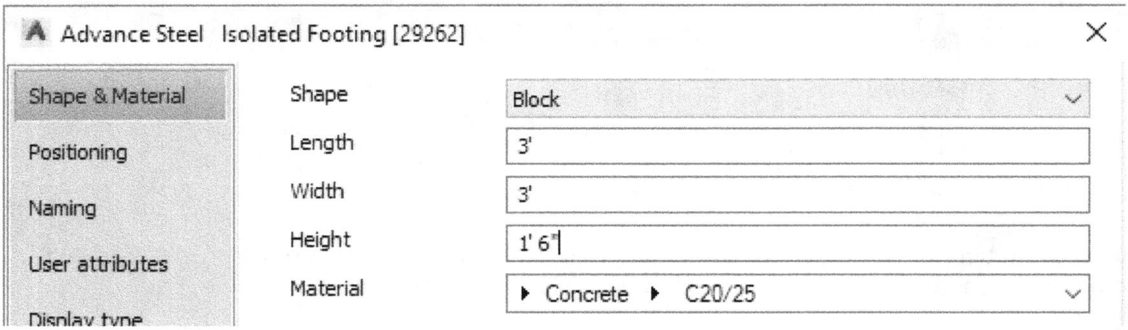

A Advance Steel Isolated Footing [29262]		✕
Shape & Material	Shape	Block ∨
Positioning	Length	3'
Naming	Width	3'
	Height	1' 6"
User attributes	Material	▶ Concrete ▶ C20/25 ∨
Display type		

Figure 4–75

8. Close the dialog box.

9. Copy the isolated foundation to the other grid intersections, as shown in Figure 4–76. Note that you can use either the AutoCAD® **Copy** or **Array** commands because the objects are not connected.

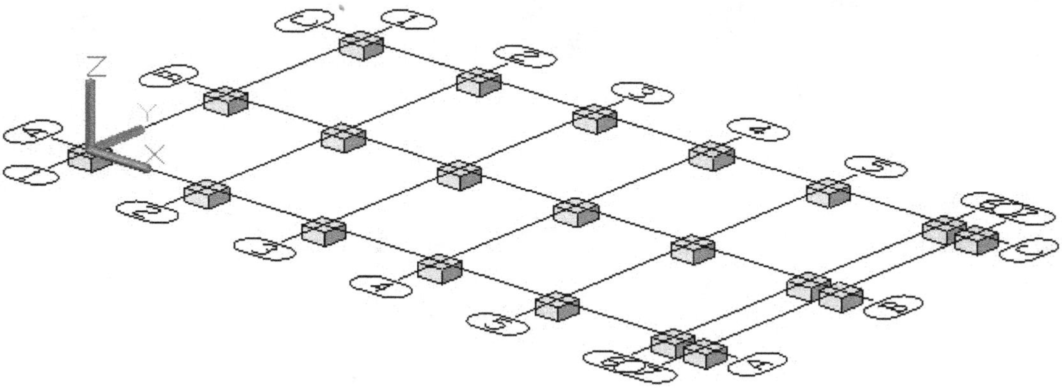

Figure 4–76

10. Save the drawing.

Task 2: Add continuous footings.

1. In the *Objects* tab>*Other objects* panel, click ✎ (Continuous footing).

2. Draw the footings from the midpoint of one isolated footing to the midpoint of the next.

3. In the *Beam* dialog box, set the *Section type* to **Rectangular>R1'6"x1'6"**.

4. In the *Positioning* tab, change the *Offset* as show in Figure 4–77.

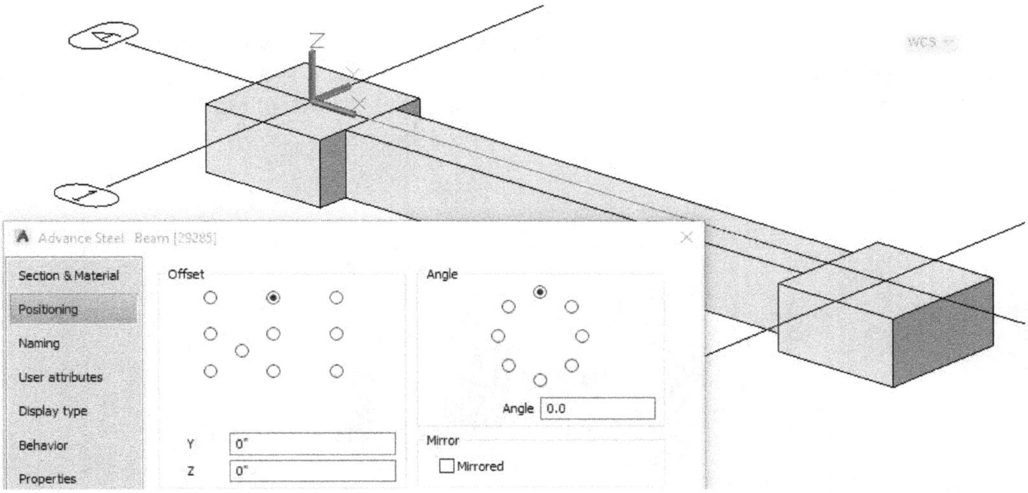

Figure 4–77

5. Add footings around the model as shown in Figure 4–78.

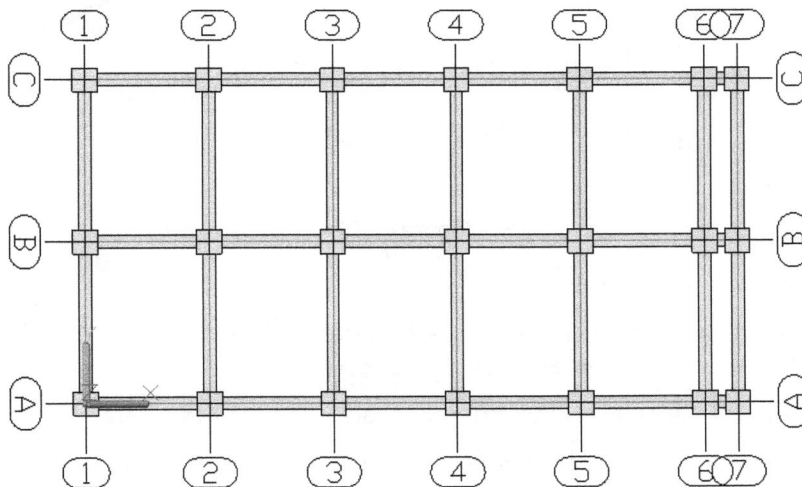

Figure 4–78

6. Unisolate the other objects.

7. Save the drawing.

End of practice

4.4 Special Parts

Sometimes it is necessary to create objects of existing buildings or equipment to avoid clashes and to help with communication. Alternatively, you may want to add special proprietary parts in your structural design and include them as special parts, as shown in Figure 4–79.

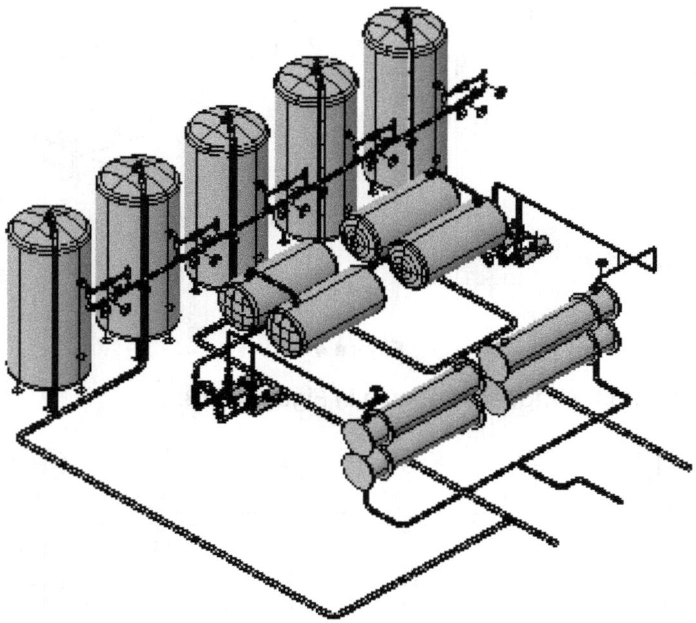

Figure 4–79

- Special parts need to be created in their own DWG file.

- Depending on the special parts' properties, they can be included in part numbering, bills of materials, weight calculations, and collision checks.

- You can create your special part at different scales, but it is best to use 1:1.

- A special part's insertion point is the only snapping point. You will need to open the special part file and add AutoCAD points, then re-insert it back into the model.

- Depending on the drawing style, special parts only display in assembly or general arrangement drawings.

- Automatic dimensions are created depending on the drawing style and based on the insertion point. You can manually add dimensions to other points as needed.

- You cannot create detail drawings of special parts.

How To: Create a Special Part

1. Start with an AutoCAD DWG file.

   - The WCS origin should be placed in a useful location because it becomes the insertion point of the part.
   - Draw your part using AutoCAD solids or convert the drawing to a solid.
   - Save your file.

2. In your Autodesk Advance Steel Tool Palette>  (Tools) category, click (Advance Steel Special Part).

3. Select a point at which to insert the part (this matches the WCS origin in the part file).

   - If the message *"The field 'Block name' cannot contain a zero-value"* displays, as shown in Figure 4–80, click **OK** to open the *Special part* dialog box.

AutoCAD   ✕

⚠ The field 'Block Name' cannot contain a zero-value. Please enter a name in the field.

OK

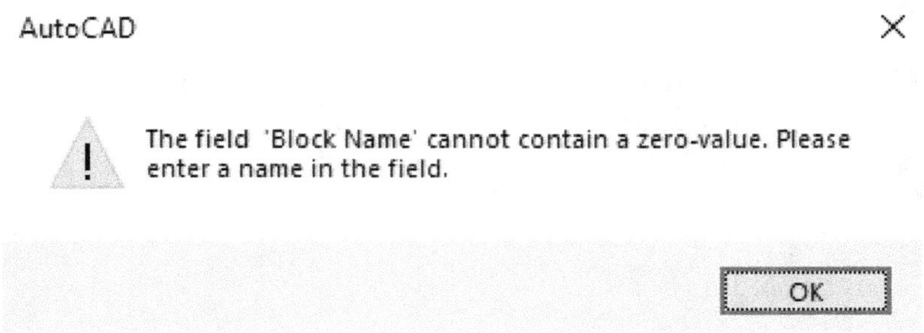

Figure 4–80

4. In the *Special part* dialog box, you can set the special part's properties using the following tabs.

Definition & Material Tab

In the *Definition & Material* tab, you can load or specify the special part's name, then add a weight, scale and material, as shown in Figure 4–81.

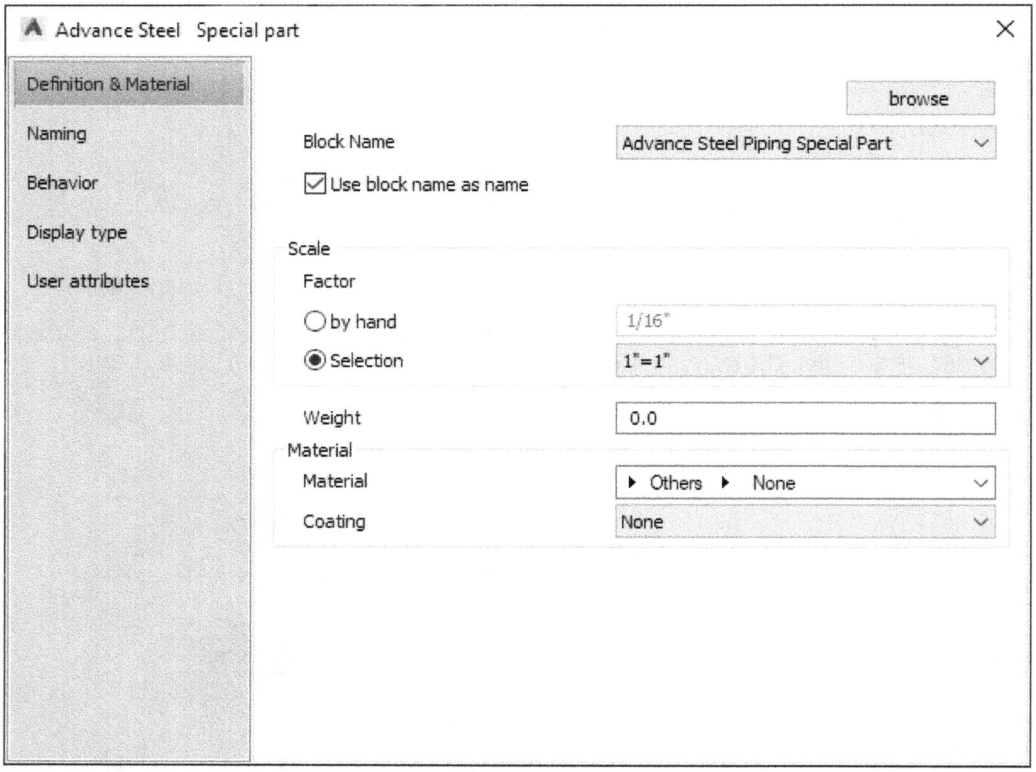

Figure 4–81

Browse	To load your special part, use the **Browse** button.
Block Name	If the special part has already been added, you can select it from the *Block Name* drop-down list.
Use block name as name	Select the **Use block name as name** check box if you want the part to keep its exact name.
Scale	*Factor:* Manually enter in the value. *Weight:* Specify what scale ratio to use.
Weight	By default, the weight of a special part is set to 0. If you want your special part to be included in calculations, enter a weight.
Material	*Material:* Select the type and subtype material. *Coating:* Specify the part's coating.

Naming Tab

In the *Naming* tab, you can specify how your information will be numbered and displayed in the BOM, as shown in Figure 4–82.

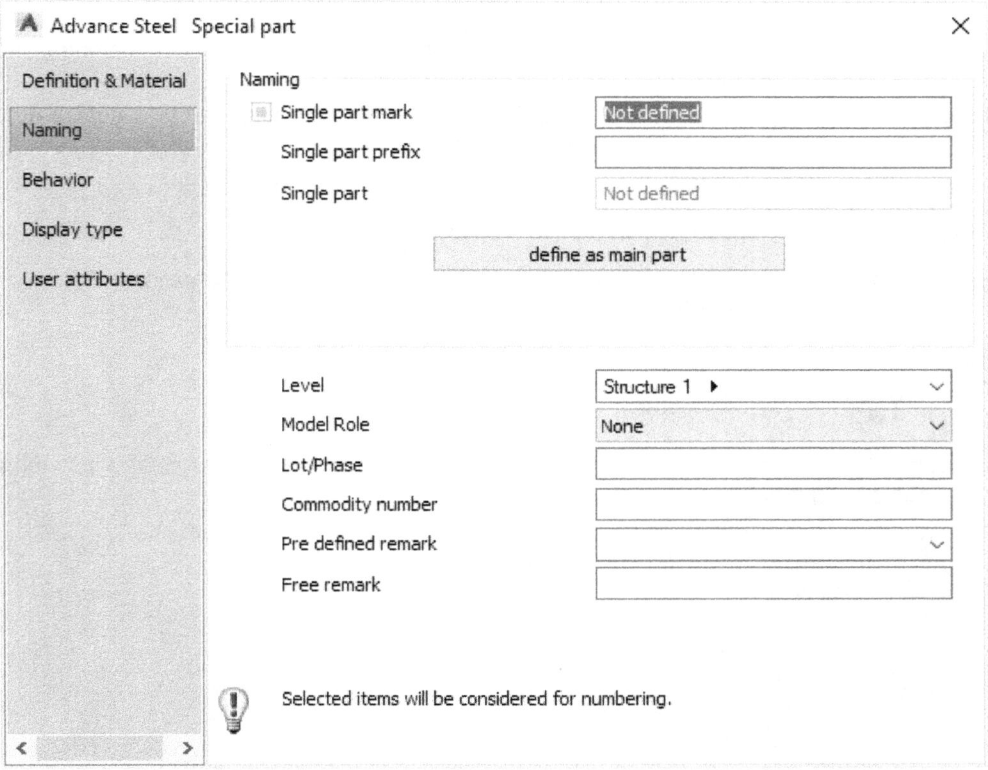

Figure 4–82

Single part mark	Specifies the elements assigned to Single part mark. You cannot modify the check box for Single part and Main part, which is indicating whether the element is a main or attached part.
Single part prefix	Specifies the prefix of the single part mark.
Single part	The single part's name.
Define as main part	Before numbering elements, this allows you to specify the selected element as a main part of an assembly. Clicking **define as main part** will show three new fields: • *Assembly mark*: The element's defined prefix assembly mark. • *Assembly prefix*: The assembly mark prefix. • *Main part*: The main part's name.

Model Role	Assigns the element's role.
	• The role is automatically assigned if the object was created by connection or structural element macro.
	• You must manually specify the model role if the object was manually created.
	• The model role is for assigning a presentation and dimensioning and labeling style.
	The model role is used by the prefix tool to assign prefixes, but it is not used directly for numbering.
Lot/Phase	Specifies if the element is assigned to the lot or the phase.
Commodity number	Specifies the element's commodity number.
Predefined remark	Element remarks saved in the database that can be used again from the *Remark* drop-down list.
Free remark	Additional element notes.

Behavior Tab

In the *Behavior* tab, you can specify the behavior for single parts and assemblies, as well as the their behavior during a collision check, as shown in Figure 4–83.

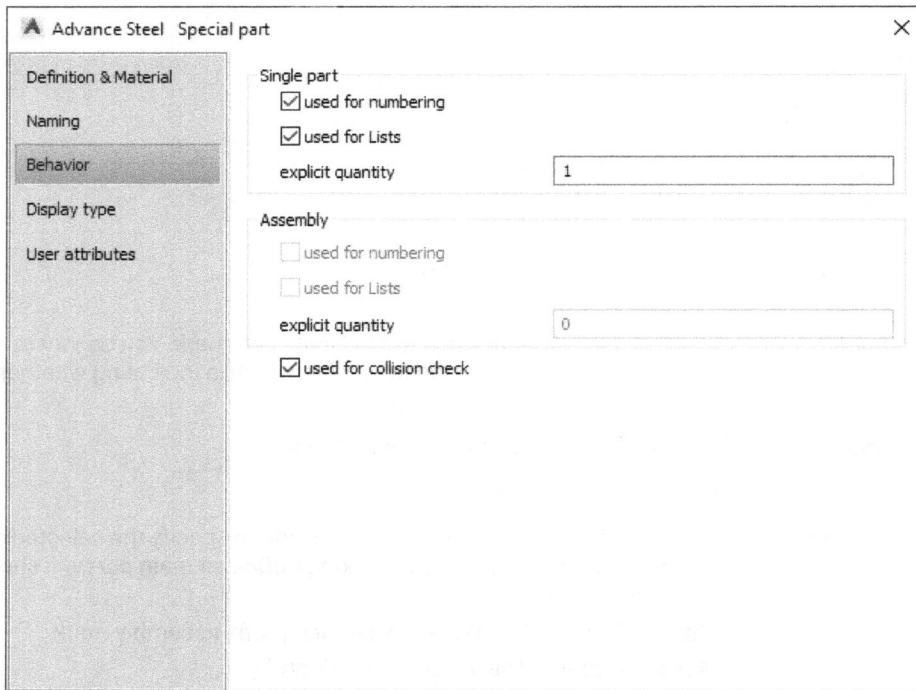

Figure 4–83

Single Part

Used for numbering	Numbering will use the single part if selected.
Used for lists	The BOM will use the single part if selected.
Explicit quantity	Specifies the single part's amount in the BOM.

Assembly

Used for numbering	Numbering will use the assembly if selected.
Used for lists	The BOM will use the assembly if selected.
Explicit quantity	Specifies the single part's amount in the BOM.
Used for collision check	Determines if the element will be used in a collision check if selected.

Display Type Tab

In the *Display type* tab, specify how special parts display in the mode: **Off** (the part will not display), **Standard** (the part is displayed), or **Box** (the part is displayed, as well as a bounding box), as shown in Figure 4–84.

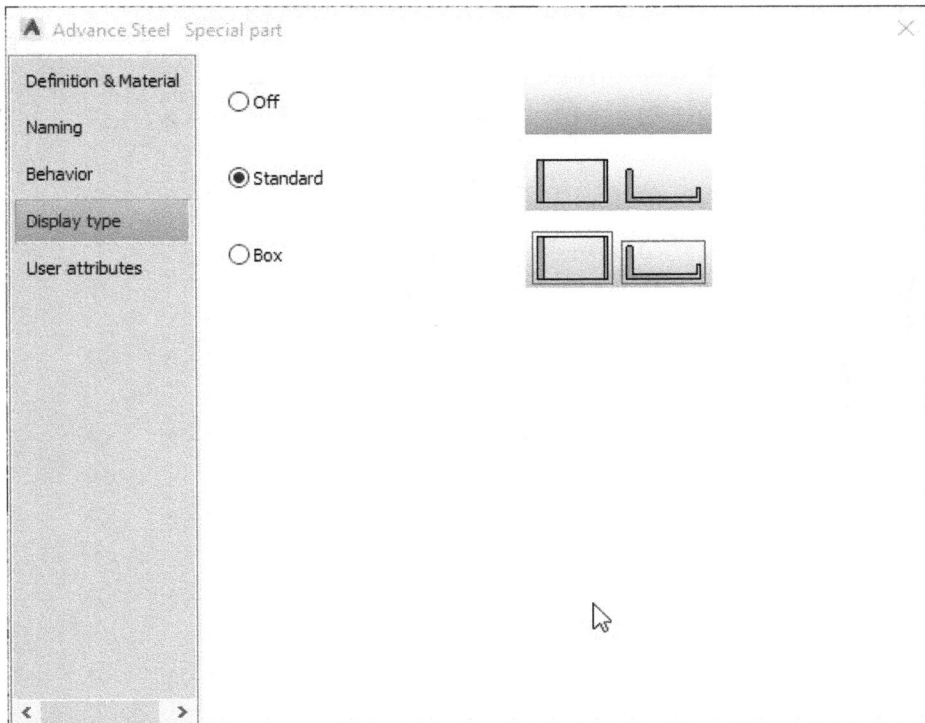

Figure 4–84

User Attributes Tab

In the *User attributes* tab, additional properties can be added if necessary to differentiate between model elements. You can specify up to ten user attributes, as shown in Figure 4–85, which can be used in the numbering process.

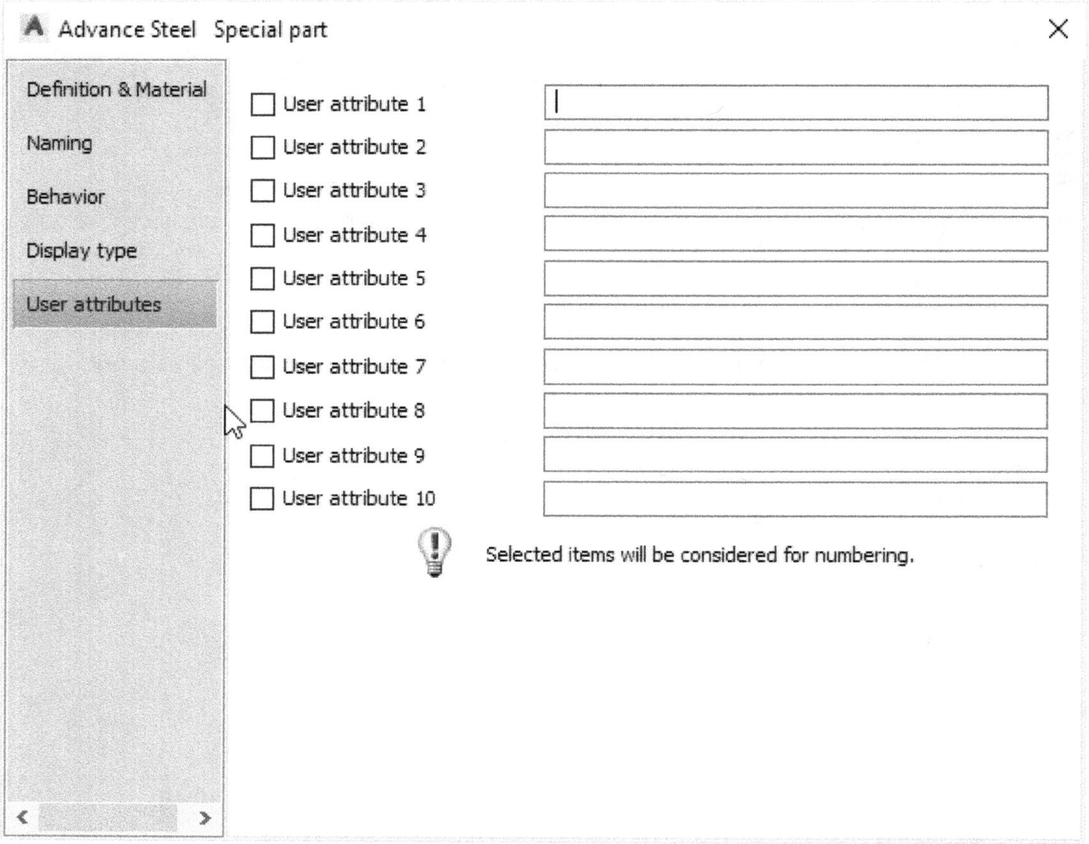

Figure 4–85

Practice 4e
Insert Special Parts

Practice Objectives

- Insert a special part.
- Modify the special part's properties.

In this practice, you will insert a special part using the Advance Steel Special Part tool and modify its properties, as shown in Figure 4–86.

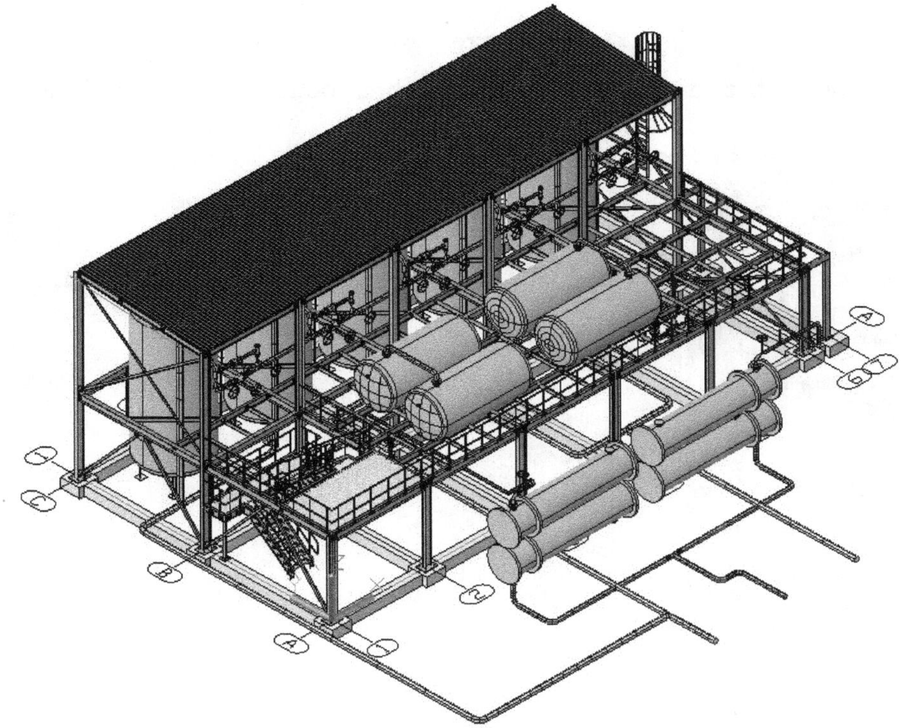

Figure 4–86

1. In the practice files folder, open **Platform-Special Parts.dwg**.

2. In the *Advance Tool Palette>* 🔧 (Tools) category, click 📋 (Advance Steel special part).

3. For the *Central Point*, type **0,0** and click **enter**.

4. Click **OK** in the *"The field 'Block name' cannot contain a zero-value"* dialog box to open the *Special part* dialog box.

5. In the *Definition & Material* tab, click **browse**.

6. Navigate to the practice files folder and choose **Advance Steel Piping Special Part.dwg**. Click **open**.

7. Select the **Use block name as name** checkbox as shown in Figure 4–87.

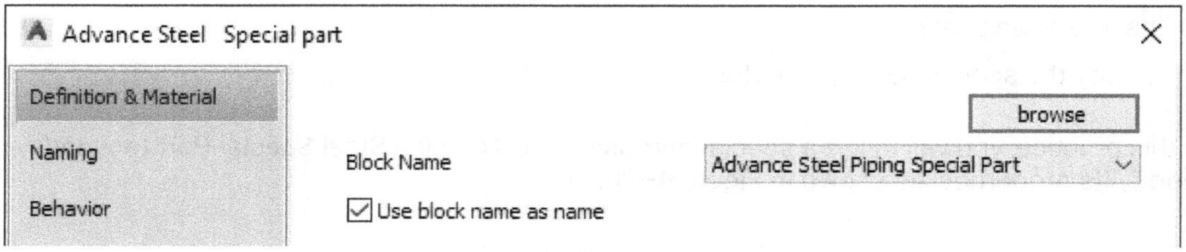

Figure 4–87

8. In the *Behavior* tab, clear the **used for collision check** option, as shown in Figure 4–88.

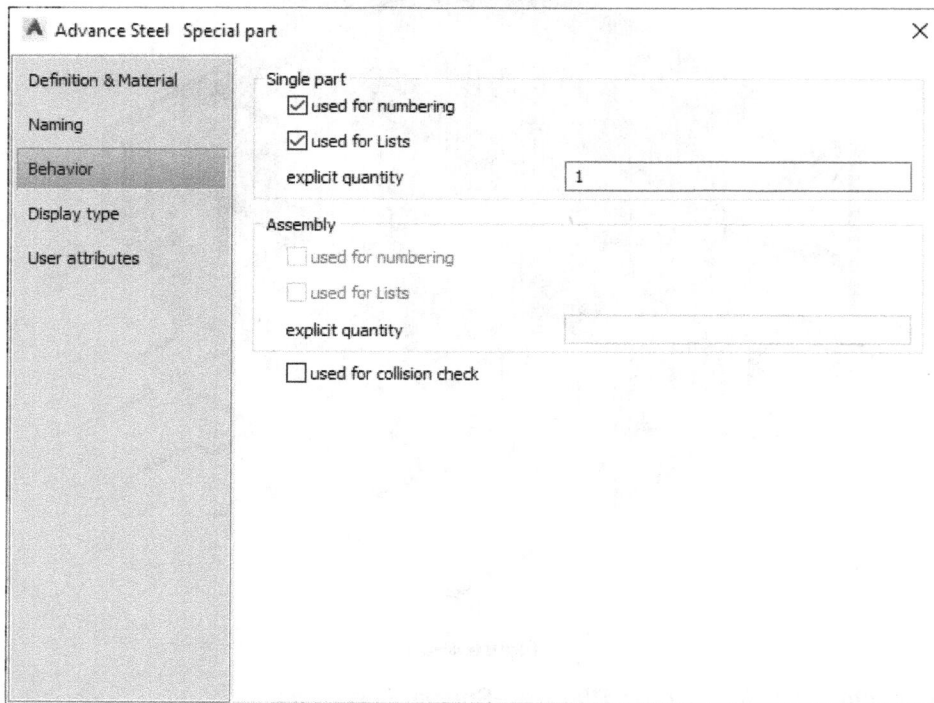

Figure 4–88

9. Close the *Special part* dialog box.

End of practice

Chapter Review Questions

1. Grating and Cladding are made of different types of objects, even though they are both designed to cover flat areas. Which of the following is true of gratings? (Select all that apply.)

   a. Used for horizontal surfaces.

   b. Used for vertical or slanted surfaces.

   c. Made from beam objects.

   d. Made from plate objects.

2. Ladders are placed along which UCS axis?

   a. X-axis

   b. Y-axis

   c. Z-axis

3. When creating stairs, there are several options that you can use to define the run of the stairs. Which of the following is used most often as a guide when you first draw lines?

   a. Start and end point

   b. Length and angle

   c. Height and angle

4. Railings can be placed anywhere in the model.

   a. True, they do not need to be attached to any other object.

   b. False, they must be placed on beams.

5. Railings are automatically attached when you model a railing continuously along a beam. However, they are not attached when two railings are made separately, as shown in Figure 4–89. Which of the following commands connects railings?

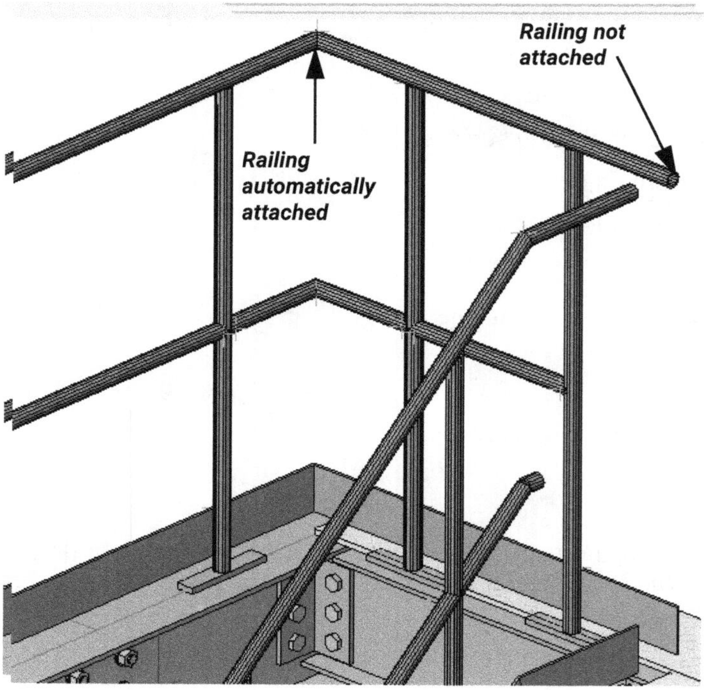

Figure 4–89

a. *Extended Modeling* tab>*Structural Elements* panel>**Railing Connect**

b. *Advance Tool Palette>* (Modify) category>**Fillet**

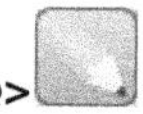

c. *Connection Vault>* (Miscellaneous) category>**Railing joint handrail**

d. *Objects* tab>*Connection objects* panel>**Link**

6. Continuous footings are made of which of the following object types?

a. Concrete

b. Beams

c. Footings

d. Steel

7. Special parts can be displayed in general arrangement drawings.

    a. True

    b. False

8. Within the *Special part* dialog box, which tab do you need to be in to select **used for collision check**?

    a. Definition & Material

    b. User attributes

    c. Behavior

    d. Positioning

Command Summary

Button	Command	Location
Grating and Cladding		
▨	Bar Grating	• **Ribbon:** *Home* tab>*Objects* panel • **Ribbon:** *Objects* tab>*Grating* panel
⌂	Create claddings	• **Ribbon:** *Home* tab>*Extended Modeling* panel • **Ribbon:** *Extend Modeling* tab>*Structural Elements* panel
△	Define cladding area	• **Ribbon:** *Home* tab>*Extended Modeling* panel • **Ribbon:** *Extend Modeling* tab>*Structural Elements* panel
⌂	Define cladding opening	• **Ribbon:** *Home* tab>*Extended Modeling* panel • **Ribbon:** *Extend Modeling* tab>*Structural Elements* panel
⌕	Grating at Polyline	• **Ribbon:** *Home* tab>expanded *Objects* panel
▨	Standard Grating	• **Ribbon:** *Home* tab>*Objects* panel • **Ribbon:** *Objects* tab>*Grating* panel
▨	Variable Grating, polygonal	• **Ribbon:** *Home* tab>expanded *Objects* panel
▨	Variable Grating, rectangular	• **Ribbon:** *Home* tab>*Objects* panel • **Ribbon:** *Objects* tab>*Grating* panel
Ladders, Stairs, and Railings		
○	Cage Ladder	• **Ribbon:** *Home* tab>*Extended Modeling* panel • **Ribbon:** *Extend Modeling* tab>*Structural Elements* panel
▥	Hand-railing	• **Ribbon:** *Home* tab>*Extended Modeling* panel • **Ribbon:** *Extend Modeling* tab>*Structural Elements* panel
▱	Straight Stair	• **Ribbon:** *Home* tab>*Extended Modeling* panel • **Ribbon:** *Extend Modeling* tab>*Structural Elements* panel
Concrete Objects		
▱	Concrete beam	• **Ribbon:** *Objects* tab>*Other objects* panel
▯	Concrete column	• **Ribbon:** *Objects* tab>*Other objects* panel
▱	Concrete curved beam	• **Ribbon:** *Objects* tab>*Other objects* panel
▱	Continuous footing	• **Ribbon:** *Objects* tab>*Other objects* panel

Button	Command	Location
	Isolated footing	• **Ribbon:** *Objects* tab>*Other objects* panel
	Polygonal slab	• **Ribbon:** *Objects* tab>*Other objects* panel
	Polygonal wall	• **Ribbon:** *Objects* tab>*Other objects* panel
	Slab	• **Ribbon:** *Objects* tab>*Other objects* panel
	Wall	• **Ribbon:** *Objects* tab>*Other objects* panel
Special Parts		
	Advance Steel Special Part	• **Advance Tool Palette:** *Tools* category

Model Verifications

Model Verification is an important part of modeling and should be done throughout the modeling process. After you have created your 3D model, minimize errors by performing a variety of checks within the model. Checking can include, but is not limited to, bolts, bolt patterns, bolt spacing, holes (root radii), beams or plates that are duplicates and modified elements that are incorrect. Advance Steel also provides a way to check a joint's strength and run a report on the findings.

Learning Objectives

- Use Clash Check on the model.
- Run a Technical Check for design issues.
- Find issues using Object Marking.
- Check the entire model for issues using **Model Check**.
- Run a joint strength report using **Joint Design**.

5.1 Clash Check

The Clash Check tool ensures that within either the entire model or a selection set, no members are interfering with another, as shown in Figure 5–1.

Figure 5–1

How To: Start a Clash Check

1. Change the *Visual Style* to **2D Wireframe**.

    • The 2D Wireframe visual style allows you to view all parts involved in a collision, including parts that would otherwise be obstructed by another object.

2. In the *Home* tab>*Checking* panel, click ▥ (Clash Check).

    • If you only want to check specific parts, select those parts first. If nothing is selected, the entire model is checked.

3. The *Clash check* dialog box displays the results, as shown in Figure 5-2.

Id	Object 1	Object 2	Coordinates	Volume
1	m1074 : FL30X10 [Ladderbrace]	m1073 : FL25X10 [Ladder band]	WCS (77' 6 3/16" in., 26' in., 18' 6 3/4" in.)	0" in.²
2	1007 : 2 M3/4"x2 1/4" [Bolt]	m1001 : L4X3X3/8 [Clip angle]	WCS (10 3/4" in., -1 1/4" in., 11' 11 5/8" in.)	0" in.²
3	1007 : 2 M3/4"x2 1/4" [Bolt]	m1001 : L4X3X3/8 [Clip angle]	WCS (14' 1" in., -1 1/4" in., 11' 11 5/8" in.)	0" in.²
4	1007 : 2 M3/4"x2 1/4" [Bolt]	m1001 : L4X3X3/8 [Clip angle]	WCS (15' 11" in., -1 1/4" in., 11' 11 5/8" in.)	0" in.²
5	1007 : 2 M3/4"x2 1/4" [Bolt]	m1001 : L4X3X3/8 [Clip angle]	WCS (29' 1" in., -1 1/4" in., 11' 11 5/8" in.)	0" in.²
6	1007 : 2 M3/4"x2 1/4" [Bolt]	m1001 : L4X3X3/8 [Clip angle]	WCS (30' 11" in., -1 1/4" in., 11' 11 5/8" in.)	0" in.²
7	1007 : 2 M3/4"x2 1/4" [Bolt]	m1001 : L4X3X3/8 [Clip angle]	WCS (44' 1" in., -1 1/4" in., 11' 11 5/8" in.)	0" in.²
8	1007 : 2 M3/4"x2 1/4" [Bolt]	m1001 : L4X3X3/8 [Clip angle]	WCS (45' 11" in., -1 1/4" in., 11' 11 5/8" in.)	0" in.²
9	1007 : 2 M3/4"x2 1/4" [Bolt]	m1001 : L4X3X3/8 [Clip angle]	WCS (59' 1" in., -1 1/4" in., 11' 11 5/8" in.)	0" in.²
10	1007 : 2 M3/4"x2 1/4" [Bolt]	m1001 : L4X3X3/8 [Clip angle]	WCS (60' 11" in., -1 1/4" in., 11' 11 5/8" in.)	0" in.²
11	1007 : 2 M3/4"x2 1/4" [Bolt]	m1001 : L4X3X3/8 [Clip angle]	WCS (74' 1 1/4" in., -1 1/4" in., 11' 11 5/8" in.)	0" in.²
12	m1074 : FL30X10 [Ladderbrace]	m1071 : FL25X10 [Ladder band]	WCS (77' 6 3/16" in., 26' in., 18' 6 3/4" in.)	0" in.²
13	m1074 : FL30X10 [Ladderbrace]	Not defined : FL25X10 [Ladder band]	WCS (77' 6 3/16" in., 26' in., 18' 6 3/4" in.)	0" in.²
14	m1074 : FL30X10 [Ladderbrace]	m1070 : FL25X10 [Ladder band]	WCS (77' 6 3/16" in., 26' in., 18' 6 3/4" in.)	0" in.²

Figure 5-2

The Clash Check Dialog Box

To locate a clash, you need to first run a (Clash Check). Performing a Clash check will display the *Clash check* dialog box, showing a list of collisions found. It is important to fix all of the collisions that are found. The following data is displayed for each clash found in the model or selection set:

- **Id** - Clashing parts are numbered.
- **Object1** - Clashing part number: section size: model role.
- **Object 2** - Part involved in the clash: section size: model role.
- **Coordinates** - Gives the center location of the clash.
- **Volume** - Shows the collision solids volume.

How To: Bring Up the Clash Check Dialog Box

1. In the *Home* tab>*Checking* panel, click (Display Clash Checking Results).
   - Double-clicking a line that corresponds to a collision zooms the model display directly to the collision, highlighting the collision solid in red.

   Note: The correction of one clash may fix several other clashes. Make sure to run a new Clash check after fixing an error.

2. Fix any errors found.

There are additional tools available in the *Clash check* dialog box that aid in reviewing collisions found, as shown in the table below.

Bottom Right Icons		
	Clash Check	Performs another Clash check of the entire model or the selection set. If an object has been temporarily ignored, running another Clash check will ignore it.
	Search Marked Objects	Within the model, the highlighted collision solid will also show an arrow pointing to the solid.
	Clear Marked Objects	Turns the marked collision solid back to its original color.
	Ignore Objects	Ignores the selected collision and temporarily removes it from the list.
	Reset Ignored Objects	Resets the list to show ignored objects.
Top Left Icons		
	Notification(s)	Clicking on this icon will show any semantic or collision (if applicable) errors found.
	Clash Check	Clicking on this icon will switch you to the *Clash check* dialog box, showing the list of collisions found for the entire model or selection set.

Collision Solids

Collision solids are the result of a Clash check. They are displayed in the *Clash check* dialog box and can be reviewed in the model by double-clicking on a clash. The collision solid is highlighted in red to make it easy to find. You can evaluate and correct the collision solid while keeping the *Clash check* dialog box opened.

How To: Find a Collision Solid

1. In the *Home* tab>*Checking* panel, click ⚏ (Clash Check). Run a Clash check on the entire model or a selection set. Alternatively, if a Clash check has already been run, in the *Home* tab>*Checking* panel, click ⚏ (Display Clash Checking Results).

2. Keep the list of collisions displayed by moving the dialog box off to one side of the screen.

3. Double-click on each error to display the collision solid in the model.

4. Review the collision and determine the best method to fix it.

5. Repeat the ⚏ (Clash Check) command on the entire model or selection set to verify the collisions have all been fixed.

Practice 5a
Run a Clash Check

Practice Objectives

- Run a Clash check.
- Resolve collision errors.

In this practice, you will check a selection set in the model using Clash Check, review the collisions found, and resolve the issues using the Advance Joint Properties. You will then run a Clash check again to verify that all the errors are resolved.

1. In the practice files folder, open **Platform-Checking.dwg**.

2. Verify your display is set to 2D Wireframe.

3. With a crossing selection, draw a window around the corner brace system at Level 2 and at grid 1C (as shown in Figure 5–3) to select the entire connection set, column, and beams as shown in Figure 5–4.

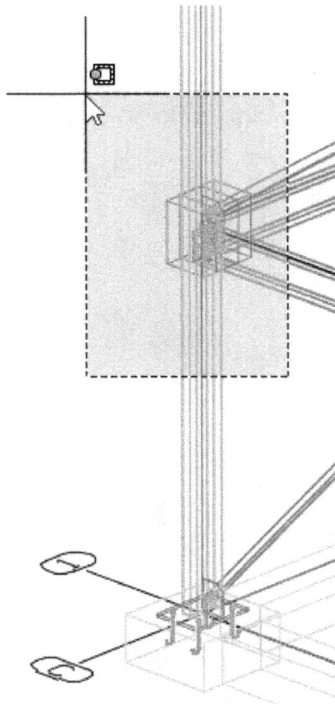

Figure 5–3

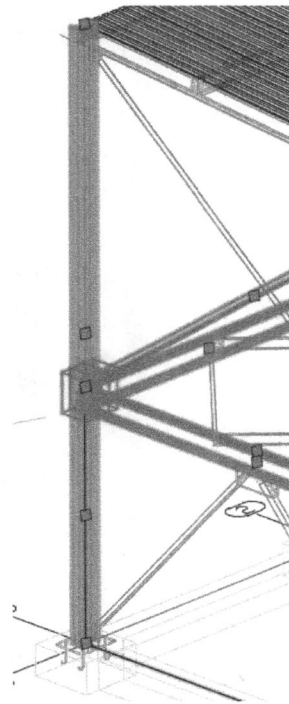

Figure 5–4

4. In the *Home* tab>*Checking* panel, click ⧉ (Clash Check).

5. In the *Clash check* dialog box, you will see the list of clashes that were found, as shown in Figure 5–5.

Id	Object 1	Object 2	Coordinates	Volume
1	b1007 : W12x30 [None]	1006 : 3 M3/4"x2" [Bolt]	WCS (7' 6" in., 40" in., 11' 5 7/8" in.)	0" in.²
2	b1007 : W12x30 [None]	1006 : 3 M3/4"x2" [Bolt]	WCS (7' 6" in., 40" in., 11' 5 7/8" in.)	0" in.²
3	b1020 : W12x30 [None]	1008 : 3 M3/4"x1 3/4" [Bolt]	WCS (0" in., 29' 11 11/16" in., 11' 5 7/8" in.)	0" in.²
4	b1020 : W12x30 [None]	1008 : 3 M3/4"x1 3/4" [Bolt]	WCS (0" in., 29' 11 11/16" in., 11' 5 7/8" in.)	0" in.²
5	b1007 : W12x30 [None]	1007 : 3 M3/4"x2 1/4" [Bolt]	WCS (7' 6" in., 40" in., 11' 5 7/8" in.)	0" in.²
6	b1020 : W12x30 [None]	1007 : 3 M3/4"x2 1/4" [Bolt]	WCS (0" in., 29' 11 11/16" in., 11' 5 7/8" in.)	0" in.²
7	1006 : 3 M3/4"x2" [Bolt]	1007 : 3 M3/4"x2 1/4" [Bolt]	WCS (5 3/4" in., 39' 9 7/8" in., 11' 5" in.)	0" in.²
8	b1007 : W12x30 [None]	m1001 : L4X3X3/8 [Clip angle]	WCS (7' 6" in., 40" in., 11' 5 7/8" in.)	1" in.²
9	b1007 : W12x30 [None]	m1001 : L4X3X3/8 [Clip angle]	WCS (7' 6" in., 40" in., 11' 5 7/8" in.)	1" in.²
10	b1020 : W12x30 [None]	m1003 : L4X4X3/8 [Clip angle]	WCS (0" in., 29' 11 11/16" in., 11' 5 7/8" in.)	1" in.²
11	b1020 : W12x30 [None]	m1003 : L4X4X3/8 [Clip angle]	WCS (0" in., 29' 11 11/16" in., 11' 5 7/8" in.)	1" in.²

Figure 5–5

6. Double-click on **Id 8** within the *Clash check* dialog box to zoom in on the collision solid highlighted in red.

7. Select the clip angle shown in Figure 5–6.

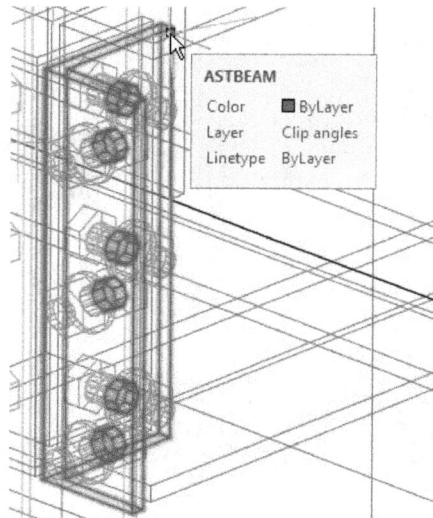

Figure 5–6

8. Right-click and select **Advance Joint Properties**.

9. In the *Clip angle* dialog box, click on the *Vertical Bolts* tab and change the **Interm. dist.** to **3"** as shown in Figure 5–7.

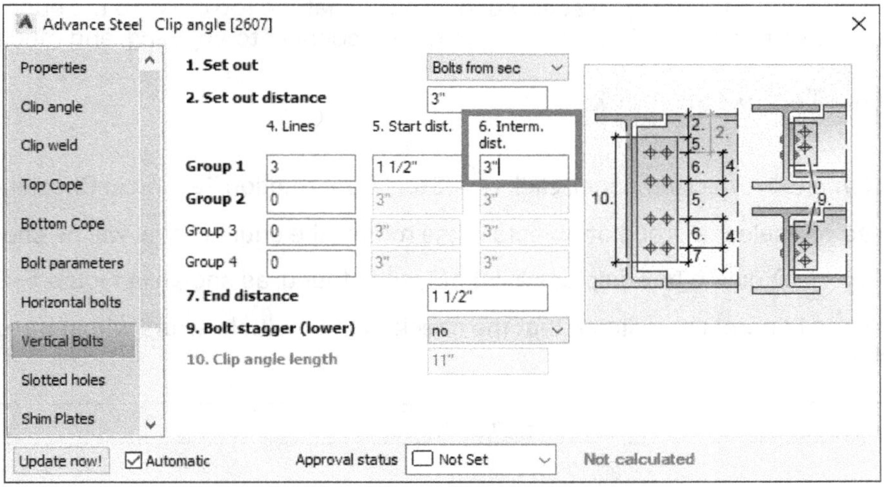

Figure 5–7

10. Repeat this for the other clip angles in the clash.

11. When they are all fixed, select the same objects again and run another Clash check.

12. If all errors have been fixed, the *Notification(s)* dialog box shows "Collision check found no errors", as shown in Figure 5–8.

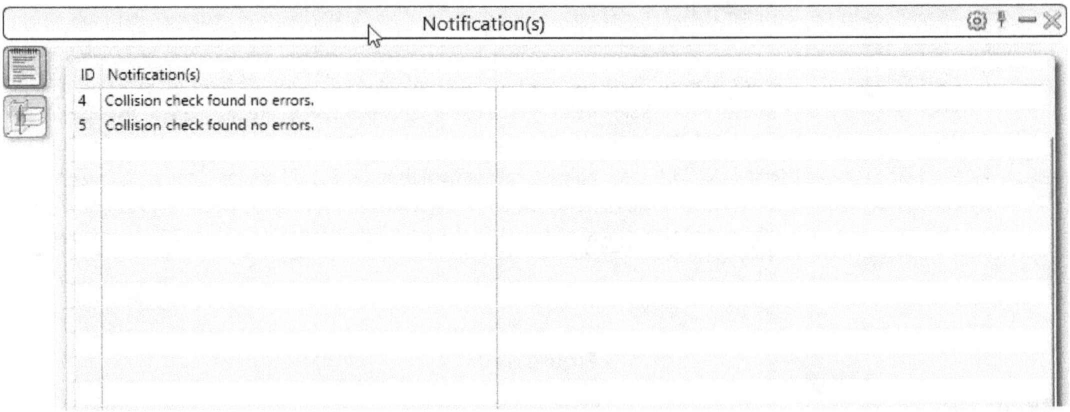

Figure 5–8

13. Save the drawing.

End of practice

5.2 Technical Check

Technical Check looks for design issues, such as bolts that are too close to the edge of a plate, bolts that are not a part of anything, welds that do not connect to anything, and much more.

How To: Use Technical Check

1. In the *Home* tab>*Checking* panel, click 🔧 (Steel Construction Technical Checking).

   - Unless you select a group of objects in the model, the entire model will be checked.

2. The *Steel check* dialog box displays a list of issues found, as shown in Figure 5–9.

   - Duplicate Id numbers indicate that the check found multiple errors within the same object.

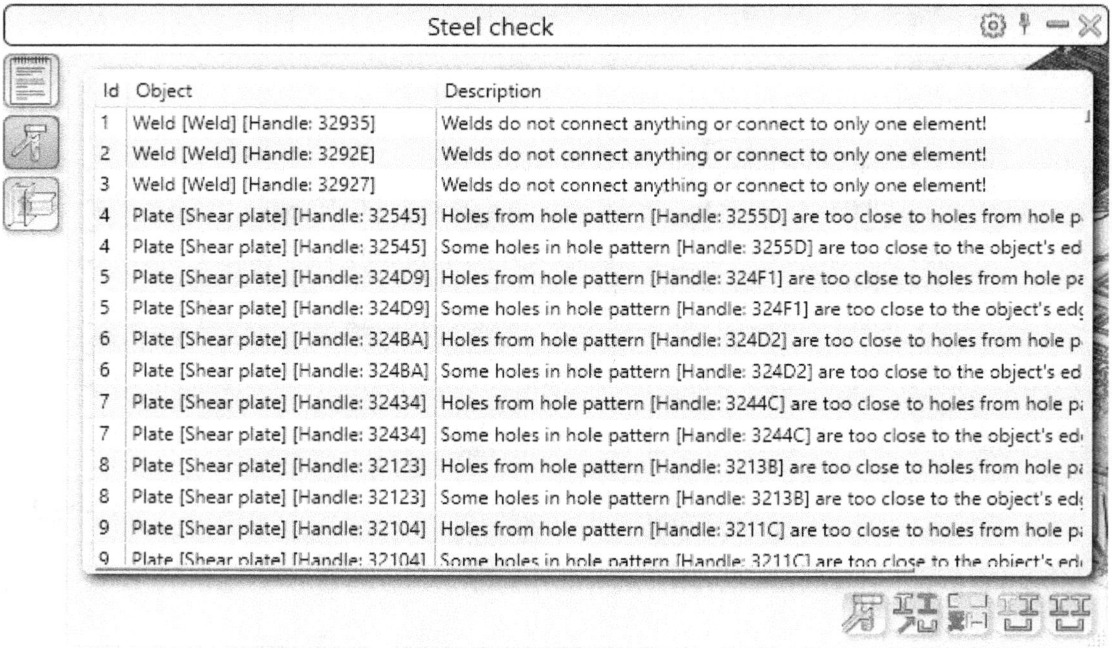

Id	Object	Description
1	Weld [Weld] [Handle: 32935]	Welds do not connect anything or connect to only one element!
2	Weld [Weld] [Handle: 3292E]	Welds do not connect anything or connect to only one element!
3	Weld [Weld] [Handle: 32927]	Welds do not connect anything or connect to only one element!
4	Plate [Shear plate] [Handle: 32545]	Holes from hole pattern [Handle: 3255D] are too close to holes from hole p
4	Plate [Shear plate] [Handle: 32545]	Some holes in hole pattern [Handle: 3255D] are too close to the object's ed
5	Plate [Shear plate] [Handle: 324D9]	Holes from hole pattern [Handle: 324F1] are too close to holes from hole pa
5	Plate [Shear plate] [Handle: 324D9]	Some holes in hole pattern [Handle: 324F1] are too close to the object's edç
6	Plate [Shear plate] [Handle: 324BA]	Holes from hole pattern [Handle: 324D2] are too close to holes from hole p
6	Plate [Shear plate] [Handle: 324BA]	Some holes in hole pattern [Handle: 324D2] are too close to the object's ed
7	Plate [Shear plate] [Handle: 32434]	Holes from hole pattern [Handle: 3244C] are too close to holes from hole pa
7	Plate [Shear plate] [Handle: 32434]	Some holes in hole pattern [Handle: 3244C] are too close to the object's edɪ
8	Plate [Shear plate] [Handle: 32123]	Holes from hole pattern [Handle: 3213B] are too close to holes from hole pa
8	Plate [Shear plate] [Handle: 32123]	Some holes in hole pattern [Handle: 3213B] are too close to the object's edç
9	Plate [Shear plate] [Handle: 32104]	Holes from hole pattern [Handle: 3211C] are too close to holes from hole pa
9	Plate [Shear plate] [Handle: 32104]	Some holes in hole pattern [Handle: 3211C] are too close to the object's edɪ

Figure 5–9

Note: *The minimum distance between hole and element is calculated with the formula, Factor*Hole diameter in Management Tools>Defaults, Joints>General, as shown in Figure 5–10.*

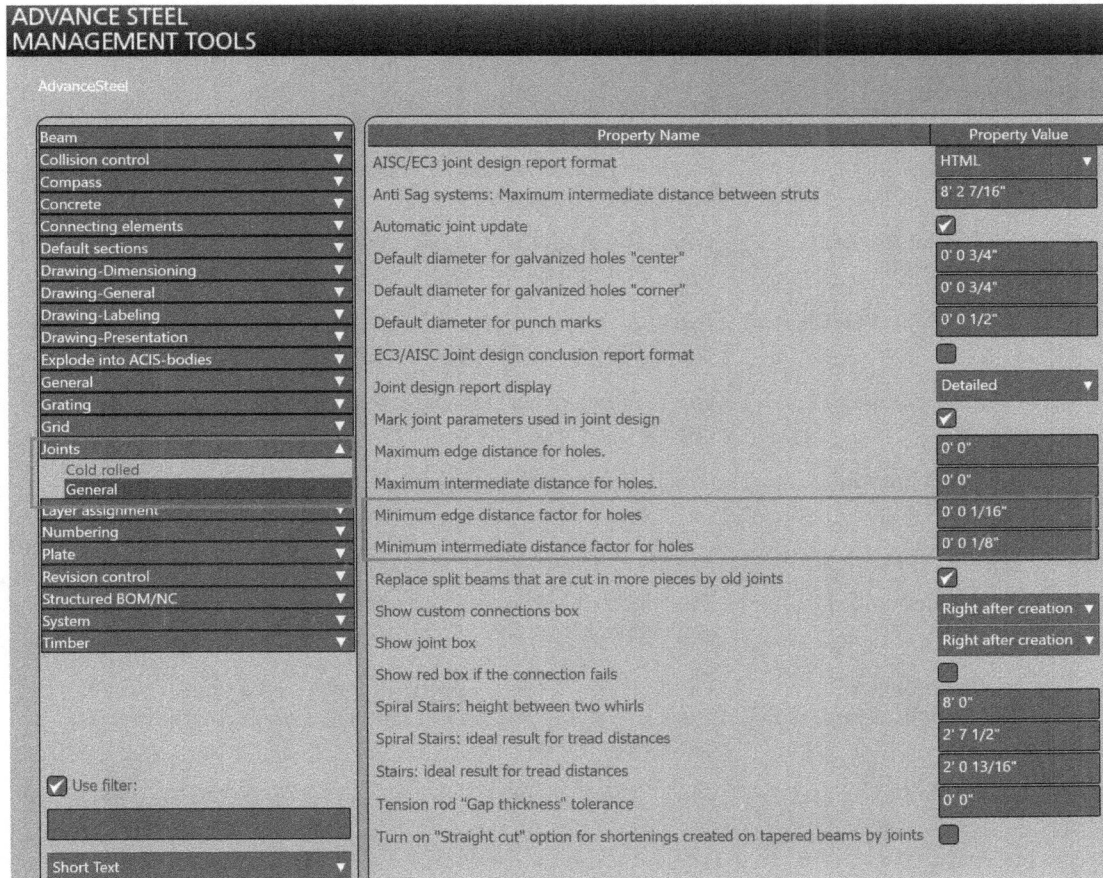

Figure 5–10

Steel Check Dialog Box

For each technical issue found, the following information is provided:

- **Id** - Each issue found is numbered. There can be more than one of the same Id number due to multiple issues found with one object.
- **Object** - Parent part type and the handle Id number.
- **Description** - The error and the handle Id of the connection element that is causing the error.

There are additional tools available in the *Steel check* dialog box that aid in reviewing issues found, as shown in the table below.

Bottom Right Icons		
	Steel Check	Performs another Steel check of the entire model or the selection set. If an object has been temporarily ignored, running another Steel check will ignore it.
	Search Marked Objects	Within the model, the highlighted collision solid will also show an arrow pointing to the solid.
	Clear Marked Objects	Turns the marked object back to its original color and turns off the red arrow pointing to the object.
	Ignore Objects	Ignores the highlighted object and temporarily removes it from the list.
	Reset Ignored Objects	Resets the list, showing ignored objects.
Top Left Icons		
	Notification(s)	Clicking on this icon will show any semantic or collision (if applicable) errors found.
	Clash Check (if applicable)	The Clash check icon appears if there was a Clash check run previously. Selecting on this icon will switch you to the *Clash check* dialog box.
	Steel Check	Clicking on this icon will switch the dialog box to the *Steel check* list.

Mark Object

Mark object allows you to find a connection element causing errors by entering its index number, found in the *Steel check* dialog box next to the object name or description, as shown in Figure 5–11.

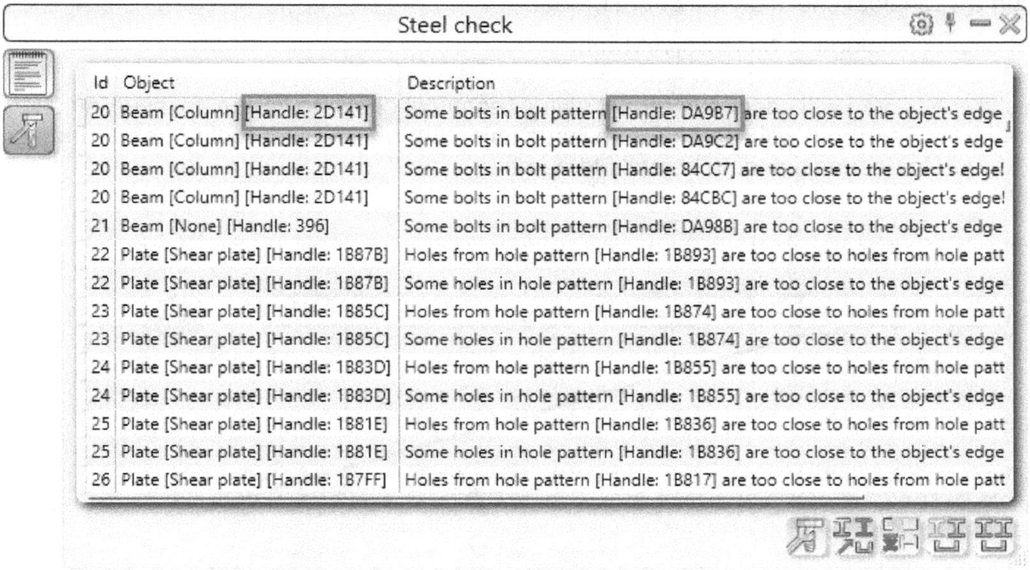

Id	Object	Description
20	Beam [Column] [Handle: 2D141]	Some bolts in bolt pattern [Handle: DA9B7] are too close to the object's edge
20	Beam [Column] [Handle: 2D141]	Some bolts in bolt pattern [Handle: DA9C2] are too close to the object's edge
20	Beam [Column] [Handle: 2D141]	Some bolts in bolt pattern [Handle: 84CC7] are too close to the object's edge!
20	Beam [Column] [Handle: 2D141]	Some bolts in bolt pattern [Handle: 84CBC] are too close to the object's edge!
21	Beam [None] [Handle: 396]	Some bolts in bolt pattern [Handle: DA98B] are too close to the object's edge
22	Plate [Shear plate] [Handle: 1B87B]	Holes from hole pattern [Handle: 1B893] are too close to holes from hole patt
22	Plate [Shear plate] [Handle: 1B87B]	Some holes in hole pattern [Handle: 1B893] are too close to the object's edge
23	Plate [Shear plate] [Handle: 1B85C]	Holes from hole pattern [Handle: 1B874] are too close to holes from hole patt
23	Plate [Shear plate] [Handle: 1B85C]	Some holes in hole pattern [Handle: 1B874] are too close to the object's edge
24	Plate [Shear plate] [Handle: 1B83D]	Holes from hole pattern [Handle: 1B855] are too close to holes from hole patt
24	Plate [Shear plate] [Handle: 1B83D]	Some holes in hole pattern [Handle: 1B855] are too close to the object's edge
25	Plate [Shear plate] [Handle: 1B81E]	Holes from hole pattern [Handle: 1B836] are too close to holes from hole patt
25	Plate [Shear plate] [Handle: 1B81E]	Some holes in hole pattern [Handle: 1B836] are too close to the object's edge
26	Plate [Shear plate] [Handle: 1B7FF]	Holes from hole pattern [Handle: 1B817] are too close to holes from hole patt

Figure 5–11

How To: Find Connection Elements Causing Errors

1. In the *Advance Tool Palette*> (Selection) category, click 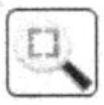 (Mark object).
2. At the command line *Handle/ID,* enter **H** for Handle and press <Enter>.
3. At the command line *Please specify the object handle*, enter the handle Id and press <Enter>.
   - It is not case sensitive.
4. The model zooms into the object, which is highlighted in red.

Practice 5b
Run a Technical Check

Practice Objectives

- Run a Technical check on the entire model.
- Find issues using the Mark object tool.

In this practice, you will check the model using Technical check to locate an issue, as shown in Figure 5–12. Then, you will search for an issue using the Mark object tool to better understand how to look for issues using an object's Id number.

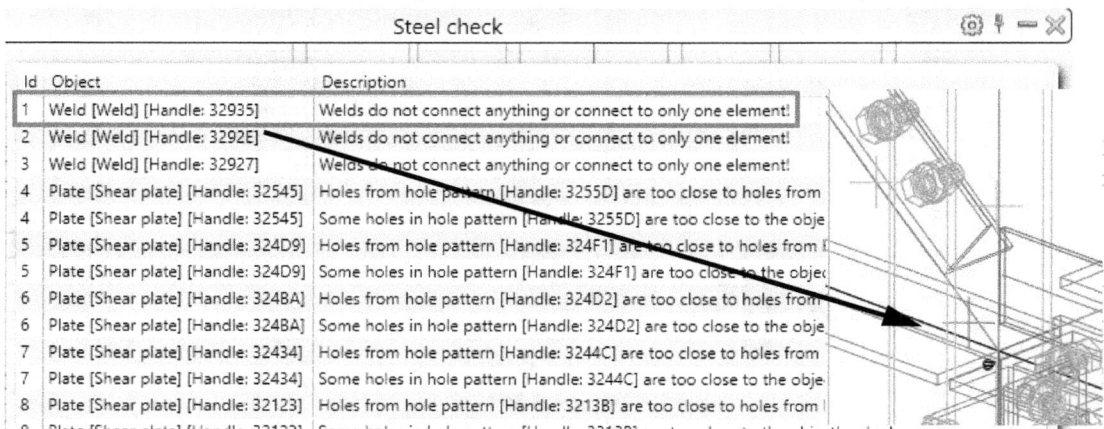

Id	Object	Description
1	Weld [Weld] [Handle: 32935]	Welds do not connect anything or connect to only one element!
2	Weld [Weld] [Handle: 3292E]	Welds do not connect anything or connect to only one element!
3	Weld [Weld] [Handle: 32927]	Welds do not connect anything or connect to only one element!
4	Plate [Shear plate] [Handle: 32545]	Holes from hole pattern [Handle: 3255D] are too close to holes from
4	Plate [Shear plate] [Handle: 32545]	Some holes in hole pattern [Handle: 3255D] are too close to the obje
5	Plate [Shear plate] [Handle: 324D9]	Holes from hole pattern [Handle: 324F1] are too close to holes from
5	Plate [Shear plate] [Handle: 324D9]	Some holes in hole pattern [Handle: 324F1] are too close to the obje
6	Plate [Shear plate] [Handle: 324BA]	Holes from hole pattern [Handle: 324D2] are too close to holes from
6	Plate [Shear plate] [Handle: 324BA]	Some holes in hole pattern [Handle: 324D2] are too close to the obje
7	Plate [Shear plate] [Handle: 32434]	Holes from hole pattern [Handle: 3244C] are too close to holes from
7	Plate [Shear plate] [Handle: 32434]	Some holes in hole pattern [Handle: 3244C] are too close to the obje
8	Plate [Shear plate] [Handle: 32123]	Holes from hole pattern [Handle: 3213B] are too close to holes from

Figure 5–12

Task 1: Run a Technical check.

1. In the practice files folder, open **Platform-Checking.dwg**.

2. In the *Home* tab>*Checking* panel, click 📐 (Steel Construction Technical Checking).

3. The *Steel check* dialog box shows a list of issues found, as shown in Figure 5–13.

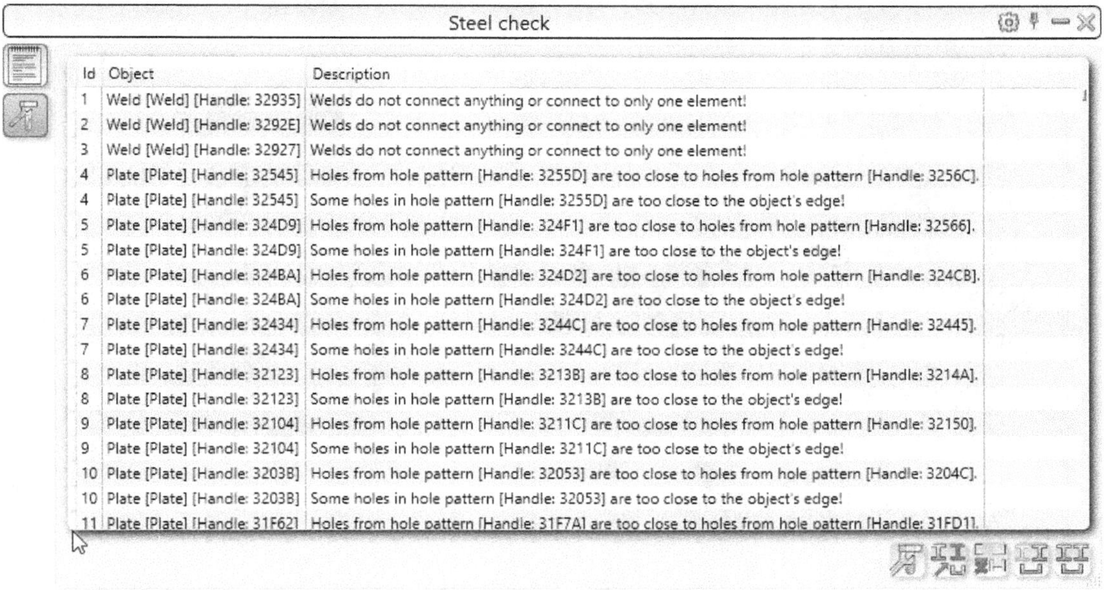

Id	Object	Description
1	Weld [Weld] [Handle: 32935]	Welds do not connect anything or connect to only one element!
2	Weld [Weld] [Handle: 3292E]	Welds do not connect anything or connect to only one element!
3	Weld [Weld] [Handle: 32927]	Welds do not connect anything or connect to only one element!
4	Plate [Plate] [Handle: 32545]	Holes from hole pattern [Handle: 3255D] are too close to holes from hole pattern [Handle: 3256C].
4	Plate [Plate] [Handle: 32545]	Some holes in hole pattern [Handle: 3255D] are too close to the object's edge!
5	Plate [Plate] [Handle: 324D9]	Holes from hole pattern [Handle: 324F1] are too close to holes from hole pattern [Handle: 32566].
5	Plate [Plate] [Handle: 324D9]	Some holes in hole pattern [Handle: 324F1] are too close to the object's edge!
6	Plate [Plate] [Handle: 324BA]	Holes from hole pattern [Handle: 324D2] are too close to holes from hole pattern [Handle: 324CB].
6	Plate [Plate] [Handle: 324BA]	Some holes in hole pattern [Handle: 324D2] are too close to the object's edge!
7	Plate [Plate] [Handle: 32434]	Holes from hole pattern [Handle: 3244C] are too close to holes from hole pattern [Handle: 32445].
7	Plate [Plate] [Handle: 32434]	Some holes in hole pattern [Handle: 3244C] are too close to the object's edge!
8	Plate [Plate] [Handle: 32123]	Holes from hole pattern [Handle: 3213B] are too close to holes from hole pattern [Handle: 3214A].
8	Plate [Plate] [Handle: 32123]	Some holes in hole pattern [Handle: 3213B] are too close to the object's edge!
9	Plate [Plate] [Handle: 32104]	Holes from hole pattern [Handle: 3211C] are too close to holes from hole pattern [Handle: 32150].
9	Plate [Plate] [Handle: 32104]	Some holes in hole pattern [Handle: 3211C] are too close to the object's edge!
10	Plate [Plate] [Handle: 3203B]	Holes from hole pattern [Handle: 32053] are too close to holes from hole pattern [Handle: 3204C].
10	Plate [Plate] [Handle: 3203B]	Some holes in hole pattern [Handle: 32053] are too close to the object's edge!
11	Plate [Plate] [Handle: 31F62]	Holes from hole pattern [Handle: 31F7A] are too close to holes from hole pattern [Handle: 31FD1].

Figure 5–13

4. Double-click on **Id 34** to zoom in on the connection in the model. The clash will be highlighted in red.

5. The object is highlighted in red. As with Clash check, you need to determine the issue and take the necessary steps to correct the issue.

6. In the *Home* tab>*Settings* panel, click **Management Tools**. Select **Defaults**. Click *Joints* tab>General, and change the *Minimum edge distance factor for holes* and *Minimum intermediate distance factor for holes* to **0**, as shown in Figure 5–14.

7. Click **Load Settings in Advance** in the upper right corner.

8. In the *Home* tab>*Settings* panel, click **Update Defaults**.

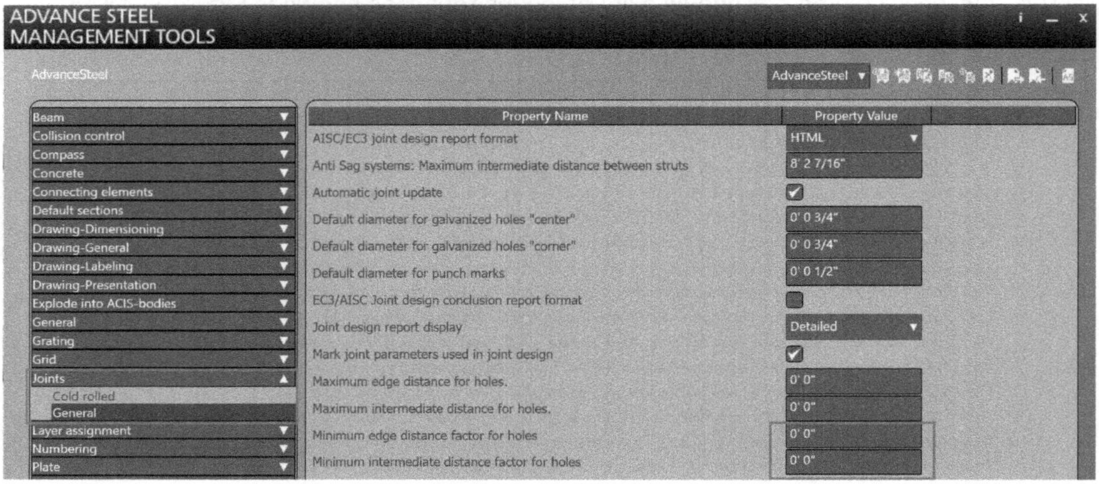

Figure 5–14

9. Run Technical Checking again.

10. Save the drawing.

End of practice

5.3 Model Check

The Model Check tool checks the entire drawing for modeling issues like overlapping shortening, as shown in Figure 5–15, and displays each finding in the *Model Check* dialog box. You can fix the errors one at a time or use the Model Check fix function to attempt to fix all or some of the errors at once.

Figure 5–15

Model Check Dialog Box

The *Model Check* dialog box shows a list of errors found in the model. For each issue found, the following information is provided:

- **Id** - Each issue found is numbered.
- **Parent** - This is the Parent number.
- **Feature** - Feature number.
- **Object Type** - Member type.
- **Error** - Description of the issue.
- **Fix Info** - Displays fix information on the member.
- **Error Code** - Error code number.

There are additional tools available in the *Model Check* dialog box that aid in reviewing errors found, as shown in the table below.

Bottom Right Icons		
	Model Check	Performs another Model check of the entire model. If an object has been temporarily ignored, running another Model check will ignore it.
	Fix All Errors	Attempt to remove one of the features triggering the problem instead of deleting the whole plate/folded plate within the entire model.
	Fix Selected Errors	Attempt to remove one of the features triggering the problem instead of deleting the whole plate/folded plate with the selected features.
	Search Marked Objects	Within the model, the highlighted collision solid will also show an arrow pointing to the solid.
	Clear Marked Objects	Turns the marked collision solid back to its original color and removes the red arrow pointing to the collision solid.

Top Left Icons		
	Notification(s)	Clicking on this icon will switch you to the *Notification(s)* dialog box showing any collision errors found.
	Model Check	Clicking on this icon will switch you to the *Model Check* dialog box.

How To: Run a Model Check

1. In the *Home* tab>*Checking* panel, click [icon] (Model Check).

    - The entire model is checked for errors.
    - If no errors are found, the *Notification(s)* dialog box displays with no errors. If errors are found, the *Model Check* dialog box displays with a list of errors, as shown in Figure 5–16.

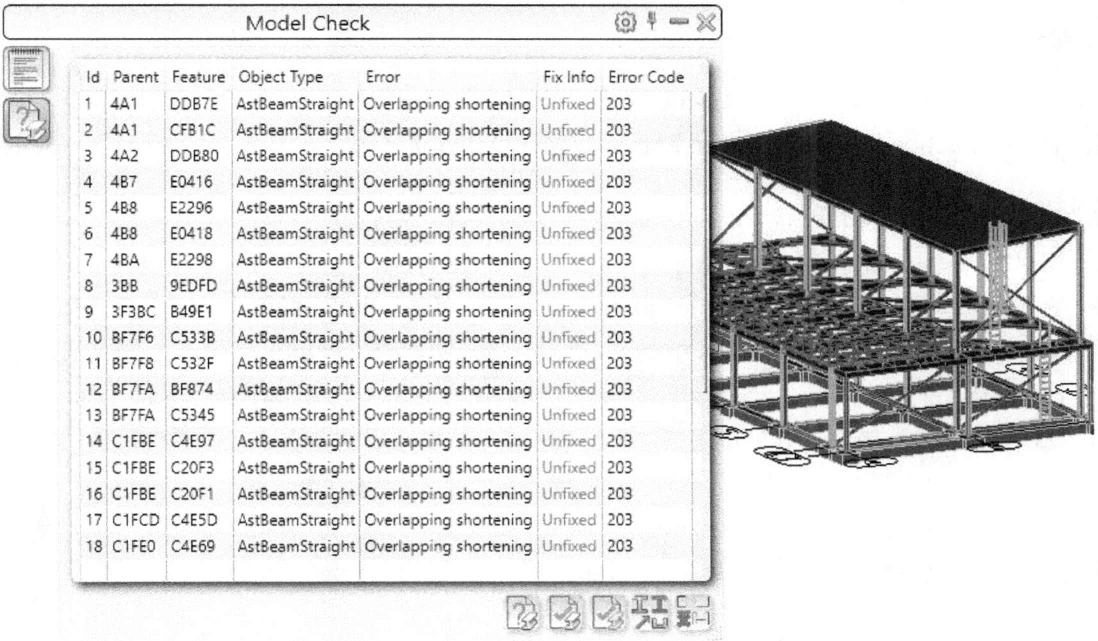

Id	Parent	Feature	Object Type	Error	Fix Info	Error Code
1	4A1	DDB7E	AstBeamStraight	Overlapping shortening	Unfixed	203
2	4A1	CFB1C	AstBeamStraight	Overlapping shortening	Unfixed	203
3	4A2	DDB80	AstBeamStraight	Overlapping shortening	Unfixed	203
4	4B7	E0416	AstBeamStraight	Overlapping shortening	Unfixed	203
5	4B8	E2296	AstBeamStraight	Overlapping shortening	Unfixed	203
6	4B8	E0418	AstBeamStraight	Overlapping shortening	Unfixed	203
7	4BA	E2298	AstBeamStraight	Overlapping shortening	Unfixed	203
8	3BB	9EDFD	AstBeamStraight	Overlapping shortening	Unfixed	203
9	3F3BC	B49E1	AstBeamStraight	Overlapping shortening	Unfixed	203
10	BF7F6	C533B	AstBeamStraight	Overlapping shortening	Unfixed	203
11	BF7F8	C532F	AstBeamStraight	Overlapping shortening	Unfixed	203
12	BF7FA	BF874	AstBeamStraight	Overlapping shortening	Unfixed	203
13	BF7FA	C5345	AstBeamStraight	Overlapping shortening	Unfixed	203
14	C1FBE	C4E97	AstBeamStraight	Overlapping shortening	Unfixed	203
15	C1FBE	C20F3	AstBeamStraight	Overlapping shortening	Unfixed	203
16	C1FBE	C20F1	AstBeamStraight	Overlapping shortening	Unfixed	203
17	C1FCD	C4E5D	AstBeamStraight	Overlapping shortening	Unfixed	203
18	C1FE0	C4E69	AstBeamStraight	Overlapping shortening	Unfixed	203

Figure 5–16

Practice 5c
Model Check

Practice Objectives

- Run a Model check on the entire model.
- Utilize Model check tools to fix all errors.
- Edit any objects that still have an error.

In this practice, you will check the entire model using Model Check and review the errors found. Using the Model check tools, you will fix as many errors as you can. For any objects that were not fixed, you will manually edit the object to resolve the issue.

1. In the practice files folder, open **Platform-Model Check.dwg**.

2. In the *Home* tab>*Checking* panel, click ▦ (Model Check).

3. A clash verification is done automatically for the entire model.

4. After the automatic verification is done, the *Notification* palette is displayed.

5. There are issues in the *Model Check* dialog box, as shown in Figure 5–17.

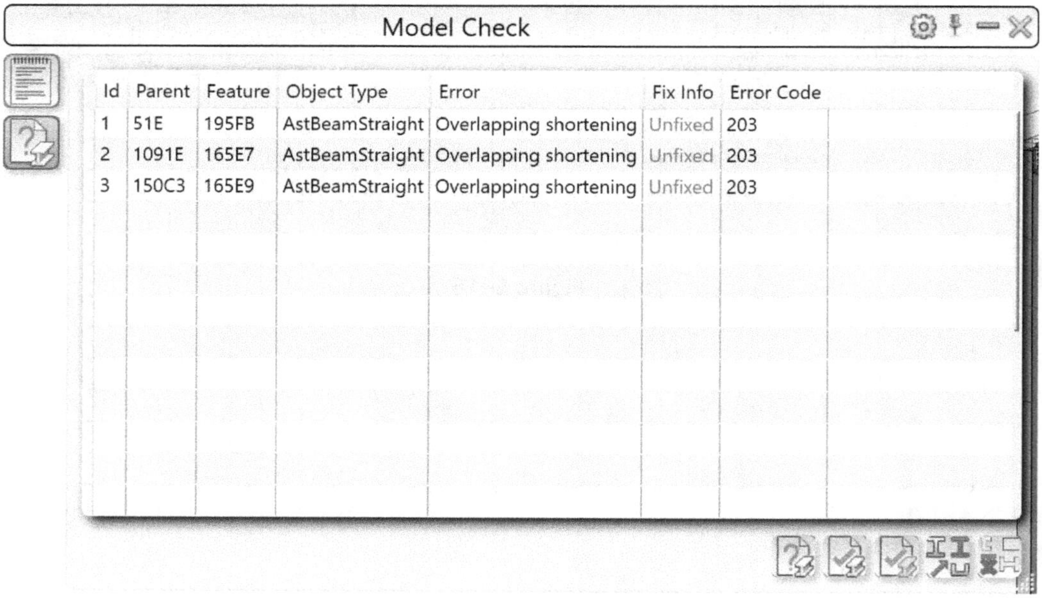

Id	Parent	Feature	Object Type	Error	Fix Info	Error Code
1	51E	195FB	AstBeamStraight	Overlapping shortening	Unfixed	203
2	1091E	165E7	AstBeamStraight	Overlapping shortening	Unfixed	203
3	150C3	165E9	AstBeamStraight	Overlapping shortening	Unfixed	203

Figure 5–17

6. Click ▦ (Fix All Errors).

7. Most of the errors have been fixed, as shown in Figure 5–18.

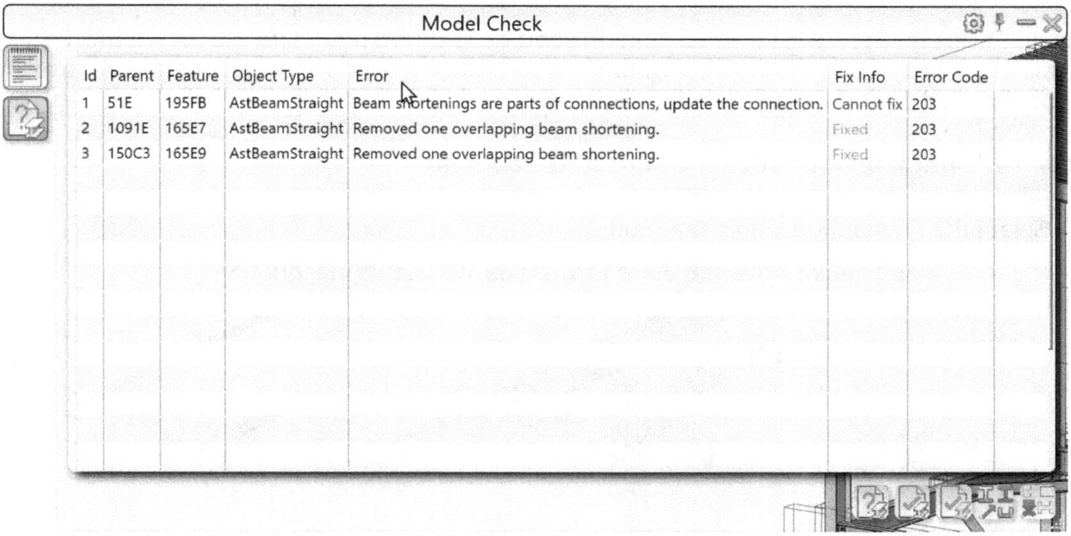

Id	Parent	Feature	Object Type	Error	Fix Info	Error Code
1	51E	195FB	AstBeamStraight	Beam shortenings are parts of connnections, update the connection.	Cannot fix	203
2	1091E	165E7	AstBeamStraight	Removed one overlapping beam shortening.	Fixed	203
3	150C3	165E9	AstBeamStraight	Removed one overlapping beam shortening.	Fixed	203

Model Check

Figure 5–18

8. Double-click on the first line to zoom into the object that cannot be fixed.

9. Delete the connections and beam, then replace it by using the **Advance Copy** tool to copy an existing beam including additional connections.

10. Save the file.

End of practice

5.4 Joint Design

Use Joint Design from the Advance Joint Properties to confirm the strength of a joint. This check is used to reduce the number of times a joint goes through the joint design process and is not intended to replace a structural engineer.

How To: Check a Connection Using Joint Design

1. Select the connection within the model.
2. Right-click and select **Advance Joint Properties**, as shown in Figure 5–19.

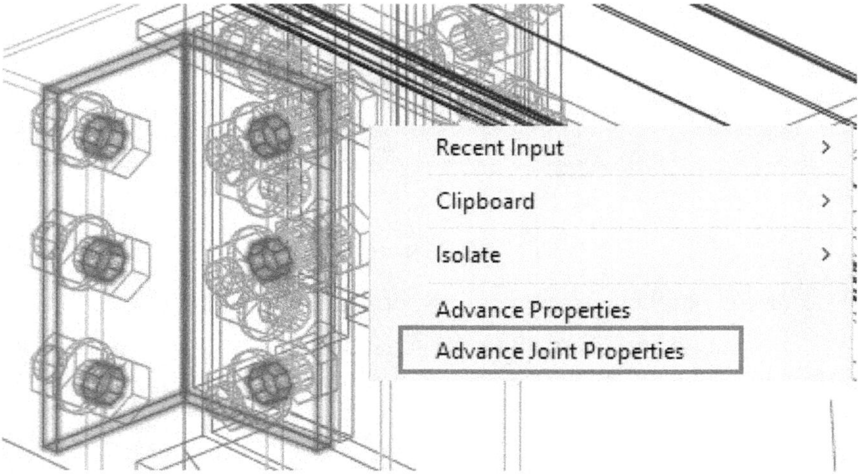

Figure 5–19

3. In the *Clip angle* dialog box, on the *Properties* tab, if the selected item was not the original connection, you can select **Upgrade to master**, as shown in Figure 5−20.

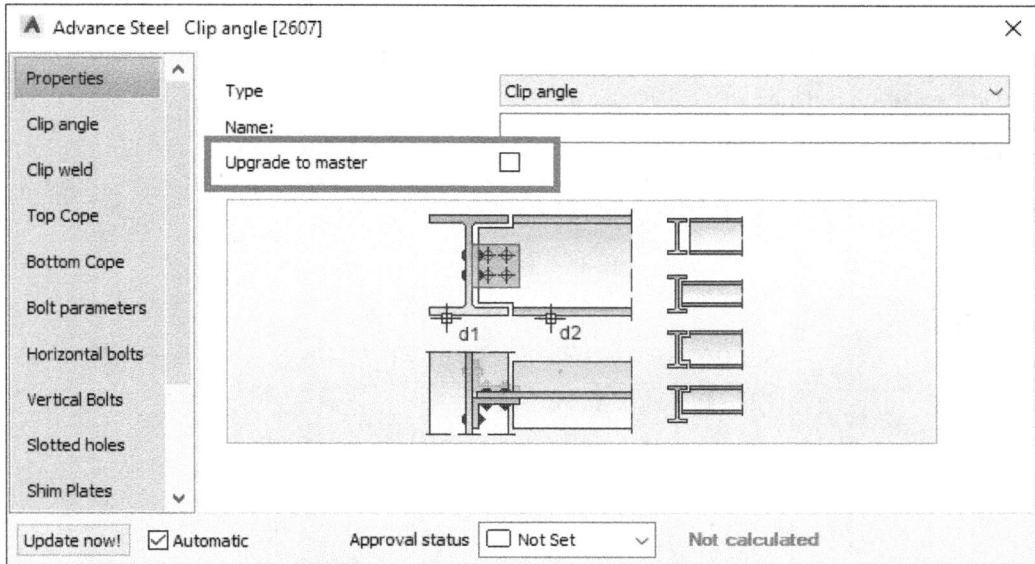

Figure 5−20

4. In the *Clip angle* dialog box, select the *Joint design* tab as shown in Figure 5−21.

Figure 5−21

5. In the *Design module* section, select the standard you are working in.

6. If required, clear the **Automatic values** checkbox to enter the necessary loads for **M**, **P**, and **V**.

   - You are only able to enter a value for M with an appropriate moment load joint type.

7. Click the **Check** button.

After the check is completed, the results are shown in the *Status* window.

   - If the joint passes, the status displays **OK Checked** in green with no failed verifications, as shown in Figure 5–22.

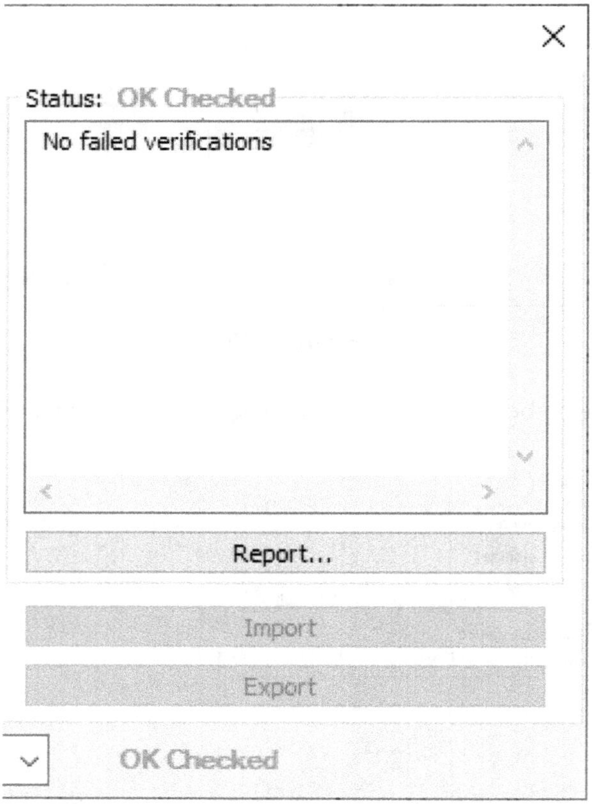

Figure 5–22

- If the joint fails, the status displays **Checking failed** in red and will display a list of failed verifications, as shown in Figure 5–23.

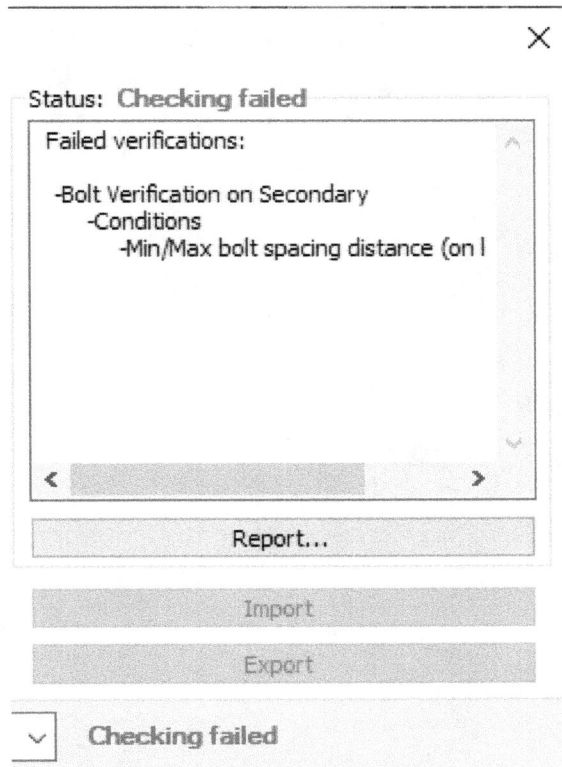

Figure 5–23

- The check result is also displayed at the bottom of each tab within the *Advance Joint Properties* dialog box.

8. The **Report** button generates an RTF or HTML report that displays each criteria that was checked, errors found, and formulas used (as shown in Figure 5–24).

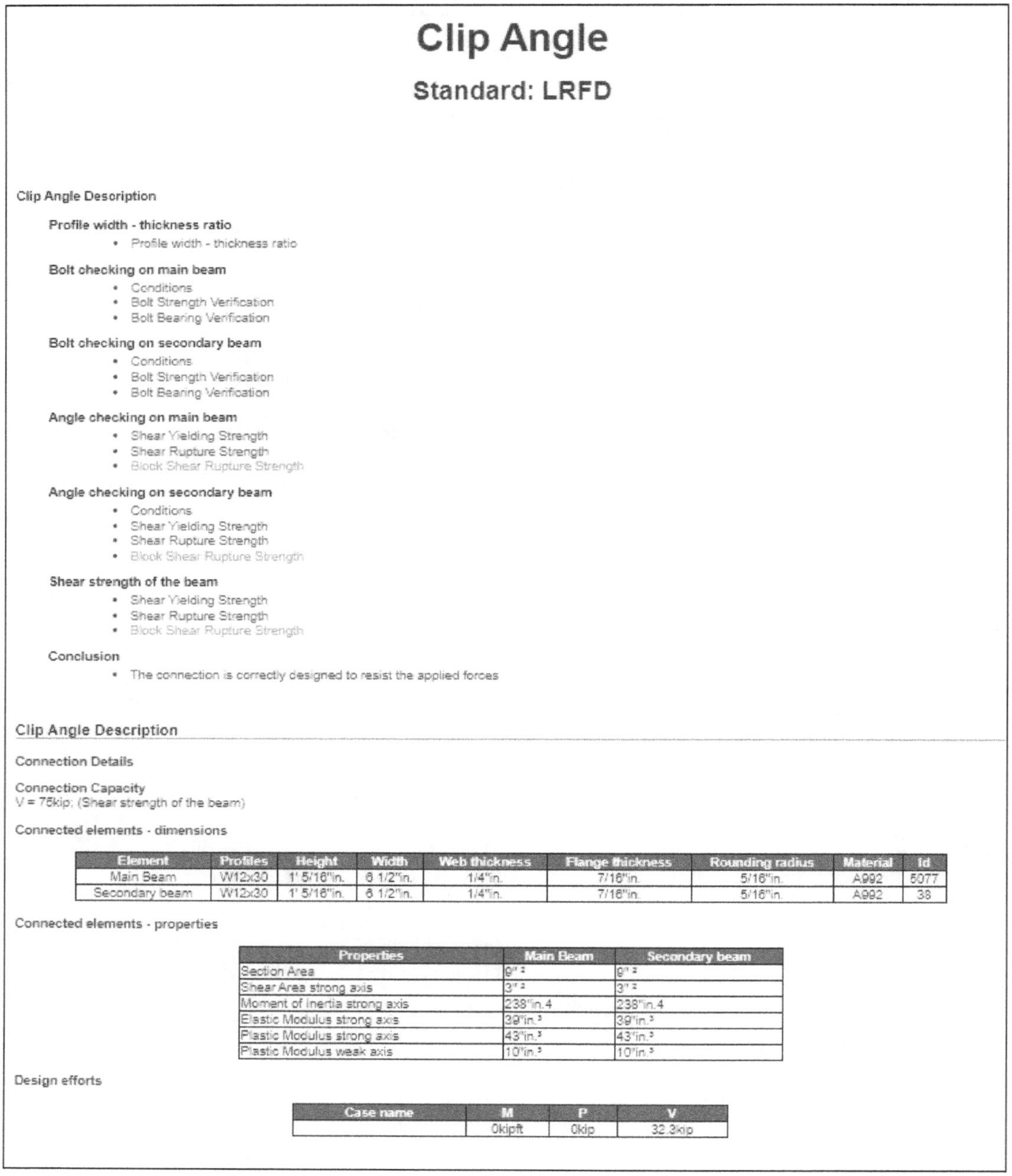

Clip Angle
Standard: LRFD

Clip Angle Description

Profile width - thickness ratio
- Profile width - thickness ratio

Bolt checking on main beam
- Conditions
- Bolt Strength Verification
- Bolt Bearing Verification

Bolt checking on secondary beam
- Conditions
- Bolt Strength Verification
- Bolt Bearing Verification

Angle checking on main beam
- Shear Yielding Strength
- Shear Rupture Strength
- Block Shear Rupture Strength

Angle checking on secondary beam
- Conditions
- Shear Yielding Strength
- Shear Rupture Strength
- Block Shear Rupture Strength

Shear strength of the beam
- Shear Yielding Strength
- Shear Rupture Strength
- Block Shear Rupture Strength

Conclusion
- The connection is correctly designed to resist the applied forces

Clip Angle Description

Connection Details

Connection Capacity
V = 75kip; (Shear strength of the beam)

Connected elements - dimensions

Element	Profiles	Height	Width	Web thickness	Flange thickness	Rounding radius	Material	Id
Main Beam	W12x30	1' 5/16"in.	6 1/2"in.	1/4"in.	7/16"in.	5/16"in.	A992	5077
Secondary beam	W12x30	1' 5/16"in.	6 1/2"in.	1/4"in.	7/16"in.	5/16"in.	A992	38

Connected elements - properties

Properties	Main Beam	Secondary beam
Section Area	9"²	9"²
Shear Area strong axis	3"²	3"²
Moment of inertia strong axis	238"in.4	238"in.4
Elastic Modulus strong axis	39"in.³	39"in.³
Plastic Modulus strong axis	43"in.³	43"in.³
Plastic Modulus weak axis	10"in.³	10"in.³

Design efforts

Case name	M	P	V
	0kipft	0kip	32.3kip

Figure 5–24

9. To change the report's format, click the **Settings** button and specify the Report criteria as shown in Figure 5–25.

    - Set the *Text format* to **Short** or **Long**.
    - The **Short** format only lists the names of the failed criteria while the **Long** format is a full report.

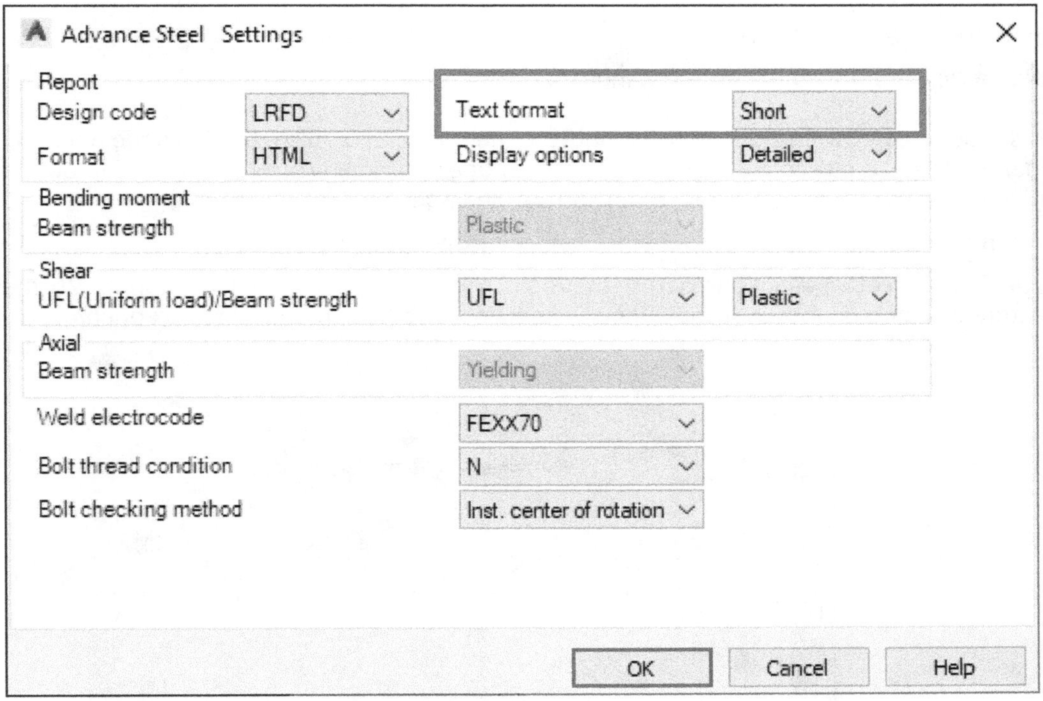

Figure 5–25

Practice 5d
Run a Joint Design Report

Practice Objectives

- Select a joint and run a strength report.
- Find the errors and fix them.
- Run a report of the joint check findings.

In this practice, you will select on a clip angle in the model and utilize the Advance Joint Properties to run a strength check on the entire model to find any errors that need to be resolved. Then, you will generate a report on the findings.

1. In the practice files folder, open **Platform-Joint Design.dwg**.

2. On Grid 1A, select a clip angle from the top of the column, as shown in Figure 5–26, then right-click and choose **Advance Joint Properties**. The *Clip angle* dialog box opens.

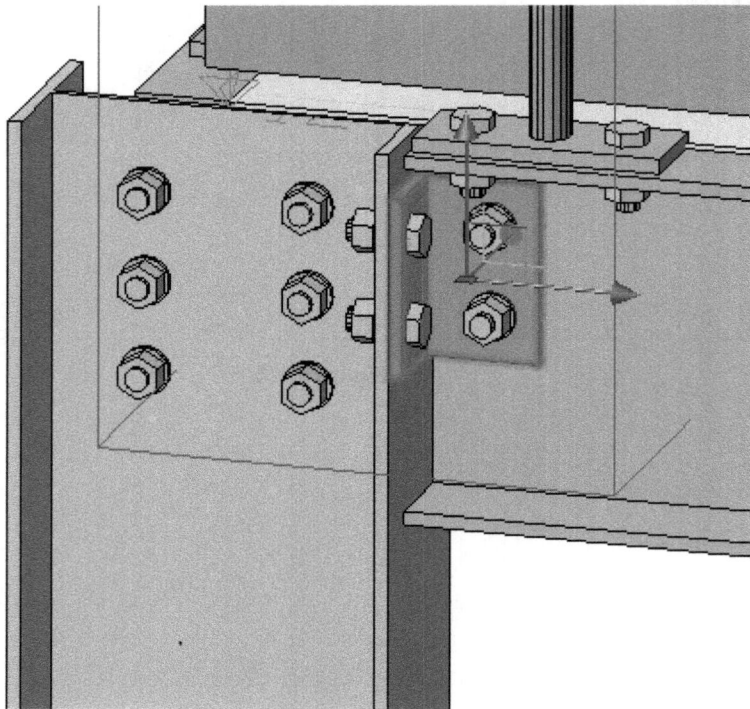

Figure 5–26

3. In the *Clip angle* dialog box>*Joint design* tab, do the following, as shown in Figure 5–27:

- Set *Design module* to **AISC**.
- Within the *Design Options* section, click **Check.**
- When the check is finished, click **Report...**.

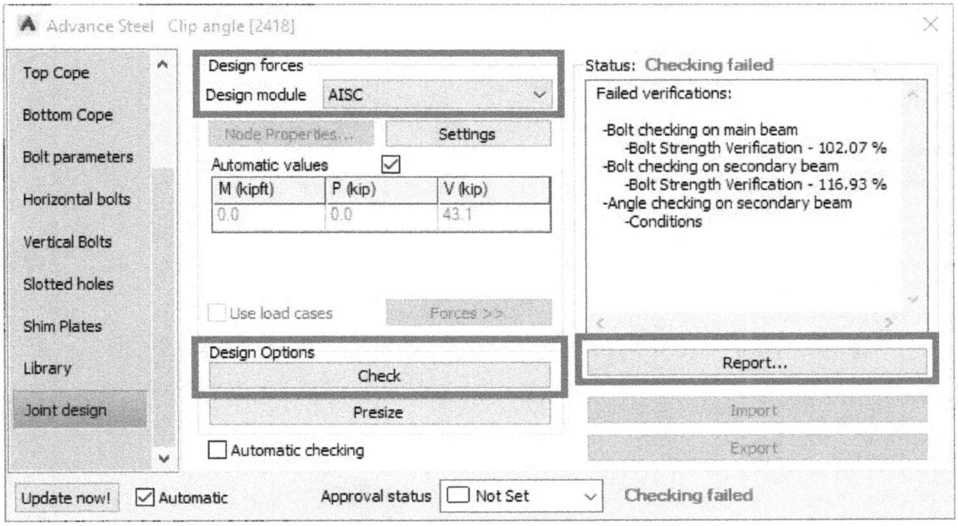

Figure 5–27

4. An HTML of the report will open as shown in Figure 5-28.

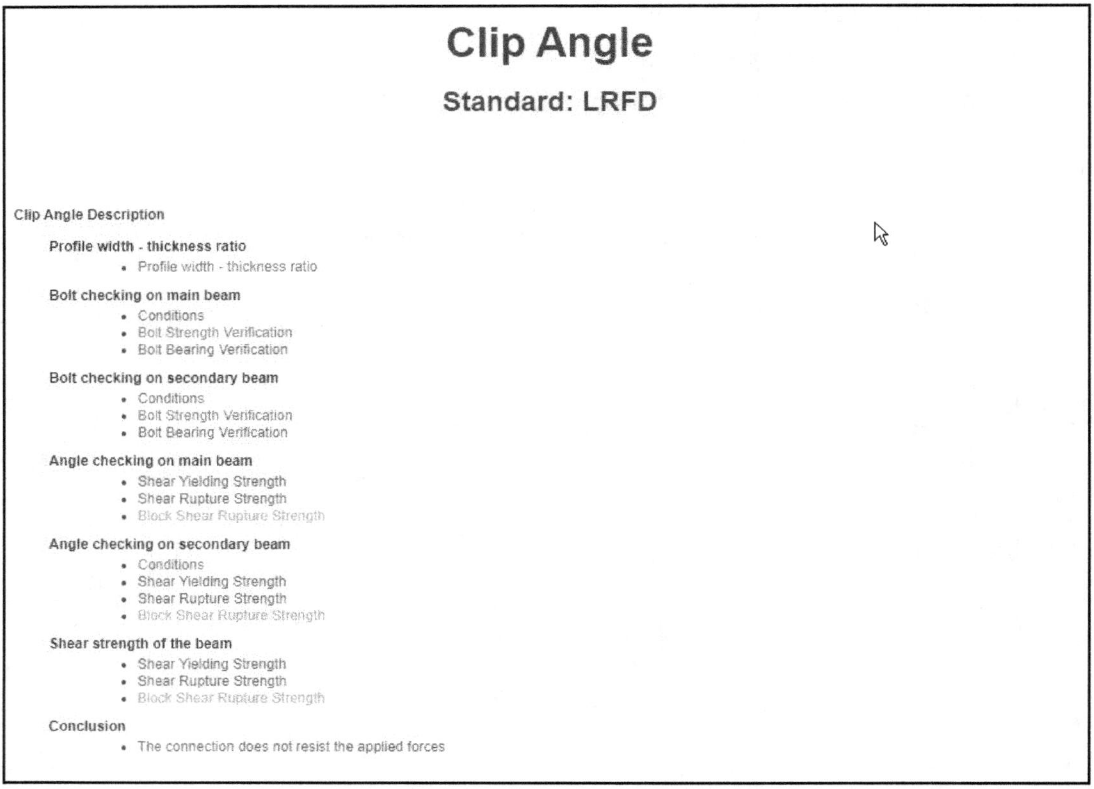

Clip Angle

Standard: LRFD

Clip Angle Description

Profile width - thickness ratio
- Profile width - thickness ratio

Bolt checking on main beam
- Conditions
- Bolt Strength Verification
- Bolt Bearing Verification

Bolt checking on secondary beam
- Conditions
- Bolt Strength Verification
- Bolt Bearing Verification

Angle checking on main beam
- Shear Yielding Strength
- Shear Rupture Strength
- Block Shear Rupture Strength

Angle checking on secondary beam
- Conditions
- Shear Yielding Strength
- Shear Rupture Strength
- Block Shear Rupture Strength

Shear strength of the beam
- Shear Yielding Strength
- Shear Rupture Strength
- Block Shear Rupture Strength

Conclusion
- The connection does not resist the applied forces

Figure 5-28

5. In the *Clip angle* dialog box>*Vertical Bolts* tab, change the **Group 1** number to **3,** as shown in Figure 5–29.

Figure 5–29

6. In the *Joint design* tab, within the *Design Options* section, click **Check.** Notice the *Status* box shows **No failed verifications**, as shown in Figure 5–30.

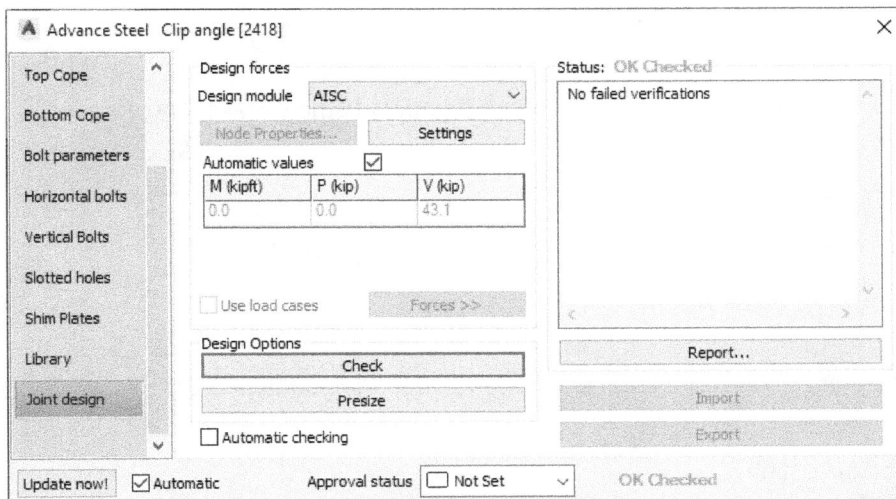

Figure 5–30

7. Save and close the drawing.

End of practice

Chapter Review Questions

1. Which of the following issues does Clash Check find?
   a. Features with the same name
   b. Hidden beams
   c. Bolt patterns
   d. Collisions

2. The correction of one clash may fix several other clashes.
   a. True
   b. False

3. What information does the *Steel check* dialog box show you?
   a. Drawing name and how to fix the issue found.
   b. Id number, parent part type, and description.
   c. Handle Id and drawing name.
   d. Description of the problem and where to find the issue.

4. Which of the following allows you to find members using the Mark object tool?
   a. Id
   b. Part number
   c. Handle/Id
   d. Part name

5. What type of model verification check do you use for design issues?
   a. Mark object
   b. Technical check
   c. Drawing check
   d. Clash check

Command Summary

Button	Command	Location
	Clash Check	• **Ribbon:** *Home* tab>*Checking* panel
	Display Clash Checking Results	• **Ribbon:** *Home* tab>*Checking* panel
	Mark Object	• **Advance Tool Palette**>*Selection* category
	Model Check	• **Ribbon:** *Home* tab>*Checking* panel
	Selection	• **Advance Tool Palette**
	Steel Check	• **Steel check dialog box:** Bottom right
	Steel Construction Technical Checking	• **Ribbon:** *Home* tab>*Checking* panel

Creating Fabrication Drawings

Once you have created a 3D model in the Autodesk® Advance Steel software, you can use the model to create 2D shop drawings. The software includes macros that automatically processes the 3D model to create 2D views, complete with dimensions and tags. These drawings are opened through the *Document Manager* and update as changes are made to the model. You can modify the detail drawings by adding callout views and cut views. You can also add and modify parametric dimensions that update if the model is changed.

Learning Objectives

- Add numbering to steel objects.
- Understand the process of turning a 3D model into 2D drawings.
- Run drawing styles and processes.
- Work with the *Document Manager*.
- Add callout views and cut views.
- Add and modify parametric dimensions.
- Add revisions.

6.1 Numbering Objects

To create the documents that are required in order to fabricate the objects in the model, you first need to number the objects. Identical parts are numbered with the same mark.

There are two types of marks for each object: single part number, and assembly (or main) part number, as shown for a column in Figure 6–1. To specify how you want the parts to be numbered, set the behavior of the parts in the Management Tools, and then run the **Numbering** command.

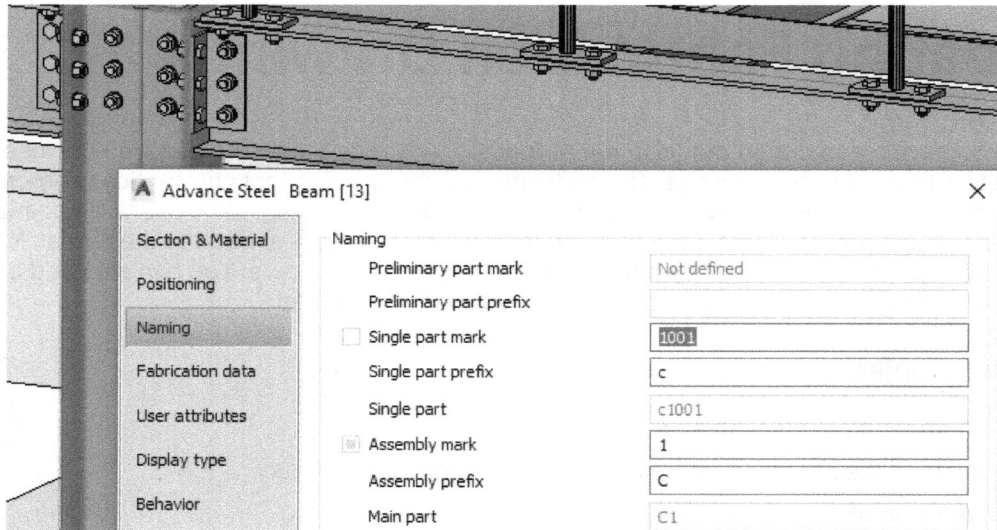

Figure 6–1

- Single parts have no other parts welded or bolted to them in the workshop.

- Main parts have connections that are welded or bolted in the workshop. These parts are considered an assembly. The heaviest part of the assembly is considered the main part.

- Like all of the documentation tools in the Autodesk Advance Steel software, the **Numbering** command has numerous ways to customize it to suit your requirements. These tools are found on the *Output* tab>*Part marks* panel, as shown in Figure 6–2.

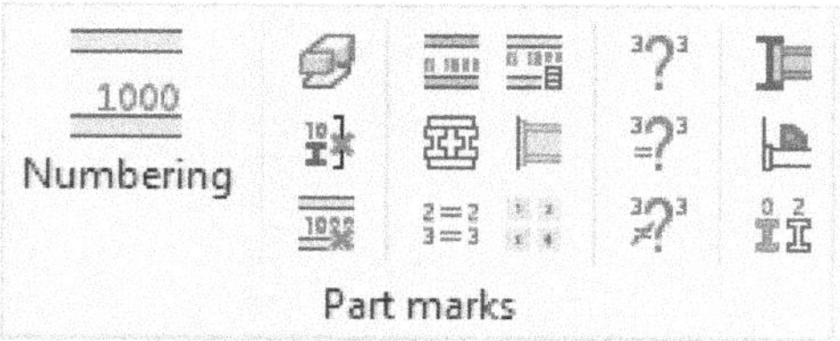

Figure 6–2

How To: Set Up a Project for Shop Drawings

1. In the *Home* tab>*Settings* panel, click (Management Tools).

2. In the *Advance Steel Management Tools* dialog box, click **Defaults**.

3. In the *Defaults* tab, expand *Drawing-General>General*. Note that the default setting for *Behavior for Main Part beams on shop drawings* and similar settings is **Create only MP detail**, as shown in Figure 6−3.

   - In this dialog box, **MP** refers to Main Parts, and **SP** refers to Single Parts.

Figure 6−3

4. If you want to create drawings that include both Main Parts and Single Parts, change each of the *Drawing-General>General* options to **Create both MP and SP details**.

5. In the upper left corner of the dialog box, click (Load Settings in Advance Steel).

6. Close the dialog box.

7. In the *Home* tab>*Settings* panel, click (Update defaults).

8. Save the drawing.

How To: Number Parts and Assemblies

1. In the *Home* tab>*Documents* panel, click (Numbering).

2. In the *Numbering* dialog box>*General* tab, select **Process single parts** and **Process assemblies**.

3. Set the *Start* and *Increment* numbers as required.

4. Change the *Method*, if required, as shown in Figure 6−4.

Figure 6−4

5. Click **OK**. A warning displays that many objects are selected. Click **Yes** to start the command.

6. The *Numbering* palette displays with the Main Parts (MP) listed, as shown in Figure 6–5.

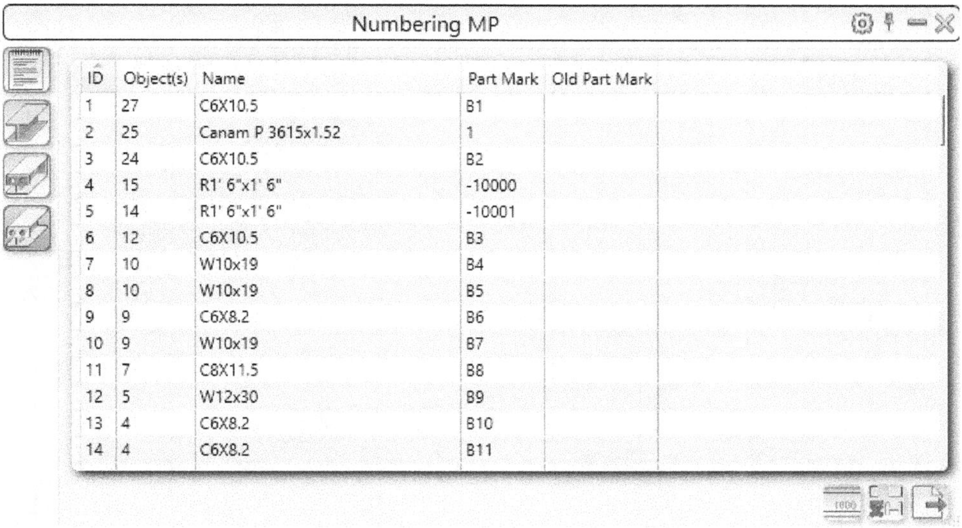

ID	Object(s)	Name	Part Mark	Old Part Mark
1	27	C6X10.5	B1	
2	25	Canam P 3615x1.52	1	
3	24	C6X10.5	B2	
4	15	R1' 6"x1' 6"	-10000	
5	14	R1' 6"x1' 6"	-10001	
6	12	C6X10.5	B3	
7	10	W10x19	B4	
8	10	W10x19	B5	
9	9	C6X8.2	B6	
10	9	W10x19	B7	
11	7	C8X11.5	B8	
12	5	W12x30	B9	
13	4	C6X8.2	B10	
14	4	C6X8.2	B11	

Numbering MP

Figure 6–5

- You can sort the information in the *Numbering* palette by clicking on the column headers.
- Click on the icons to the side to display the SP (Single Part) and PP (Process Preliminary mark) lists.
- Default Part Marks have the following designations:

Mark Prefix	Designation
- (minus symbol)	Concrete objects
(none)	Grating
B	Beams
C	Columns
D	Bracing
M	Miscellaneous (parts of railing and ladders)
P	Plates
S	Stair assemblies

- If you want to display the *Numbering* palette again, rerun the **Numbering** command.
- If you add additional objects to the model that require numbering, or if you want to change the numbering sequences, rerun the **Numbering** command.

- For another view of numbering, in the *Advance Tool Palette>* ☐ (Selection) category, click
  ☐ (Modify Browser). In the Model Browser, you can sort the list by main part, single part, and preliminary part, as shown in Figure 6-6.

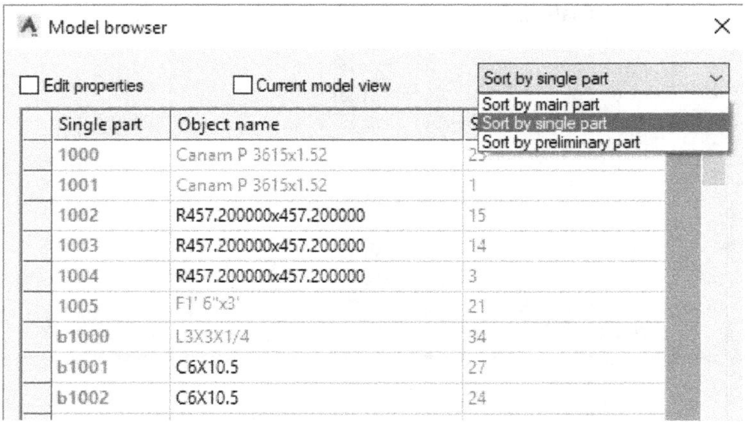

	Single part	Object name	S...
	1000	Canam P 3615x1.52	25
	1001	Canam P 3615x1.52	1
	1002	R457.200000x457.200000	15
	1003	R457.200000x457.200000	14
	1004	R457.200000x457.200000	3
	1005	F1' 6"x3'	21
	b1000	L3X3X1/4	34
	b1001	C6X10.5	27
	b1002	C6X10.5	24

Figure 6-6

- You can control if you want to use holes to drive part numbering. Select a hole, then right-click to open the *Advance Steel Hole pattern* dialog box. In the *Hole definition* tab, select or deselect **Use hole(s) for numbering** accordingly, as shown in Figure 6-7.

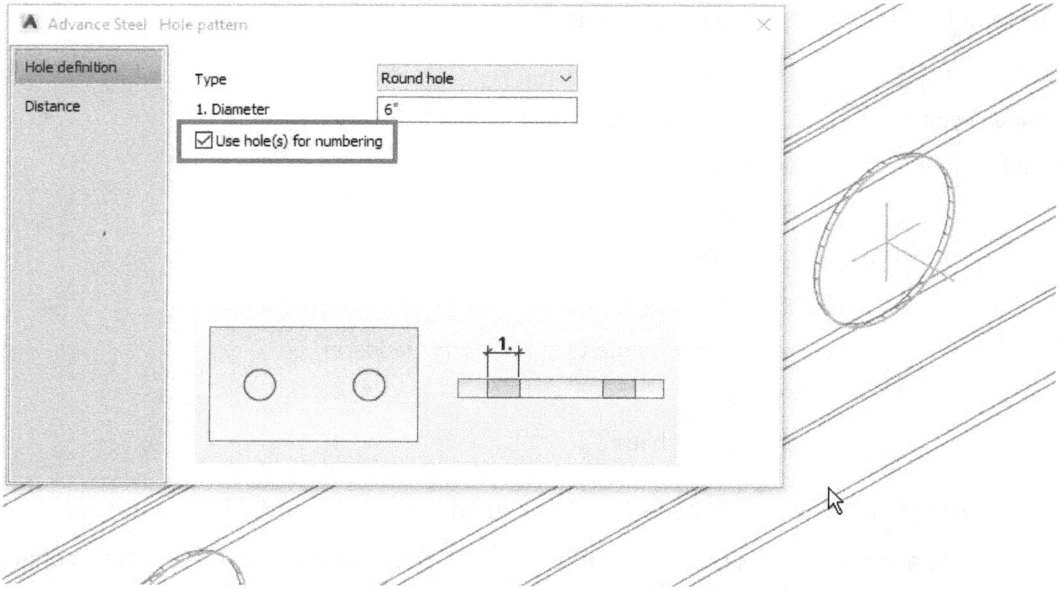

Figure 6-7

- You can also use beams to control if you want to use holes to drive part numbering. Right-click on a beam to open the *Advance Steel Beam* dialog box. In the *Behavior* tab, select or deselect **Use hole(s) for numbering** accordingly, as shown in Figure 6–8.

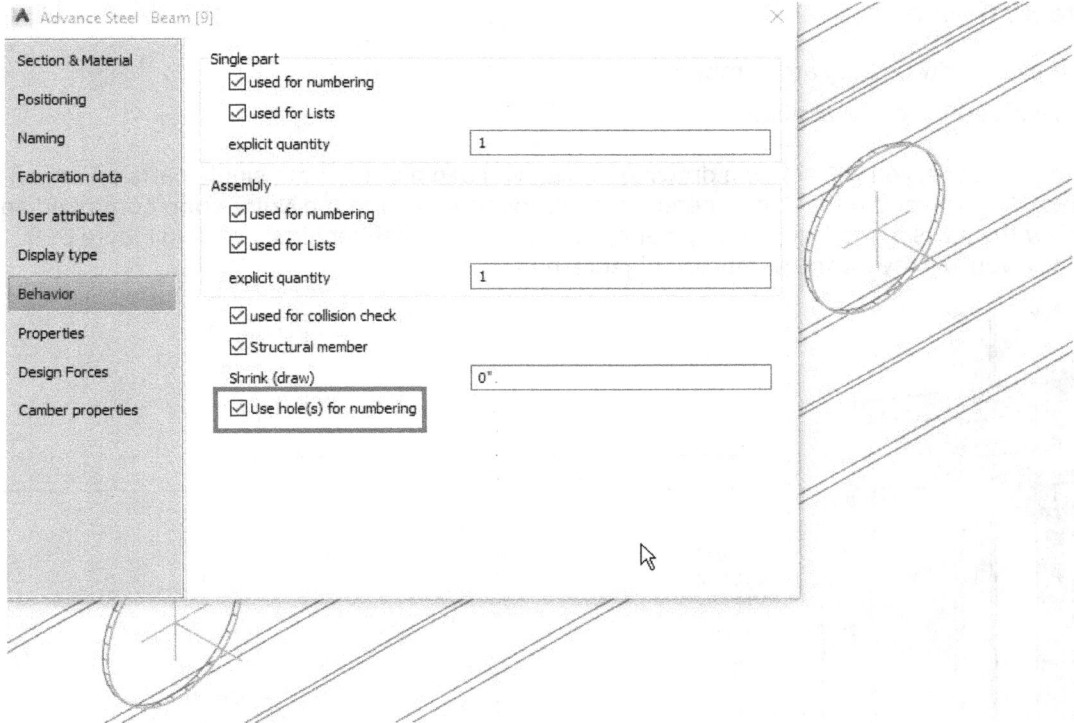

Figure 6–8

Practice 6a
Number Objects

Practice Objectives

* Review how objects are numbered.
* Number objects in a model.

In this practice, you will set up a drawing so that you can number both single parts and main parts. You will review the unnumbered objects, you will then run the **Numbering** command and review the parts in the *Numbering* palette, as shown with a ladder highlighted in Figure 6–9. Finally, you will review the numbered objects in the model.

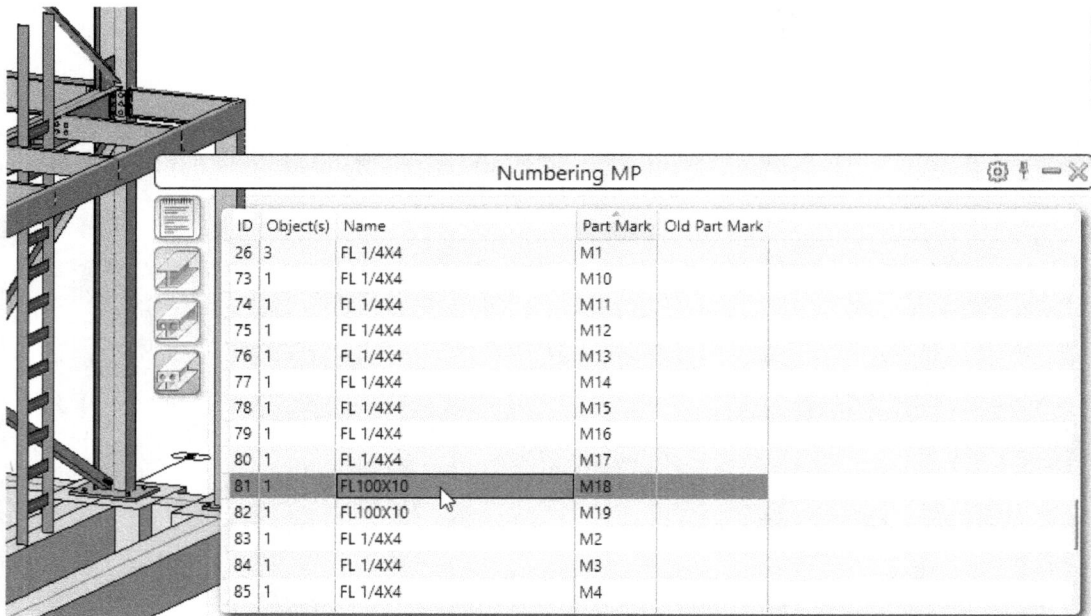

ID	Object(s)	Name	Part Mark	Old Part Mark
26	3	FL 1/4X4	M1	
73	1	FL 1/4X4	M10	
74	1	FL 1/4X4	M11	
75	1	FL 1/4X4	M12	
76	1	FL 1/4X4	M13	
77	1	FL 1/4X4	M14	
78	1	FL 1/4X4	M15	
79	1	FL 1/4X4	M16	
80	1	FL 1/4X4	M17	
81	1	FL100X10	M18	
82	1	FL100X10	M19	
83	1	FL 1/4X4	M2	
84	1	FL 1/4X4	M3	
85	1	FL 1/4X4	M4	

Figure 6–9

Task 1: Set up the drawing.

1. In the practice file folder, open **Platform-Numbering.dwg**.

2. Double-click on Column A2 to open the *Advance Properties* dialog box.

3. Open the *Naming* tab. Note that the *Single part mark* is listed as **Not defined** (as shown in Figure 6–10), and that it is not set as a main part.

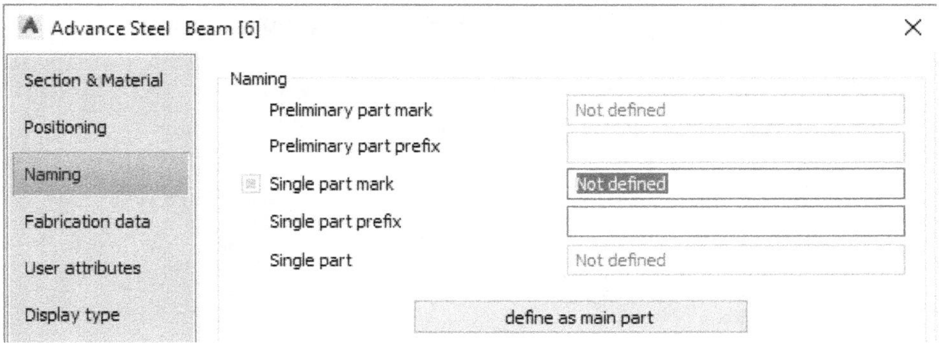

Figure 6–10

4. Zoom in on the stairs.

5. Select the left stringer and open the **Advance Properties**. In the *Naming* tab, note that this part has not been defined.

6. In the *Home* tab>*Settings* panel, click ⊛ (Management Tools).

7. In the *Advance Steel Management Tools* dialog box, click **Defaults**.

8. In the *Defaults* tab, expand *Drawing-General>General*. Change the **Property Values** for the Behavior for Main Parts and Standalone Parts to **Create both MP and SP details**, as shown in Figure 6–11.

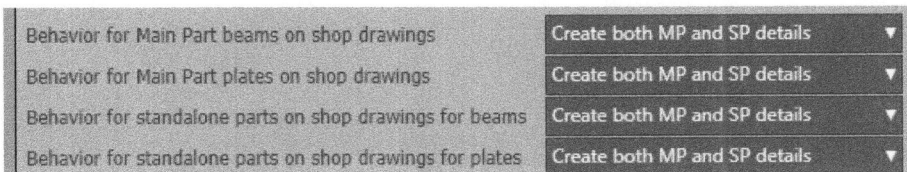

Figure 6–11

9. In the upper right corner of the dialog box, click 🔲 (Load Settings in Advance Steel).

10. Close the dialog box.

11. In the *Home* tab>*Settings* panel, click 🗔 (Update defaults).

12. Save the drawing.

Task 2: Number and review the parts and assemblies.

1. In the *Home* tab>*Documents* panel, click ⬚ (Numbering).

2. In the *Numbering* dialog box>*General* tab, select both **Process single parts** and **Process assemblies**.

3. Leave the *Start* and *Increment* numbers to their defaults. In the *Single Part* and *Assembly* areas, change Method to **SP: 1000, 1001...;MP: 1,2,3**, as shown in Figure 6–12.

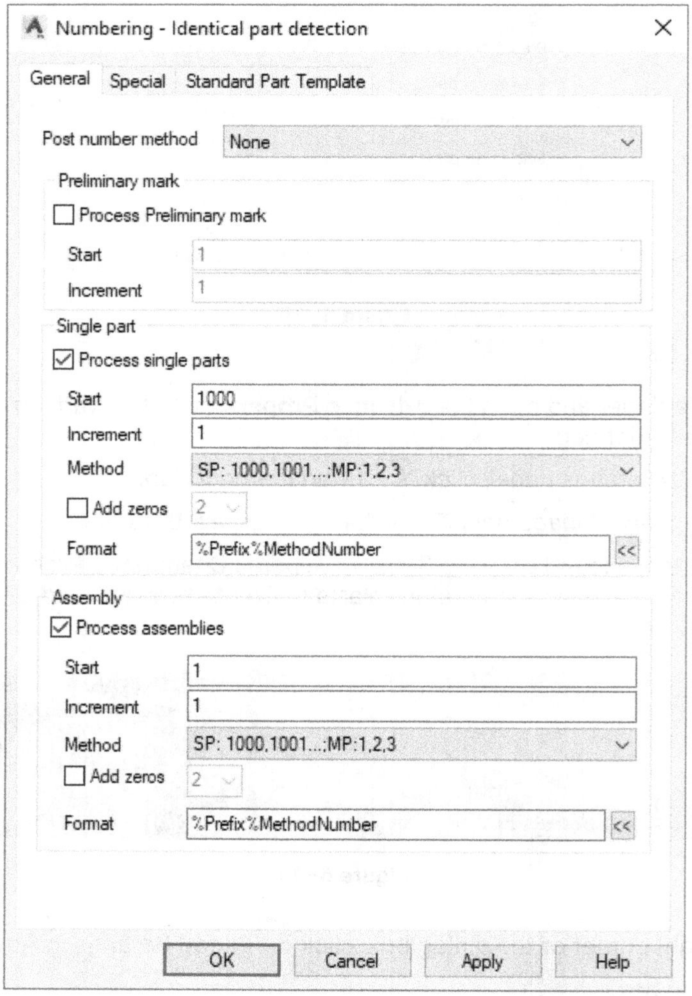

Figure 6–12

4. Click **OK**.

5. A warning about many objects being selected might display. Note that it might take time to process all of the parts.

6. When its finished, select **Yes**.

7. When the processing is complete, the Numbering tool palette and the Main Part (MP) numbers display, as shown in Figure 6–13.

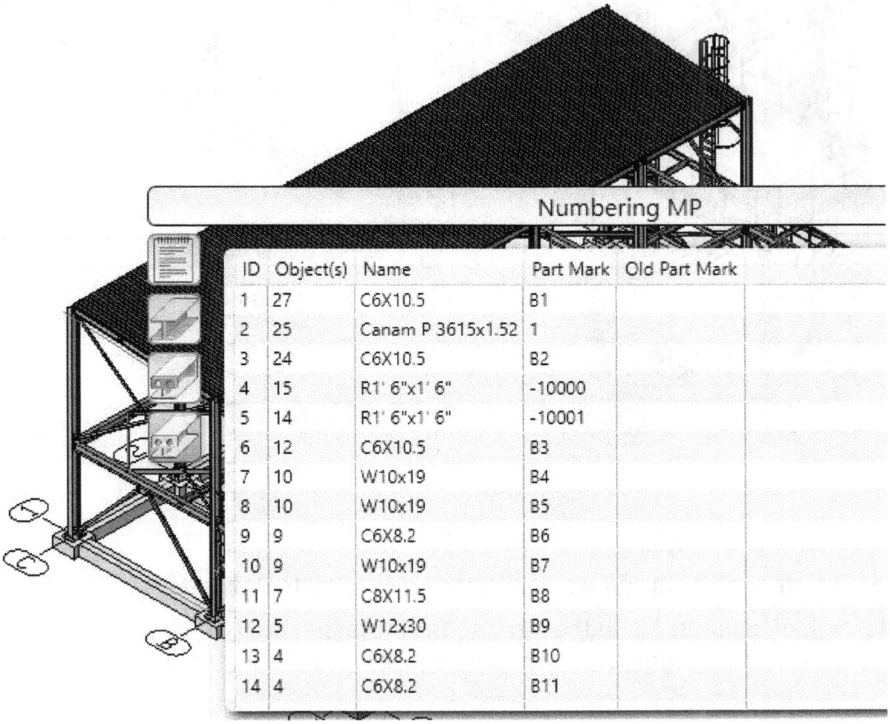

Numbering MP

ID	Object(s)	Name	Part Mark	Old Part Mark
1	27	C6X10.5	B1	
2	25	Canam P 3615x1.52	1	
3	24	C6X10.5	B2	
4	15	R1' 6"x1' 6"	-10000	
5	14	R1' 6"x1' 6"	-10001	
6	12	C6X10.5	B3	
7	10	W10x19	B4	
8	10	W10x19	B5	
9	9	C6X8.2	B6	
10	9	W10x19	B7	
11	7	C8X11.5	B8	
12	5	W12x30	B9	
13	4	C6X8.2	B10	
14	4	C6X8.2	B11	

Figure 6–13

8. Click on the *Part Mark* column heading to sort the list.

9. Scroll down to the columns and double-click on each column set to review which assemblies are selected. For example, column C1 includes the four columns on A 2-5 (as shown in Figure 6−14), while C2 includes the four columns on C 2-5.

Figure 6−14

10. Review any other objects you want, and then close the tool palette.

11. Double-click on column A2. In the *Naming* tab, note that the part now has both a *Single part mark* and an *Assembly mark*, as shown in Figure 6−15.

Figure 6−15

12. Double-click on the left stringer of the stair and review the *Naming*. Note that it is considered a single part, but that it is also part of an Assembly.

13. Double-click on the left stringer of the stair and review the *Naming*. Note that *Assembly mark* is selected (as shown in Figure 6–16) and that this is considered the Main part of the assembly.

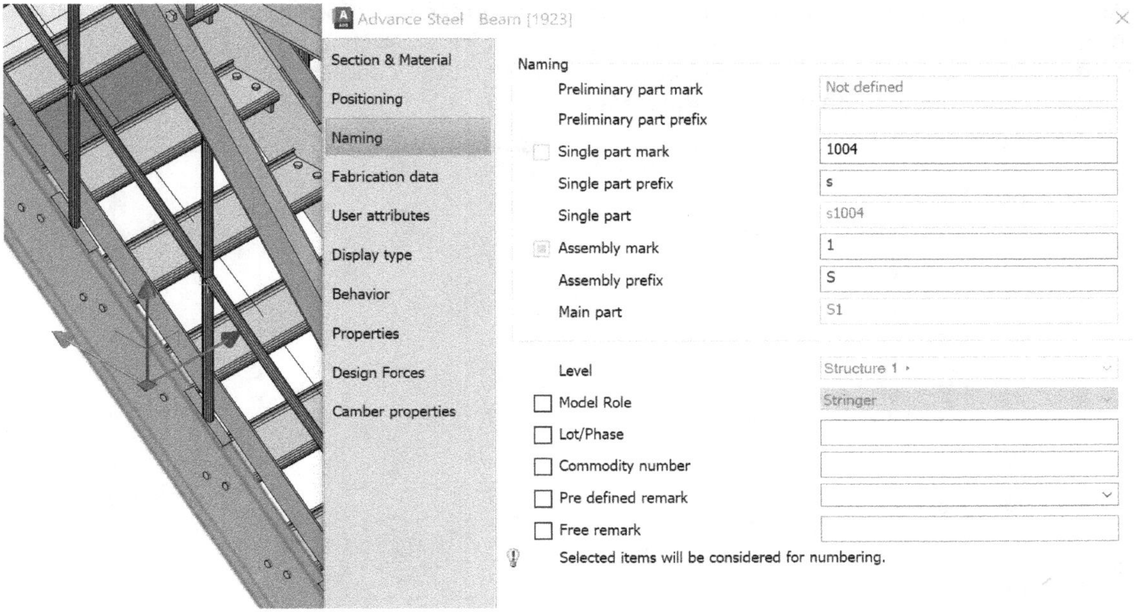

Figure 6–16

14. If you have time, check other object names.

15. Save the project.

End of practice

6.2 Tools for Creating Drawings

Instead of creating 2D fabrication drawings from scratch, the drawings in the Autodesk Advance Steel software are created using macros that take the information stored in the 3D model, and create a layout with a title block, as shown for an anchor plan in Figure 6–17. Once created, these separate drawings are accessed through the *Document Manager*, are linked to the 3D model, and update when the model is modified.

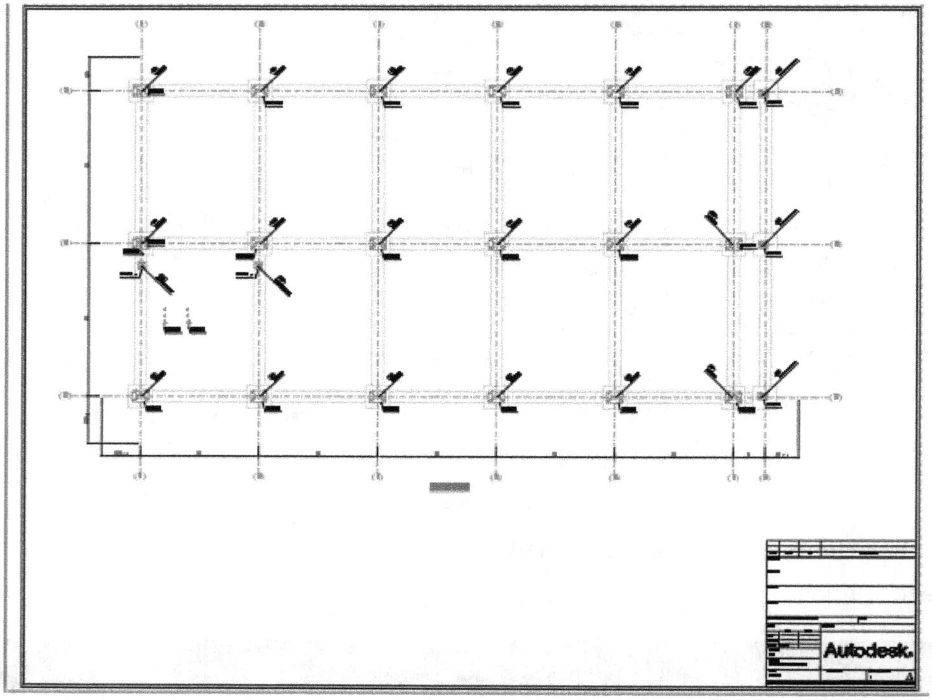

Figure 6–17

- The size of the selected title block (also called a prototype) controls the size of the sheet.

- Many annotations are automatically added through the macros, but you can also make changes and add annotations, symbols, and dimensions.

Quick Document Palettes

The macros that you use to produce the fabrication drawings are accessed using three palettes. They are found on the *Home* tab>*Documents* panel or *Output* tab>*Documents* panel, as shown in Figure 6–18.

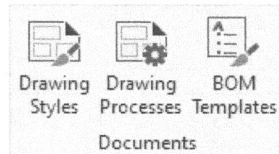

Figure 6–18

- **Drawing Styles:** Create individual 3D isometrics and 2D plans (such as the anchor plan in Figure 6–17), elevations, and details for single parts, assemblies, ladders, stairs, and railings. Each view is created separately, and the UCS direction impacts the view.

- **Drawing Processes:** Create part drawings in multiple drawing styles, as well as Bills of Materials (BOMs) (as shown in Figure 6–19) for all or selected single parts or assemblies.

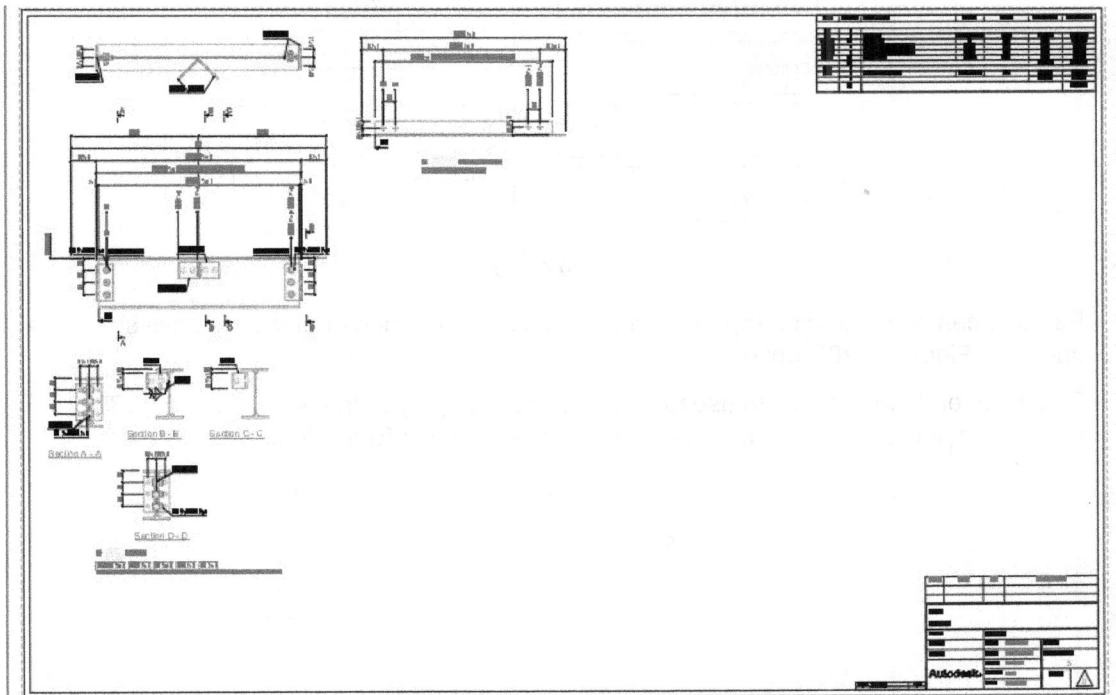

Figure 6–19

- **BOM Templates:** Create separate drawings with BOMs for Assemblies (as shown in Figure 6–20), Drawings, Fasteners, and Parts.

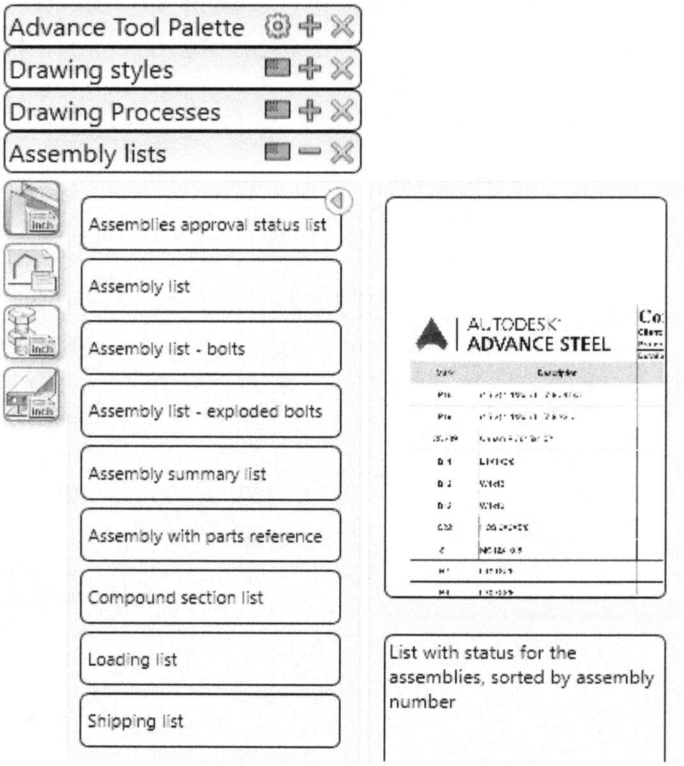

Figure 6–20

- Palettes can be docked on top of each other and then opened or closed to save space, as shown in Figure 6–20, above.

- Examples and tips on how to use the selected tool display to the right of the tool. The arrowhead at the top of the tool list opens or closes this information.

Using the Document Manager

To access drawings once they are created, you need to open them through the *Document Manager*, as shown in Figure 6–21. In the *Home* tab>*Documents* panel or *Output* tab>*Document Manager* panel, click 🖳 (Document Manager).

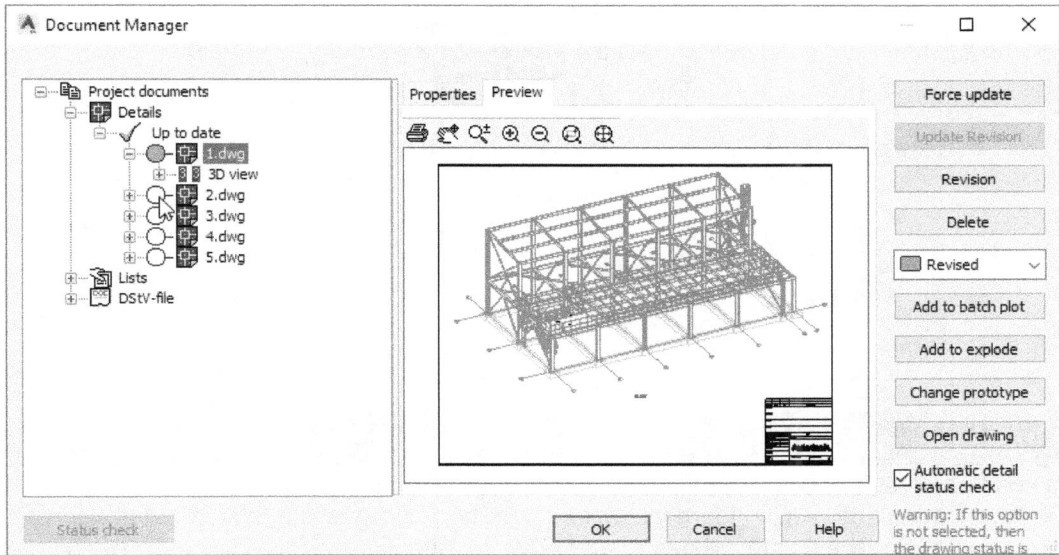

Figure 6–21

- The **Add to explode** button will explode drawings. They are then copied to a new branch, named *batch explodes*.

- Each of the drawings that are connected with the model are listed on the left in the *Details* node. Click the plus sign to expand the list and the minus sign to close each level of the node.

- The *Preview* tab displays a selected drawing without opening it.

- To open a drawing, double-click on the drawing name or select the drawing name and click **Open drawing**.

- You can delete a drawing if you no longer need it.

- When you make a change to a model, the detail drawings need to be updated, as shown in Figure 6–22. If you do not want to record the change, select the *Update required* node and click **Force Update**.

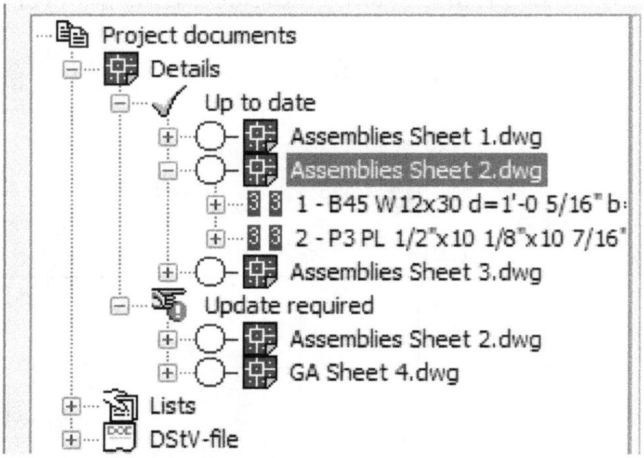

Figure 6–22

Hint: Opening Detail Drawings from the Model

You can open related detail drawings directly from the model. Select a part of the model (such as the stair shown in Figure 6–23) and right-click. If there is a related detail view created, click **Show assembly detail** to open the drawing.

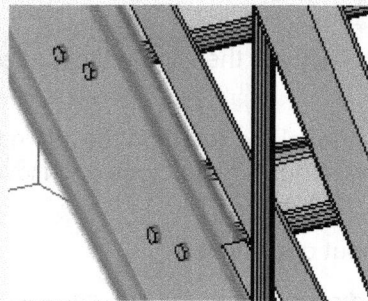

Figure 6–23

Working with Documentation Drawings

When you open a drawing, note that the views created by the drawing style or process are automatically placed in Paper Space on a layout using the title block size that is specified in the macro, as shown in Figure 6–24.

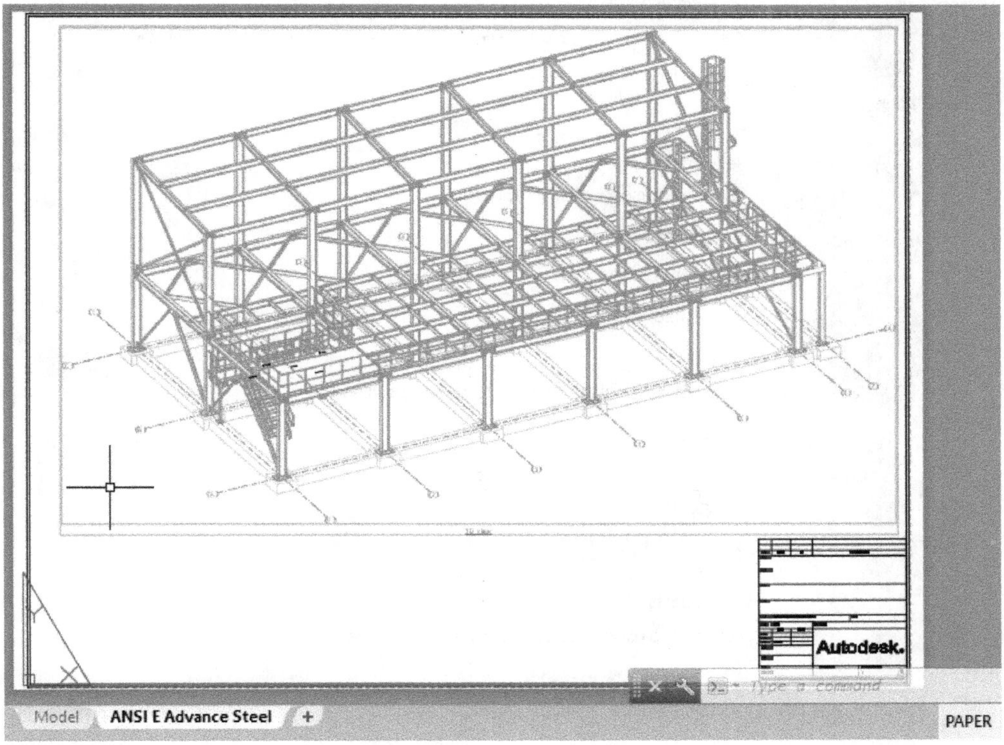

Figure 6–24

- When you hover the cursor over a view, two or more green boxes display. The outer box is the detail frame, which might surround one or more inner boxes that are view frames.

- Double-click on the inner green box to open the *View* dialog box (shown in Figure 6–25), where you can change the scale or other view-related features.

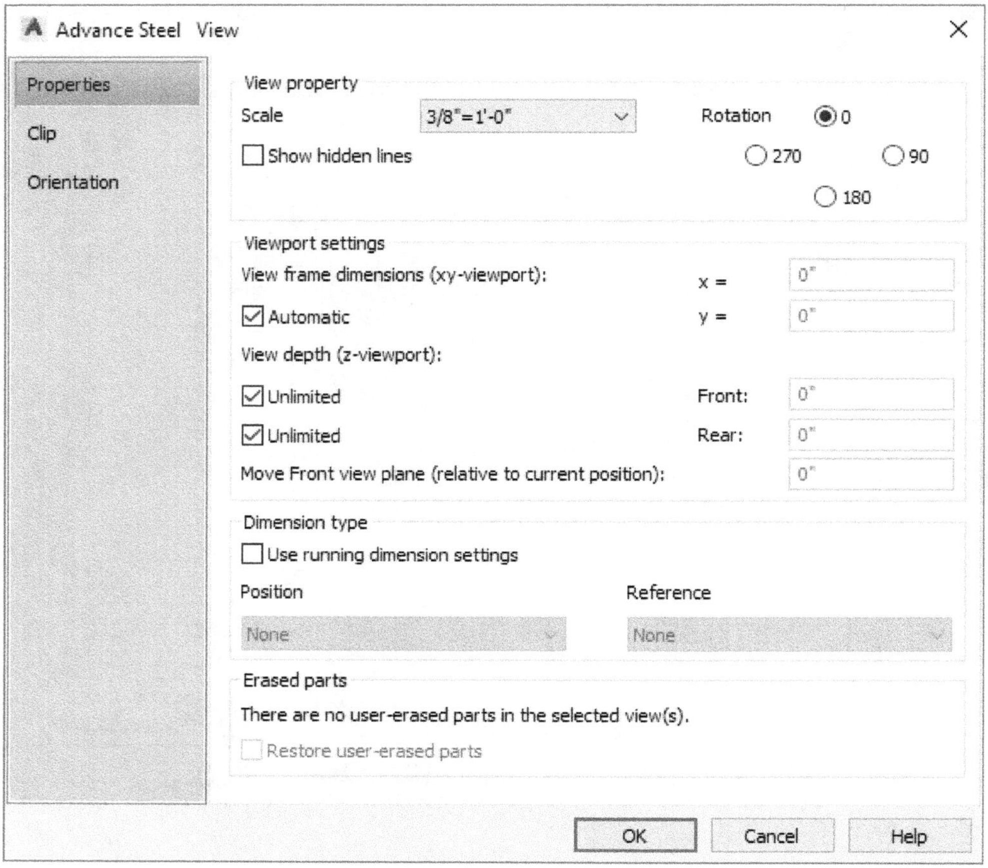

Figure 6–25

- Double-click on the outer green box to open the *Edit Detail* dialog box, shown in Figure 6–26. This enables you to modify the information for all of the inner views. In this example, five different views are connected together in an assembly drawing.

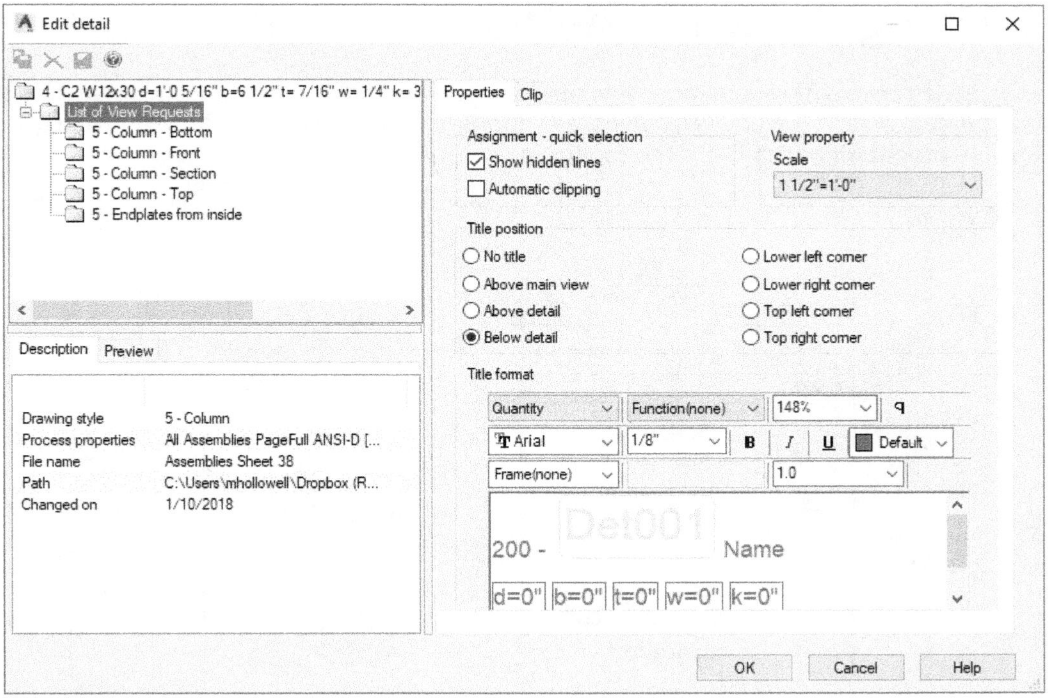

Figure 6-26

- You can drag and drop views to move them on the layout as required.

- Annotation elements in a view can be moved with grips. For example, when a label overlaps another label or lines, you can move it using the left grip, as shown in Figure 6-27. To rotate a label, use the right grip.

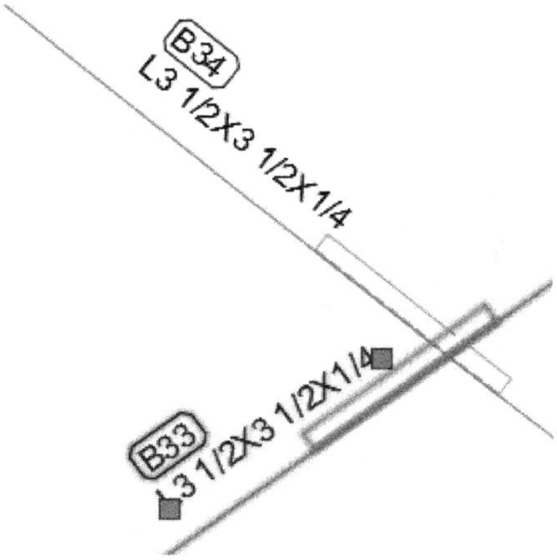

Figure 6-27

6.3 Using Drawing Styles

Drawing Styles create individual drawings for specified parts of the model. The categories and tools in the tool palette include 3D, details, elevations, and plan views, as shown in Figure 6–28. Review the process information in the tooltip.

> **Note:** *Take time to review all of the different categories and macros you can access in this tool palette.*

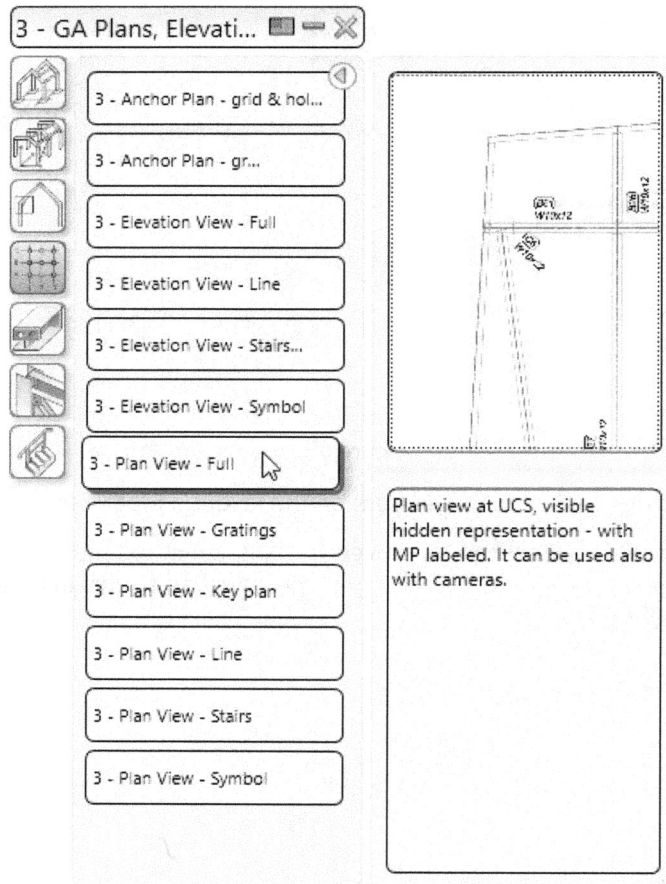

3 - GA Plans, Elevati...

- 3 - Anchor Plan - grid & hol...
- 3 - Anchor Plan - gr...
- 3 - Elevation View - Full
- 3 - Elevation View - Line
- 3 - Elevation View - Stairs...
- 3 - Elevation View - Symbol
- 3 - Plan View - Full
- 3 - Plan View - Gratings
- 3 - Plan View - Key plan
- 3 - Plan View - Line
- 3 - Plan View - Stairs
- 3 - Plan View - Symbol

Plan view at UCS, visible hidden representation - with MP labeled. It can be used also with cameras.

Figure 6–28

- It is important to set the UCS before you run any of the drawing styles.

  - For 3D views, arrange the view the way you want it, and then in the *Advance Tool Palette>* [icon] (UCS) category, click [icon] (UCS View). The text will be placed on the face of the view, rather than the model.

  - For 2D plan views, set the UCS to **World** for base plan views, or move the UCS up to specify other levels.

  - When creating elevation views, orient the UCS so that the XY-axis is on the face of the elevation.

How To: Create a Detail Drawing Using Drawing Styles

1. In the *Home* tab>*Documents* panel or *Output* tab>*Documents* panel, click [icon] (Drawing Styles) to open the *Drawing Styles* palette.

2. Set the UCS as required.

3. In the *Drawing Styles* palette, select the type of view you want to create.

4. In the *Create detail* dialog box (shown in Figure 6–29), select the method you want to use.

   - You can modify the prototype (title block size) and scale after you create the drawing.

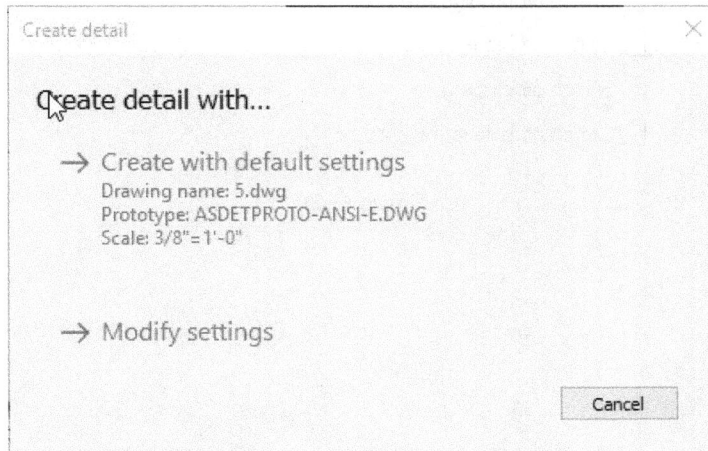

Create detail ✕

Create detail with...

→ Create with default settings
    Drawing name: 5.dwg
    Prototype: ASDETPROTO-ANSI-E.DWG
    Scale: 3/8"=1'-0"

→ Modify settings

                    Cancel

Figure 6–29

5. Follow any prompts to select objects as required.

6. The drawing is created and placed in the correct location in the folder structure.

- If you select options that include labels, the labels use the numbers specified using the **Numbering** command.

💡 Hint: Modifying Settings

When you click **Modify Settings**, the *Select destination file* dialog box displays. You can change the drawing information (such as the scale, as shown in the *New drawing* tab in Figure 6–30), the size of the viewport on the *Detail box* tab, and the title block on the *Prototype (template)* tab. If other drawings have already been created in the set, they are listed on the *Existing drawings* tab.

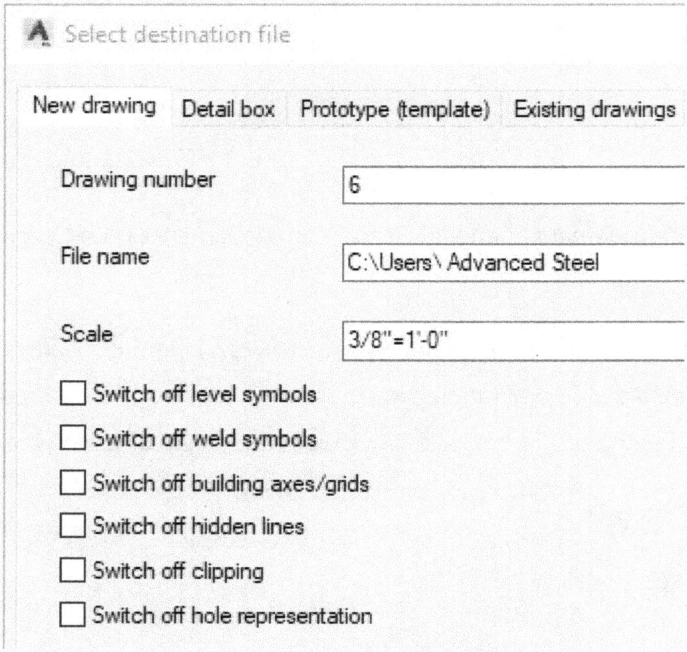

Figure 6–30

Practice 6b
Use Drawing Styles

Practice Objectives

- Use drawing styles to create 3D, plan, and elevation views and the corresponding documents.
- Modify UCS locations as required to correctly orient the views.
- Review the new drawings in the *Document Manager*.

In this practice, you will create a 3D view, two plan views (including the anchor plan shown in Figure 6–31), and an elevation view using the Drawing Styles and the Autodesk Advance Steel software defaults. You will modify the location of the UCS to create these views as required. You will also review the new drawings you have created in the *Document Manager*.

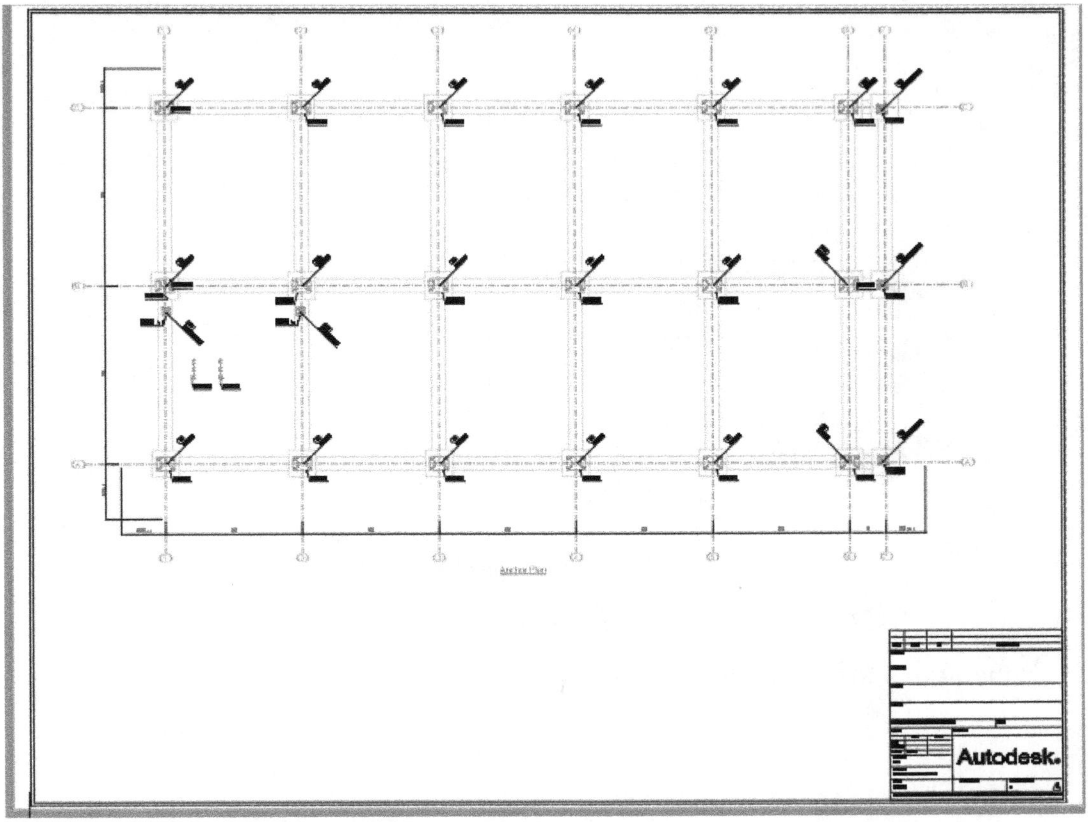

Figure 6–31

Task 1: Create a 3D view document.

1. In the practice files folder, open **Platform-Documents.dwg**.

2. In the *Output* tab>*Documents* panel, click (Drawing Styles).

3. In the *Drawing Styles* palette, click  (0 - Engineering).

4. Click on the play icon and hover the cursor over **0 - 3D View - Model - w/o Labels** to display the description and text shown in Figure 6–32. Note in the description this tool needs the UCS set to the view.

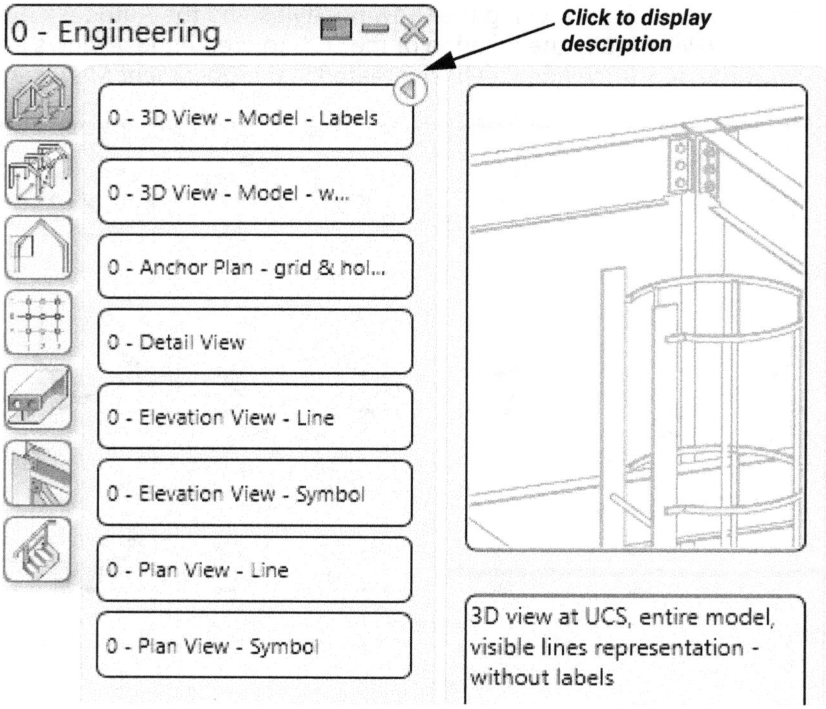

Figure 6–32

5. Set the view to **SW Isometric**.

6. In the *Advance Tool Palette>* ![UCS icon] (UCS) category, click ![UCS View icon] (UCS View). The UCS icon rotates so that the X-axis is to the right, the Y-axis is pointing up, and the Z-axis is pointing toward you, as shown in Figure 6–33.

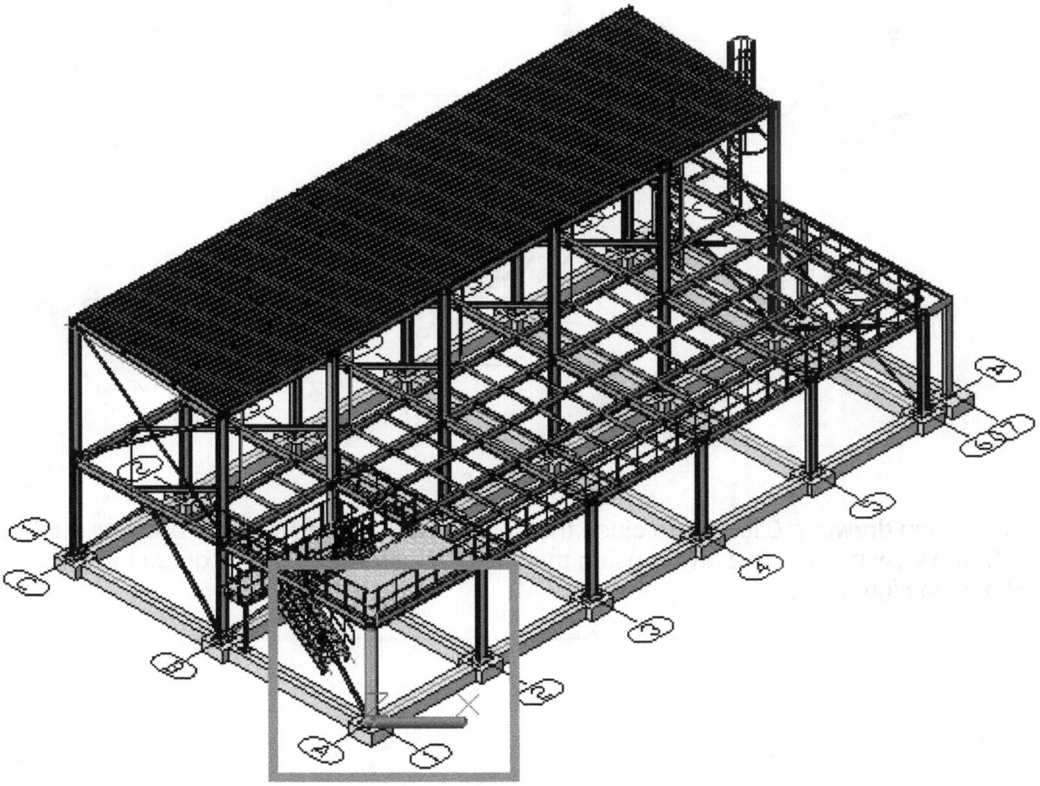

Figure 6–33

7. In the *Drawing Styles* palette, click **0 - 3D View - Model - w/o Labels.**

8. In the *Create detail* dialog box, click **Create with default settings**. The new drawing begins processing.

9. In the *Output* tab>*Document Manager* panel, click ![Document Manager icon] (Document Manager).

10. In the *Document Manager*, in the left pane, expand **Details> Up to date** and click on **1.dwg** (the new document). In the right pane, select the *Preview* tab to display the document, as shown in Figure 6–34.

Figure 6–34

11. Click **Open drawing**. Click **OK** to close the *Document Manager*.The new drawing displays with the view placed on a layout using the title block specified by the default settings, as shown in Figure 6–35.

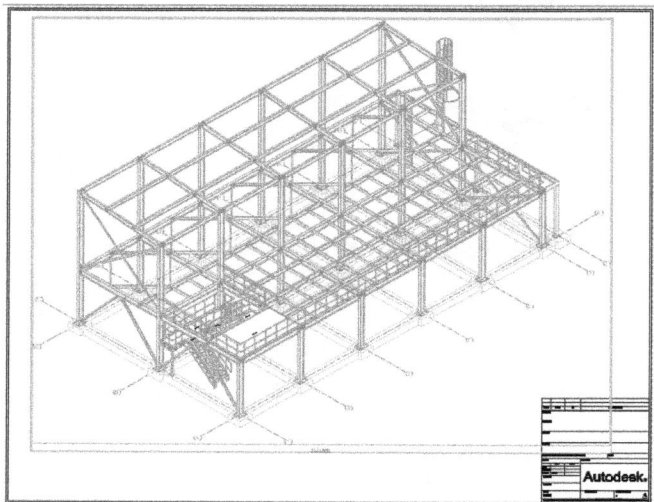

Figure 6–35

12. Review the drawing and then close it without making any changes.

Task 2: Create plan views for Level 0 and Level 1.

1. Set the UCS to **World**.

2. In the *Drawing Styles* tool palette> (3 - GA Plans, Elevations / Sections) category, click **3 - Anchor Plan - grid dimensioned**, as shown in Figure 6−36.

Figure 6−36

3. In the *Create detail* dialog box, click **Create with default settings**.

4. Draw a box around the entire model. This is what will display in the detail window. A new drawing is created that can be reviewed in the *Document Manager*.

5. Move the UCS to the top node of Column A1.

6. In the *Drawing Styles* tool palette> (3 - GA Plans, Elevations / Sections) category, click **Plan View - Full**. Select *Create with default settings* and create a window around the entire model.

7. Open the *Document Manager* and review the new drawings, as shown in Figure 6–37. If you open the drawings, do not make any changes at this time.

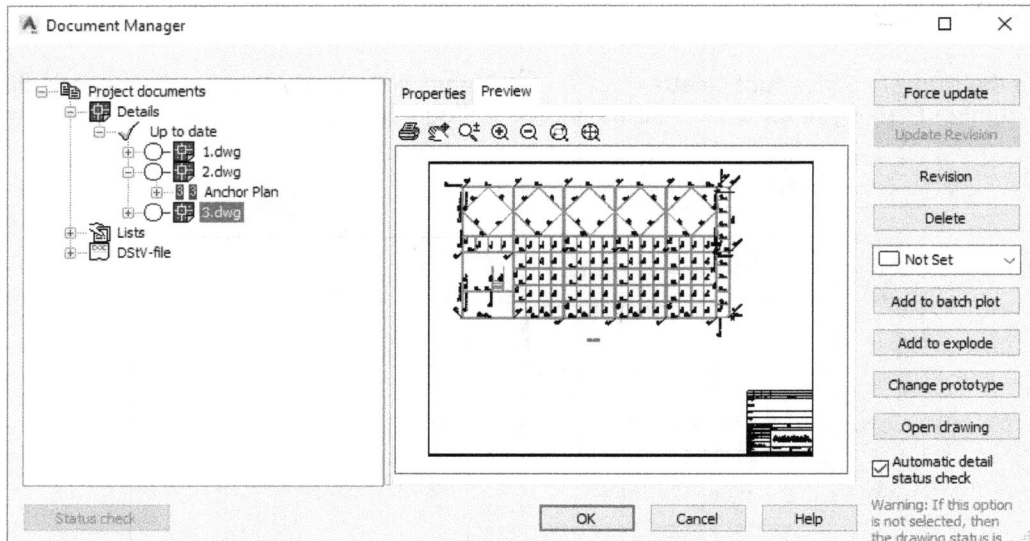

Figure 6–37

8. Close the *Document Manager*.

9. Save the drawing.

Task 3: Create an elevation view.

1. Rotate the view and set the UCS to **World** and then reorient it to the midpoint of the concrete footing at Column C1, as shown in Figure 6–38.

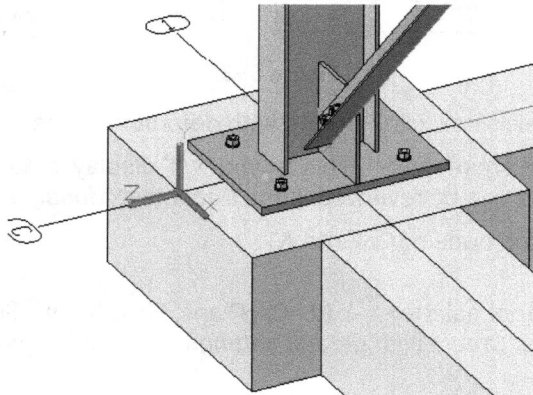

Figure 6–38

2. In the *Drawing Styles* tool palette> ▦ (3 - GA Plans, Elevations / Sections) category, click **3 - Elevation View - Symbol**.

3. Create the detail using the default settings and window the area as shown in Figure 6–39.

   Note: Toggle off object snaps as they might get in the way when you make the window selection.

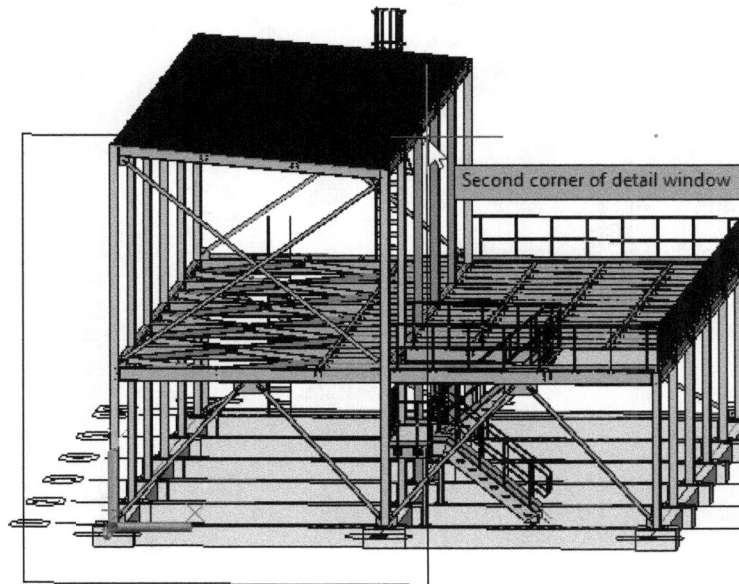

Second corner of detail window

Figure 6–39

4. Open the *Document Manager* and review **4.dwg**. Note that the file is in symbolic mode, rather than the full model view, as shown in Figure 6–40.

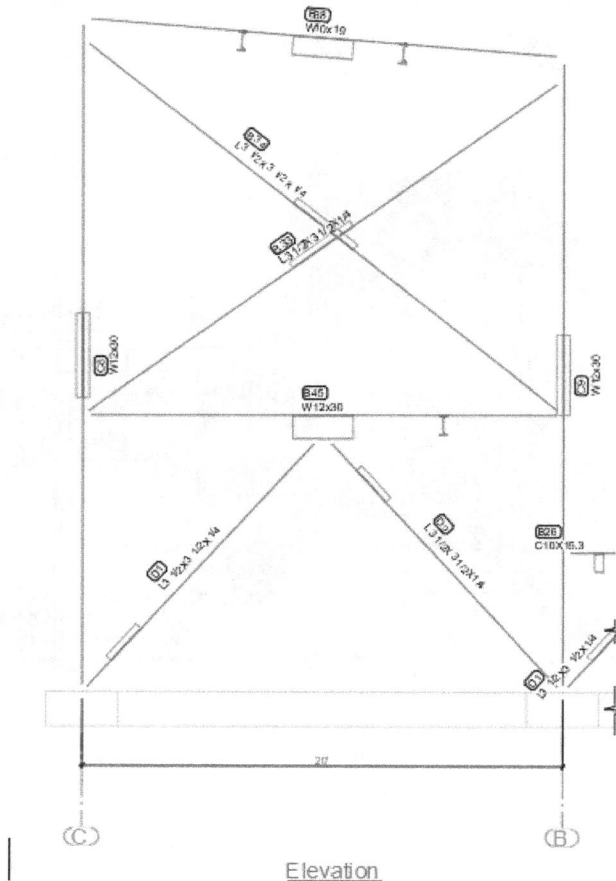

Figure 6–40

5. Set the UCS to **World** and the view to S**W Isometric**.

6. Save the main model.

7. If you have time, make other elevations and plans using the tools found in the *Drawing Styles* tool palette.

End of practice

6.4 Running Drawing Processes

Drawing Processes are rules that automatically create multiple drawing styles that are centered around camera views, parts, or assemblies and which frequently include the related bill of materials, as shown in Figure 6–41.

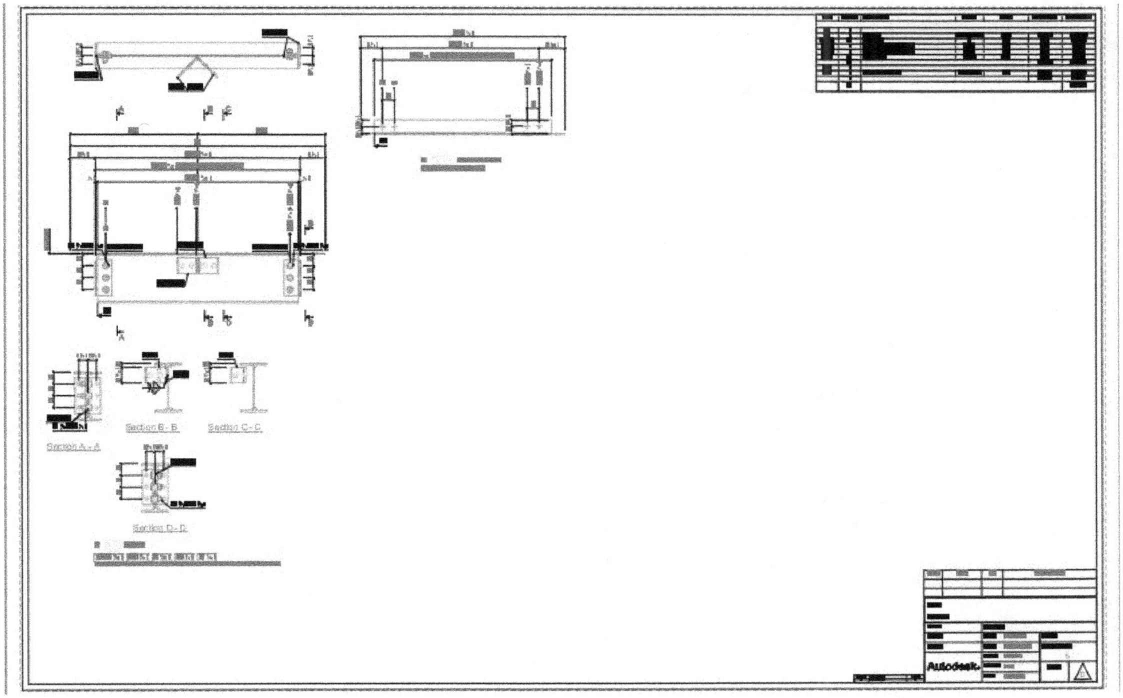

Figure 6–41

- The *Drawing Processes* tool palette is divided into three types of processes made by cameras, single parts, and assemblies.

How To: Run Processes

1. In the *Home* tab>*Documents* panel or the *Output* tab>*Document* panel, click 🖼️ (Drawing Processes) to open the *Drawing Processes* tool palette.

2. In the *Drawing Processes* tool palette, select the category and method you want to use.

3.   The *Process Properties* dialog box displays, as shown in Figure 6–42. In most cases you can use the defaults.

  * The selection options vary for each tool. You can limit the options that display using the drop-down list.

  Note: *The default drawing number varies depending on how many other drawings you have created using drawing styles or drawing processes by this point in the model.*

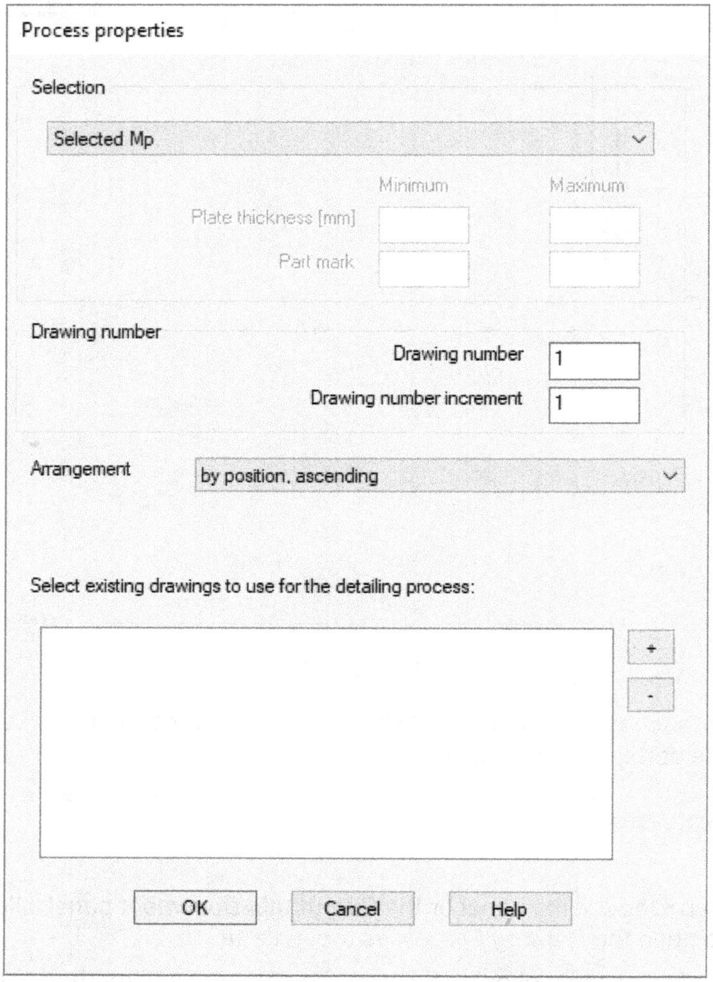

Figure 6–42

4.   Click **OK**. The drawings are processed.

5.   Open the *Document Manager* to review the new drawings.

Single Parts and Assemblies

The *Single Parts* and the *Assemblies* categories are divided into similar groupings. For example, the [icon] (Singleparts - All) category (shown in Figure 6–43) includes tools that create views of all of the single parts in a model. The [icon] (Assemblies - Selected) category (shown in Figure 6–44) requires you to first select the part of the model you want to detail before running the processes.

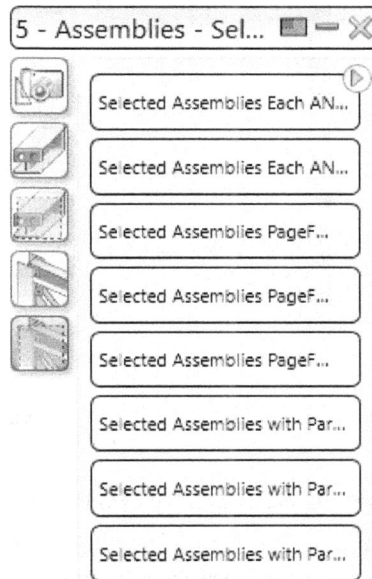

| Figure 6–43 | Figure 6–44 |

Each category of tools is sub-divided into template sizes (based on the size of the title block) and options for how many details are placed on a sheet.

* The **Each** options place one detail on one sheet.

* The **PageFull** options place as many details on one sheet as possible and then create additional sheets as required.

- For example, the **All Sp PageFull ANSI-D** tool creates details of all of the single parts in a model, and then places them on multiple ANSI-D sized sheets, as shown in the *Document Manager* in Figure 6-45.

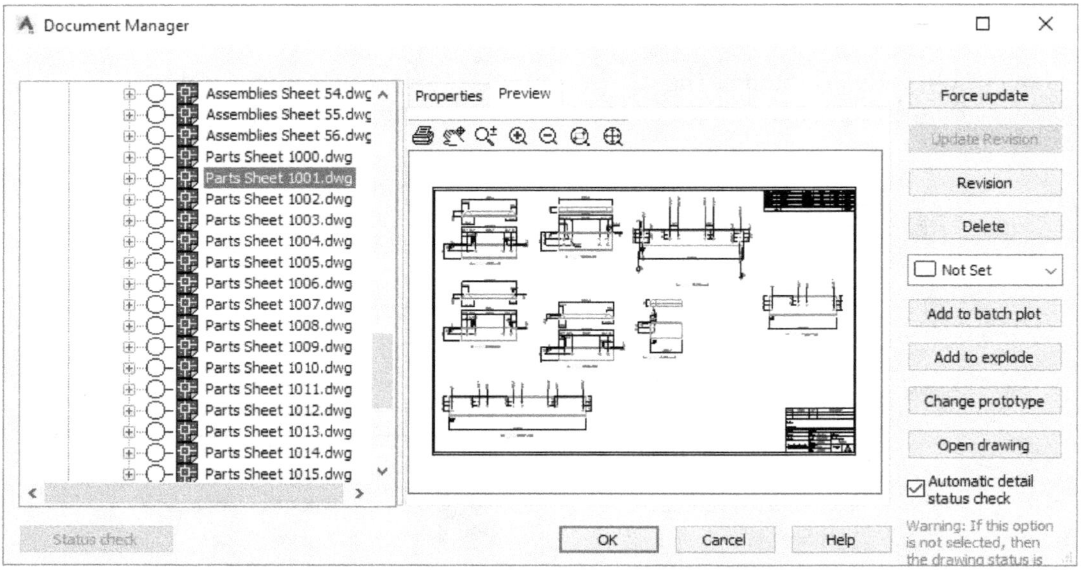

Figure 6-45

- You can use the *Advance Tool Palette*> (Selection) category> (Search filter) or use (Create new query) from the *Project Explorer* to find parts for which the single part shop drawings and assembly drawings have been created. The search can be saved in the *Project Explorer* where you can show or hide parts and assemblies. The searches can be highlighted in different colors in the 3D model.

Hint: Moving Views Between Sheets

The Drawing Processes tools typically place the appropriate number of views on a sheet, but you can move the views between sheets if required. In the *Document Manager*, expand the sheet node to display the views, as shown in Figure 6–46. Then, drag and drop one view at a time to the new sheet location.

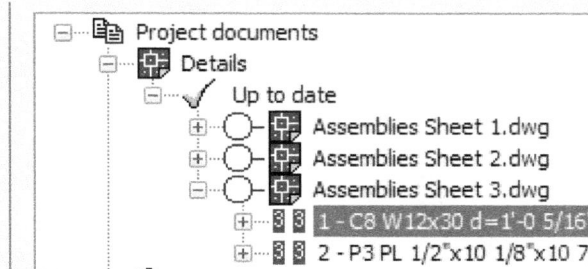

Figure 6–46

In the example shown in Figure 6–47, the two views from the original **Assemblies Sheet 3.dwg** have been moved to **Assemblies Sheet 2.dwg** and an update is required. If this is not a revision, click **Force Update**.

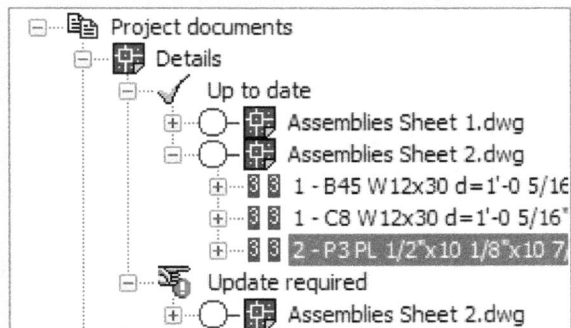

Figure 6–47

- Once the views have been reorganized, open the modified sheet and move the views as required.

Creating Camera Views

When you want to specify a certain portion of the model to detail, rather than objects such as parts and assemblies, you can create a camera view (as shown in Figure 6–48) and then run a drawing process based on the camera. There are two methods for creating camera views: **Create camera at UCS** and **Create camera(s) at node**.

How To: Create a Camera at the UCS

1. Set the UCS in the XY plane in which you want the camera.

2. In the *Advance Tool Palette>* (Tools) category, click  (Create camera, UCS).

3. Click a point where you want the camera.

4. In the *Camera* dialog box>*Detail box* tab, change *xy-Viewport* to **Fixed** and set a size for the camera box. You can also modify the *Z-Viewport* size, as shown in Figure 6–48.

   • You can use grips on the camera box to modify its size.

   • Once you have placed the camera box, use the **Move** command to move it.

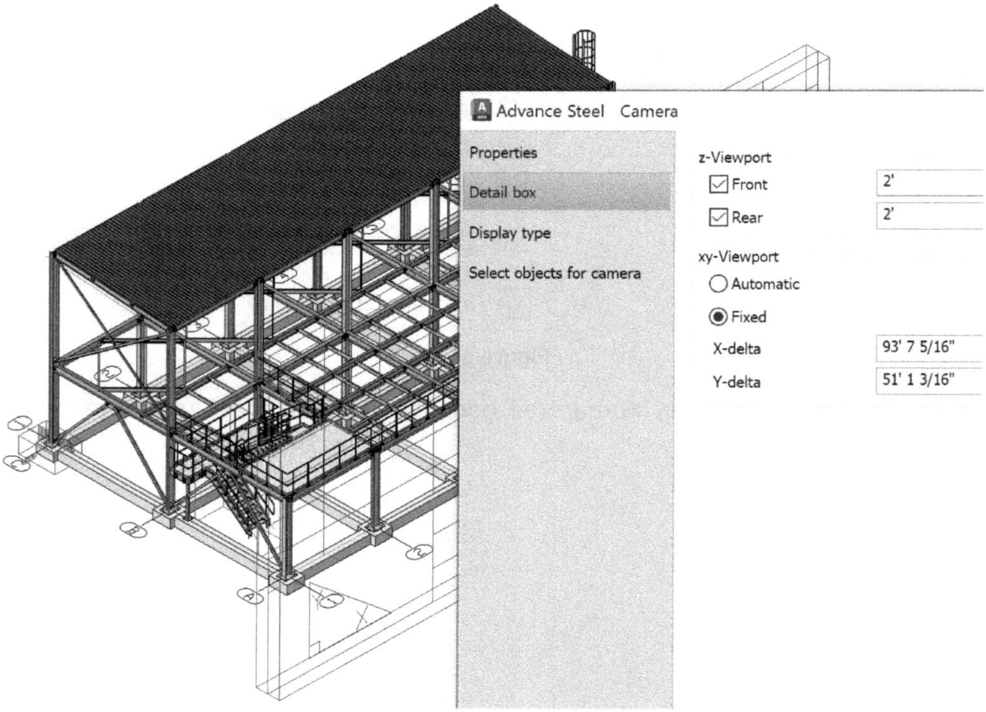

Figure 6–48

How To: Create a Camera at a Node

1. Toggle on the joint boxes where you want the cameras.

    *   **Hint:** Right-click on a part of the connection object and select **Advance Joint Properties**. Close the dialog box and the joint box displays.

    Note: The node is a joint box.

2. In the *Advance Tool Palette>* (Tools) category, click (Create camera(s) at node).
3. Select a joint box and press <Enter>.
4. Select the arrow in the direction you want to show the detail, as shown in Figure 6–49.
5. Modify the size of the camera box using grips (as shown in Figure 6–50), or double-click on the box to open the *Camera* dialog box and modify the size on the *Detail box* tab.

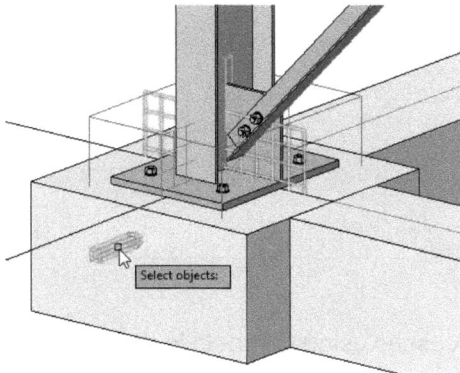

Figure 6–49

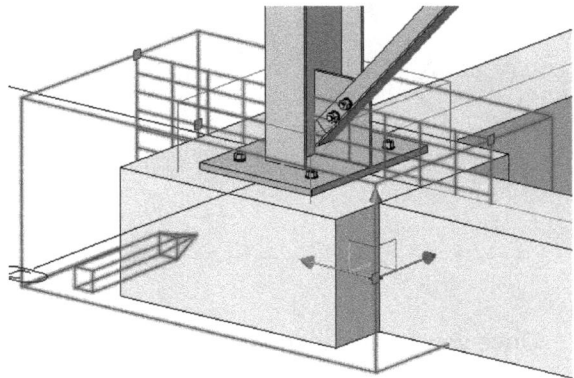

Figure 6–50

6. Select the camera, then right-click and select **Advance Properties**, or double-click on the box to open the *Camera* dialog box.

7. Select the *Select objects for camera* sub-tab, as shown in Figure 6–51. You can then select either the **Consider all objects inside the viewport** option or the **Select objects for camera detailing** option.

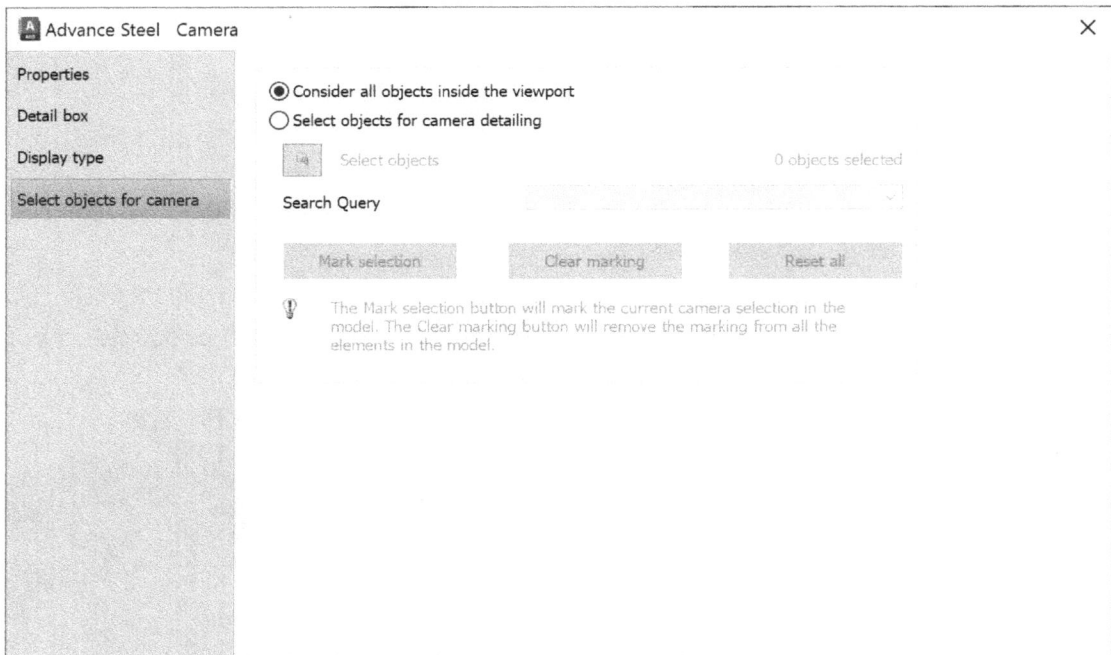

Figure 6–51

8. Once you have placed the cameras, you can run processes using the tools in the *Drawing Processes* tool palette> (Cameras) category.

Practice 6c
Run Drawing Processes

Practice Objectives

- Create drawings from a selected assembly.
- Create stair assembly documentation.
- Create section details using cameras.
- Review the new drawings through the *Document Manager*.

In this practice, you will create detail drawings using Drawing Processes for a connection assembly and a stair assembly. You will create camera views and run drawing processes to extract the details from those views, as shown in Figure 6–52. You will also review the new drawings created in the *Document Manager* and move views from one sheet to another.

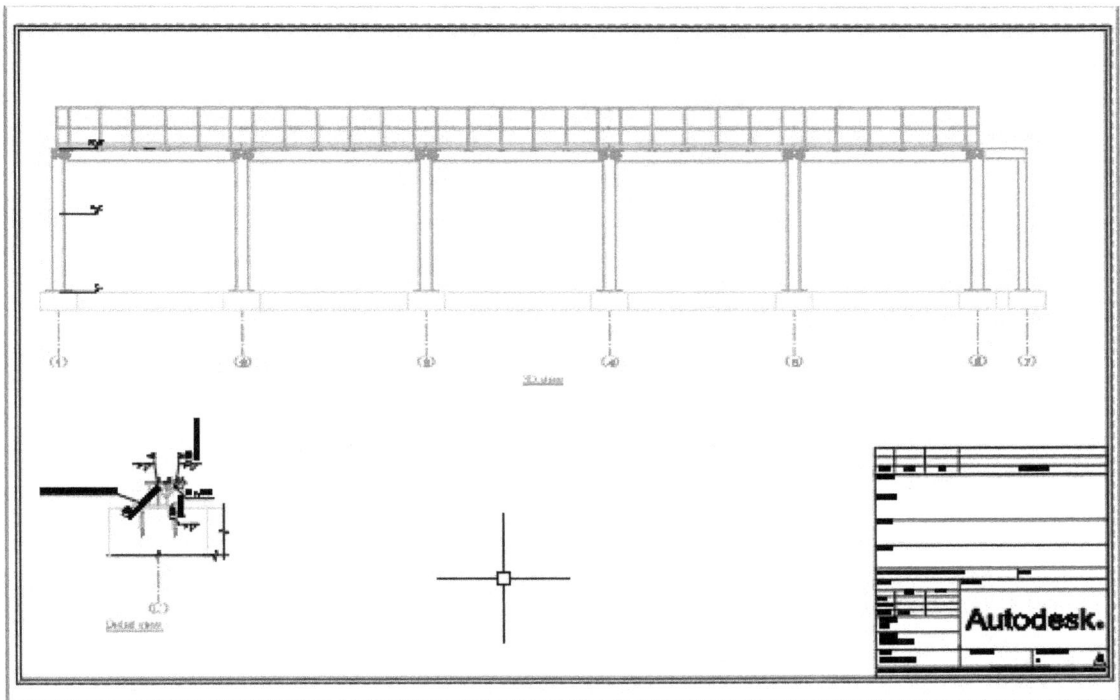

Figure 6–52

Task 1: Create drawings from a selected assembly.

1. In the practice files folder, open **Platform-Documents-Processes.dwg**.

2. In the *Output* tab>*Documents* panel, click ▦ (Drawing Processes). Review the categories available.

3. Select the objects shown in Figure 6−53.

Select these objects

Figure 6−53

4. In the *Drawing Process* palette, click ▦ (5 - Assemblies - Selected), and then select **Selected Assemblies PageFull ANSI-D**.

5. In the *Process properties* dialog box, accept the defaults.

6. Open the *Document Manager*. Note that three sheets have been created based on the selected assemblies, as shown in Figure 6–54.

   • The number of sheets that are connected to the drawing depend on how many times you have run other Drawing Styles or Drawing Processes.

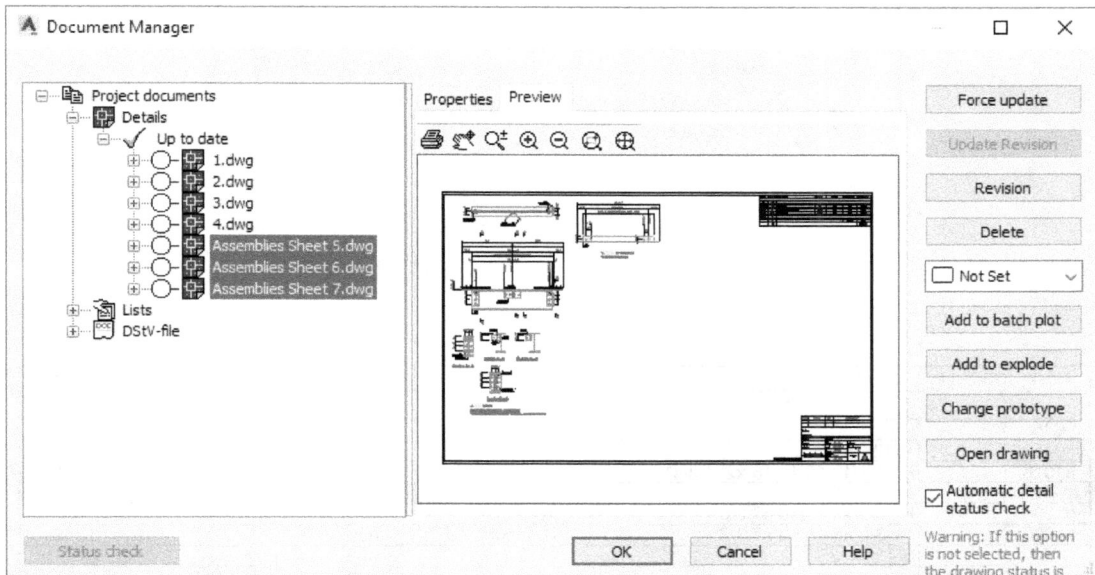

Figure 6–54

7. Open each sheet and review the information. Note that all of the detail sections display (as shown in Figure 6–55), as well as the related BOM (as shown in Figure 6–56).

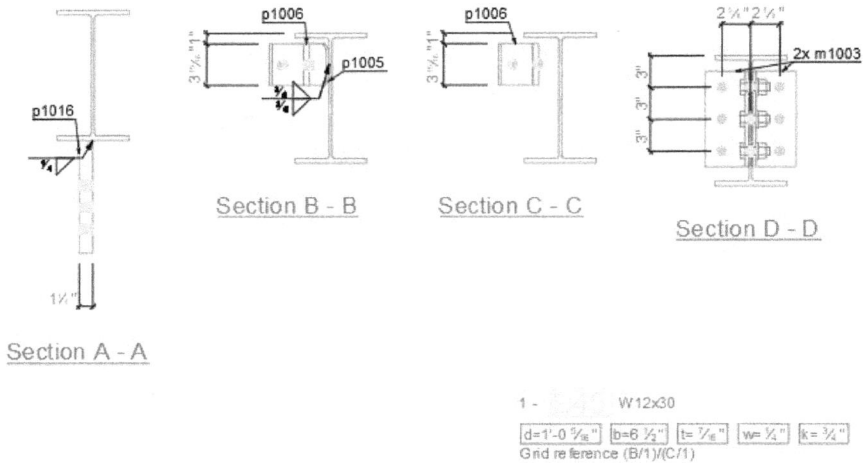

Figure 6–55

Mark	Quantity	Description	Length	Grade	Part weight	Total weight
B45	1					
B45	1	W12x30	19'-11 3/8"	A992	598.42	598.42
p1016	1	PL 1 1/4"x10 11/16"x2'-4 3/4"	2'-4 3/4"	A36	106.59	106.59
m1003	2	L4X4X3/8	9"	A36	7.35	14.70
p1005	1	PL 3/8"x3 15/16"x7 3/8"	7 3/8"	A36	3.08	3.08
p1006	1	PL 3/8"x3 15/16"x7 3/8"	7 3/8"	A36	3.08	3.08
1007	3	3/4"ø A325	2 1/4"	10.9	0.86	2.57
	9				728.43	728.43
	9					**728.43**

Figure 6–56

Task 2: Create a search and mark objects.

1. In the *Project Explorer*, click **?₊** (Create new query), as shown in Figure 6–57.

Figure 6–57

- Alternatively, you can use the *Advance Tool Palette>* (Selection) category> (Search filter).

2. In the *Search and mark objects* dialog box>*General* tab, select **Assign color** and select a color from the drop-down list, as shown in Figure 6–58.

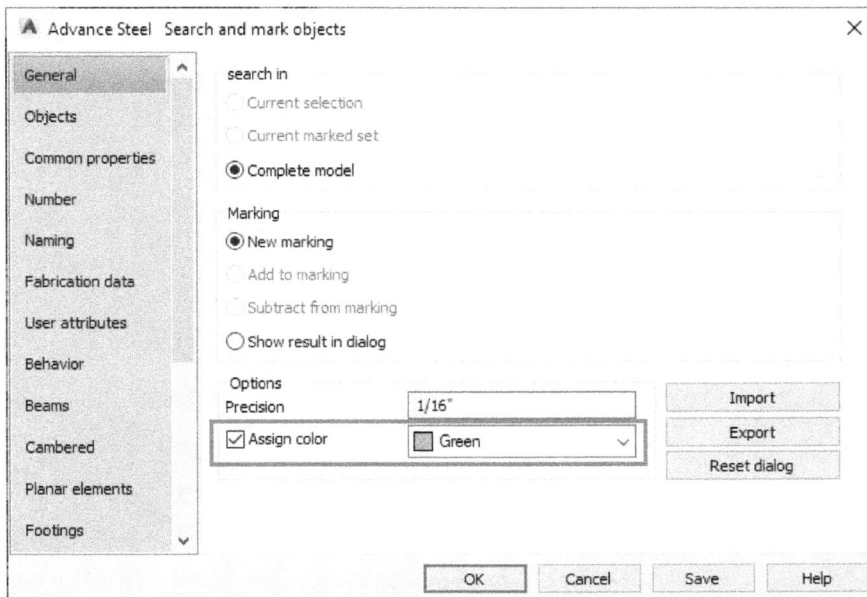

Figure 6–58

3. In the *Objects* tab, select **Steel beam**, as shown in Figure 6−59.

Figure 6−59

4. In the *Behavior* tab, in the *Created drawings* area, select **Has Assembly drawing** and select the checkbox under the blank column, as shown in Figure 6−60.

Figure 6−60

5. Click **Save**. Name the search **Assembly drawings** and click **OK** twice to get back to the drawing.

6. The objects found in the search are now displayed in the model in green and the search is listed in the *Project Explorer* under *Queries*, as shown in Figure 6–61.

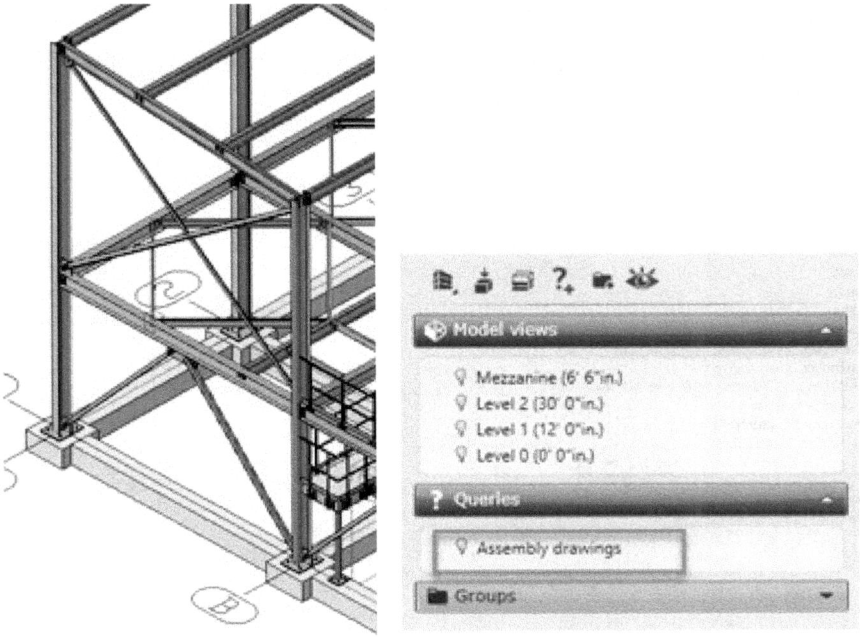

Figure 6–61

7. Close the detail drawings. You do not need to save them.

Task 3: Modify details on sheets.

1. In the model drawing, open the *Document Manager*.

2. Expand **Assemblies Sheet 7.dwg** and select the detail, then drag and drop it to **Assemblies Sheet 6.dwg**, as shown in Figure 6–62.

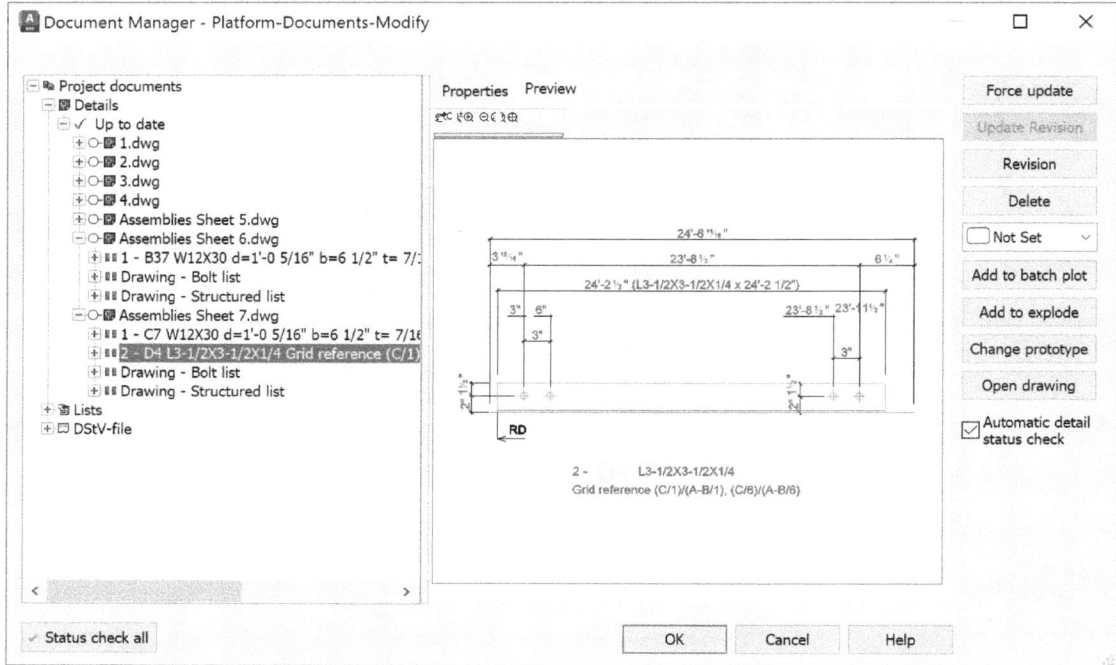

Figure 6–62

3. **Assembly Sheet 7.dwg** is removed and an update is required. Click **Force Update**. This fixes the BOM schedule by adding the materials from the moved details.

4. Select **Assembly Sheet 6.dwg** and click **Open drawing**.

5. Note that one of the sets of details is overlapping the BOM schedule. Select the green box surrounding the group of details and drag it down.

6. In the *Labels & Dimensions* tab>*Management* panel, select 🖼 (Update Lists). This will update the BOM.

7. Save and close the detail drawing.

Task 4: Create stair assembly documentation.

1. Zoom in on the stair objects.

2. Double-click on the left stringer, as shown in Figure 6–63.

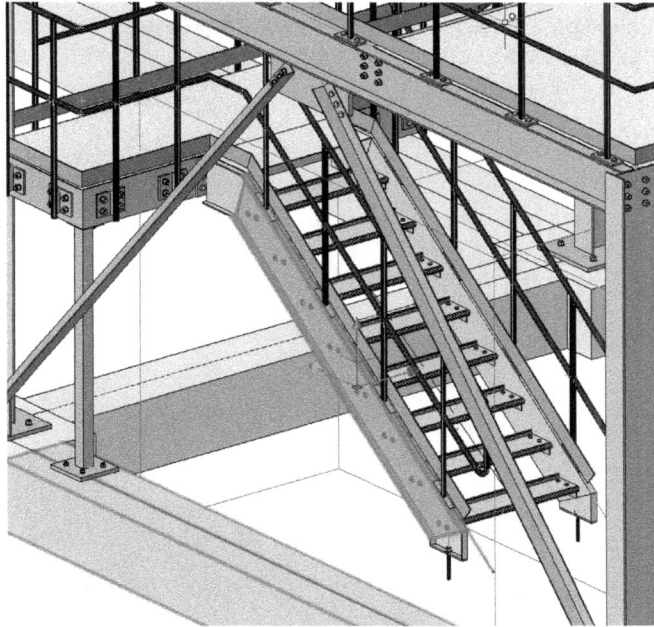

Figure 6–63

3. In the related *Advance Properties* dialog box, select the *Naming* tab and note that the *Assembly mark* option is selected, as shown in Figure 6–64.

Naming	
Preliminary part mark	Not defined
Preliminary part prefix	
☐ Single part mark	1004
Single part prefix	s
Single part	s1004
☒ Assembly mark	1
Assembly prefix	S
Main part	S1

Figure 6–64

4. Close the dialog box and select the right stringer again.

5. In the *Drawing Processes* tool palette> (Assemblies - Selected) category, click **Selected Assemblies PageFull ANSI-D**.

6. In the *Process properties* dialog box, keep the defaults and click **OK**.

7. When the assembly is processed, open the *Document Manager* and open the new drawing. Zoom in so that the plan and section of the stringer displays as shown in Figure 6–65.

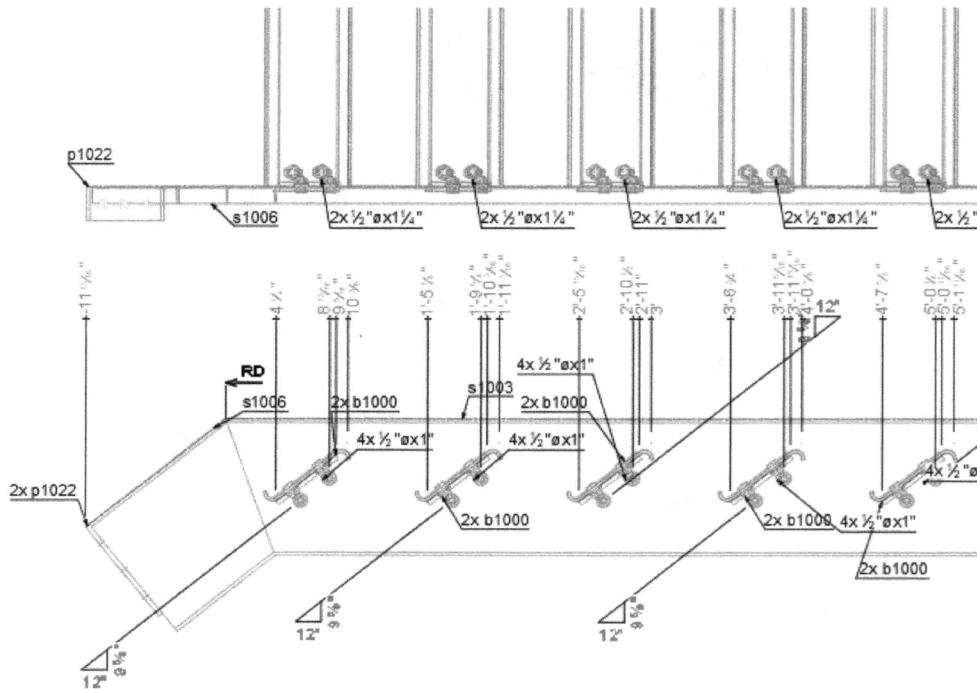

Figure 6–65

8. If you have time, create assembly drawings of the ladders using the process outlined in this Task.

9. Close any open detail drawings.

Task 5: Create section details using cameras.

1. Zoom in on the base of column C1.

2. Select one of the connector objects, right-click and select **Advance Joint Properties**. Close the dialog box. The joint box displays as shown in Figure 6–66.

Figure 6–66

3. In the *Advance Tool Palette*> (Tools) category, click (Create Camera at node).

4. Select the joint box and press <Enter>. Select the arrow pointing toward the model. The new camera view is created as shown in Figure 6–67.

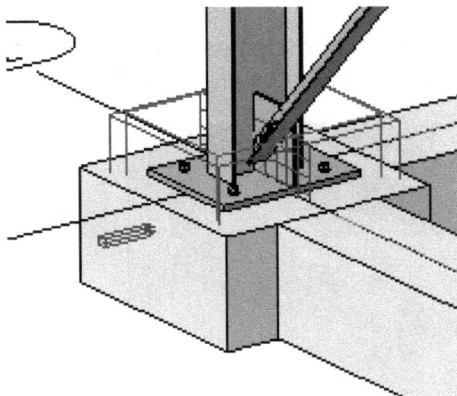

Figure 6–67

5. Zoom in on the camera box and modify it using grips, or by double-clicking on the box and modifying the size in the *Camera* dialog box>*Detail box* tab. When you are finished, the entire footing should be inside the box, as shown in Figure 6–68.

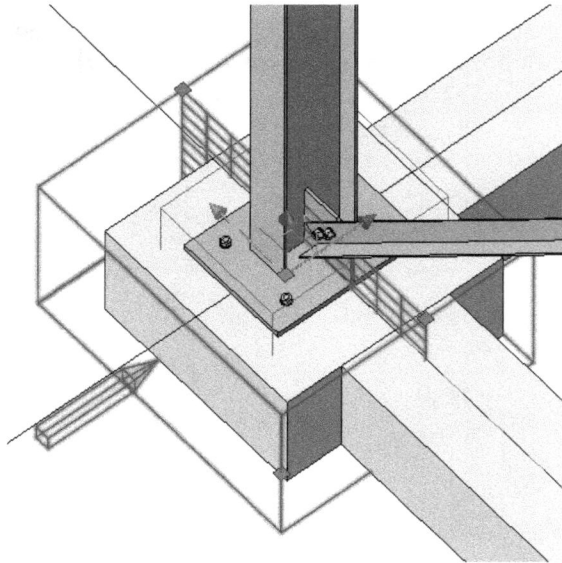

Figure 6–68

6. Zoom out that the front of the platform displays.

7. Move the UCS to the midpoint of concrete beam A3-4 and rotate the UCS so the XY plane is on the surface and the Z-axis is pointing away from the model, as shown in Figure 6–69.

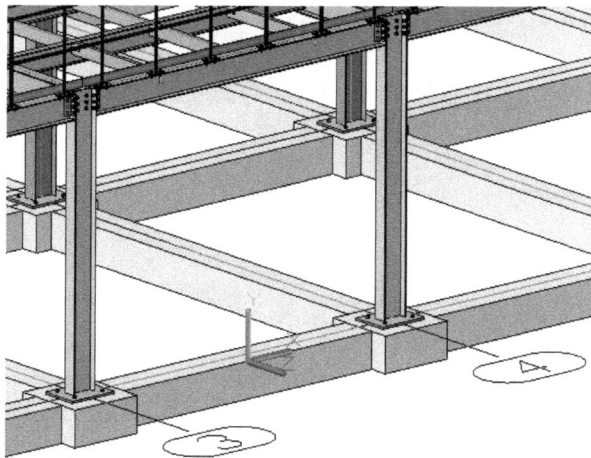

Figure 6–69

8. In the *Advance Tool Palette>* (Tools) category, click (Create camera, UCS).

9. At the *Please locate point* prompt, select the same midpoint as the UCS.

10. In the *Camera* dialog box>*Detail box* tab, set the sizes similar to those shown in Figure 6–70. Close the dialog box.

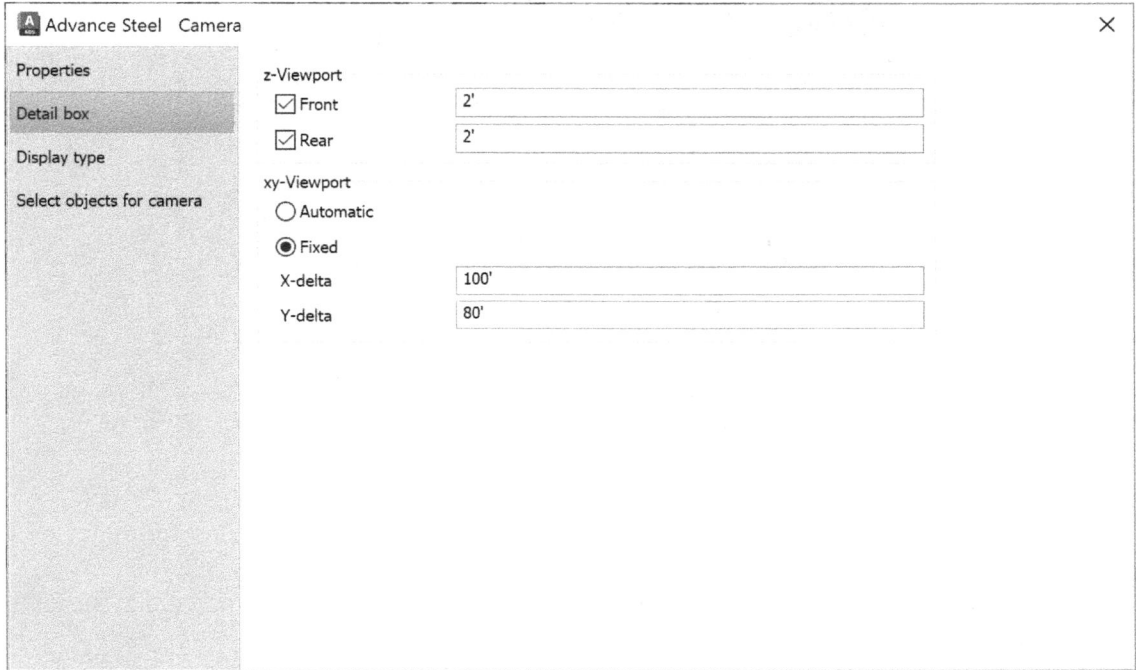

Advance Steel Camera		✕
Properties	z-Viewport	
Detail box	☑ Front	2'
	☑ Rear	2'
Display type	xy-Viewport	
Select objects for camera	◯ Automatic	
	⦿ Fixed	
	X-delta	100'
	Y-delta	80'

Figure 6–70

11. Modify the camera box by adjusting the X-delta and Y-delta so that only the first level is covered. You can also use the camera box grips as shown in Figure 6–71.

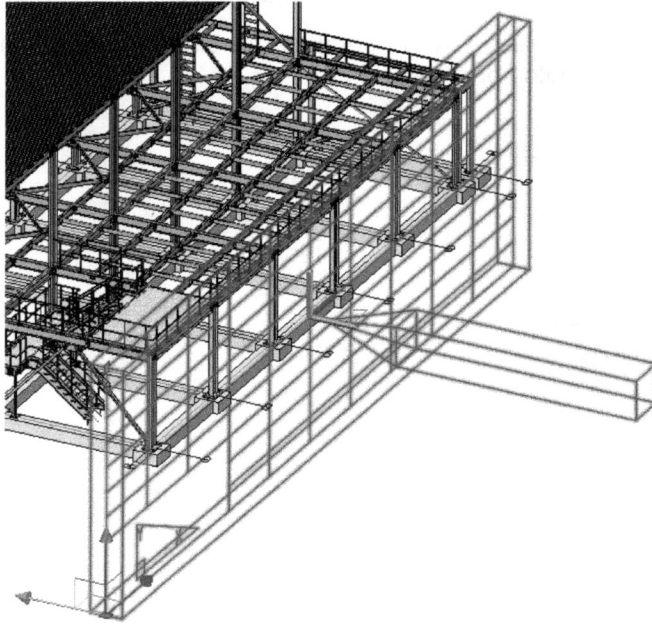

Figure 6–71

12. Ensure that you still have this camera selected and then also select the camera at the joint box.

13. In the *Drawing Processes* toolbar> ▣ (Cameras) category, click **Selected Cameras PageFull ANSI-D**.

14. In the *Process properties* dialog box, click **OK**.

15. Open the *Document Manager* and view the new sheet.

16. Save and close the drawings.

End of practice

6.5 Modifying Detail Drawings

While most detail drawings are created using Drawing Styles and Drawing Processes, you can make modifications to the detail drawings. Some typical changes that you might make include adding cut views (sections), callout views (as shown in Figure 6–72) and modifying or adding parametric dimensions.

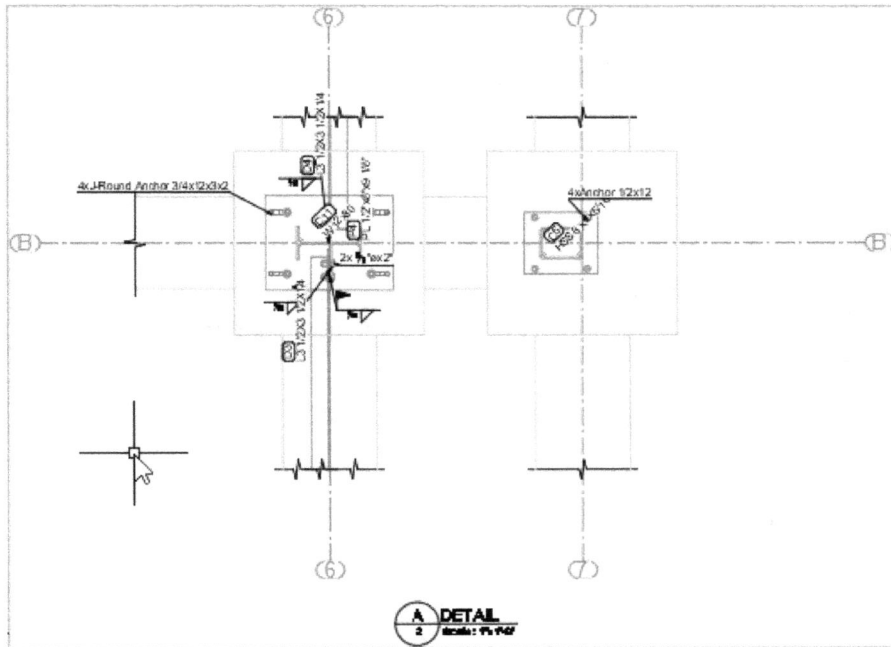

Figure 6–72

- These tools only work inside detail drawings that are opened through the *Document Manager*.

- To display or hide the green frame around views, in the *Labels & Dimensions* tab>*Parametric views* panel, click ▨ (Toggle the display of green frames around details).

How To: Create a Cut View

1. In the *Labels & Dimensions* tab>*Parametric views* panel, click ▨ (Create cut view).
2. Select an object in the view where you want to place the section.
3. Draw a line in front of the area you want to cut.

---

4. Select a point that defines the cut depth, as shown in Figure 6-73.

Figure 6-73

5. Cut symbols are placed in the primary detail and the new section is added to the same sheet, as shown in Figure 6-74.

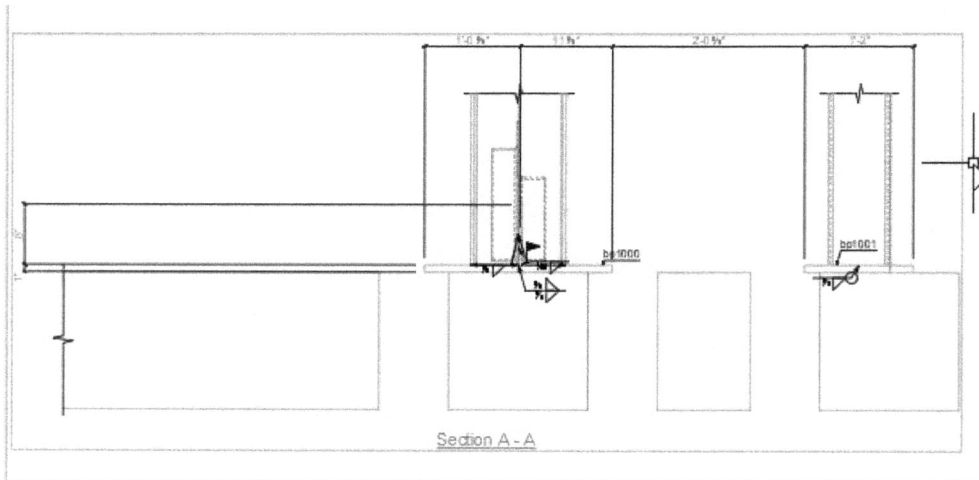

Figure 6-74

- To modify the view (such as by changing the scale, clip, or orientation), double-click on the inner green frame.

How To: Create a Callout View

1. In the *Labels & Dimensions* tab>*Parametric views* panel, click ⬚ (Create Callout View).

2. Select an object in the view where you want to place the callout.

3. Pick two diagonal points that define the area of the callout.

4. The new callout view is placed on the same sheet.

5. If required, modify the callout box and label using grips, as shown in Figure 6–75.

Figure 6–75

*   If a detail does not fit on the sheet, you can move it by dragging it by the green frame, as shown in Figure 6–76. Alternatively, in the *Labels & Dimensions* tab>*Parametric views* panel,

    click ▦ (Move view) and follow the prompts to move the view.

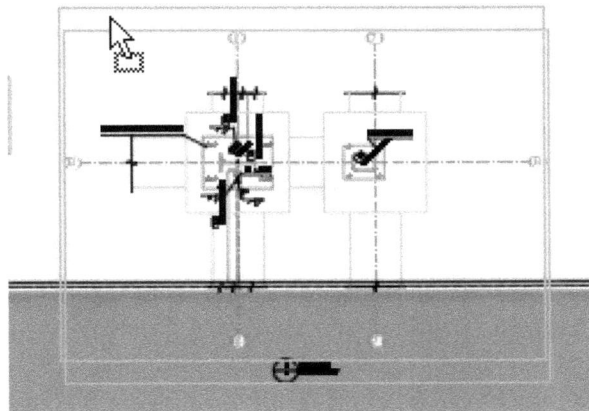

Figure 6–76

- To remove a view from the sheet, select the green frame and press <Delete>. Alternatively, in the *Labels & Dimensions* tab>*Parametric views* panel, click ▣ (Delete view) and follow the prompts to move the view.

- If the size of the title block is not appropriate for the views, you can change it in the *Labels & Dimensions* tab>*Parametric views* panel by clicking ▣ (Change prototype file). In the *New drawing* dialog box (shown in Figure 6–77), select the Prototype template you want to use and click **OK**.

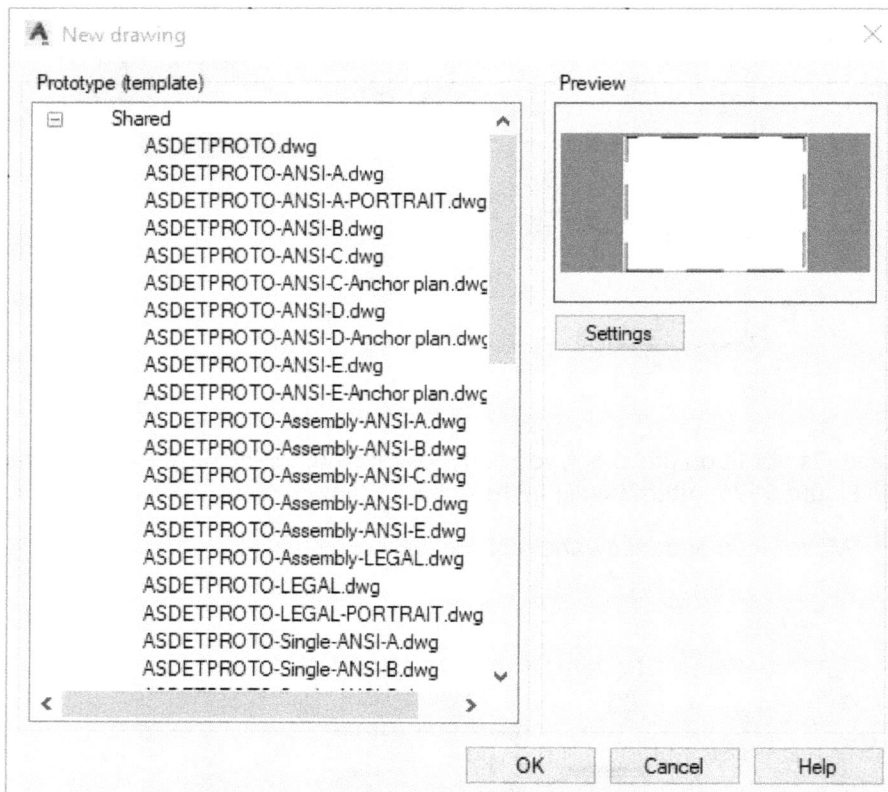

Figure 6–77

Working with Parametric Dimensions

Parametric dimensions in detail drawings automatically update if changes are made to the model, as shown in Figure 6–78 and Figure 6–79. To add and modify these dimensions, you must use very specific tools to ensure that they update correctly. Many of these dimensions are automatically added when you run drawing styles.

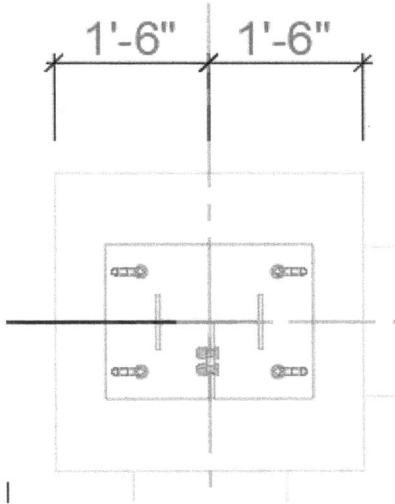

Figure 6–78

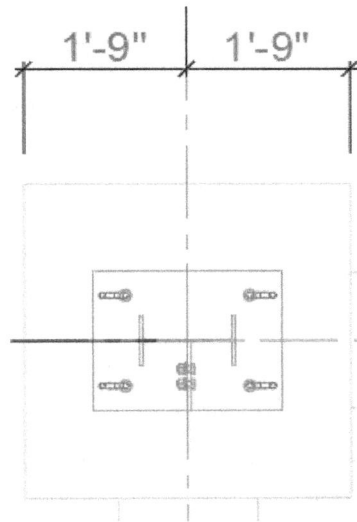

Figure 6–79

- When a change is made to the model, you need to update the drawing in the *Document Manager*. Do not make changes to model objects in the detail drawings.

How To: Add Parametric Dimensions

1. In the *Labels & Dimensions* tab>*Parametric Dimensions* panel, click ⊢⊣ (Parametric Dimensions).

2. Note that the default command is ⊢⊣ (Linear). You can select others from the drop-down list, shown in Figure 6–80.

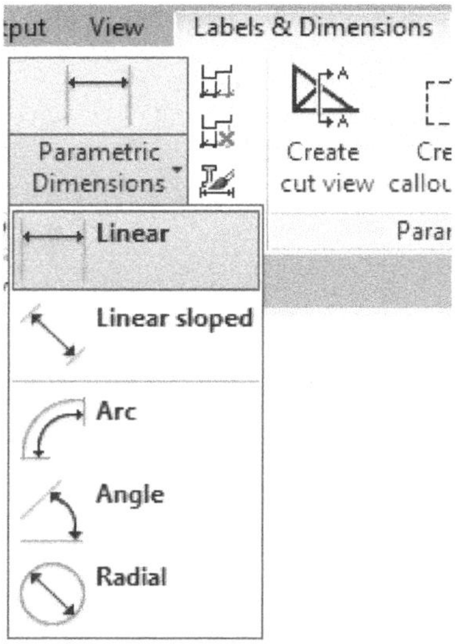

Figure 6–80

3. At the *Please select a view* prompt, select an item in the view where you want to add the dimensions.

4. You are prompted for the first two dimension points, and then to specify the placement of the dimension line.

5. If you want to add a string of dimensions based on the first selection, continue selecting points. The additional points are added to the first dimension, as shown in Figure 6-81. When you are finished, press <Enter>.

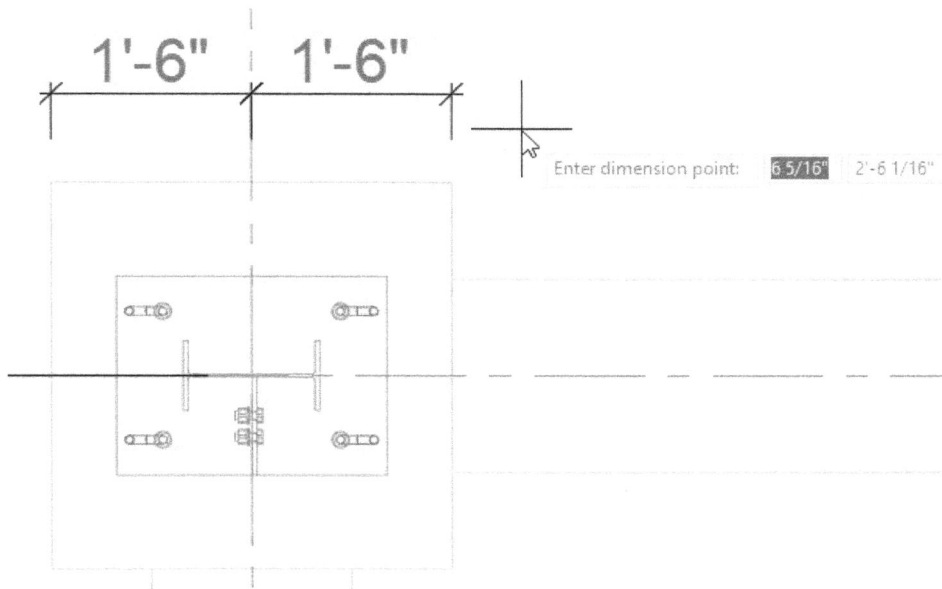

Figure 6-81

6. Note that you are still in the command. If you want to add another string of dimensions in a different location, start by selecting the view again and then selecting the dimension points following the prompts.

7. When you are finished adding dimensions, press <Enter> twice. This opens the *Dimensions* dialog box (shown in Figure 6–82), where you can make modifications to all of the dimensions in the set.

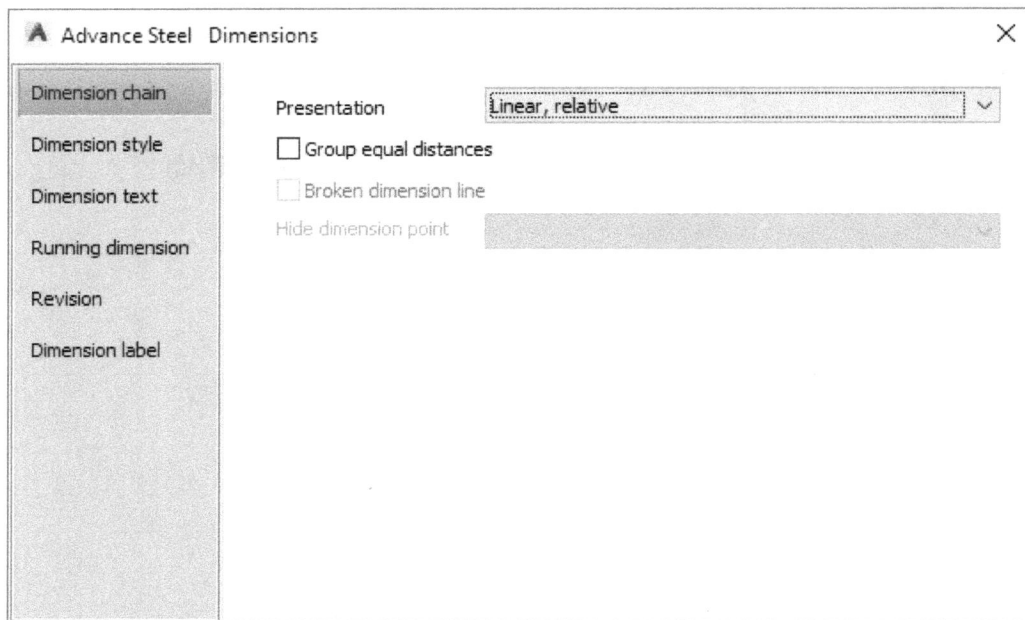

Figure 6–82

How To: Add Dimension Lines to Existing Dimensions

1. In the *Labels & Dimensions* tab>*Parametric Dimensions* panel, click 🔲 (Insert Points).
2. Select the dimension where you want to insert points and press <Enter>.
3. Select the points where you want to add the dimension lines.
4. Press <Enter> to complete the command.

How To: Delete Dimension Lines from Existing Dimensions

1. In the *Labels & Dimensions* tab>*Parametric Dimensions* panel, click 🔲 (Delete Point).
2. Select the dimension line you want to remove and press <Enter>.
3. Select other lines as required and press <Enter> to complete the command.

• To delete an entire dimension, select it and press <Delete>, or use the AutoCAD® **Erase** command.

Weld Symbols

You can remove and add leaders from weld symbols. Select a weld symbol, right-click, and select **Add leader line** or **Remove leader line** from the menu, as shown in Figure 6−83. When adding a leader, click a reference point on the drawing. When removing a leader, select the leader line you want to remove.

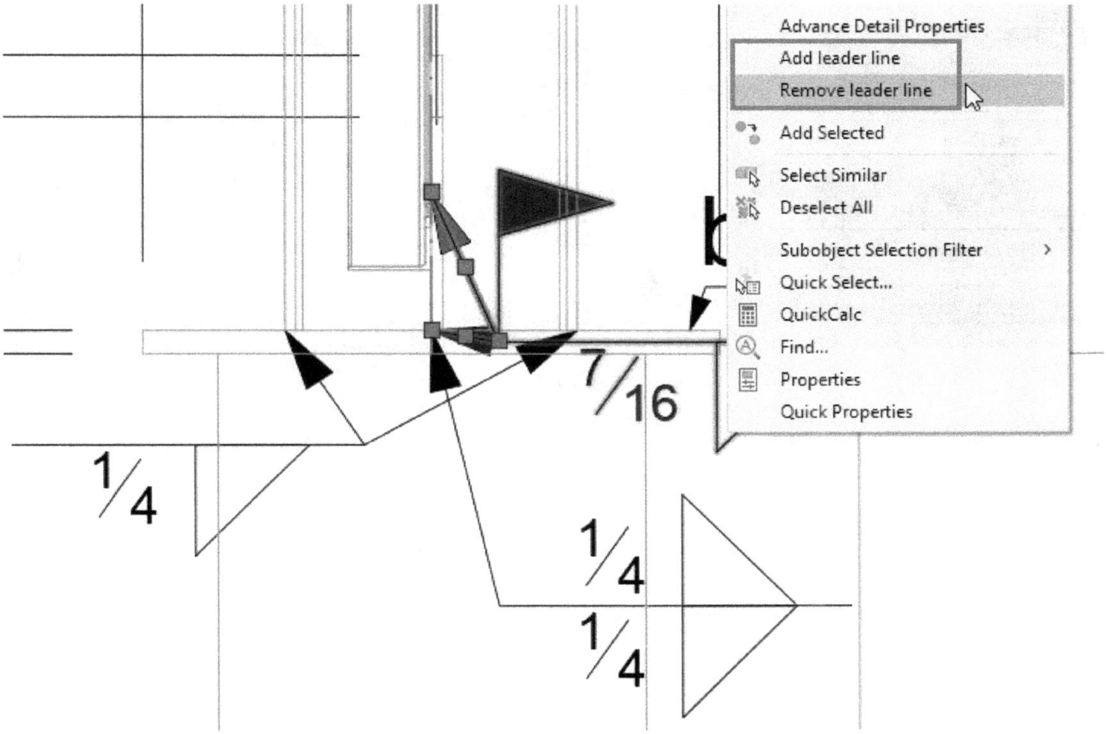

Figure 6−83

In Management Tools, there is an option under the Defaults called **Weld symbols combine only if the weld point connects the same objects in the model** (as shown in Figure 6–84). Once selected, this setting combines weld symbols corresponding to welds connecting elements with the same single part numbers.

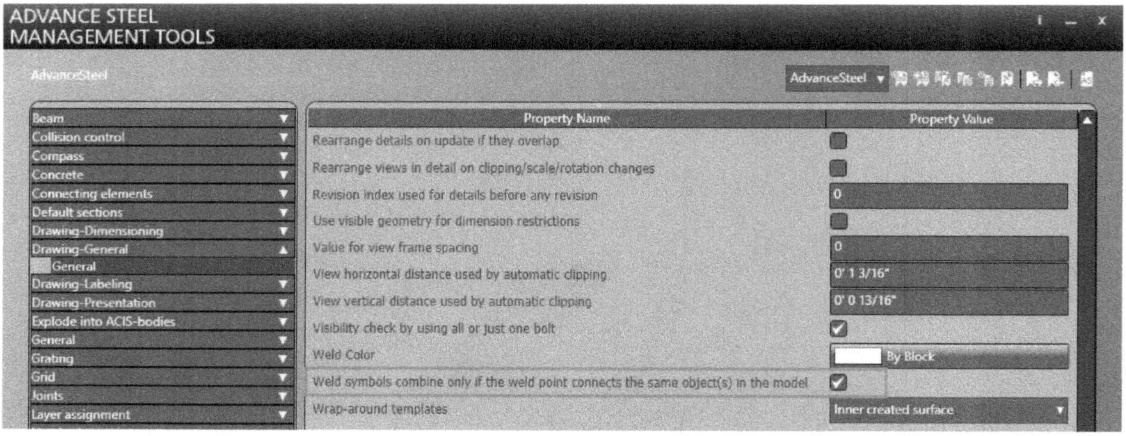

Figure 6–84

6.6 Revising Models and Drawings

When you change a model, all of the related drawings created using the Autodesk Advance Steel processes are also updated. You can force an update or you can manually apply revisions. When you apply revisions, a revision table is added to the title block and the parts of the drawing that have been updated are circled by a revision cloud, as shown in Figure 6–85.

Note: Make changes to the model, which copies changes to the drawings.

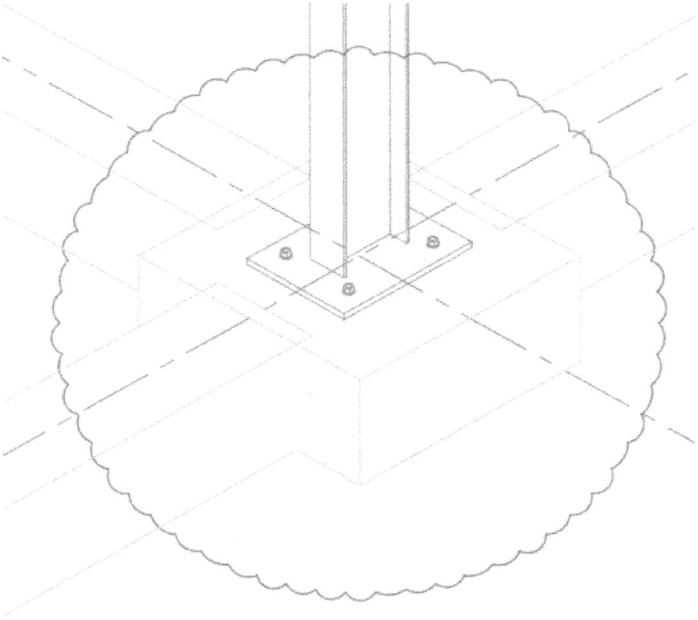

Figure 6–85

How To: Apply Revisions

1. In the model drawing, make changes to the model.

2. Open the *Document Manager*. The drawings that need revisions are grouped in the **Update required** node, as shown in Figure 6–86.

    • Ensure that **Automatic detail status check** is selected so that the updates display.

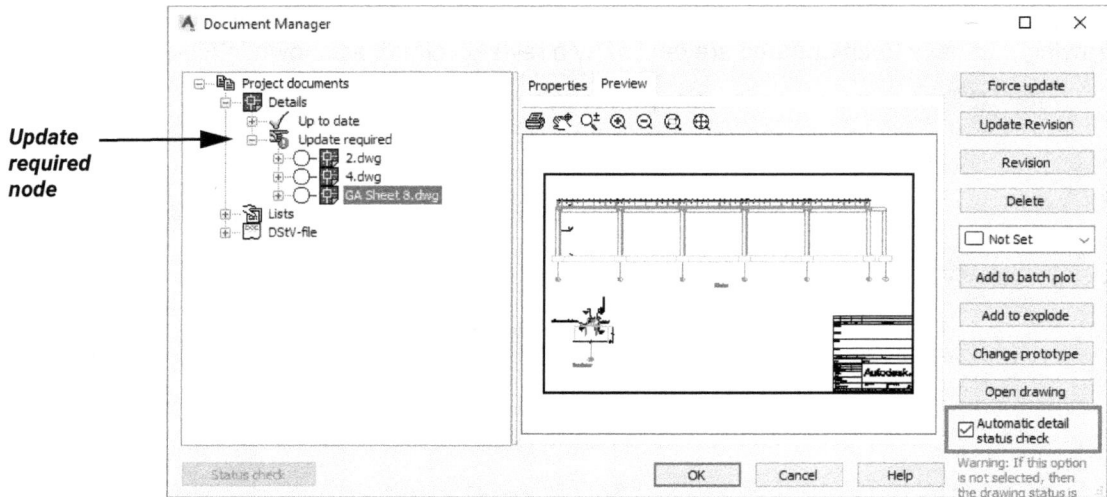

Update required node

Figure 6–86

3. Select the drawings that require updating and click **Update Revision**.

    • If you click **Revision**, the revision is added to the title block, but revision clouds are not automatically added.

4. In the *Add revision mark to* dialog box, click **Add**.

5. In the *Revision mark* dialog box, insert the required information and click **OK**.

    - If you select more than one drawing to be revised, the *Add revision mark to* dialog box prompts you to specify the index method for the drawings, as shown in Figure 6−87.

    - Add information about what changes were made to the original drawing on the *Backup* tab.

Figure 6−87

6. Click **OK** to add the revision.

Practice 6d
Modify Detail Drawings

Practice Objectives

- Add and modify parametric dimensions.
- Create callout views and cut views.
- Modify the scale of views.
- Change the prototype (title block) drawing.
- Apply revisions.

In this practice, you will open an anchor plan drawing and remove extra dimensions. You will then add overall dimensions to the plan. You will add a callout view and a cut view and modify the scale of the new views. Finally, you will change the prototype drawing to a larger title block so that the new views fit better on the sheet, as shown in Figure 6–88. You will also modify the size of footings in the model and apply revisions.

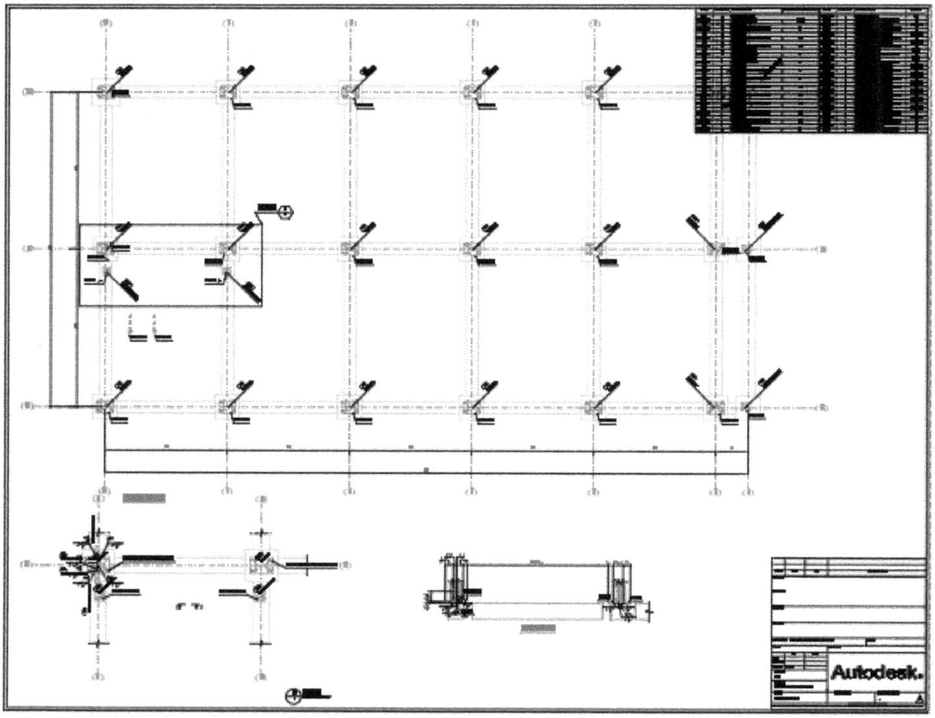

Figure 6–88

Task 1: Delete and add parametric dimensions.

1. In the practice files folder, open **Platform-Documents-Modify.dwg**.

2. Open the *Document Manager*.

3. Expand *Details>Up to date*.

4. Select **2.dwg**. Click **Change prototype**.

5. Select **ASDETPROTO-ANSI-E-Anchor plan.dwg**. Click **OK**.

6. Double-click on **2.dwg** to open it.

7. Double-click on the green frame around the view. Change the *Scale* to **1/4"=1'-0"**.

8. Zoom in on the upper left corner. Note that there is an extra dimension that is not required, as shown in Figure 6–89.

9. In the *Labels & Dimensions* tab>*Parametric dimensions* panel, click 🔲 (Delete Point).

10. Select the dimension line you want to delete and press <Enter>. The line is removed (as shown in Figure 6–90), and you stay in the command.

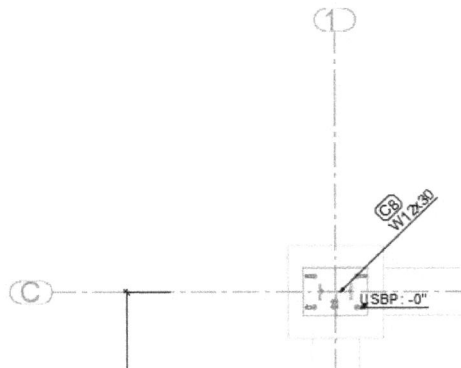

Figure 6–89   Figure 6–90

11. Pan around the plan and delete the other extra dimensions at the corners of the grids, pressing <Enter> after each selection. When you are finished, with nothing selected, press <Enter> again to end the command.

12. Select one of the dimensions and one of the grid lines.

13. Right-click and select **Select Similar**.

14. In the Status Bar, expand 🔲 (Isolate Objects) and select **Isolate Objects**. Note that only the dimensions and grid lines now display. This will make it easier to add overall dimensions to the grid.

15. Move the existing dimensions closer to the model area to make room for overall dimensions, as shown in Figure 6–91.

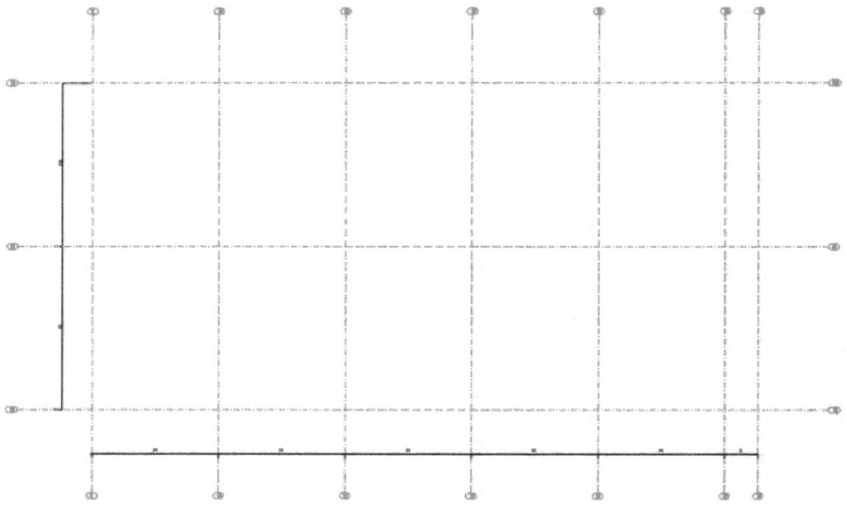

Figure 6–91

16. In the Status Bar, set the *Object Snaps Settings* to **Intersection** and **Preferred for manual dimensions**. Ensure that **Object Snaps** are on.

- The **Preferred for manual dimensions** object snap is for reference only. It displays the points on the Advance Steel objects that you can dimension to.

Note: GRID Intersection Points does not work in detail documents.

17. In the *Labels & Dimensions* tab>*Parametric Dimensions* panel, click (Parametric Dimensions).

18. At the *Please select a view* prompt, click on one of the grid lines.

19. For the first dimension point, select the intersection of Grid C1.

20. For the second dimension point, select the intersection of Grid A1.

21. Place the dimension to the left of the existing dimensions, as shown in Figure 6–92.

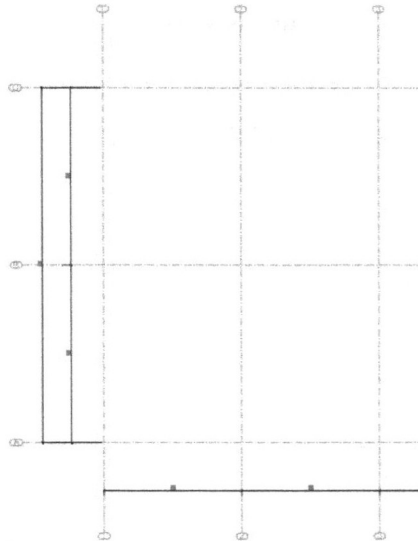

Figure 6-92

22. Press <Enter>. Note that you are still in the command, but are prompted to specify the view again. Select another grid line and create an overall dimension at the bottom of the model, as shown in Figure 6-93.

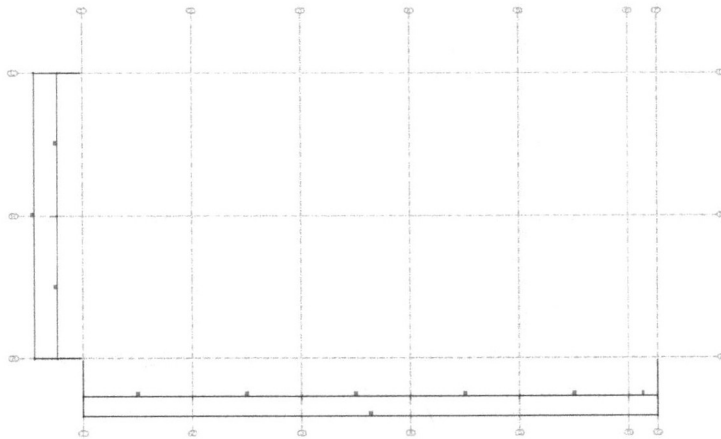

Figure 6-93

23. Press <Enter> twice to finish the command and open the *Dimensions* dialog box.

24. Close the dialog box.

25. End object isolation.

26. Save the drawing.

Task 2: Create a cut view and a callout view.

1. Continue working in **2.dwg**, which you opened through the *Document Manager*.
2. Zoom in on the area around Grids A, B, 1, and 2.

3. In the *Labels & Dimensions* tab>*Parametric views* panel, click ⌑ (Create Callout View).
4. Select an object inside the view.
5. Pick two diagonal points around the columns that support the stair landing, as shown in Figure 6−94.

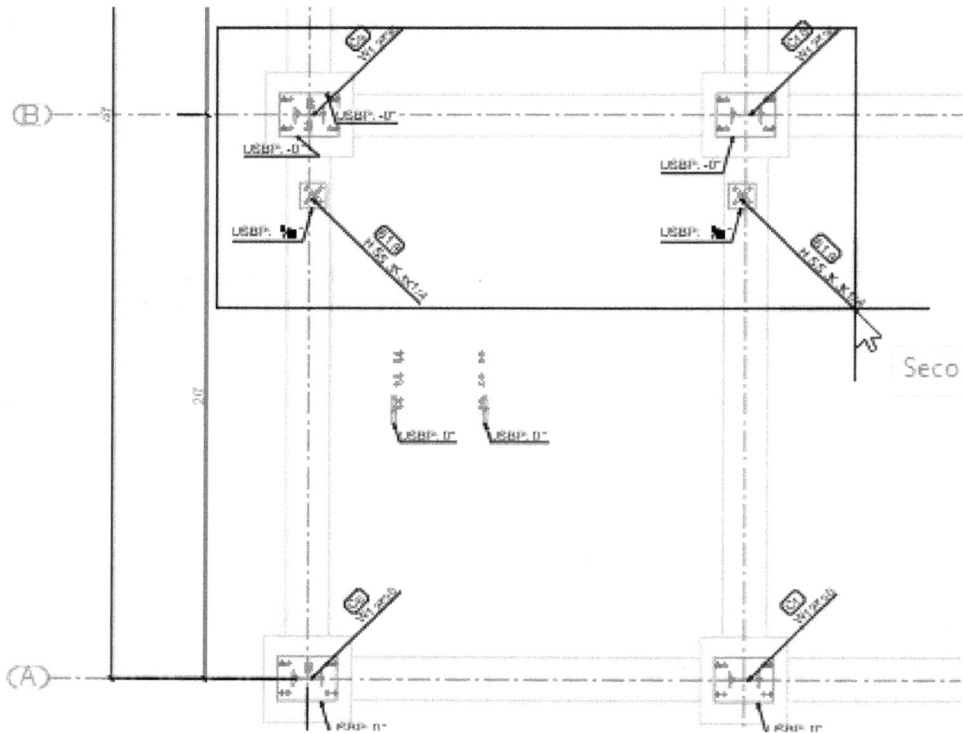

Figure 6−94

6. Note that the callout is placed on the same sheet, but that it is too big, as shown in Figure 6–95.

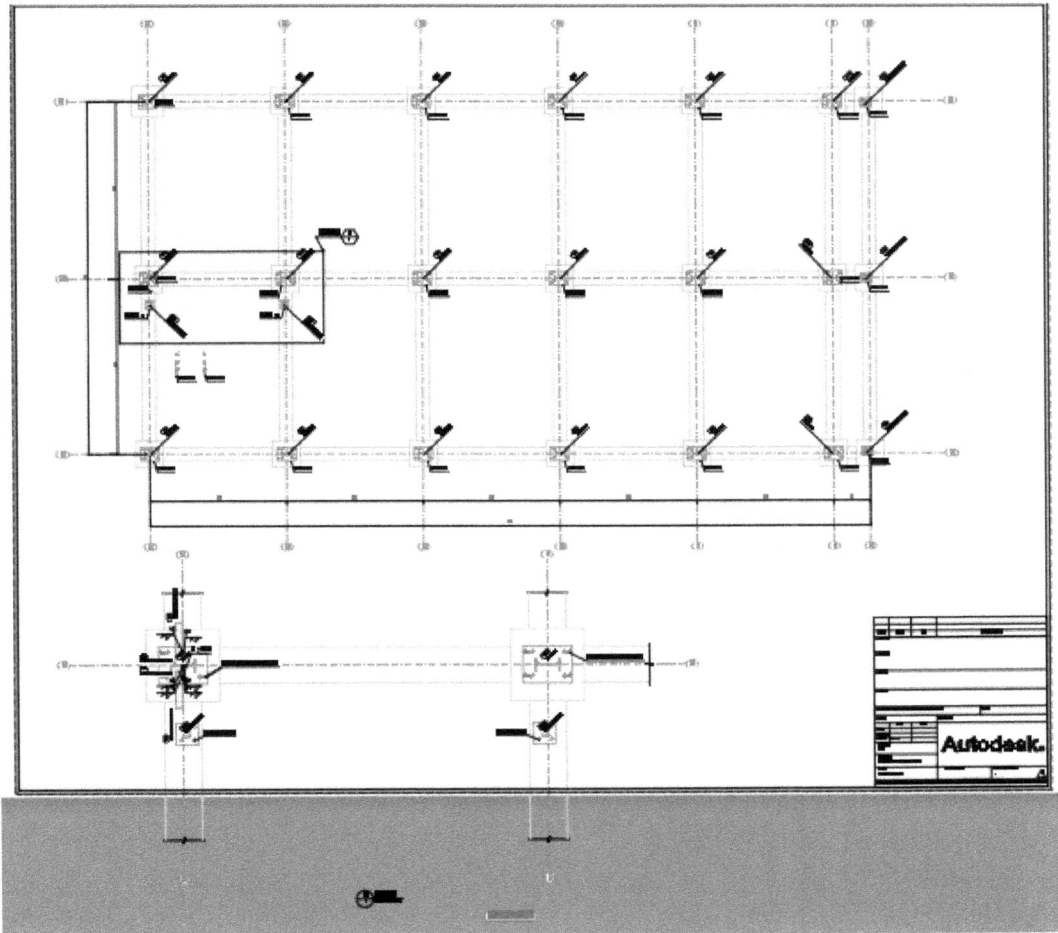

Figure 6–95

7. Hover the cursor over the edge of the new callout view and double-click on the green frame.

8. In the *View* dialog box, change the *Scale* to **1/2" =1'-0"** and click **OK**. Note that the view now fits better. Move the titles as required.

9. In the *Labels & Dimensions* tab>*Parametric views* panel, click ⊠ (Create cut view).

10. Select an object in the callout view.

11. Draw a line (which displays in a dashed yellow) across the front of the view, and then specify the cut depth as shown in Figure 6–96.

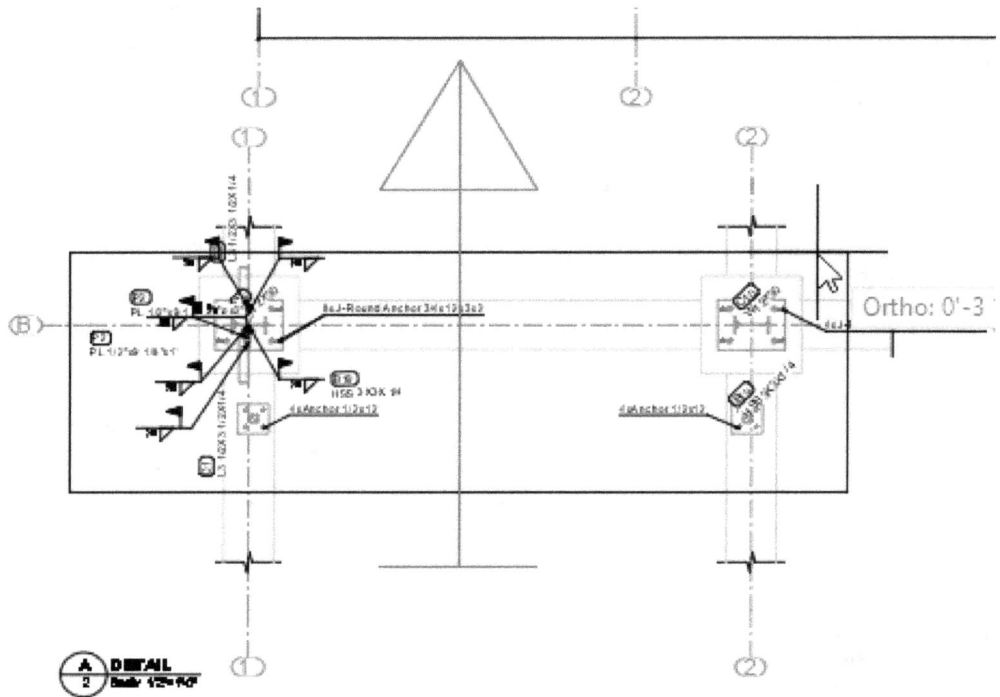

Figure 6–96

12. If this view is too large, change the *Scale* to **1/2" =1'-0**".

13. Move the new view onto the sheet.

14. Select a weld symbol, right-click, and select **Remove leader line** to remove unnecessary leaders, as shown in Figure 6–97.

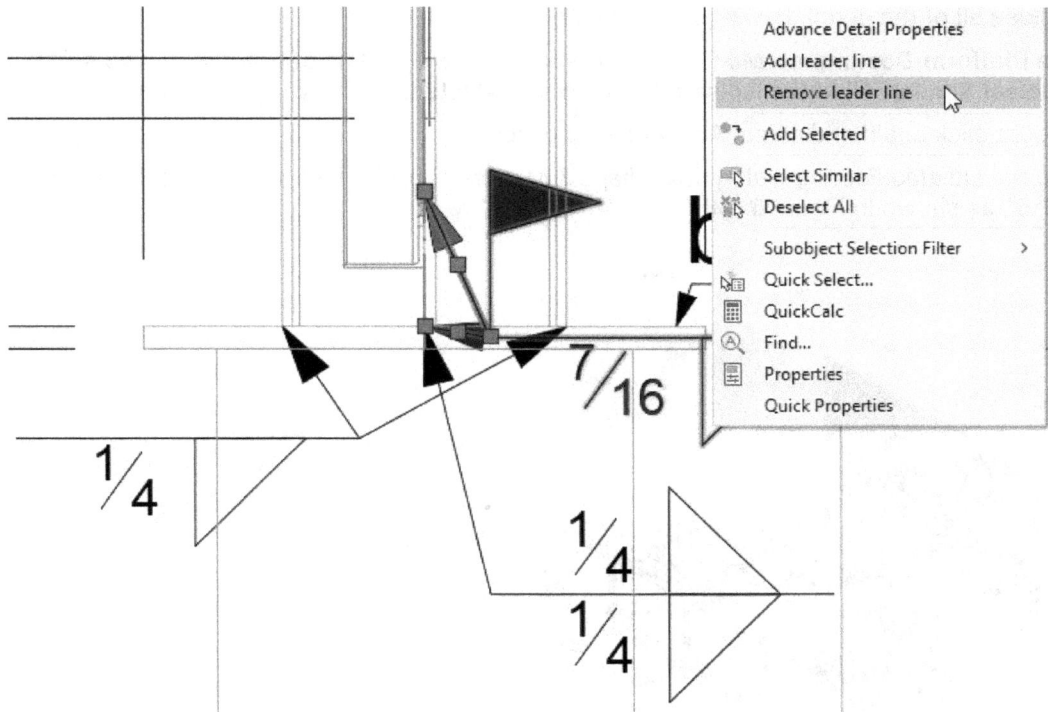

Figure 6–97

15. Save and close the drawing.

Task 3: Make changes to the model and apply revisions.

1. Close all of the detail drawings that you have open.

2. In **Platform-Documents-Modify.dwg,** right-click on the footing at Column A1 and select **Select Similar**. All of the Isolated footings are selected, as shown in Figure 6–98.

3. Right-click again and select **Advance Properties**.

4. In the *Isolated Footing* dialog box>*Shape & Material* tab, change the *Length* and *Width* to **3' 6"**, as shown in Figure 6–98.

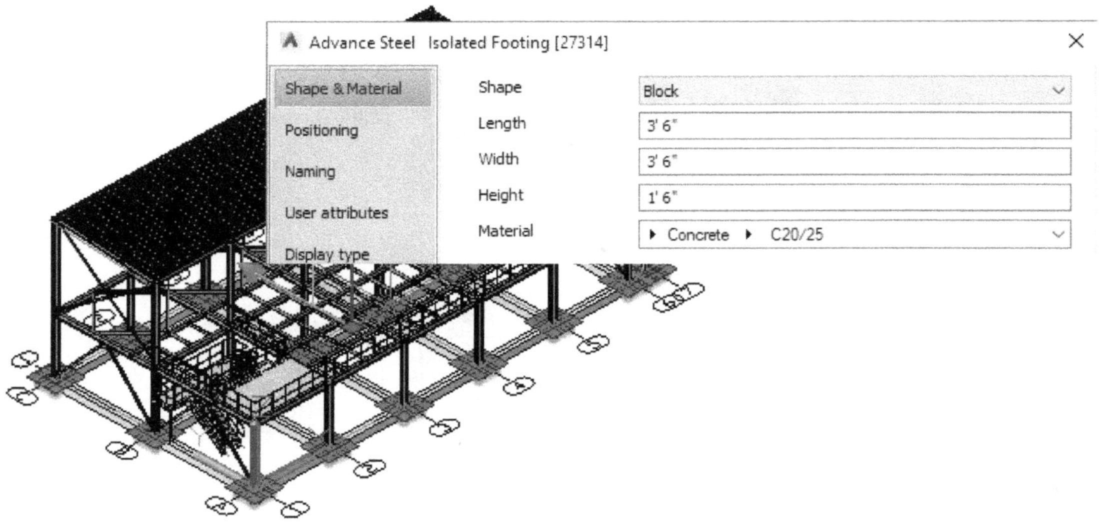

Figure 6–98

5. Close the dialog box.

6. Save the drawing.

7. Open the *Document Manager* and expand the *Update required* node, as shown in Figure 6–99. Note that each of the drawings have details that are impacted by the change to the footings.

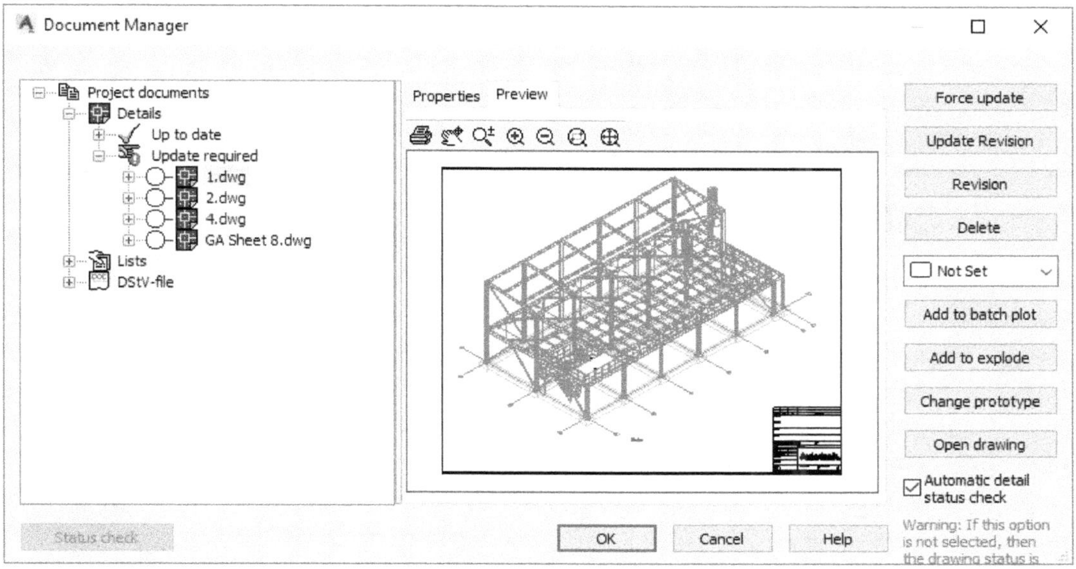

Figure 6–99

8. Select **1.dwg** and click **Update Revision**.

9. In the *Add revision mark to* dialog box, click **Add**.

10. In the *Revision mark* dialog box, add the information shown in Figure 6–100. You can use your own initials as the *Author*. Click **OK**.

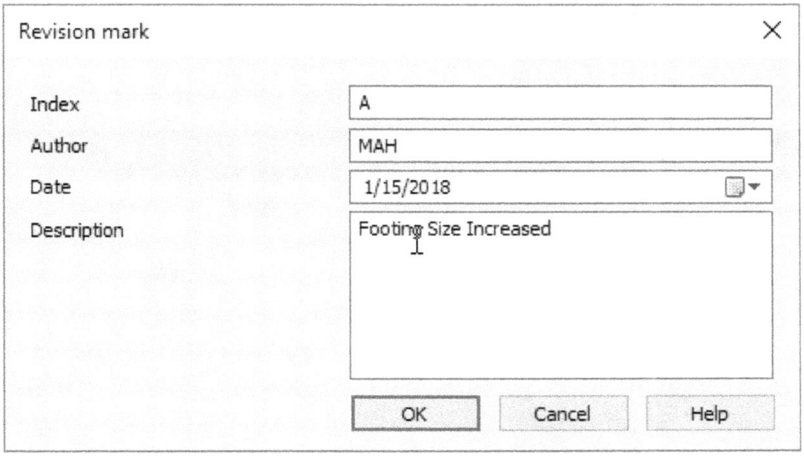

Figure 6–100

11. The program processes the revision and a new drawing is created using the *Index* number. The changes are also noted in the drawing with revision bubbles, as shown in Figure 6–101.

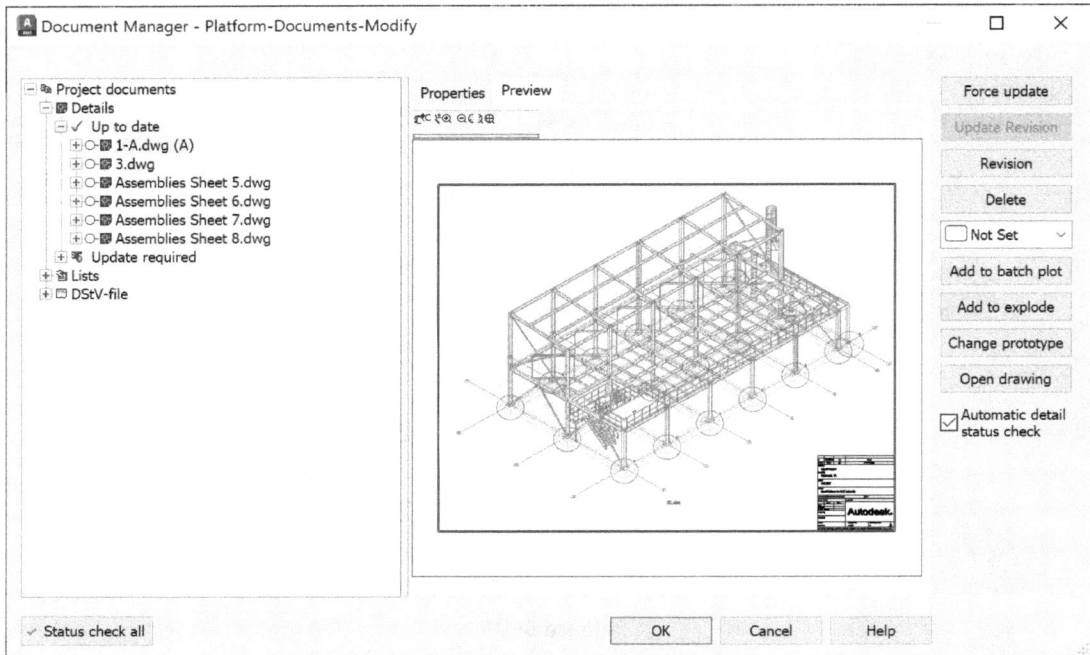

Figure 6–101

12. Open the new **1-A.dwg (A)**.

13. Zoom in on the title block and review the revision information.

14. Close the detail drawing.

15. In the model drawing, add revisions to the rest of the drawings.

16. Save and close the model drawing.

End of practice

Chapter Review Questions

1. At what point in the project do you need to number steel objects?

   a. They are automatically numbered as they are added.

   b. As you finish working on a level.

   c. Before you produce the fabrication drawings.

   d. As you produce the fabrication drawings.

2. Which of the following statements is true about creating fabrication drawings in the Autodesk Advance Steel software?

   a. To create 2D drawings, you need to create layouts, add title blocks, and then make viewports of different parts of the model.

   b. To create 2D drawings, you need to use the tools on the *Quick Document* palettes to create the drawings.

   c. To create 2D drawings, you need to open the *Document Manager* and assign parts of the model to different drawings.

   d. To create 2D drawings, you need to export the model to the *Document Manager*.

3. What is the difference between Drawing Styles and Drawing Processes?

   a. Drawing Styles create the title block layouts, while Drawing Processes create the views.

   b. Drawing Styles are impacted by the UCS, while Drawing Processes are not.

   c. Drawing Styles are used primarily for overall views, while Drawing Process are used for detail views.

   d. Drawing Styles assign the title block, scale, and rules that create a detail. Drawing Processes are macros that apply a series of drawing styles to parts.

4. The *Document Manager* is where you create detail drawings.

   a. True

   b. False

5. Where are callout views and cut views created?

   a. In the detail drawings.

   b. In the model drawing.

6. To change the size of a parametric dimension in a detail drawing, you make the changes in the model drawing.

   a. True

   b. False

7. Which of the following statements are true about revisions? (Select all that apply.)

   a. You can force an update without applying revisions.

   b. Revisions must be applied before you can update drawings.

   c. Revisions are only added to detail drawings.

   d. Revision clouds display in detail drawings and in the model.

Command Summary

Button	Command	Location
	Change prototype file	• **Ribbon:** *Labels & Dimensions* tab>*Parametric views* panel
	Create callout view	• **Ribbon:** *Labels & Dimensions* tab>*Parametric views* panel
	Create camera, UCS	• **Advance Tool Palette**>*Tools* category
	Create camera(s) at node	• **Advance Tool Palette**>*Tools* category
	Create cut view	• **Ribbon:** *Labels & Dimensions* tab>*Parametric views* panel
	Delete point	• **Ribbon:** *Labels & Dimensions* tab>*Parametric dimensions* panel
	Delete view	• **Ribbon:** *Labels & Dimensions* tab>*Parametric views* panel
	Document Manager	• **Ribbon:** *Home* tab>*Documents* panel • **Ribbon:** *Output* tab>*Document Manager* panel
	Drawing Processes	• **Ribbon:** *Home* tab>*Documents* panel • **Ribbon:** *Output* tab>*Documents* panel
	Drawing Styles	• **Ribbon:** *Home* tab>*Documents* panel • **Ribbon:** *Output* tab>*Documents* panel
	Insert points	• **Ribbon:** *Labels & Dimensions* tab>*Parametric dimensions* panel
	Management Tools	• **Ribbon:** *Home* tab>*Settings* panel
	Model Browser	• **Advance Tool Palette**>*Selection* category
	Move view	• **Ribbon:** *Labels & Dimensions* tab>*Parametric views* panel
	Numbering	• **Ribbon:** *Home* tab>*Documents* panel
	Parametric Dimensions	• **Ribbon:** *Labels & Dimensions* tab>*Parametric dimensions* panel

Button	Command	Location
	Toggle the display of green frames around details	• **Ribbon:** *Labels & Dimensions* tab>*Parametric views* panel
	Update defaults	• **Ribbon:** *Home* tab>*Settings* panel

Bills of Materials and Numerical Control Files

In addition to the typical fabrication drawings, there are other types of files that can be created from the data stored in the Autodesk® Advance Steel 3D model. Bill of Materials (BOM) files can be exported and saved to .PDFs or to files that can be reviewed in the *Document Manager*. These BOMs can provide you with many types of lists, including lists of all of the bolts in the model, to all of the items that need to be shipped to a site. Additionally, Numerical Control (NC) machines can take data from the Autodesk Advance Steel model and fabricate the materials without requiring someone to read the drawings. You can export .DXF or .NC files from the model through simple commands.

Learning Objectives

- Extract BOM lists.
- Export data to .DXF or .NC files.

7.1 Extract BOM Lists

Some BOM lists are added to detail drawings when you run drawing processes on the model. Other BOM lists are created as export-ready files that can be reviewed in the *Document Manager*. The easiest way to create these files is to use the options on the *BOM Templates* palette, shown in Figure 7–1.

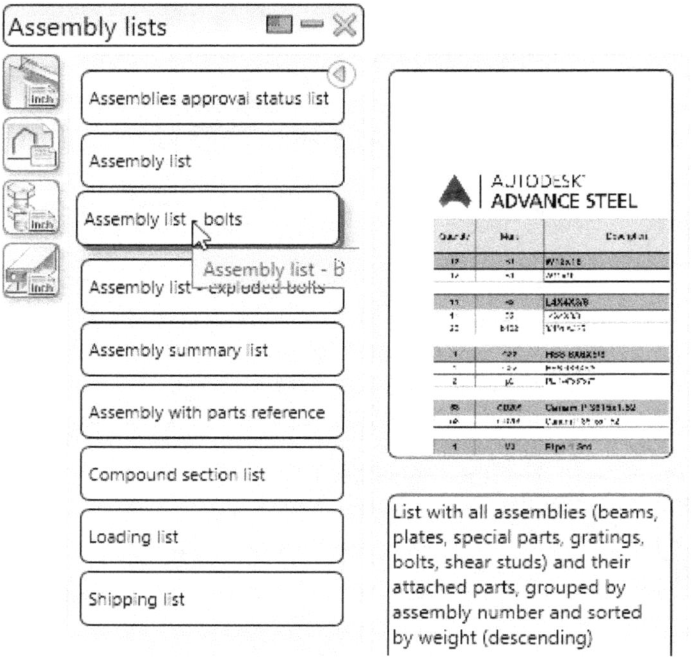

Figure 7–1

How To: Use the BOM Templates Palette

1. In the *Output* tab>*Documents* panel, click (BOM Templates).
2. Select one of the categories, including:

    *   (Assembly lists)
    *   (Drawing lists)
    *   (Fasteners lists)
    *   (Parts lists)

3. Select the type of BOM list that you want to create.
4. The model is searched and the BOM is created in a separate window, as shown in Figure 7–2.

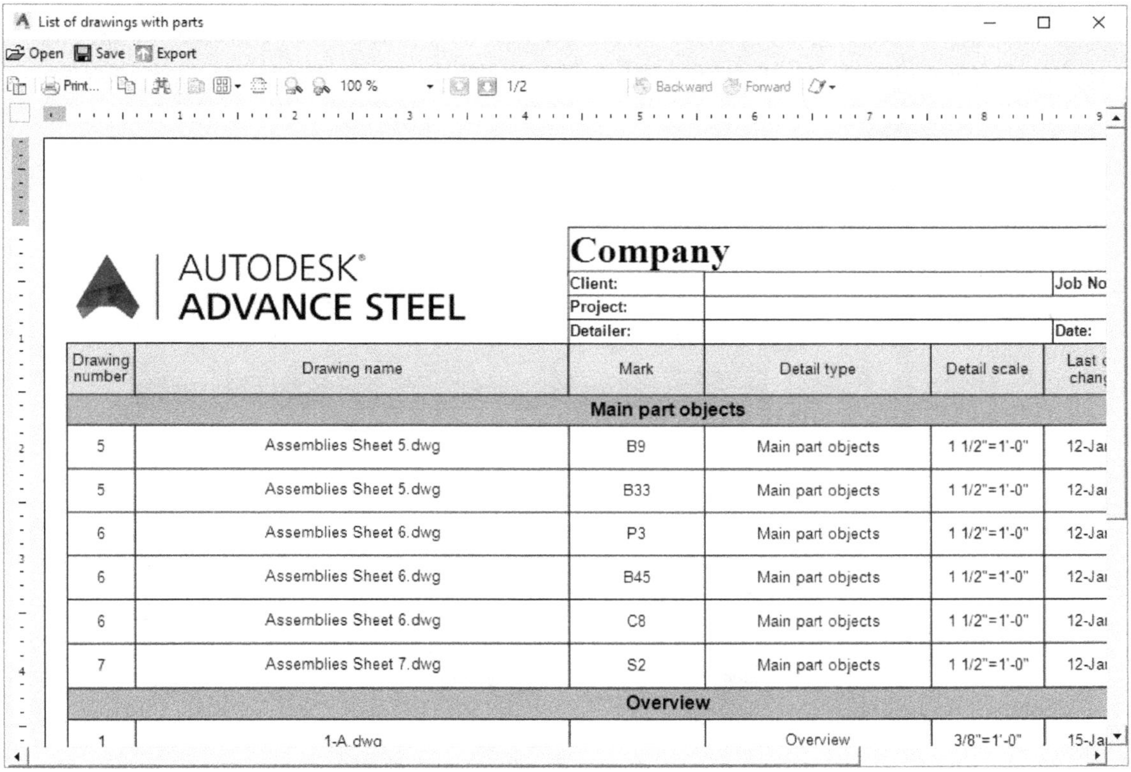

Drawing number	Drawing name	Mark	Detail type	Detail scale	Last c chan
Main part objects					
5	Assemblies Sheet 5.dwg	B9	Main part objects	1 1/2"=1'-0"	12-Jai
5	Assemblies Sheet 5.dwg	B33	Main part objects	1 1/2"=1'-0"	12-Jai
6	Assemblies Sheet 6.dwg	P3	Main part objects	1 1/2"=1'-0"	12-Jai
6	Assemblies Sheet 6.dwg	B45	Main part objects	1 1/2"=1'-0"	12-Jai
6	Assemblies Sheet 6.dwg	C8	Main part objects	1 1/2"=1'-0"	12-Jai
7	Assemblies Sheet 7.dwg	S2	Main part objects	1 1/2"=1'-0"	12-Jai
Overview					
1	1-A.dwg		Overview	3/8"=1'-0"	15-Ja

Figure 7–2

5. In the *List* window, you can review, print, export, and save the sheets, as required.

- To export the BOM:

   a. In the *List* window, click ⬆ (Export).

   b. In the *Report export* dialog box, select the type of *Export Format* you want to use (as shown in Figure 7–3) and set the *Export Options*.

   c. Click **OK**. The file is saved in the model's drawing folder in the *BOM* subfolder.

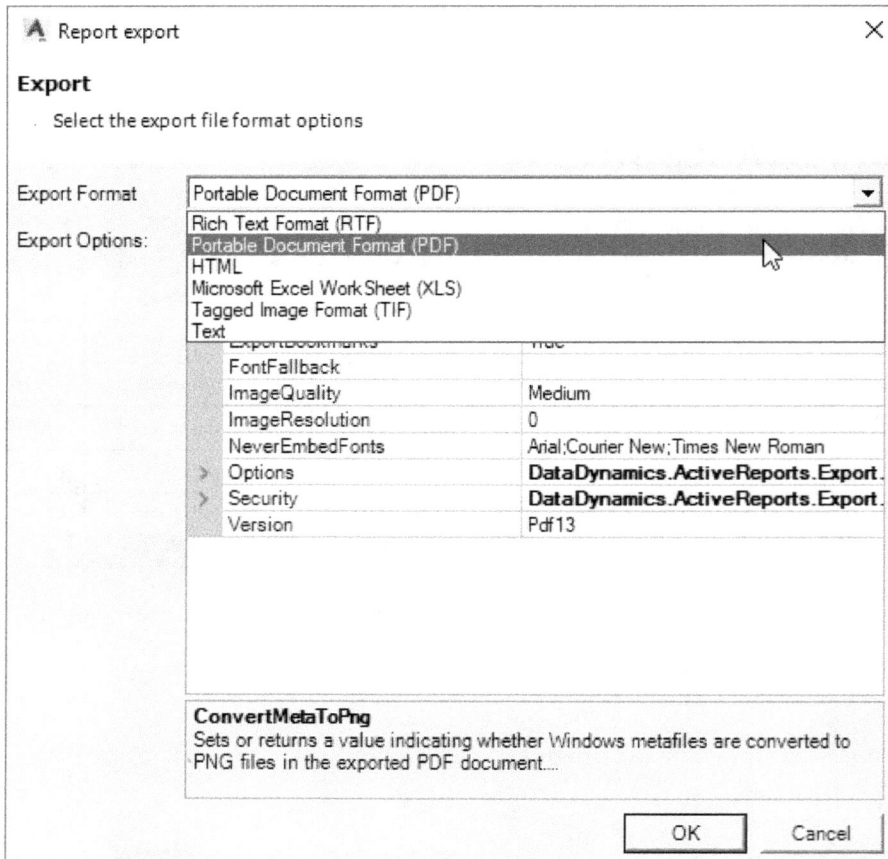

Figure 7–3

Note: As with other automatically created drawings, the BOM lists need to be saved in the correct location to be viewed and updated through the Document Manager.

- To save the BOM, click 💾 (Save). The file is saved as an RDF file in the model's drawing folder in the BOM subfolder. You can view the file through the *Document Manager*.

- To view saved BOM lists, open the *Document Manager*, and expand the *Lists* node, as shown in Figure 7−4. You can view the information In the *Preview* tab and print the sheet.

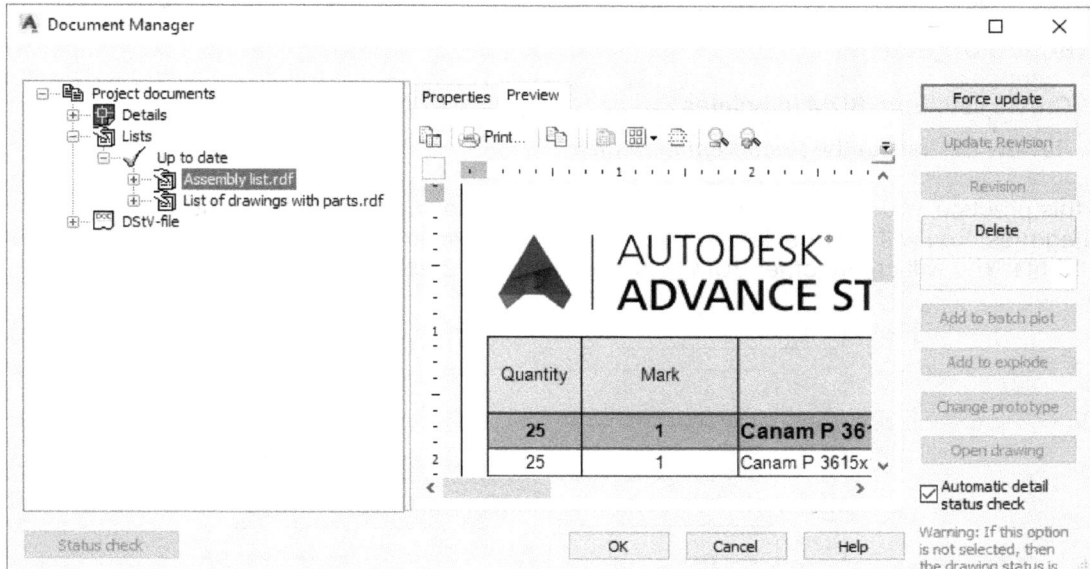

Figure 7−4

Practice 7a
Extract BOM Lists

Practice Objectives

- Create lists from BOM templates.
- Review the lists in the *Document Manager*.

In this practice, you will open the BOM templates palette and explore the options in the categories. You will use the **Saw list pictures** tool (shown in Figure 7–5) to create and save a .RDF file. You will create other BOM lists and then view them in the *Document Manager*.

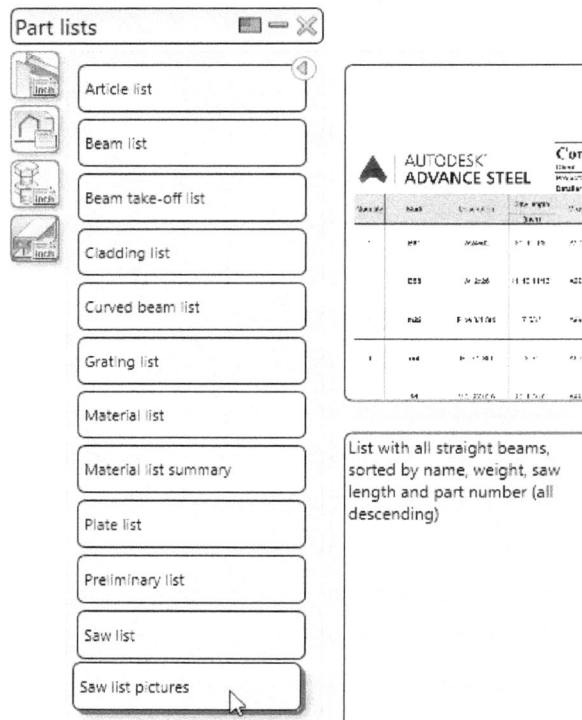

Figure 7–5

Task 1: Extract BOM lists.

1. In the practice files folder, open **Platform-Documents-BOM.dwg**.

2. In the *Output* tab>*Documents* panel, click (BOM Templates).

3. In the *BOM templates* palette, review the various options.

4. In the (Parts lists) category, click **Saw list pictures**. This creates a list as outlined in the information in the tooltip, shown in Figure 7–6.

> List with all straight beams, sorted by name, weight, saw length and part number (all descending)

Figure 7–6

5. When the program finishes processing, the list opens in a separate window, as shown in Figure 7–7.

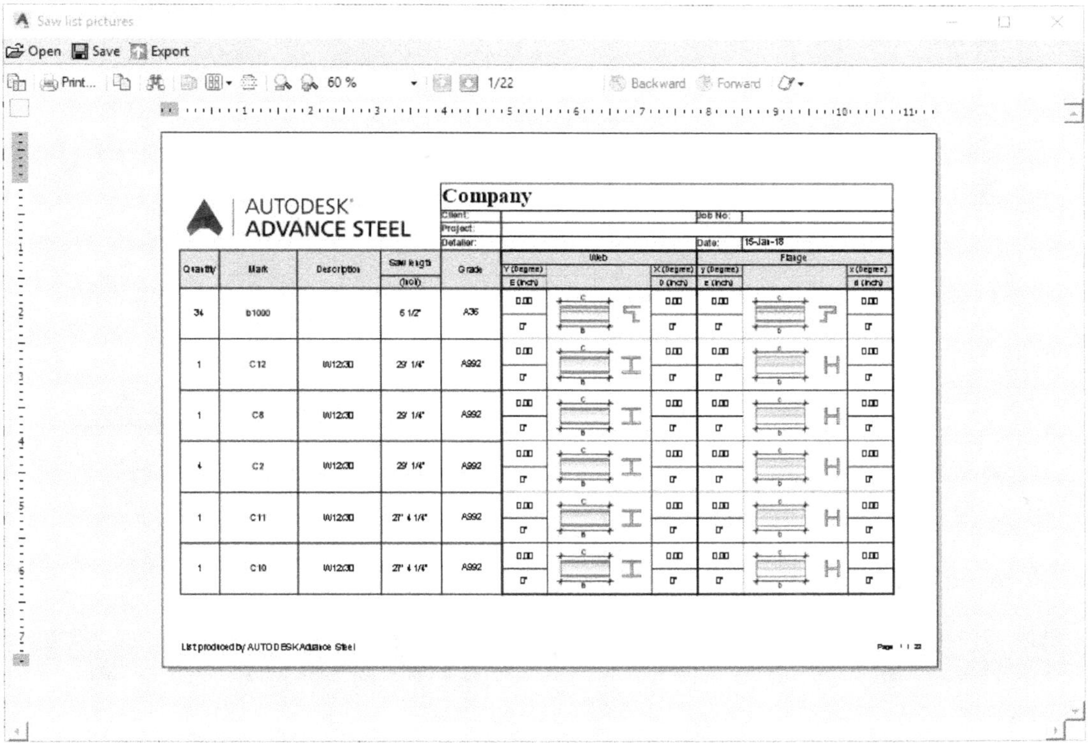

Figure 7–7

6. Review the information by scrolling through the list.

7. Click (Save).

8. In the *Save As* dialog box, ensure that the file is being saved to the practice files>*Platform-Documents-BOM>BOM* folder and click **Save**.

9. Create several other lists from other categories in the BOM templates palette and save them to the same folder.

10. Close all of the extra windows.

Task 2: Review the saved list in the Document Manager.

1. In **Platform-Documents-BOM.dwg**, open the *Document Manager*.

2. Expand the **Lists>Up to date** node and review the lists within the *Preview* tab, as shown in Figure 7–8.

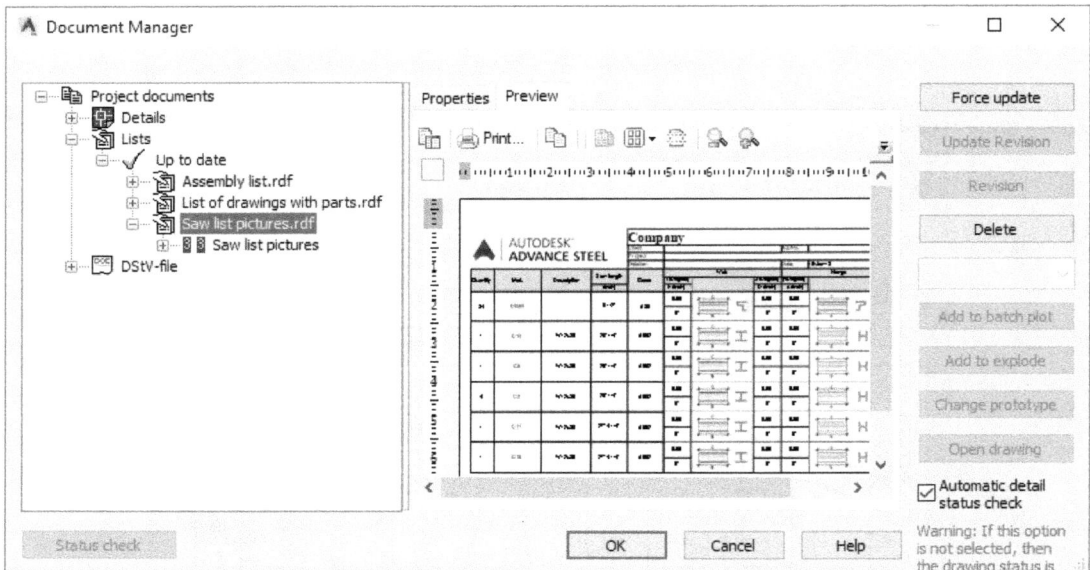

Figure 7–8

3. Close the *Document Manager*.

End of practice

7.2 Exporting Data to .NC and .DXF Files

Many fabrication shops include CNC machines. Autodesk Advance Steel geometry can be exported out as .DXF drawings (as shown in Figure 7–9), which are viewed through the *Document Manager*, and numerical control specification data that runs on these machines.

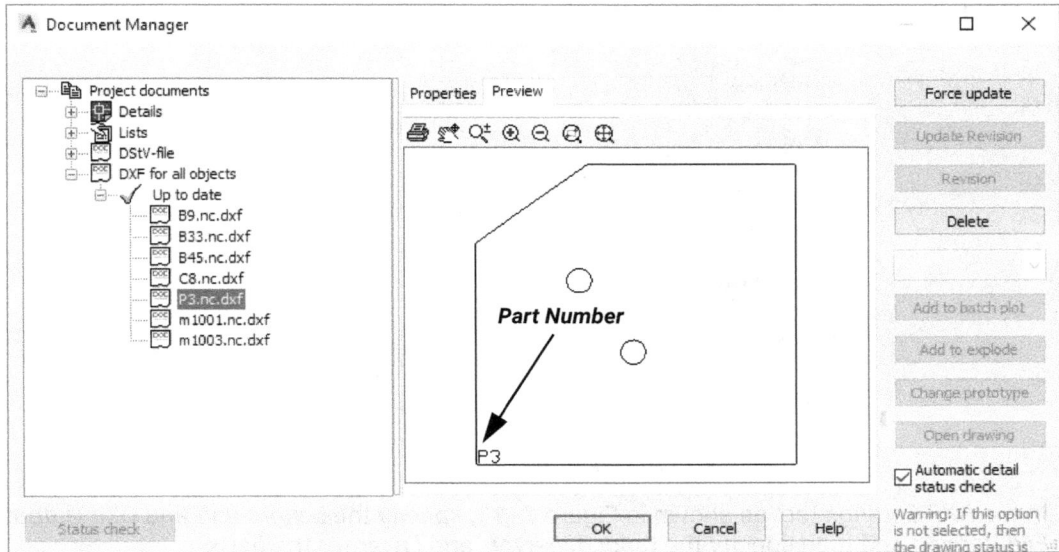

Figure 7–9

- The .DXF export only works for plates and beams. The part number is engraved on the object.

How To: Modify .NC and .DXF Settings

1. In the *Home* tab>*Documents* panel or the *Output* tab>*NC&DXF* panel, click ⬛ (NC Settings).

2. In the *NC Settings* tab (shown in Figure 7–10), specify whether the files are created using *Main Part Numbers* or *Single Part Numbers*. Assign the other settings as required.

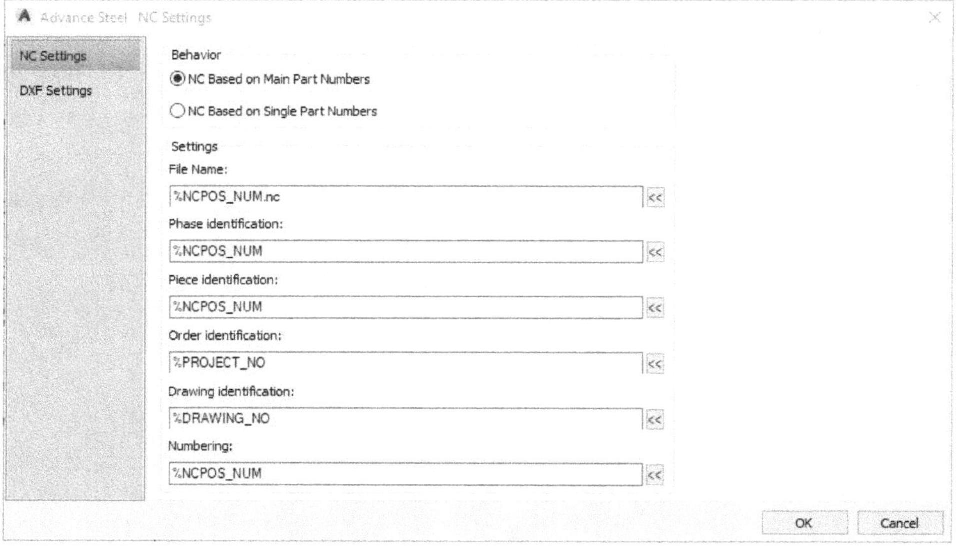

Figure 7–10

3. In the *DXF Settings* tab, as shown in Figure 7–11, specify the *Length unit* and *DXF Output File Version* and then specify the *Color, Linetype,* and *Layer* for the parts.

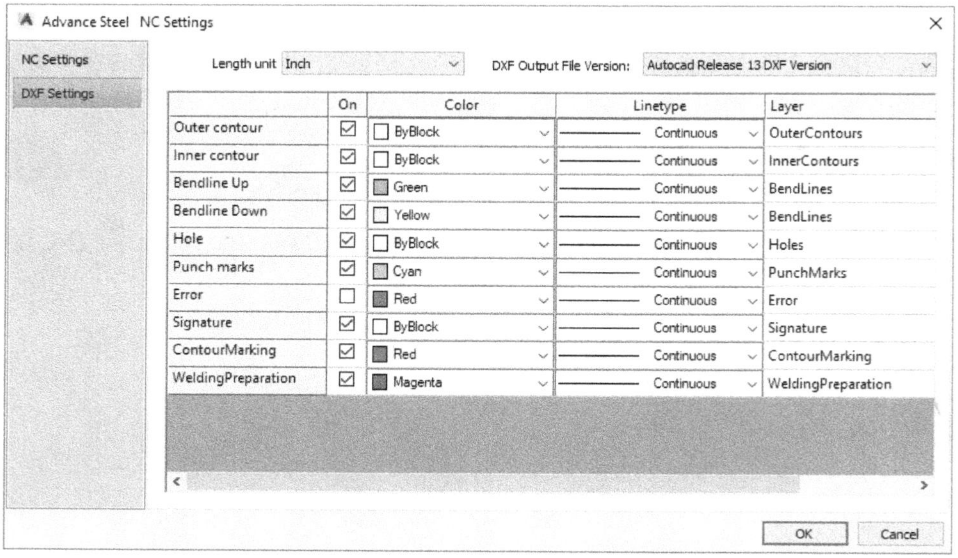

Figure 7–11

4. Click **OK**.

How To: Run .NC or .DXF processes

1. In the *Home* tab>*Documents* panel or *Output* tab>*NC&DXF* panel, click 🔲 (NC) or 🔲 (DXF (all objects)).

   - If you do not want to process the entire model, you can select the objects you want to export before starting the command.

2. The model is processed and the data is extracted.

3. The files are saved in the model's drawing folder in the *DStv>NC* subfolder. This is where you go to select the files to send to your NC machine.

- You can view, but not open, the information from the *Document Manager*, as shown for an .NC file in Figure 7–12.

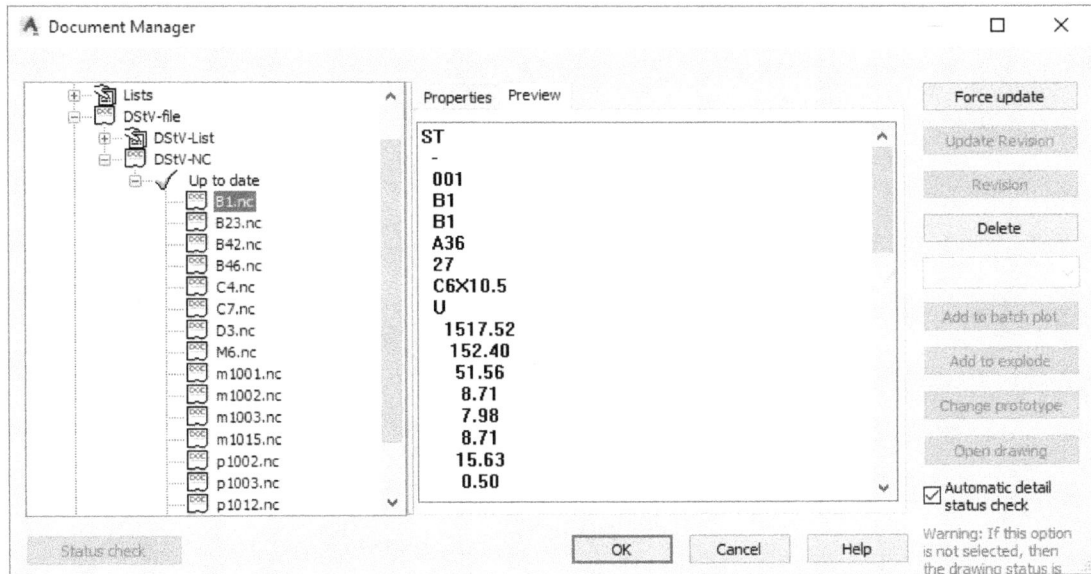

Figure 7–12

- Many NC machines read numbers, rather than drawings.

Practice 7b
Export Data to .NC and .DXF Files

Practice Objectives

- Review .NC and .DXF Settings.

- Export model information to .NC and .DXF files.

- Review the .NC and .DXF files in the *Document Manager* and Windows Explorer.

In this practice, you will review the .NC and .DXF settings. You will then run the .NC and .DXF processes on the model and review the information created in the *Document Manager*, as shown in Figure 7–13.

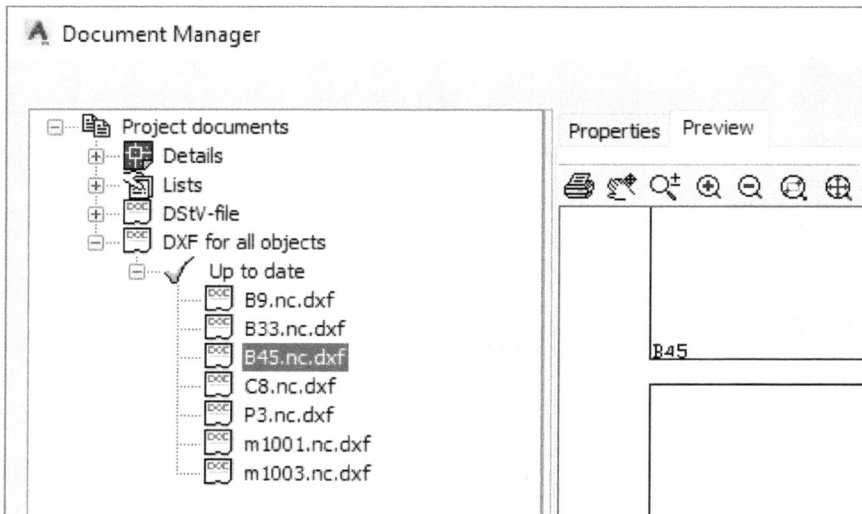

Figure 7–13

Task 1: Review .NC and .DXF settings and run the processes.

1. In the practice files folder, open the project **Platform-Documents-BOM.dwg**.

2. In the *Home* tab>*Documents* panel or *Output* tab>*NC&DXF* panel, click 🔲 (NC Settings).

3. In the *NC Settings* dialog box, review the settings on the *NC Settings* tab and the *DXF Settings* tab.

4. Close the dialog box.

5. In the model, select the objects shown in Figure 7−14.

    - It is OK if you select the grids - only the appropriate objects will be processed.

Figure 7−14

6. In the *Home* tab>*Documents* panel or the *Output* tab>*NC&DXF* panel, click 🔲 (DXF (all objects)). The selected objects are processed.

7. With the same selected objects, in the *Home* tab>*Documents* panel or the *Output* tab>*NC&DXF* panel, click 🔲 (NC). All of the selected objects in the model are processed.

8. Save the drawing.

Task 2: Review the files in the Document Manager and Windows Explorer.

1. Open the *Document Manager*.

2. Expand the **DStv-file>DStV-NC>Up to date** node and review the list of .NC files, as shown in Figure 7–15.

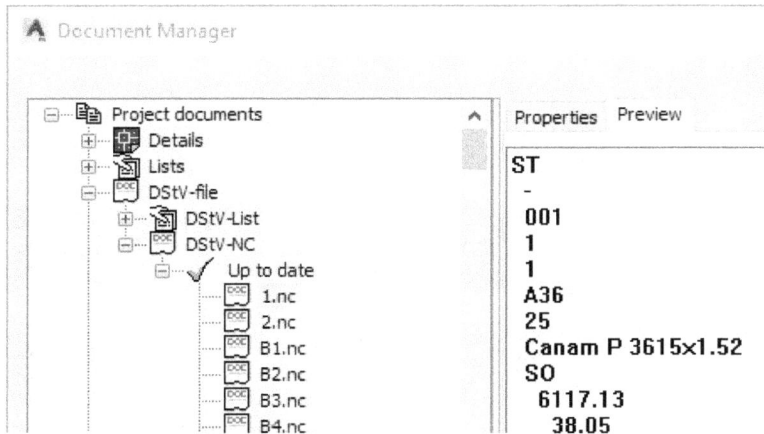

Figure 7–15

3. Close the **DStV-file** nodes and expand the **DXF for all objects>Up to date** nodes. Review the much shorter list.

4. Select one of the .DXF files, as shown in Figure 7–16. In the *Preview* tab, zoom in to view the data.

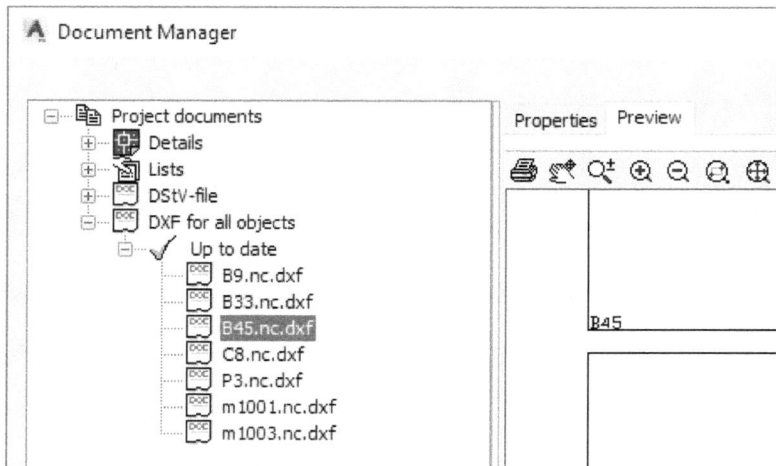

Figure 7–16

5. Close the *Document Manager*.

6. Open Windows Explorer and navigate to the practice files folder. Find the **Platform-Modify-Documents.dwg** file and then open the related *Platform-Modify-Documents* folder.

7. Navigate to the *DStV>NC* folder and review its contents, shown in Figure 7–17. Note that both the .NC files and the .DXF files are stored here.

   Note: *The .UPD files link the .NC and .DXF files to the model. Do not remove them.*

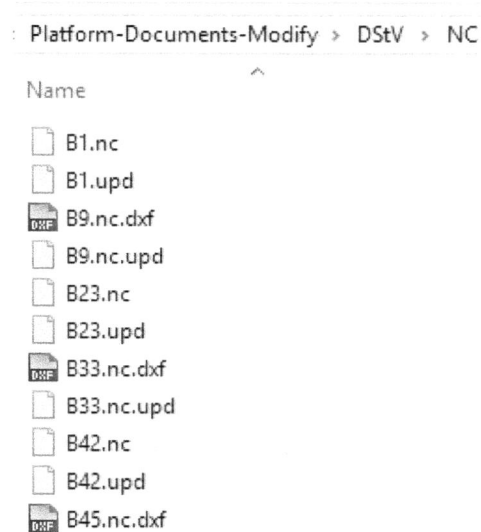

Platform-Documents-Modify > DStV > NC

Name

☐ B1.nc
☐ B1.upd
▨ B9.nc.dxf
☐ B9.nc.upd
☐ B23.nc
☐ B23.upd
▨ B33.nc.dxf
☐ B33.nc.upd
☐ B42.nc
☐ B42.upd
▨ B45.nc.dxf

Figure 7–17

8. Close Windows Explorer.

9. Save the model drawing.

End of practice

Chapter Review Questions

1. What do you need to run to create a BOM file as a separate document?
   a. Drawing Processes
   b. Drawing Styles
   c. BOM Templates
   d. Create Lists

2. What file type do you need to save a BOM file (such as the Bolt List shown in Figure 7–18) as to read it in the *Document Manager*?

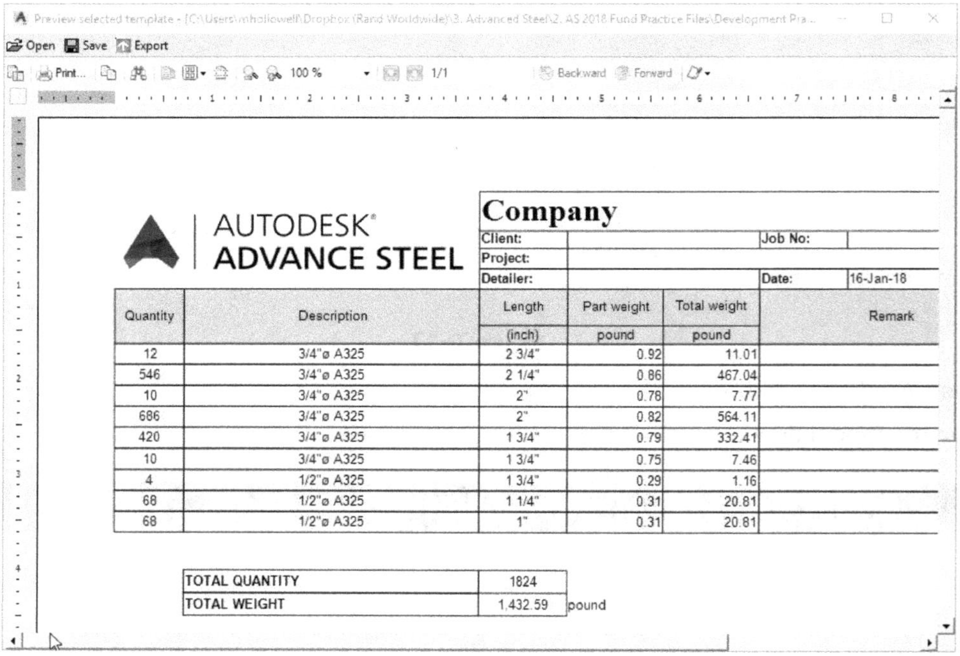

Company			
Client:		Job No:	
Project:			
Detailer:		Date:	16-Jan-18

Quantity	Description	Length (inch)	Part weight pound	Total weight pound	Remark
12	3/4"ø A325	2 3/4"	0.92	11.01	
546	3/4"ø A325	2 1/4"	0.86	467.04	
10	3/4"ø A325	2"	0.78	7.77	
686	3/4"ø A325	2"	0.82	564.11	
420	3/4"ø A325	1 3/4"	0.79	332.41	
10	3/4"ø A325	1 3/4"	0.75	7.46	
4	1/2"ø A325	1 3/4"	0.29	1.16	
68	1/2"ø A325	1 1/4"	0.31	20.81	
68	1/2"ø A325	1"	0.31	20.81	

TOTAL QUANTITY	1824	
TOTAL WEIGHT	1,432.59	pound

Figure 7–18

   a. .DOC
   b. .XLS
   c. .RDF

3. If you want to create a .DXF file from only part of a model, you need to select the model objects first and then run the ☐ (DXF (all objects)) command.
   a. True
   b. False

4. .NC files can be copied to a server where an NC machine can read it.
   a. True
   b. False

Command Summary

Button	Command	Location
	BOM Templates	• **Ribbon**: *Home* tab>*Documents* panel • **Ribbon**: *Output* tab>*Documents* panel
	DXF (all objects)	• **Ribbon**: *Home* tab>*Documents* panel • **Ribbon**: *Output* tab>*NC&DXF* panel
	NC	• **Ribbon**: *Home* tab>*Documents* panel • **Ribbon**: *Output* tab>*NC&DXF* panel
	NC Settings	• **Ribbon**: *Home* tab>*Documents* panel • **Ribbon**: *Output* tab>*NC&DXF* panel

Advance Steel Prototypes

Autodesk® Advance Steel provides drawing prototypes, in both imperial and metric, that act as templates for drawing sheets and are used to create drawing views. Drawing views can include drawing frame, title block, revision table, BOM, dimension, text, and line styles. You can use the Autodesk AutoCAD editing tools to customize and build your prototypes to suite company needs.

Learning Objectives

- Edit prototypes.
- Set up prototype units and project settings.
- Define paper size.
- Edit title block page frame and page header.
- Modify BOM list position.
- Add and adjust a reserved area.
- Define the detail arrangement settings.
- Configure dimension styles during prototype editing.
- Review the *Management Tools* dimension styles.
- Load linetypes.

8.1 Editing a Prototype Drawing

Advance Steel prototype drawings can be edited as needed and reused in various projects. It is recommended that you start with the files that are provided by the Advance Steel software, as shown in Figure 8–1, or copy files from an existing drawing.

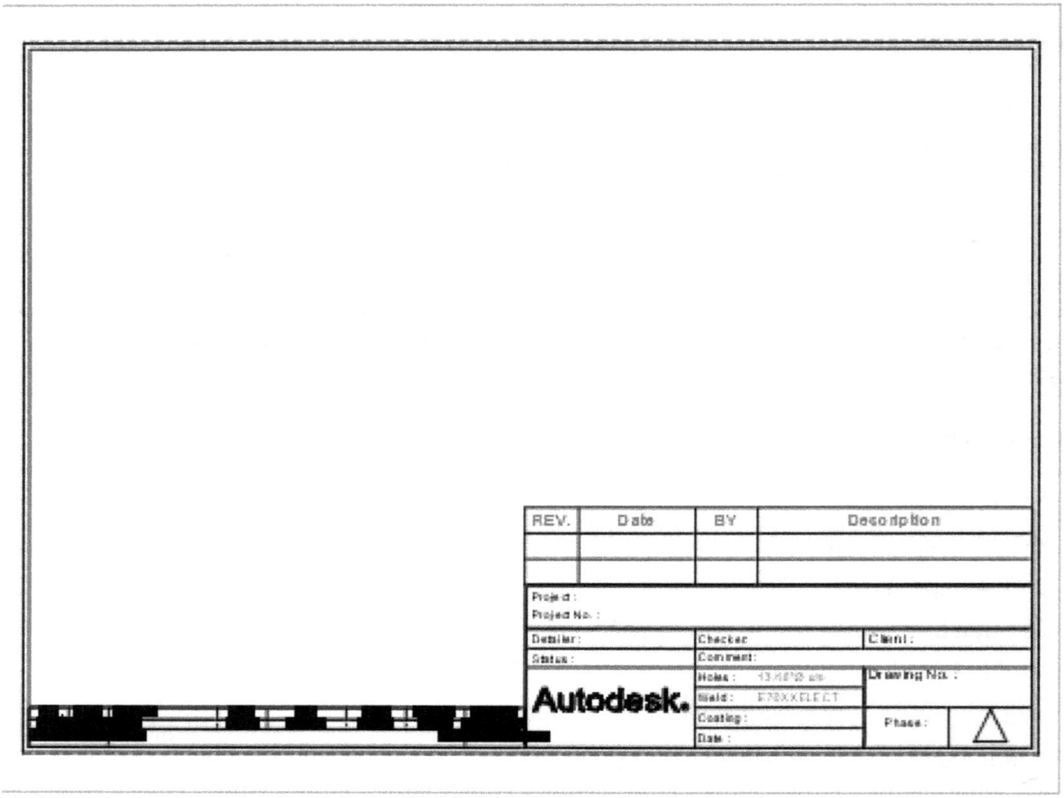

Figure 8–1

How To: Edit a Drawing Prototype

1. In the *Output* tab>*Document Manager* panel, click <img> (Edit Prototypes). The *Edit default prototypes* dialog box displays as shown in Figure 8–2.

2. Select the appropriate file and click **Open**.

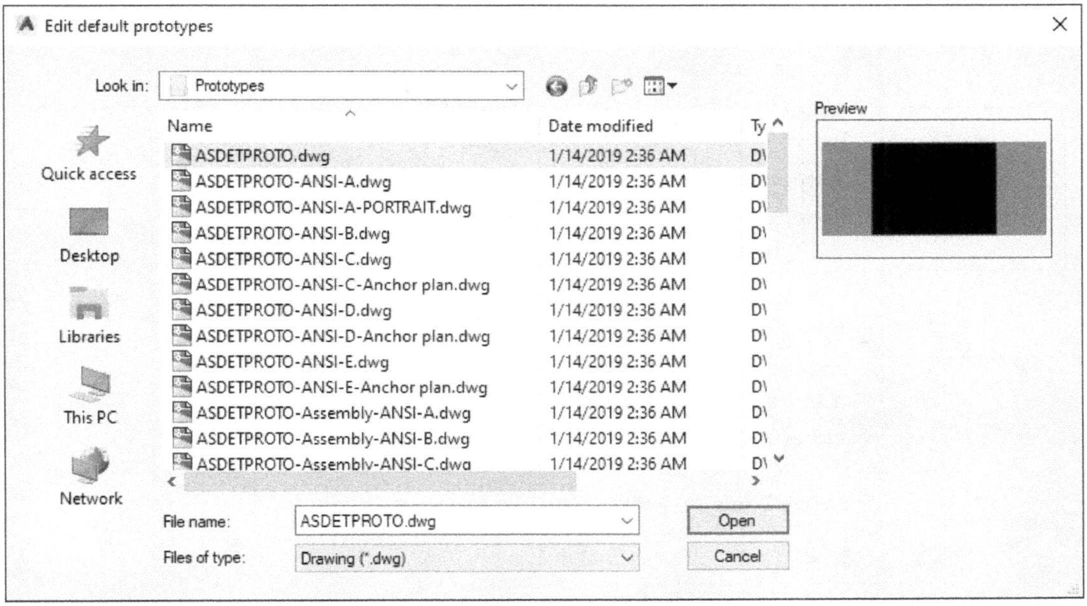

Figure 8–2

3. In the Quick Access Toolbar or Application Menu, click <img> (Save As) and save the file with a different name to avoid overwriting the original file.

Prototype Units

The *Drawing Units* dialog box in Advance Steel determines what precision is used, in either Imperial or Metric units. With the *Drawing Units* dialog box, you can control the **Type** and **Precision** of the *Length* and *Angle*, whether the angle will dimension clockwise, the **Units to scale inserted content**, and the **Units for specifying the intensity of lighting** as shown in Figure 8–3.

Figure 8–3

How To: Set the Drawing Units

1. In the ▲ (Application Menu), select ✎ (Drawing Utilities)> `0.0` (Units). Alternatively, at the command line, type **UNITS** and press <Enter>.

    Note: It is important to verify that the insertion scale matches the units of the drawing to avoid incorrect scaling.

2. In the *Length* section, select an option in the *Type* and *Precision* drop-down lists.

3. In the *Angle* section, select an option in the *Type* and *Precision* drop-down lists and set the angle rotation to be **Clockwise** from **0** as required.

4. In the *Insertion scale* section, select an option in the *Units to scale inserted content* drop-down list.

5. In the *Lighting* section, select an option in the *Units for specifying the intensity of lighting* drop-down list (this is only required if you are working in 3D).

Edit Project Settings

Editing the Project Settings allows you to customize the prototype drawings project data, including the *Project Info, Preferences, Weight, Length, Angle, Area*, and *Volume unit* and user attributes.

How To: Modify Project Data

1. In the *Home* tab>*Settings* panel, click 🖽 (Project Settings...). The *Project data* dialog box opens, as shown in Figure 8–4.

2. Set the units as required.

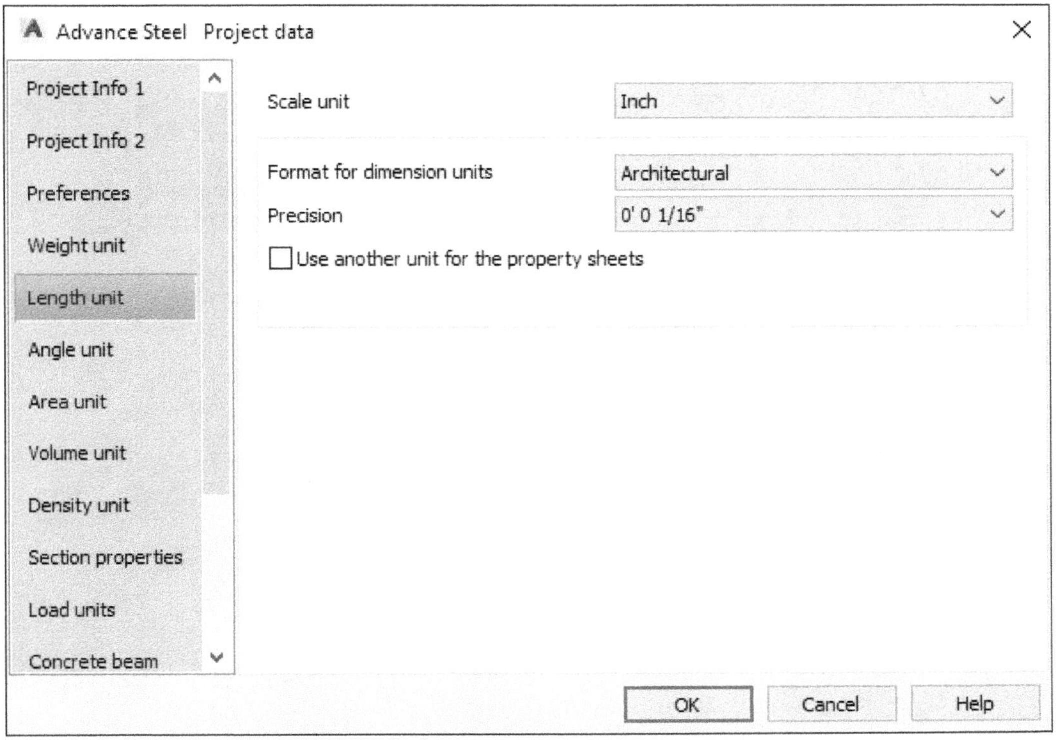

Figure 8–4

8.2 Working with the Page Setup Manager

Part of editing a drawing prototype is setting up the paper size. You will use the same *Page Setup Manager* as the Autodesk AutoCAD software.

You can simplify your day-to-day work by creating layouts in your template files to match the printers and paper sizes you normally use, as shown in Figure 8−5. These layouts are then ready to use in new drawings based on the templates.

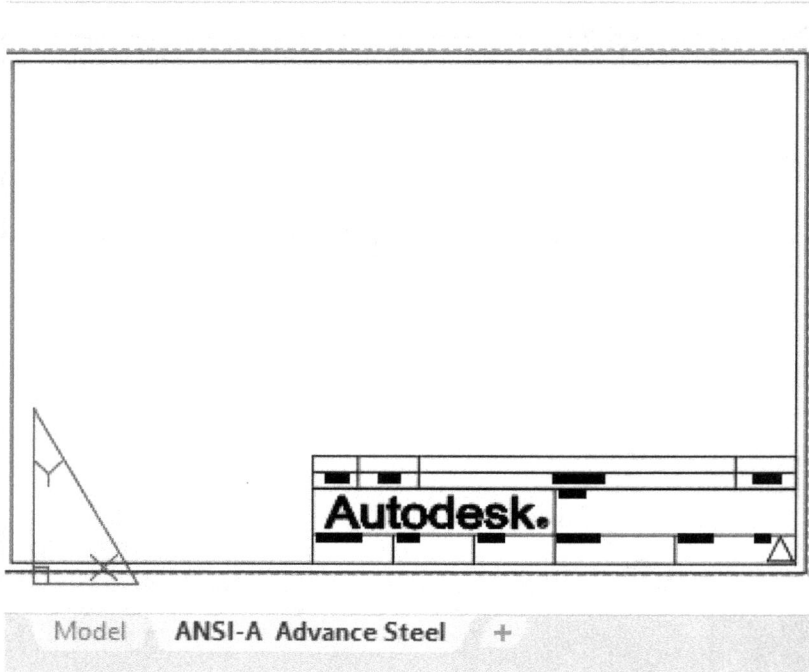

Figure 8−5

Define Paper Size

The *Page Setup Manager* allows you to define printer settings, paper size, and standards as shown in Figure 8-6. You can assign an existing page setup to the current layout, create new page setups, modify existing page setups, and import page setups from another file.

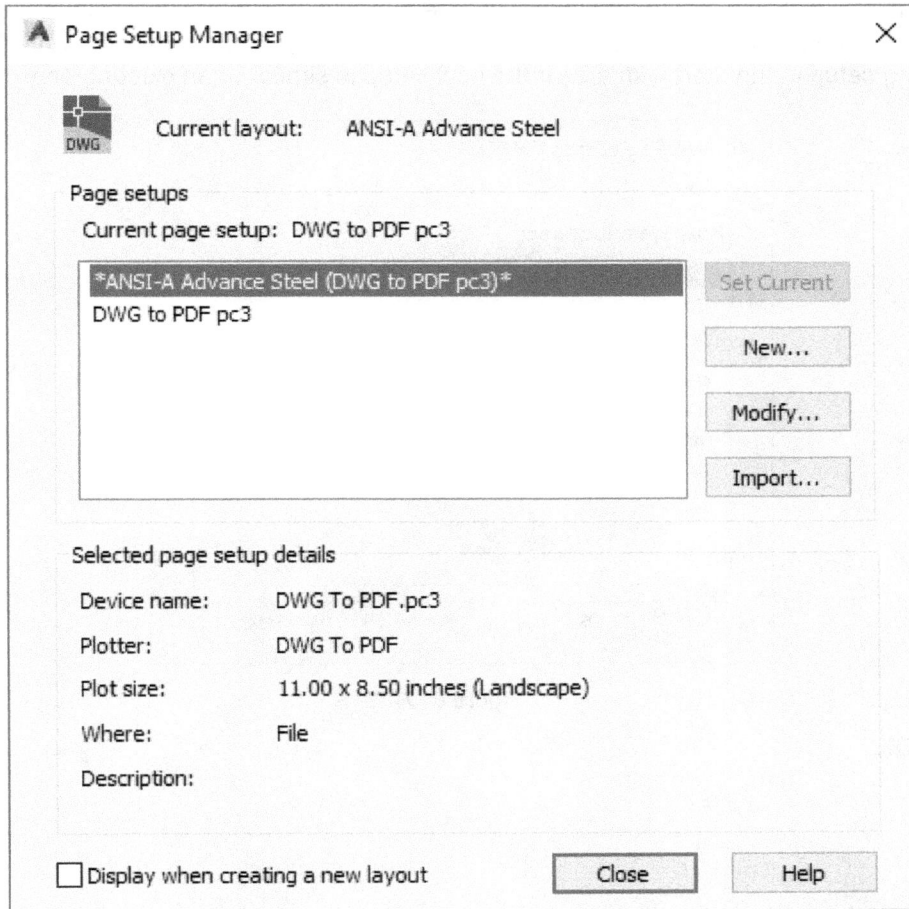

Figure 8-6

How To: Create a Page Setup

1. In the *Output* tab>*Plot* panel, click ▣ (Page Setup Manager). Alternatively, in the Application Menu, select **Print>Page Setup**.

2. In the *Page Setup Manager* dialog box, click **New**.

3. In the *New Page Setup* dialog box, shown in Figure 8–7, type a name for the setup. Select an existing setup in the *Start with* area if the new setup is similar to an existing one.

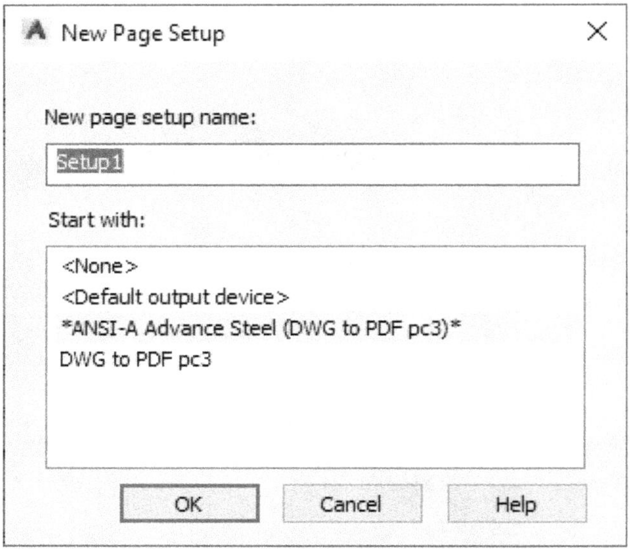

Figure 8–7

4. Click **OK**.

5. In the *Page Setup* dialog box, shown in Figure 8-8, select the printer or plotter that you want to use. This determines the paper sizes from which you can select.

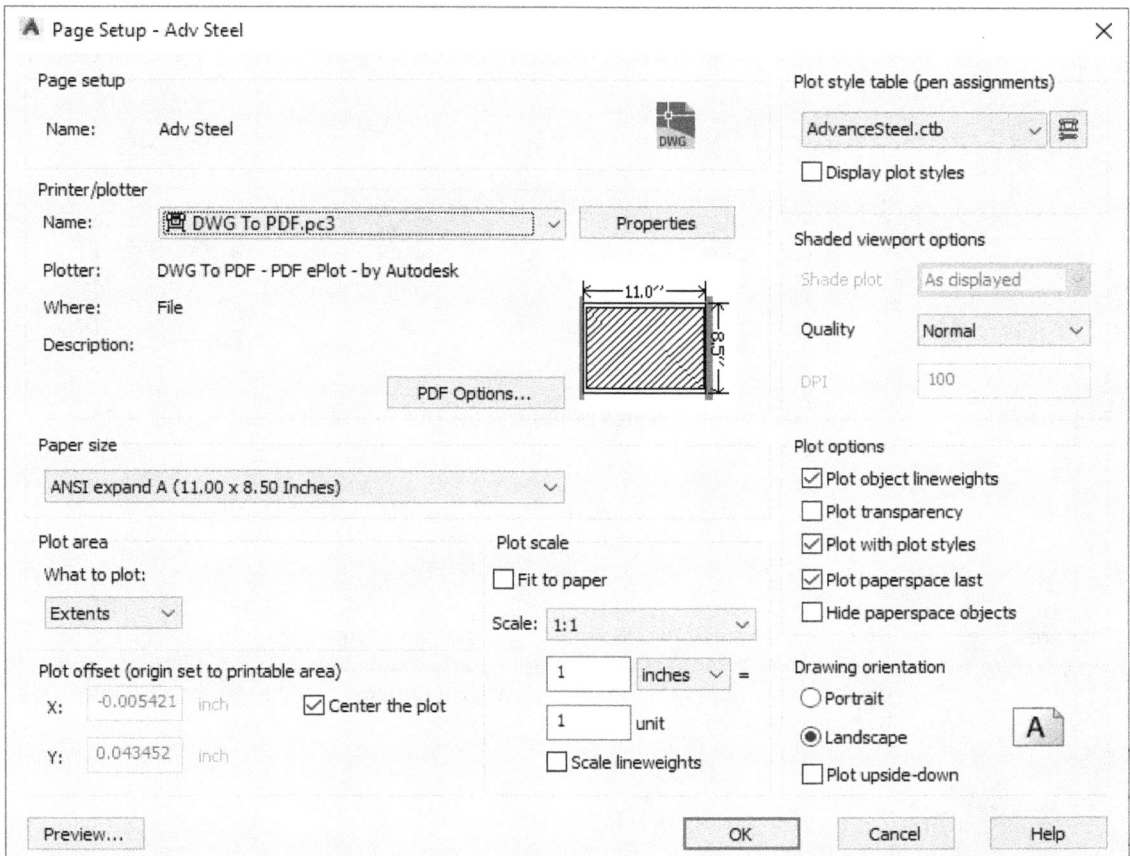

Figure 8-8

6. Specify the *Paper size, Plot area, Plot offset, Plot scale, Plot style table, Shaded viewport options, Plot options,* and *Drawing orientation.*

7. Click **Preview...** to display a preview of how the setup is going to print on the sheet.

8. Right-click in the preview and select **Exit**.

9. When the page setup is finished, click **OK** to return to the *Page Setup Manager.* The new page setup can now be applied to a layout.

Page Setup Options

Printer/ plotter Enables you to select from the list of available printing devices. Check with your CAD manager if the printer/plotter you want to use is not in the list. The AutoCAD software includes several predefined plotter configurations, such as DWF6 e-plot.

Paper size Enables you to set the size of the layout. The available sizes depend on the selected plotter.

Plot area Sets the printable area. Normally, you use the **Layout** option to plot the entire layout to the extents of the printable area. You can also print the **Extents** of the drawing, the **Display** in the drawing area, a **Window** that you select, or a **View** that has been defined in the drawing.

Plot offset Controls where the drawing starts to plot on the paper. Depending on your plotter, you might need to set this so that the drawing fits correctly on the paper. The **Center the plot** option is not available when the *Plot area* is set to **Layout**.

Plot scale Sets the scale when you are printing from a layout. The default scale is 1:1. However, you can set a different scale if you are creating a check plot on a smaller piece of paper.

Plot scale

☐ Fit to paper

Scale: 1:1

1 inches ∨ =

1 unit

☐ Scale lineweights

IMPORTANT: The *Plot scale* for a layout is almost always 1:1 because the layout is created at the actual size required to fit on the piece of paper. The scaling of the model is done using the Viewport Scale.

Plot style table Coordinates the layer color to pen weight or sets up other special effects for plotted output. Consult your CAD manager about which one you should use.

Plot style table (pen assignments)

AdvanceSteel.ctb

☐ Display plot styles

Shaded viewport options For 3D models, this enables you to set viewports to be hidden or rendered and to control the image quality.

Shaded viewport options

Shade plot As displayed

Quality Normal

DPI 100

Plot options	Enables you to plot using lineweights or plot styles and to specify how to treat paperspace objects.

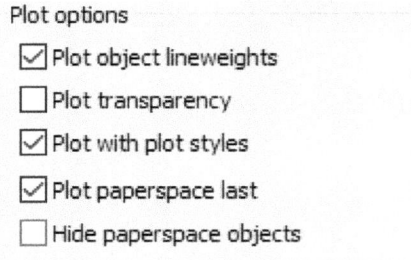

Plot options

☑ Plot object lineweights

☐ Plot transparency

☑ Plot with plot styles

☑ Plot paperspace last

☐ Hide paperspace objects

Drawing orientation	Sets the paper orientation to **Portrait** (the short edge of the paper is at the top of the page) or **Landscape** (the long edge of the paper is at the top of the page).

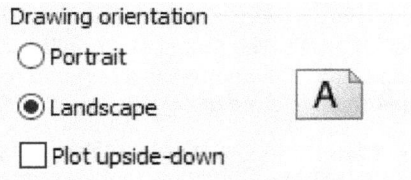

Drawing orientation

○ Portrait

◉ Landscape

☐ Plot upside-down

How To: Apply a Page Setup to a Layout

1. Right-click on the layout that you want to set and select **Page Setup Manager...**.
2. In the *Page Setup Manager*, select a page setup with the required plotter and paper size.
3. Click **Set Current** to apply it to the layout.
4. Click **Close** to close the *Page Setup Manager*.

How To: Import a Page Setup from Another File

1. Open the *Page Setup Manager*.

2. Click **Import...**.

3. In the *Select Page Setup From File* dialog box, select the file that contains the page setup you want to use and click **Open**.

4. In the *Import Page Setups* dialog box, select the setup that you want to import, as shown in Figure 8-9.

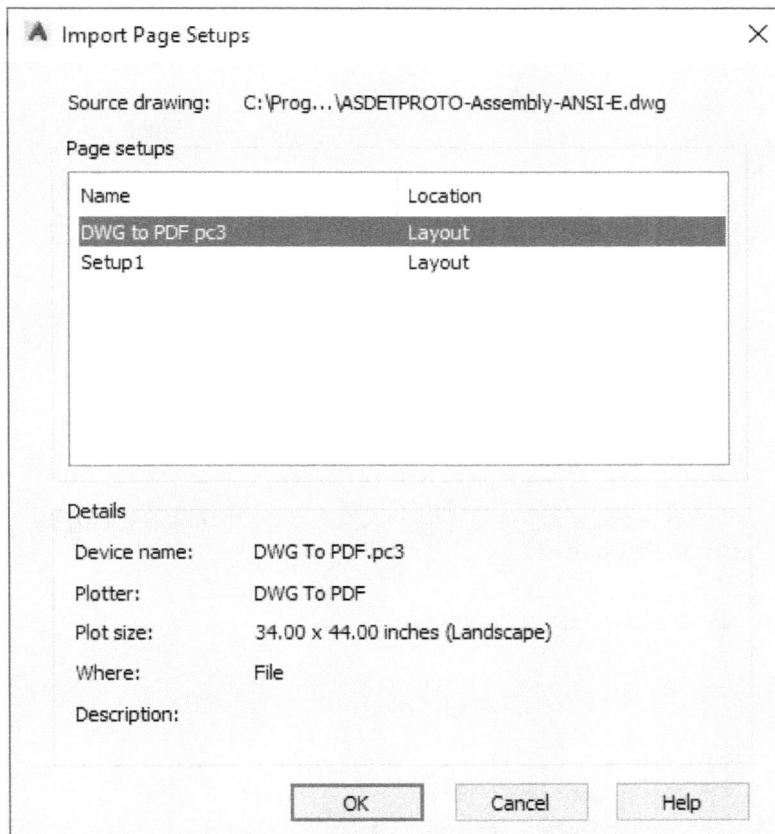

Figure 8-9

Note: If names are not displayed in the list, the drawing might have layouts but they have not been saved as page setups.

5. Click **OK** to complete the process. The imported page setup can now be used in your current drawing.

How To: Modify the Page Setup Manager

1. Right-click on the *Layout* tab and select **Page Setup Manager...** from the contextual menu, as shown in Figure 8–10.

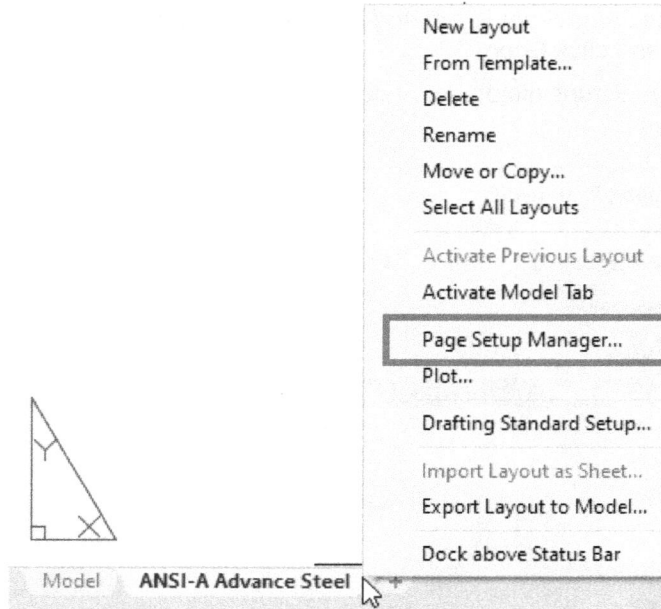

New Layout
From Template...
Delete
Rename
Move or Copy...
Select All Layouts

Activate Previous Layout
Activate Model Tab

Page Setup Manager...
Plot...

Drafting Standard Setup...

Import Layout as Sheet...
Export Layout to Model...

Dock above Status Bar

Model   **ANSI-A Advance Steel**

Figure 8–10

2. Add or edit the existing page settings and set it as current, which will mark it with an asterisk as shown in Figure 8–11.

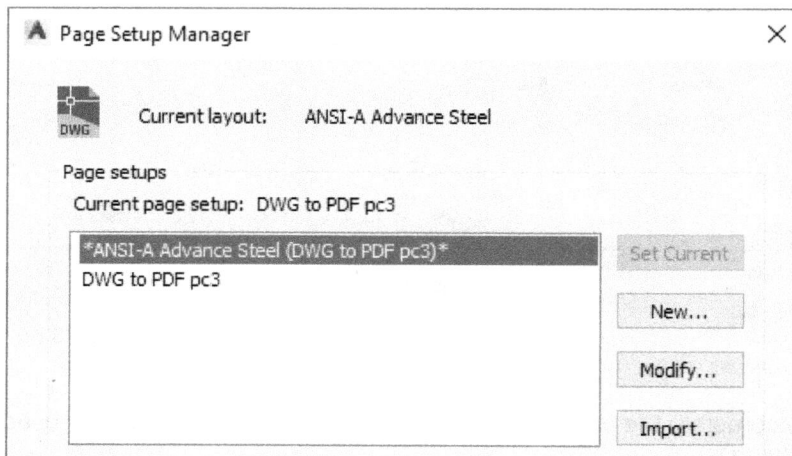

Page Setup Manager   ✕

Current layout: ANSI-A Advance Steel

Page setups
Current page setup: DWG to PDF pc3

ANSI-A Advance Steel (DWG to PDF pc3)
DWG to PDF pc3

Set Current

New...

Modify...

Import...

Figure 8–11

8.3 Title Block Drawing Frame

The prototype title block border is an AutoCAD block called **HYPERSTEELPAGEFRAME**, highlighted in Figure 8–12. The title block page frame could also contain a BOM, revision list, structure list, and additional notes depending on company needs.

Note: Do not change the name of the block page frame or copy the block to other files.

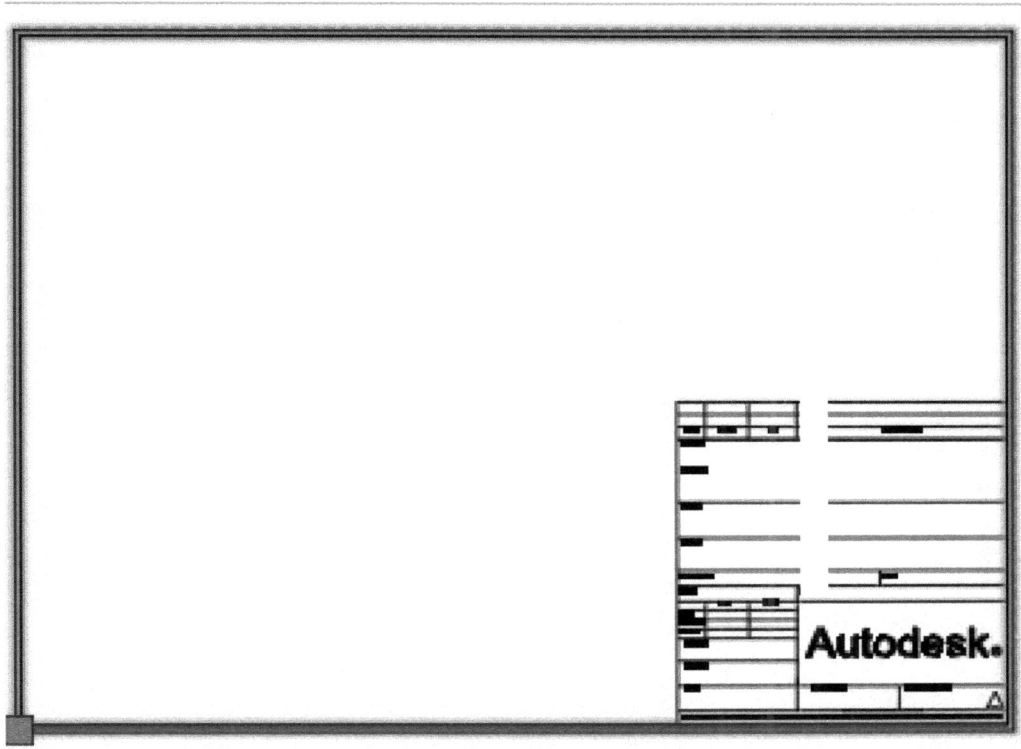

Figure 8–12

How To: Edit the Drawing Frame

1. Select the frame and at the command line, type **EXPLODE**.

   - Edit the frame using basic AutoCAD tools.

2. To redefine the page frame, use the AutoCAD **BLOCK** command to bring up the *Block Definition* dialog box, as shown in Figure 8–13.

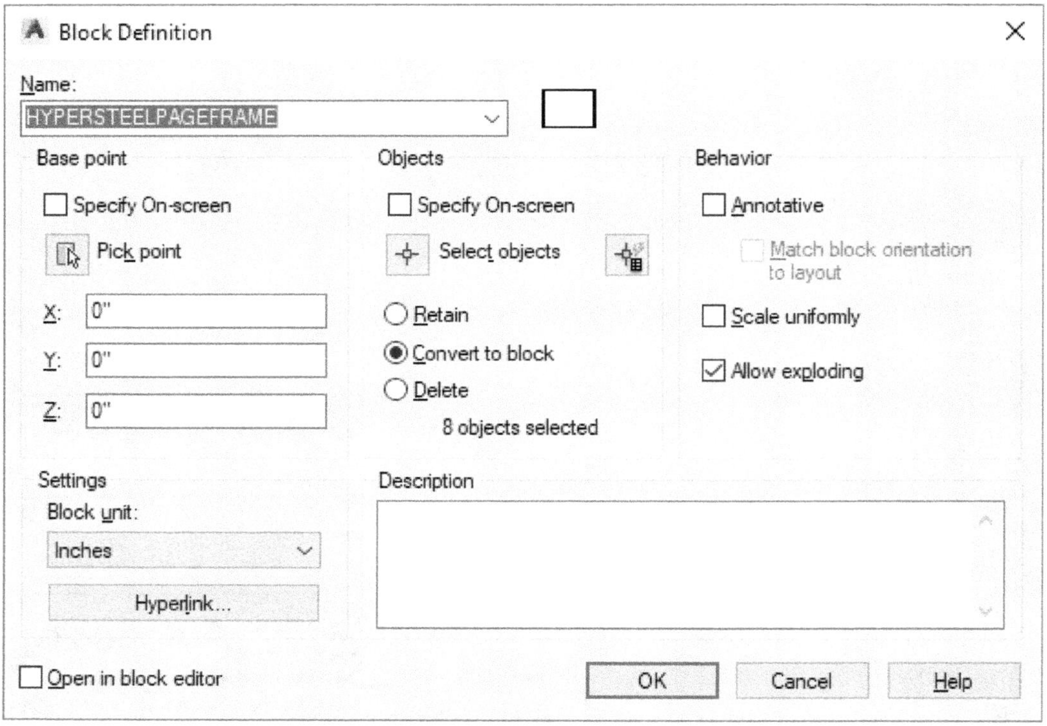

Figure 8–13

3. In the *Name* drop-down list, select **HYPERSTEELPAGEFRAME**, as shown in Figure 8–13.

4. Within the *Base point* section, leave the coordinates at 0,0,0.

5. In the *Objects* section, click ⊕ (Select objects) to select the objects that define the page frame.

6. Click **OK**.

7. When prompted with the *Block - Redefine Block* message, as shown in Figure 8–14, choose the appropriate option.

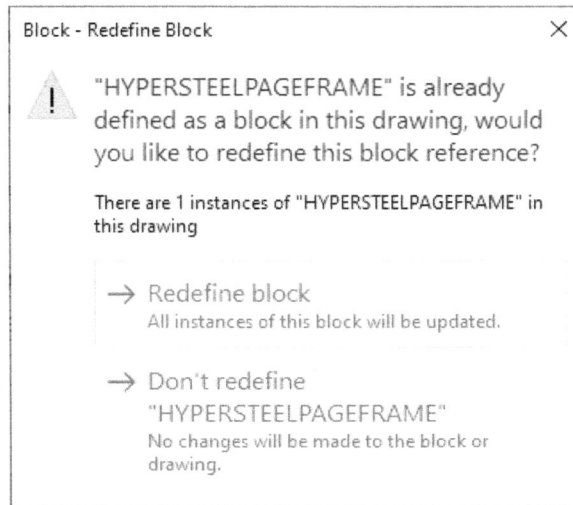

Figure 8–14

Note: In the Settings section, the Block units must match the prototype's units.

8. Click **OK** and save.

Title Block Page Header

The prototype title block page header is an AutoCAD block called **HYPERSTEELPAGEHEADER**.

Drawing information can be added to the title block page header using tokens. These attribute tags are intelligent text that automatically pull project information from the drawing file. Here is an example of some of the token attributes. To see a complete list, go to the Advance Steel help menu and search "title block token," then choose **title block token/attribute list**.

Note: Do not change the name of the block page frame or copy the block to other files.

PROJECT	Project Name
PROJECT NO	Project Number
CLIENT	Client Name
BUILDING	Building Name
BUILDING LOCATION	The Location of the Building

How To: Modify the Title Block

1. In the prototype drawing, select the title block page header as shown in Figure 8–15.

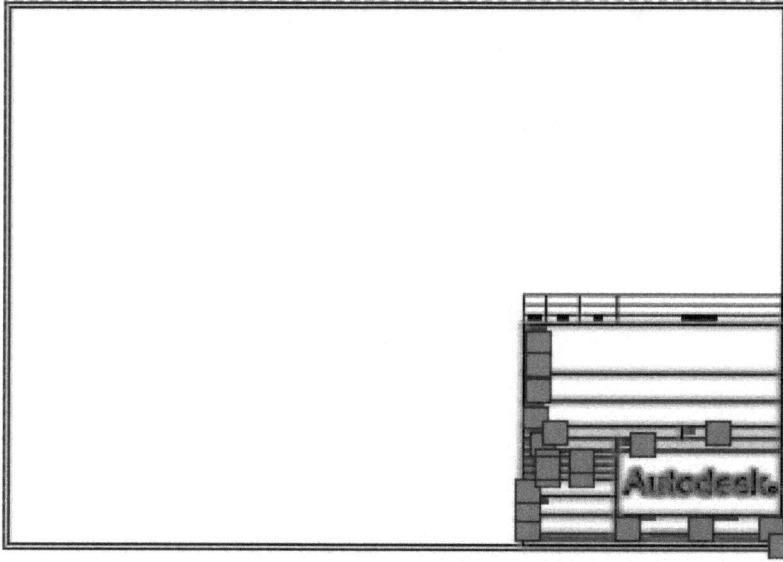

Figure 8–15

2. At the command line, type **EXPLODE**. This explodes the title block so the header displays as lines and tags that can be modified and edited using AutoCAD commands.

3. You can use the **STRETCH** command to reshape the border, create new lines, create or move text, or add tokens using the **ATTDEF** command.

4. When you have modified the title block, you can select all the objects that represent the page header, excluding any BOM's, frame, or tables.

5. Use the **BLOCK** command to recreate the header block.

6. In the *Block Definition* dialog box>*Name* section, select **HYPERSTEELPAGEHEADER**.

7. Leave the base point coordinates at 0,0,0.

8. Verify the units are correct.

9. Click **OK**.

10. Click **Redefine** when prompted.

11. Click **OK**.

12. Save the drawing.

How To: Add a Revision Table

1. In the *Labels & Dimensions* tab, expand the *Management* panel and select ⊞ (Revision Table).

2. Pick two points to indicate the position of the revision table, as shown in Figure 8–16.

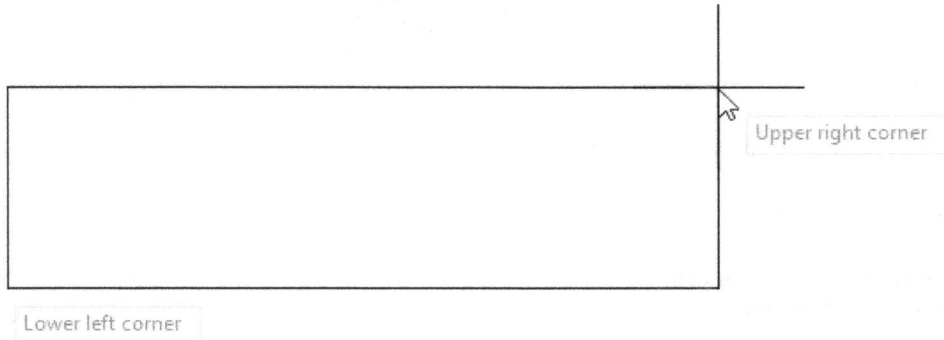

Upper right corner

Lower left corner

Figure 8–16

3. In the *Revision Control* dialog box>*Layout* tab, select the style of table.

- The first and second options create the revision table only as a header, as shown in Figure 8–17.

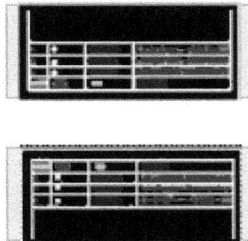

Figure 8–17

- The third and fourth options create an empty revision table with lines, as shown in Figure 8–18.

Figure 8–18

4. Review the other tabs available in the *Revision Control* dialog box.

Edit the BOM

In order for a BOM to generate in a drawing, you must place a BOM template in the prototype drawing. You can then set the size, position, and direction to extend when adding more lines.

How To: Edit the BOM

1. In the *Labels & Dimensions* tab>*Management* panel, click (Insert a drawing list).
2. Select a position on the prototype for the BOM with a first and second pick, as shown in Figure 8–19. The *List* dialog box displays.

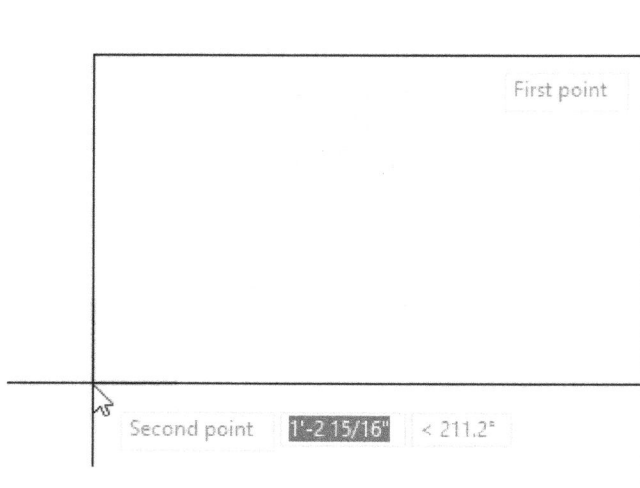

First point

Second point   1'-2 15/16"   < 211.2°

Figure 8–19

3. In the *List* dialog box>*Layout* tab, you can lock the BOM into position where you drew your rectangle by clicking on one or more of the four check boxes (but not all of them), as shown in Figure 8–20. Clicking on the top two check boxes will allow the BOM to expand down the page if any new lines are added.

Note: It is not recommended to check all four boxes, as this will force the BOM to scale and fit in the rectangle.

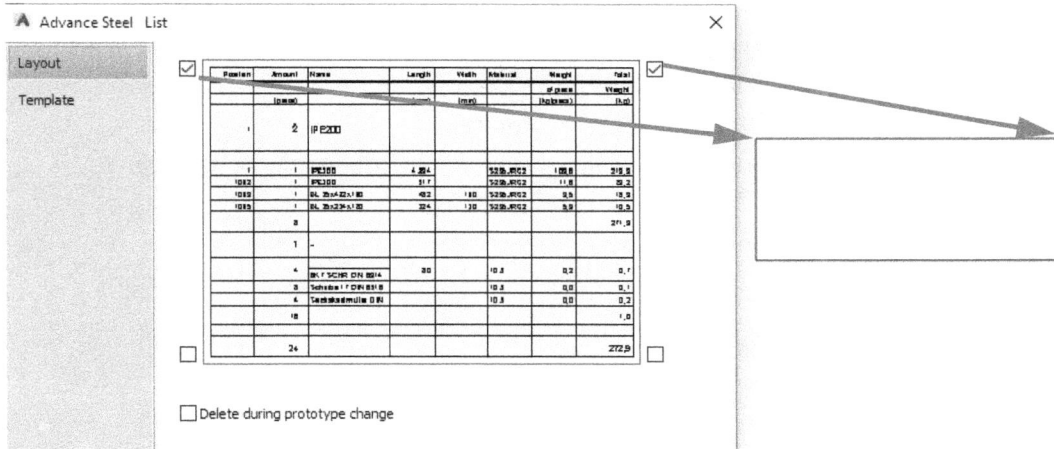

Figure 8–20

4. Switch to the *Template* tab, then click on the node next to *Advance Templates* and select a BOM template.

- The template will be inserted and will automatically update if associated with a specific part. Otherwise, it will display the token attributes as shown in Figure 8–21.

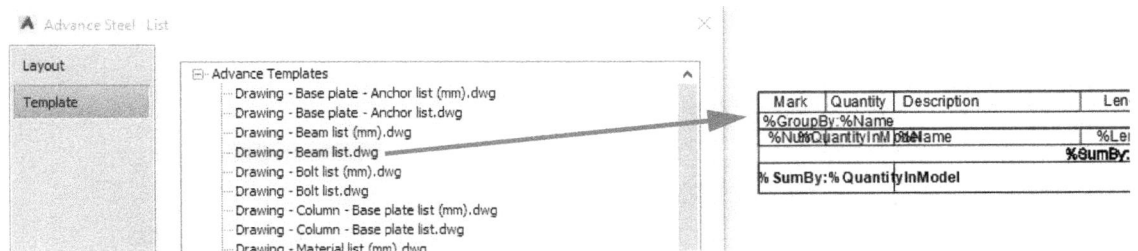

Figure 8–21

Adding a Reserved Area

The remaining area on the prototype drawing is for placing your details, drawings, etc. If you need more space than what is given, you can add a block called **ADVANCESTEEL_RESERVED_AREA** to the prototype drawing.

How To: Add a Reserved Area

1. At the command line, type **RECTANGLE** and press <Enter>, then draw a rectangle in the area you want to reserve for your drawing.

2. At the command line, type **BLOCK** and press <Enter>.

3. In the *Block Definition* dialog box>*Name* area, select **ADVANCESTEEL_RESERVED_AREA**, as shown in Figure 8–22.

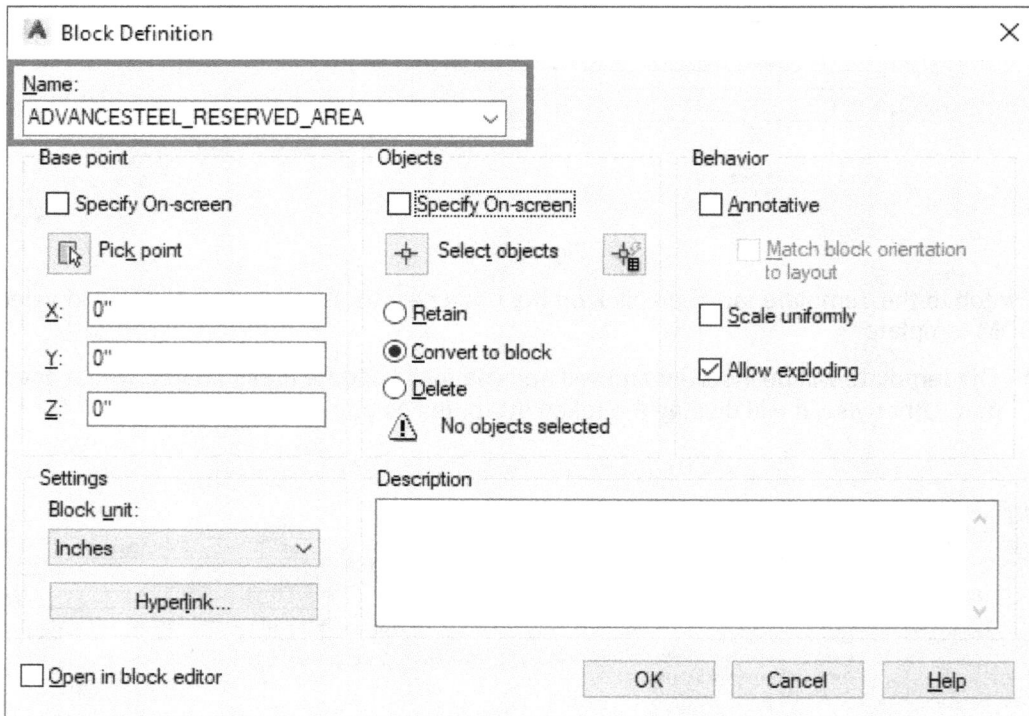

Figure 8–22

4. Set the *Base point* coordinates to **0,0,0**.

5. In the *Objects* section, click ⊕ (Select objects) and select the rectangle. Click **OK**.

6. When prompted with the *Block - Redefine Block* dialog box as shown in Figure 8–23, select the appropriate option.

   Note: In the Settings section, the Block units must match the prototype's units.

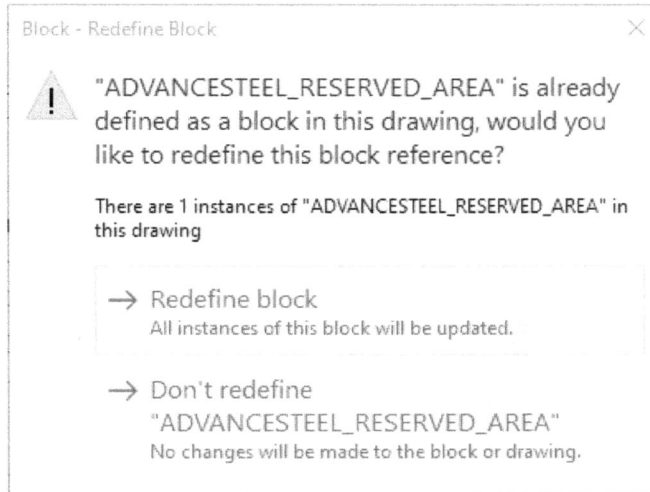

Figure 8–23

7. Save the drawing.

Define the Detail Arrangement Settings

To position your details correctly within the prototype drawing, you can use the *Layout Detail* prototype settings.

How To: Define the Detail Arrangement

1. In the *Output* tab>*Document Manager* panel, click 🔳 (Define drawing layout).
2. The *Layout detail prototype* dialog box displays, as shown in Figure 8–24.

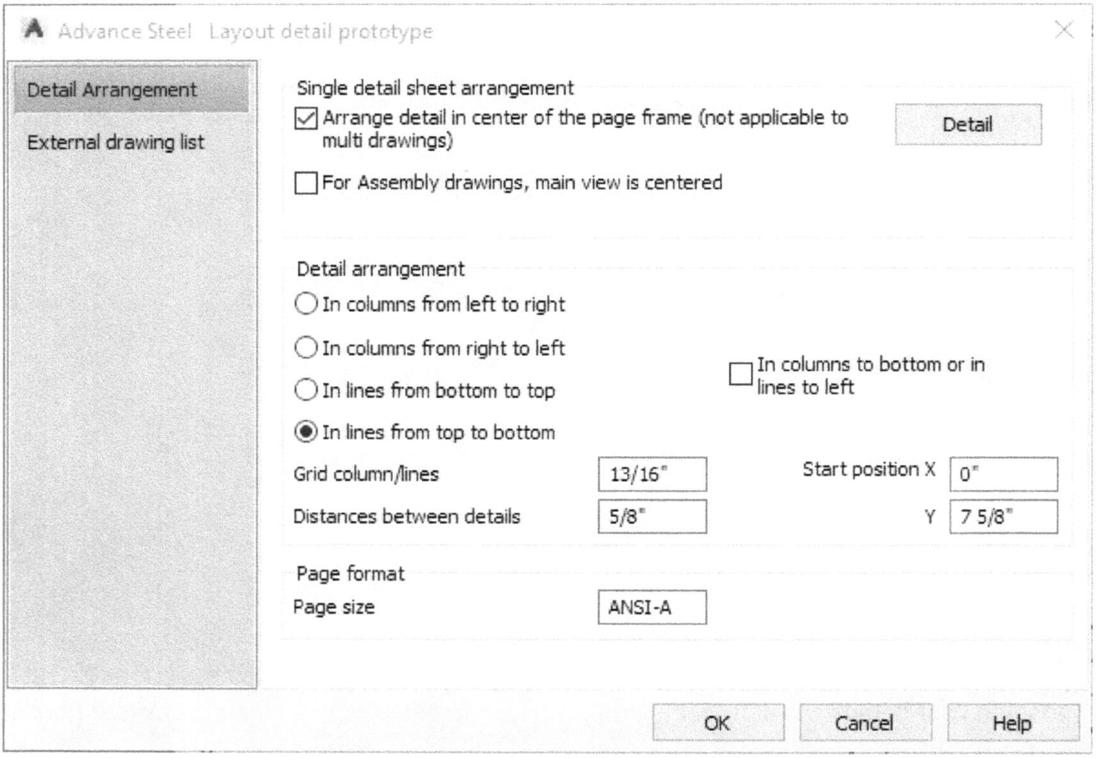

Figure 8–24

Detail Arrangement Tab

- **Arrange detail in the center of the page frame:** This is for single drawings only and the drawing needs to be centered in the frame.
- **For Assembly drawings, main view is centered:** Used for assembly drawings using the front view as the centered drawing or main view.
- **Detail arrangement**: To specify the direction of the frame if multiple drawings will be placed: *In columns from left to right, In columns from right to left, In lines from bottom to top*, or *In lines from top to bottom*.
- **Start position:** Enter X and Y values.

Configure Dimension Styles

During the prototype drawing editing, dimension styles are stored with a set name. They can be modified in the *Dimension Style Manager* and the *Management Tools* dialog box.

How To: Configure Dimension Styles

In the *Labels & Dimensions* tab>*Parametric Dimensions* panel, click ⛏ (Change Dimension Style). The *Dimension Style Manager* dialog box displays, as shown in Figure 8-25.

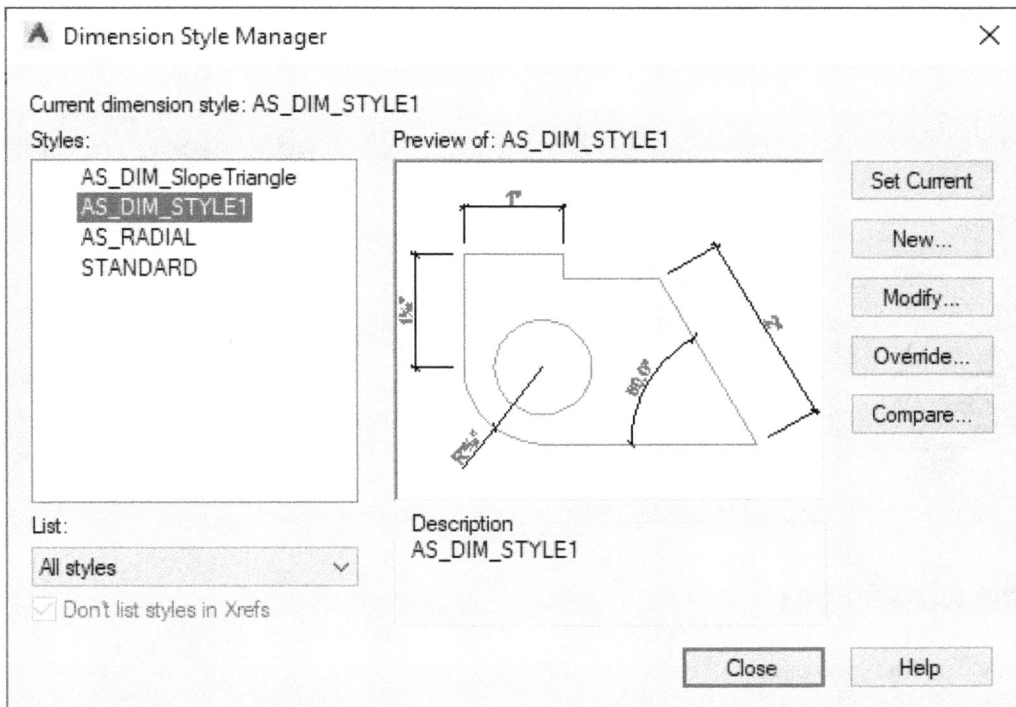

Figure 8-25

- The *Dimension Style Manager* lists the dimension styles stored in the prototype drawing.

- You can create a new style or modify an existing style.

- When you are finished making modifications or creating a new style, you can click **Close** to save the changes.

Management Tools Dimension Styles

In the Management Tools Defaults, you can change the software's Dimension Style defaults to match the dimension styles in the prototype drawing.

How To: Access the Management Tools Dimension Styles

1. In the *Home* tab>*Settings* panel, click (A) (Management Tools), then click [0.0] (Defaults).

2. Expand the *Drawing-Dimensioning* tab and select **Dimension styles**, as shown in Figure 8–26.

    *   For each dimension style, there is a specific default **Property Value**.

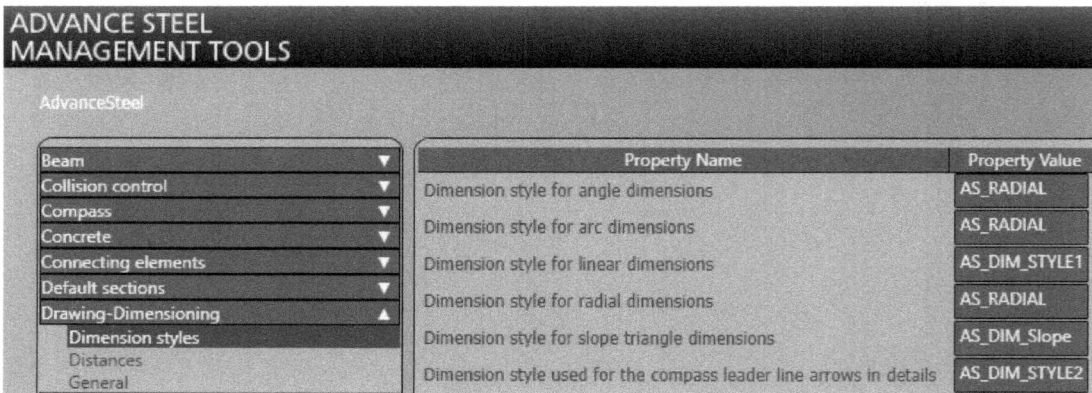

	Property Name	Property Value
Beam ▼		
Collision control ▼	Dimension style for angle dimensions	AS_RADIAL
Compass ▼	Dimension style for arc dimensions	AS_RADIAL
Concrete ▼		
Connecting elements ▼	Dimension style for linear dimensions	AS_DIM_STYLE1
Default sections ▼	Dimension style for radial dimensions	AS_RADIAL
Drawing-Dimensioning ▲		
Dimension styles	Dimension style for slope triangle dimensions	AS_DIM_Slope
Distances	Dimension style used for the compass leader line arrows in details	AS_DIM_STYLE2
General		

ADVANCE STEEL MANAGEMENT TOOLS — AdvanceSteel

Figure 8–26

Load Linetypes

Add custom linetypes or reload existing linetypes to your prototype drawing by opening the *Linetype Manager* and loading the linetypes.

How To: Load Linetypes

1. In the *Tools* tab, expand the *Linetype* drop-down list and select **Other...**, as shown in Figure 8–27

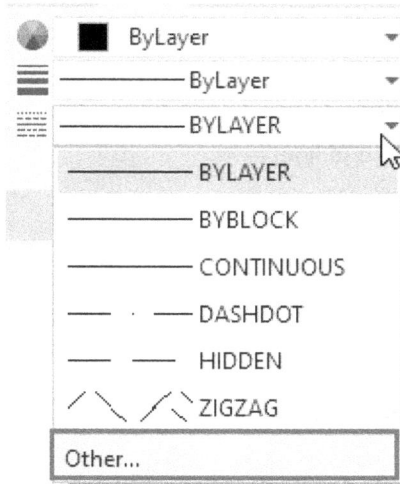

Figure 8–27

2. In the *Linetype Manager* dialog box, select **Load...**, as shown in Figure 8–28.

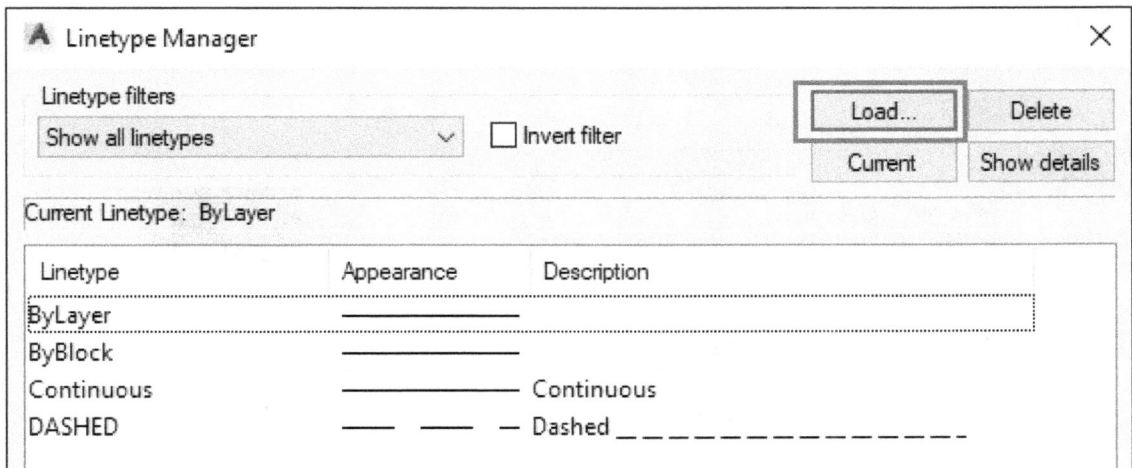

Figure 8–28

3. Select a linetype to load and click **OK**.

4. Click **OK** to exit the *Linetype Manager*.

Practice 8a
Edit a Prototype Drawing

Practice Objectives

- Open a template file.
- Edit a prototype drawing.
- Modify the default page header.
- Modify units, dimension style, and linetypes.
- Learn about assembly prototype drawings.

A couple of different scenarios can happen with prototype drawings. Sometimes the fabricator will supply a DWG file with a title block, or they will send a PDF of what they want it to look like. Depending on company needs, you will have to create the title block by customizing an existing prototype drawing. In this practice, you will modify one of Autodesk's prototypes.

Task 1: Open a prototype drawing.

1. In the Quick Access Toolbar or Application Menu, click ☐ (New).
2. In the *Select template* dialog box, select **ASTemplate.dwt** and click **Open**.
3. In the *Output* tab>*Document Manager* panel, click 🖨 (Edit Prototypes).
4. Open **ASDETPROTO-ANSI-D.dwg** as shown in Figure 8–29.

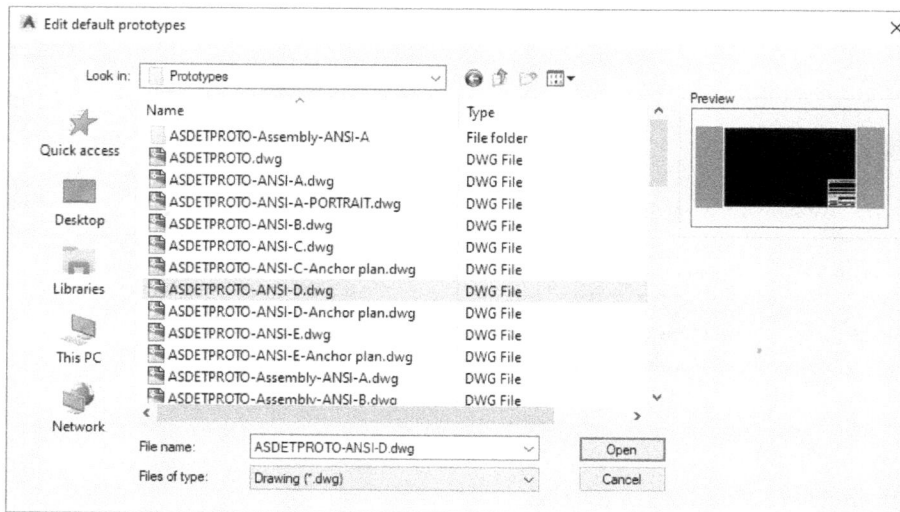

Figure 8–29

Task 2: Edit the prototype drawing.

1. In the Quick Access Toolbar or Application Menu, click ▣ (Save As) and save the drawing to the practice files folder as **CLASSPROTO_ANSI-D.dwg**.

2. Explode the HYPERSTEELPAGEHEADER block using the AutoCAD **EXPLODE** command, as shown in Figure 8–30.

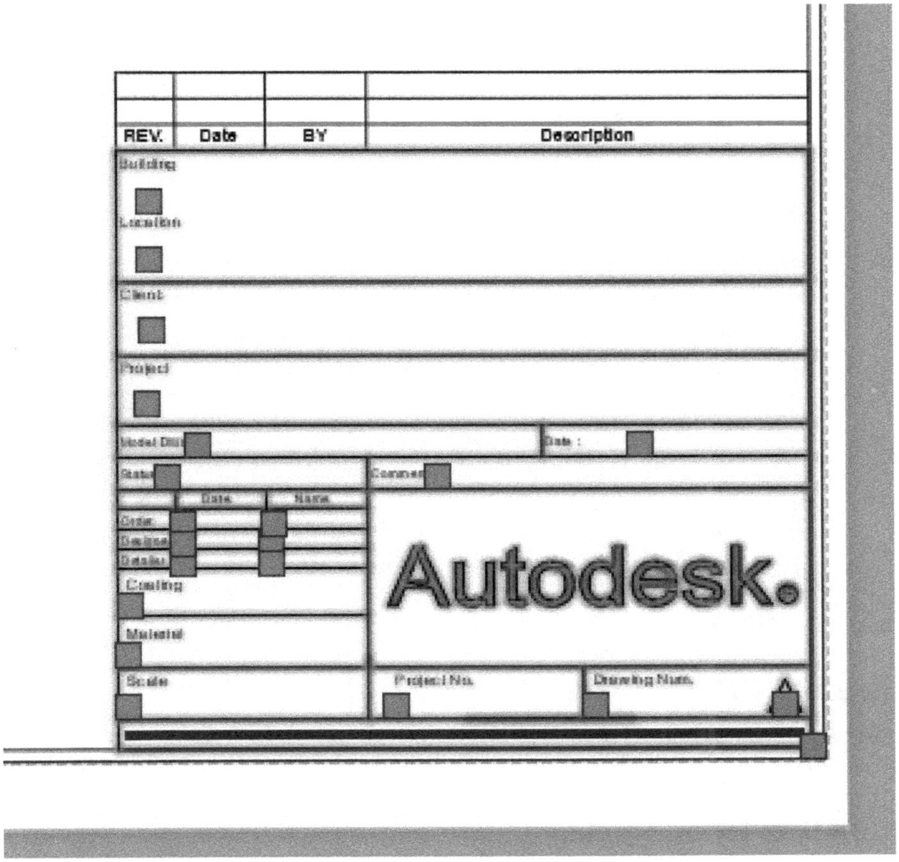

Figure 8–30

3. Remove the Autodesk logo, *Status*, *Comment*, and text at the bottom of the title block as shown in Figure 8–31.

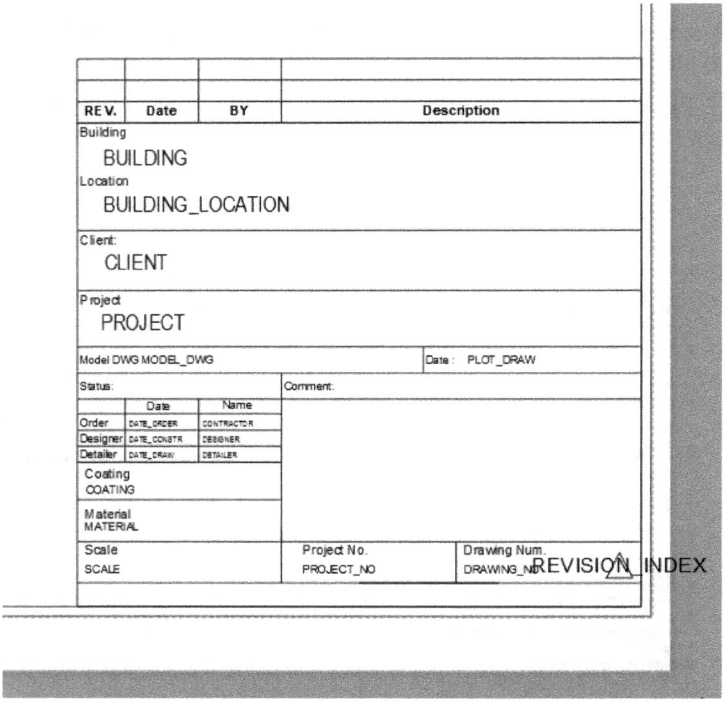

Figure 8–31

Task 3: Modify the existing page header.

1. Start the **BLOCK** command.

2. In the *Block Definition* dialog box>*Name* section, select **HYPERSTEELPAGEHEADER**, as shown in Figure 8–32.

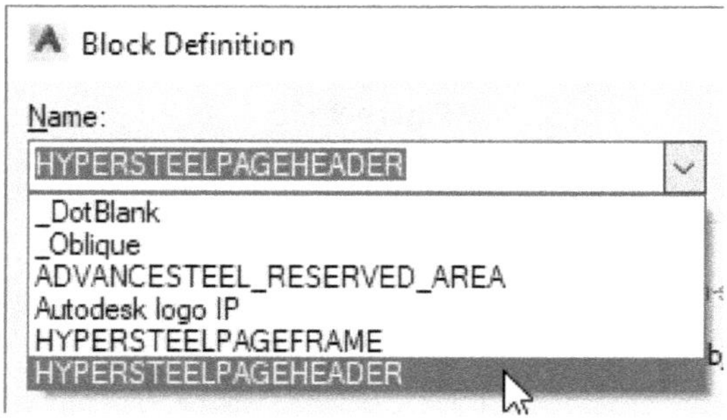

Figure 8–32

3. Use (Pick point) to manually select the lower right corner of the title block, as shown in Figure 8-33.

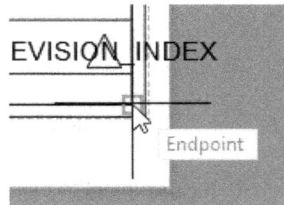

Figure 8-33

4. In the *Objects* section, use (Select objects) to select the page header information using a window selector. Do not include the Revision Table, as shown in Figure 8-34.

Figure 8-34

5. Press <Enter> to accept the selection.
6. In the *Block Definition* dialog box, click **OK**.

7. In the *Blocks - Redefine Block* dialog box, click **Redefine** as shown in Figure 8–35.

Blocks - Redefine Block

The block definition has changed. Do you want to redefine it ?

Redefine   No

Figure 8–35

8. Click **OK** in the *Edit Attributes* dialog box. The page header displays with no attributes, as shown in Figure 8–36.

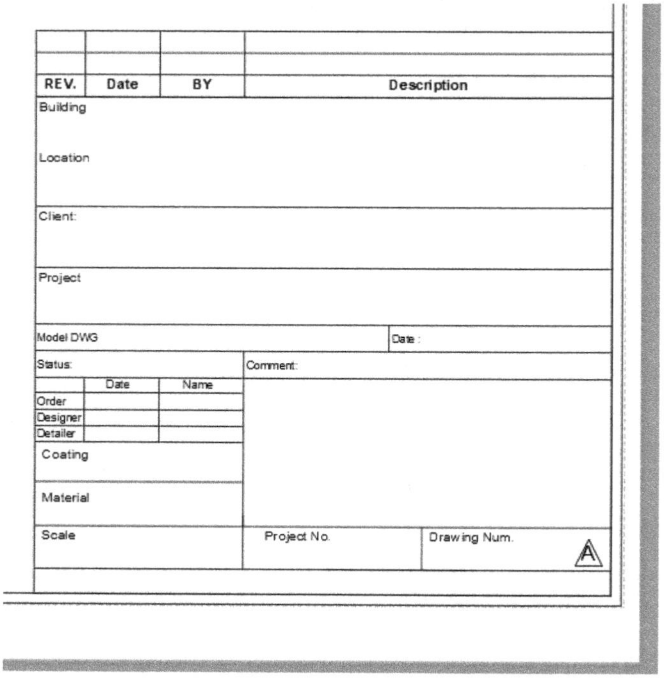

Figure 8–36

Task 4: Modify units, dimension style, and linetypes.

1. At the command line, type **UNITS** and press <Enter>.

2. In the *Drawing Units* dialog box, in the *Length* section, set the *Type* to **Architectural** and the *Precision* to **0'-0 1/16"**.

3. Click **OK**.

4. In the *Labels & Dimensions* tab>*Parametric Dimensions* panel, click ⊥̶ (Change Dimension Style).

5. Select **AS_DIM_STYLE1** and click **Modify**.

6. In the *Modify Dimension Style* dialog box, click on each tab to review the settings.

7. Click **OK** to close the *Modify Dimension Style* dialog box.

8. Click **Close** to close the *Dimension Style Manager*.

9. At the command line, type **LINETYPE** and press <Enter>.

10. In the *Linetype Manager*, click **Load**.

11. Hold down <Ctrl> and select **CENTER**, **DASHED** and **HIDDEN** as shown in Figure 8–37.

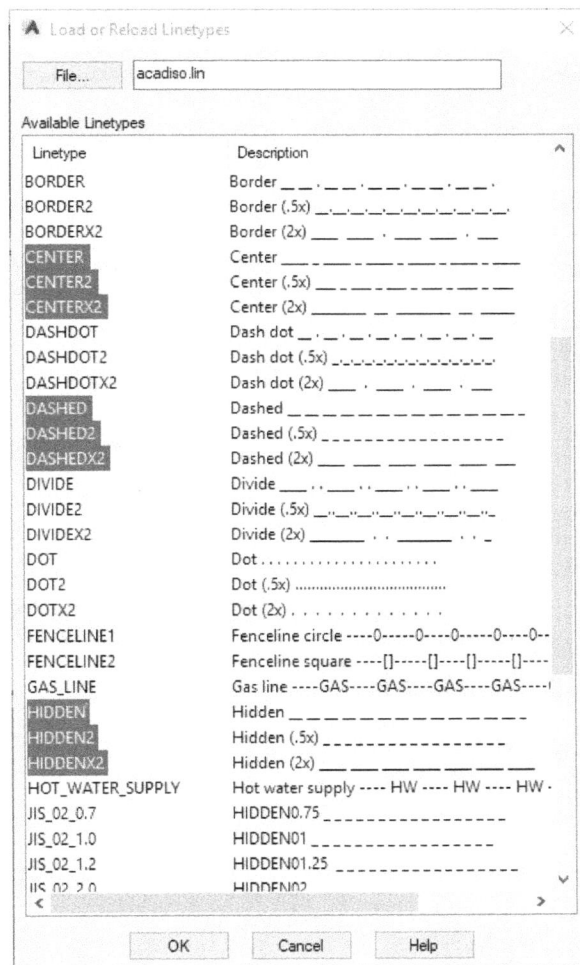

Figure 8–37

12. Click **OK** to load the linetypes.

13. Click **OK** in the *Linetype Manager*.

14. Save the drawing and keep it open.

Task 5: Assembly prototype drawing.

1. In the Quick Access Toolbar or Application Menu, click 💾 (Save As) and save the drawing as **CLASSPROTO-Assembly-ANSI-D.dwg**.

2. In the *Labels & Dimensions* tab>*Management* panel, click 📋 (Insert a drawing list).

3. Click the upper right of the title block for the first point, as shown in Figure 8−38.

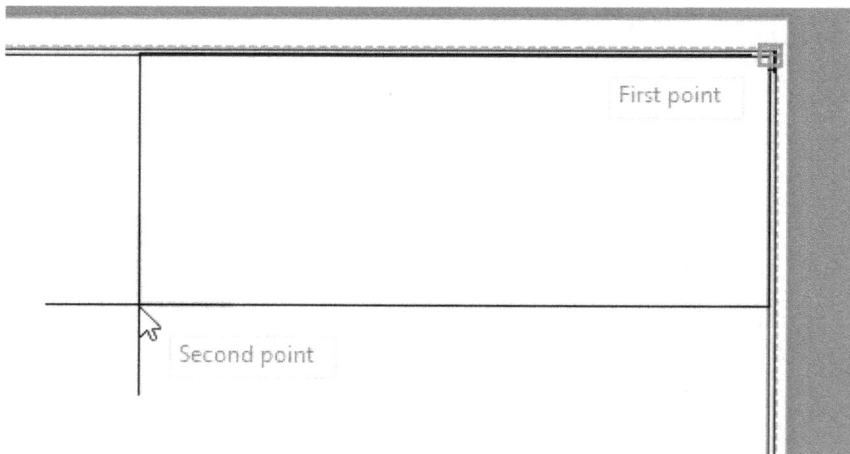

First point

Second point

Figure 8−38

4. In the *List* dialog box>*Layout* tab, select the two top check boxes.

5. Switch to the *Template* tab, expand *Advance Templates* and select **Drawing - Structured list.dwg**.

   • It will appear in the drawing area behind the *List* dialog box.

6. Close the dialog box by clicking on the **X** in the upper right corner.

7. Repeat Steps 5 and 6 to add **Drawing-Bolt list.dwg**.

8. Add the *Bolt* list near the lower, left area next to the Header, as shown in Figure 8–39.

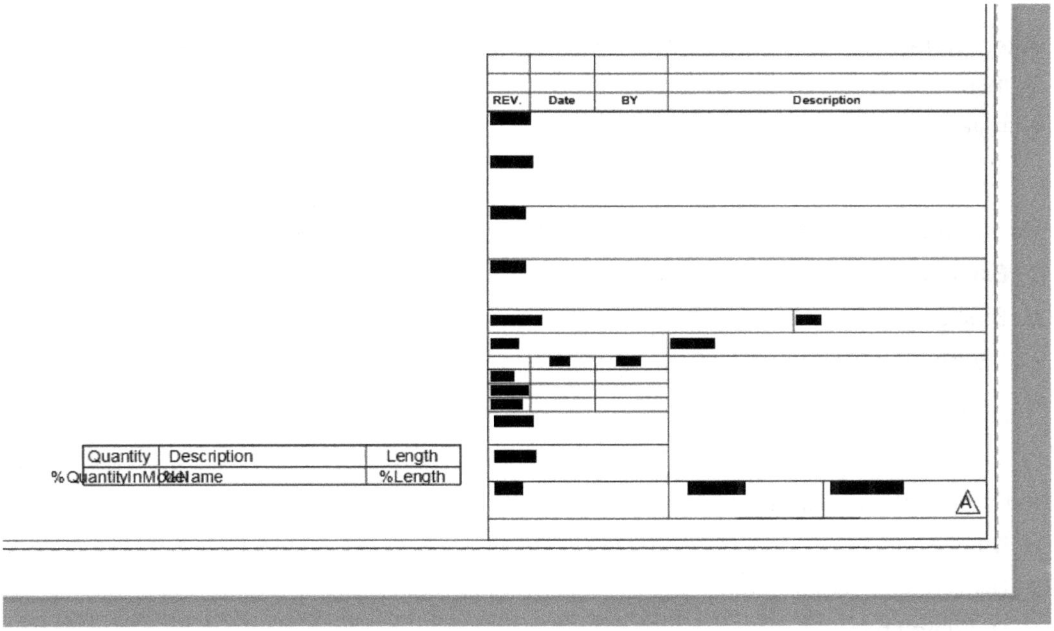

REV.	Date	BY	Description

Quantity	Description	Length
%QuantityInModel	%Name	%Length

Figure 8–39

End of practice

Chapter Review Questions

1. A template drawing is the same as a prototype drawing.
    a. True
    b. False

2. To edit a page header, you need to use which AutoCAD command?
    a. Explode
    b. Block
    c. Insert
    d. Xref

3. In the title block, attributes are used for what?
    a. Units
    b. UCS
    c. Drawing styles
    d. Drawing information

4. The Project Data is where you can set which settings?
    a. Assembly type
    b. Dimension style
    c. Length unit
    d. Template file

5. Where can you configure default values for object properties and define application behavior?
    a. Management Tools
    b. Template file
    c. Unit
    d. Project Styles

Command Summary

Button	Command	Location
	Change Dimension Style	• **Ribbon:** *Labels & Dimensions* tab>*Parametric Dimensions* panel
	Define Drawing Layout	• **Ribbon:** *Output* tab>*Document Manager* panel
	Edit Prototypes	• **Ribbon:** *Output* tab>*Document Manager* panel
	Insert a drawing list	• **Ribbon:** *Labels & Dimensions* tab>*Management* panel
	Project Settings	• **Ribbon:** *Home* tab>*Settings* panel
	Revision Table	• **Ribbon:** *Labels & Dimensions* tab>*Management* panel
0.0	**Units**	• **Application Menu: Drawing Utilities**

Drawing Style Manager

Autodesk Advance Steel provides several drawing styles relevant to general manufacturing needs to automatically create various layouts, single parts, shop drawings, general arrangement drawings, and assemblies. Sometimes the styles need to be modified to suit company needs and this can be done through the *Drawing Style Manager*, a multifaceted tool that contains predefined drawing styles to assist in the vast requirements of steel detailing. Each style is set up with several sub-styles that can be modified as needed.

Learning Objectives

- Explore the *Drawing Style Manager* user interface.
- Modify categories and drawing styles.
- Navigate the drawing style structure and view properties.

9.1 Drawing Style Manager User Interface

The *Drawing Style Manager* user interface gives you several drawing styles and sub-styles that can be modified to specific standards. Modification can be done using the tools from the toolbar or by right-clicking on the style to get a contextual menu with modification tools. The *Drawing Style Manager* is broken down in Figure 9–1.

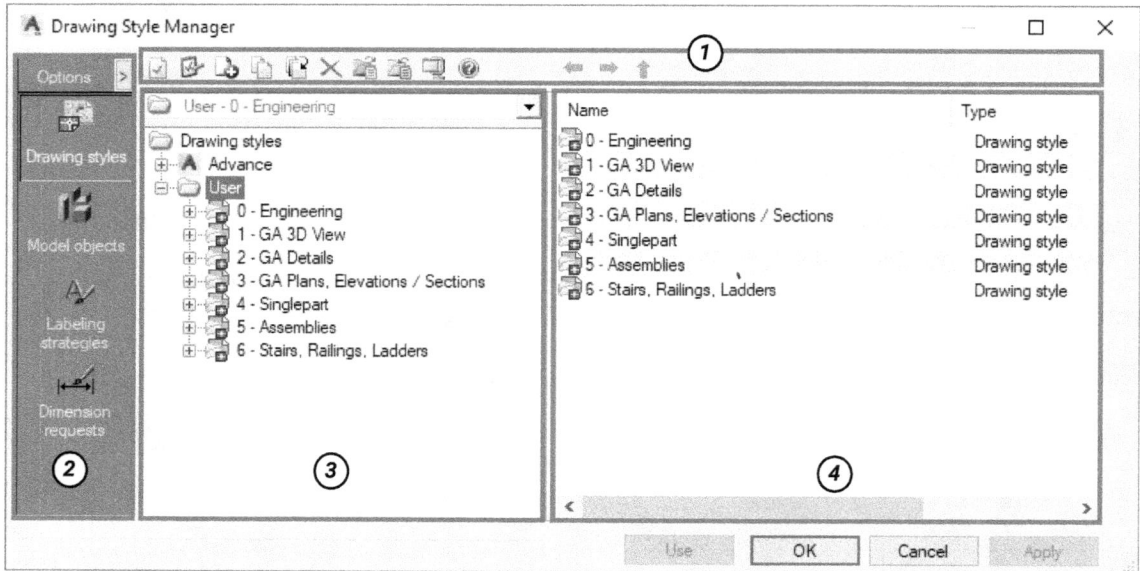

Figure 9–1

1. Toolbar	3. Tree Panel
2. Component Panel	4. Properties Panel

1. Toolbar

The toolbar provides editing tools to manage your styles. The three arrows aid in navigating through individual styles and switch to previously viewed styles in the order they were reviewed.

	Properties	Change a selected drawing's Name, Category, and Description.
	Use	Use the selected drawing style to create a new detail.
	New	Create a new category.
	Copy	Copy an existing style.
	Deep Copy	Copies a drawing's style and sub-styles.
	Delete	Delete a style from the tree panel.
	Import	Import available drawing styles.
	Export	Export a drawing style.
	Compact	Decrease the size of the database.
	Context Help	Opens the Autodesk Advance Steel help menu.
	Back	Return to a previously displayed sheet.
	Forward	Go to the next sheet.
	Up	Go up one level.

2.Component Panel

The component panel displays the sub-styles for each of the drawing styles. You can modify these sub-styles to suit company needs.

	Drawing Style	Displays the Advance and User drawing styles in the tree panel.
	Model Objects	Displays the Advance and User model objects, model roles, and group.
	Labeling Strategies	Displays the Advance and User group section name and label arrangement and strategy.
	Dimension Requests	Displays the Advance and User dimension requests for dimension definitions, group, and chain direction.

3. Tree Panel

The tree panel displays two main categories: *Advance* and *User*. By clicking the nodes, you can expand or collapse each section to review all the drawing styles.

• **Advance** – Contains all drawing styles, as shown in Figure 9–2. Drawing styles in this category can only be deep copied.

• **User** – All the *Advance* drawing styles are copied within the *User* category, as shown in Figure 9–3, and can be modified, copied, or deleted.

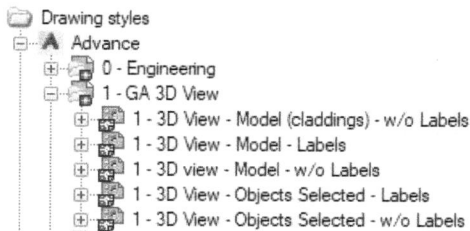

Figure 9–2

Figure 9–3

⌕ Hint: Options Fly-Out

The *Options* fly-out provides different ways to view the tree panel. You can select **Show full tree** or **Full expand** or both. The *Drawing Style Manager* does not indicate which drawing styles work with certain settings so it is recommended to keep the **Warning on modifying a shared entity** checked as shown in Figure 9–4.

Switching these options requires the *Drawing Style Manager* dialog box to be restarted.

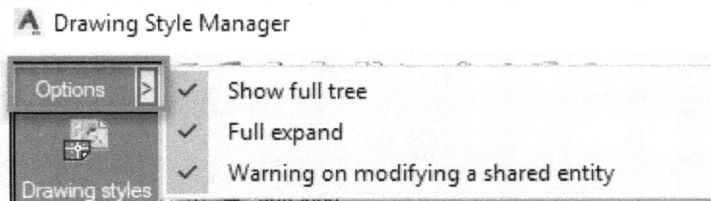

Figure 9–4

In the tree panel, you can right-click on a drawing style, view, or model object to bring up the contextual menu, which provides all the tools that you find on the toolbar (as shown in Figure 9–5).

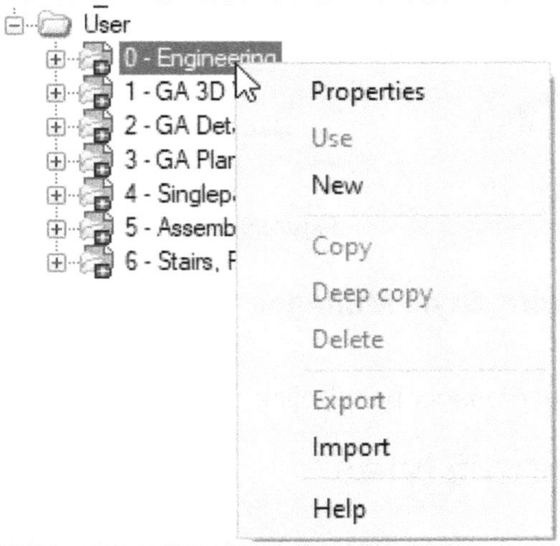

Figure 9–5

4. Properties Panel

When a drawing style, view, or model object is selected, its properties and all its settings will show in the *Properties* panel as shown in Figure 9–6.

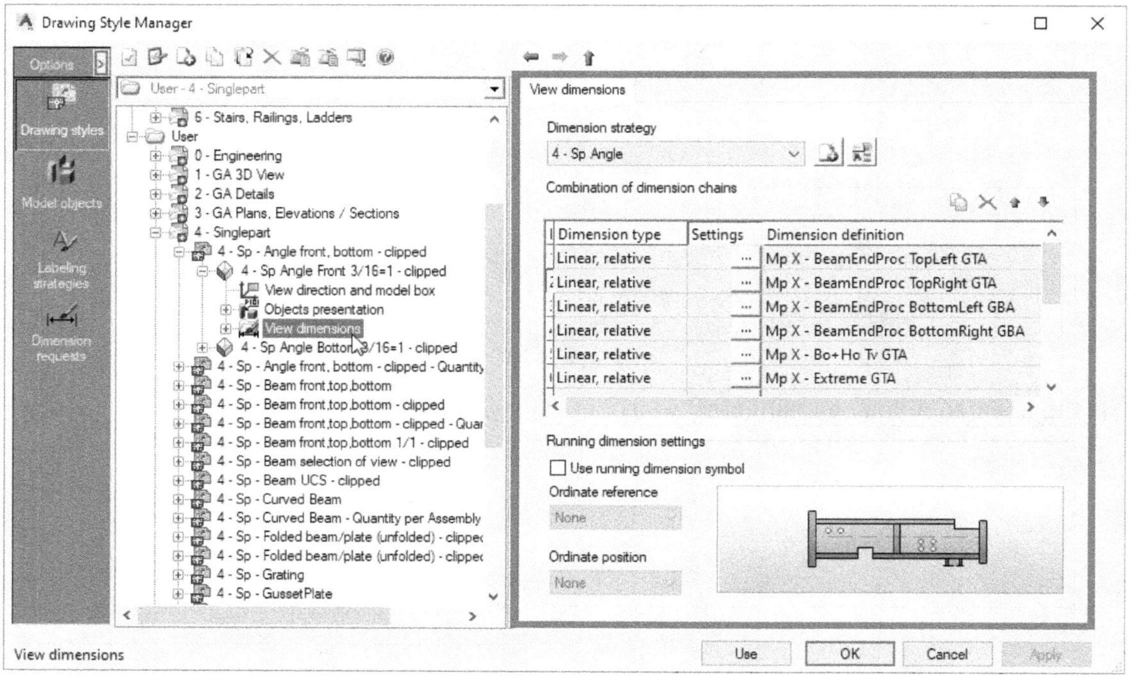

Figure 9–6

How To: Open the Drawing Style Manager

In the *Output* tab>*Document Manager* panel, click (Drawing Style Manager).

Modifying Drawing Styles

The Drawing Style Manager editing tools, from the toolbar or contextual menu, allow you to create new styles and copy or modify the properties of existing categories and styles.

How To: Create a New Category

1. In the tree panel, select the **User** drawing style.

2. On the toolbar, or alternatively by right-clicking on the style, click (New).

3. In the *New Drawing style category* dialog box, shown in Figure 9–7, enter a unique category name.

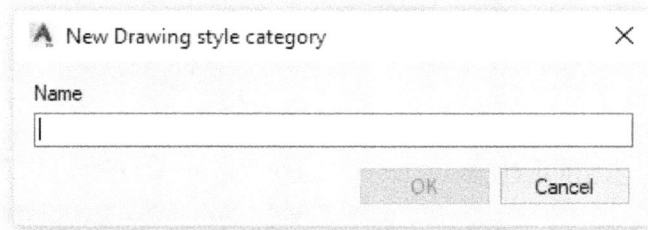

Figure 9–7

4. The new category appears in the tree panel and is ready to be modified.

How To: Move a Style to Another Category

1. In the tree panel>*User* folder, select a drawing style.

2. On the toolbar, or alternatively by right-clicking on the style, select ☑ (Properties).

3. In the *Drawing style* dialog box>*Category* section, choose a new category you would like to move your drawing style to, as shown in Figure 9–8.

   Note: You can also use drag and drop in the tree panel to move styles to different categories.

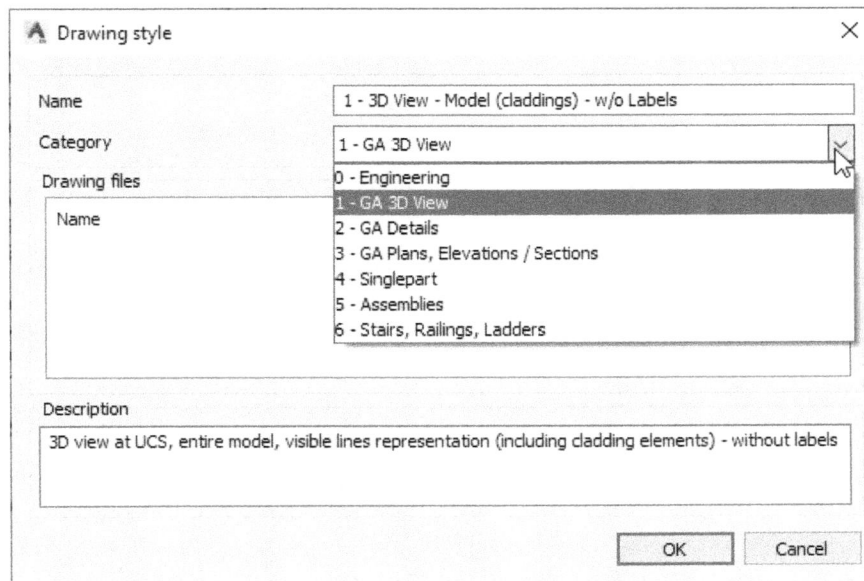

Figure 9–8

How To: Use an Existing Style

1. In the tree panel, select a drawing style.

   Note: In order to use a drawing style that contains columns or special and main parts, the object must first be numbered.

2. On the toolbar, or alternatively by right-clicking on the style, select ⬚ (Use).

3. In the *Create detail* dialog box, select either **Create with default settings** or **Modify settings** as shown in Figure 9–9.

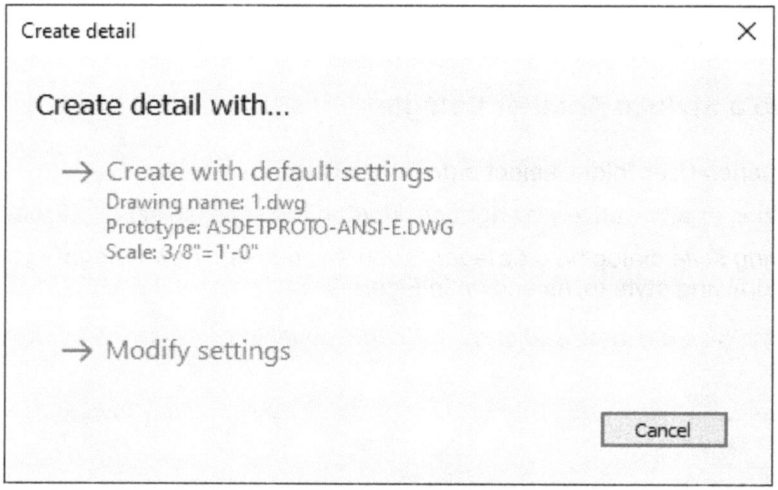

Figure 9–9

- Choosing **Modify settings** brings up the *Select destination file* dialog box, as shown in Figure 9–10.

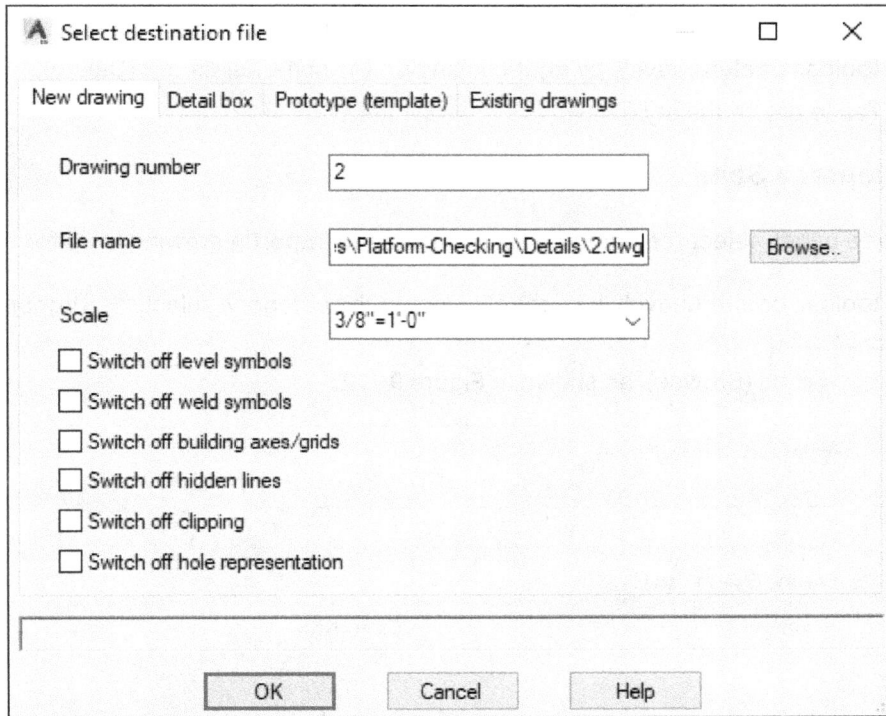

Figure 9–10

How To: Copy a Style

1. In the tree panel, select the drawing style you want to copy.

2. On the toolbar, or alternatively by right-clicking on the style, select ⧉ (Copy).

3. Enter the name for the drawing style, as shown in Figure 9–11.

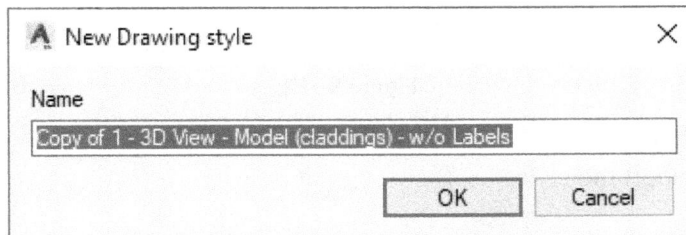

Figure 9–11

How To: Delete a Style

1. In the tree panel, select a drawing style.

2. On the toolbar, or alternatively by right-clicking on the style, select ✕ (Delete).

3. Select **Yes** to delete the style.

How To: Import a Style

1. In the tree panel, select the category that you want to import a drawing style to.

2. On the toolbar, or alternatively by right-clicking on the category, select 📑 (Import).

3. Click [...] (Browse), as shown in Figure 9–12.

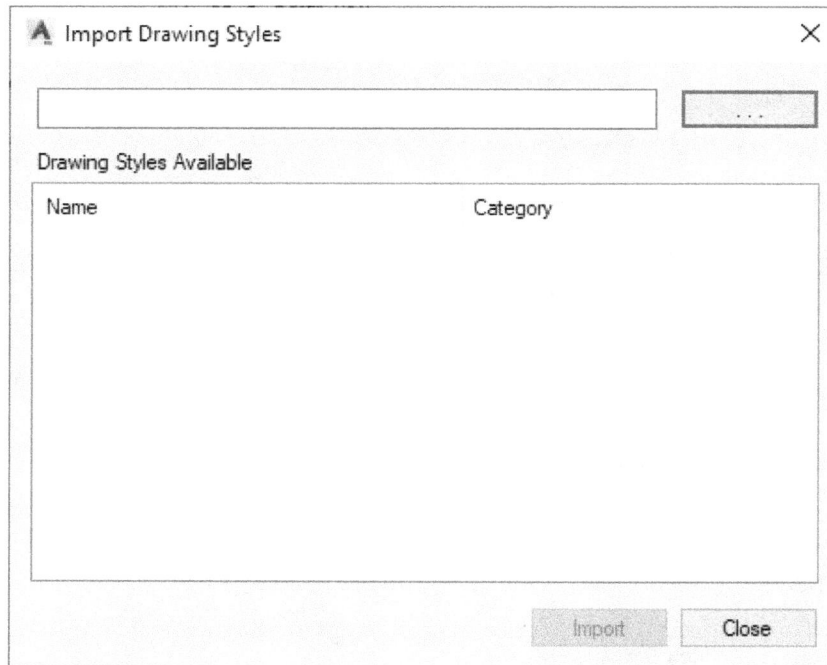

Figure 9–12

4. In the *Open* dialog box, select a database file (.mdf or .mdb).

5. Click **Open** to import the style.

6. In the *Import Drawing Styles* dialog box, select from the *Drawing Styles Available* list and click **Import**, as shown in Figure 9–13.

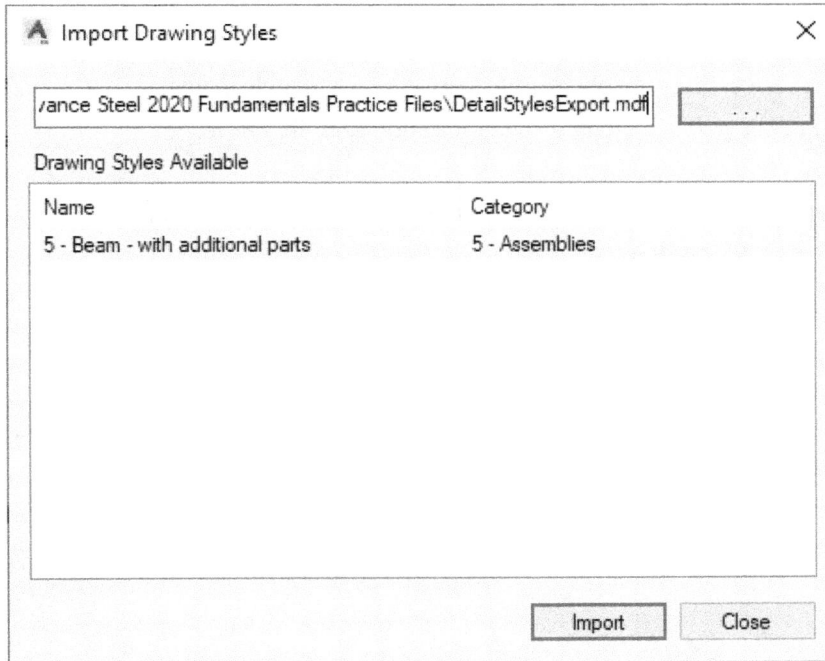

Figure 9–13

7. In the *Import Drawing Styles* dialog box, you can change:

   - The name of the imported drawing style.
   - The category that the imported drawing style will fall under.

8. Click **Next** to start the import.

9. The *Import Drawing Styles* dialog box will display, indicating that the style was successfully imported, as shown in Figure 9–14.

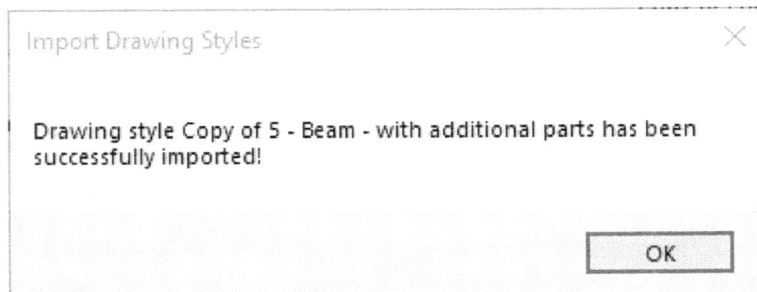

Figure 9–14

How To: Export a Style

1. In the tree panel, select a drawing style that you want to export.

2. On the toolbar, or alternatively by right-clicking on the style, select 📑 (Export).

3. In the *Open* dialog box, specify a folder location and file name.

4. Click **Open**.

5. The *Export Drawing Styles* dialog box will indicate if the export was successful, as shown in Figure 9−15.

Export Drawing Styles   ✕

Drawing style 5 - U and C has been successfully exported!

OK

Figure 9−15

How To: Compact the Database

1. On the toolbar, click (Compact) to decrease the size of the database.

2. In the *Compact the database* dialog box, specify what you want to compact, which database to compact, and the compact behavior: **Compact** or **Compact with purge**, as shown in Figure 9–16.

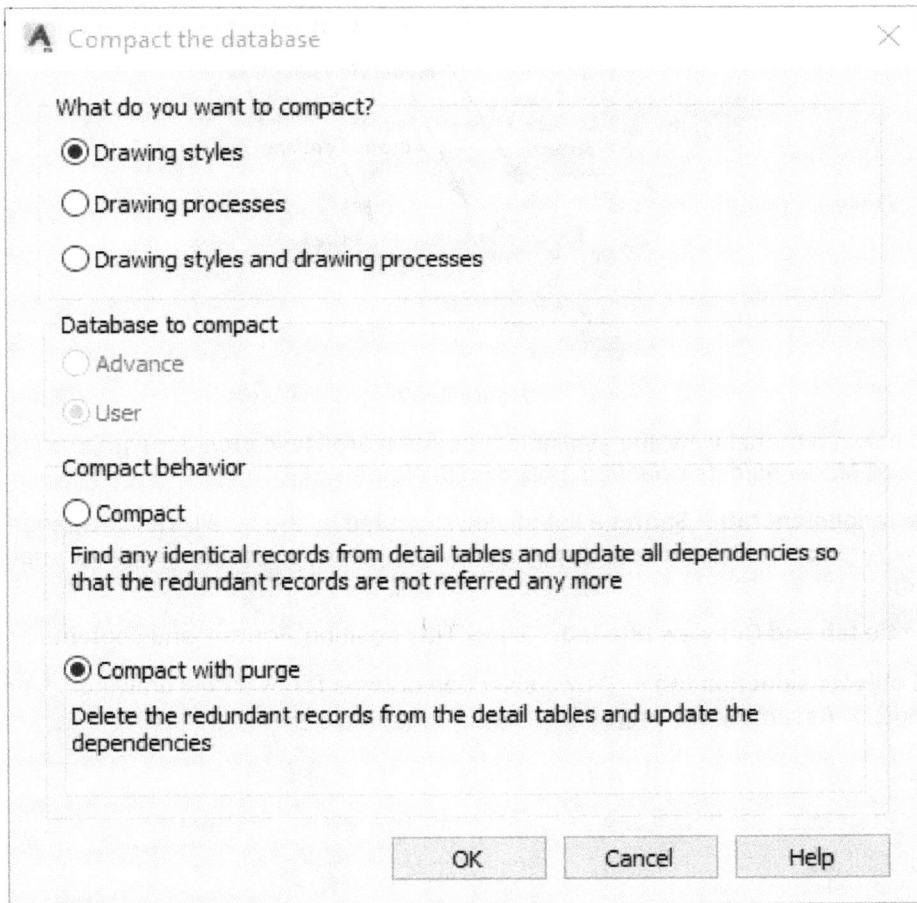

Figure 9–16

3. Click **OK** to start compacting.

Drawing Style Structure

Within each category, there are various drawing styles, and each style can have multiple views. Each style's view has advance options for defining the view presentation, dimensions, and labels, as shown in Figure 9–17.

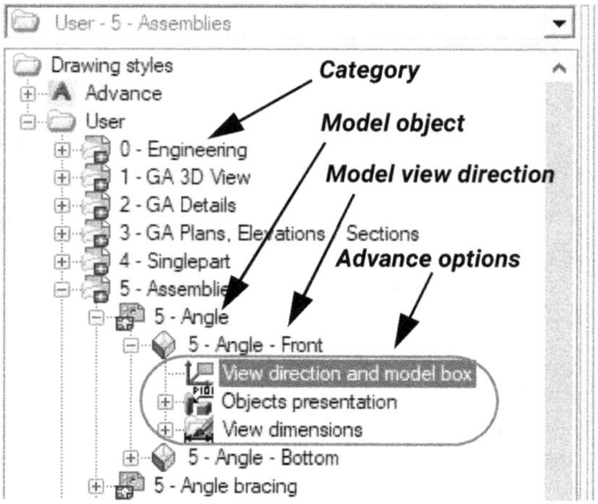

Figure 9–17

When you select a model view, the available properties are *View arrangement*, *Detail title*, *Cut view title*, and *Model objects selection* as shown in Figure 9–18.

- **View arrangement tab** – Shows a list of views created by the detail, Views' arrangement preview, Arrangement definition, distance between views, orientation, and automatic clipping.

- **Detail title tab and Cut view title tab** – Show Title position, Format, and contents of the title.

- **Model objects selection tab** – Shows selection options for overview drawings, Single part drawings, or Assembly drawings.

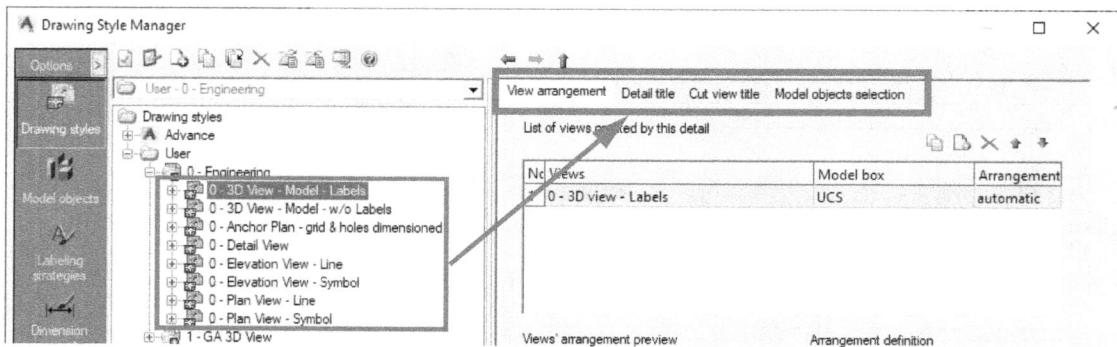

Figure 9–18

View Direction and Model Box

To define how an object's view direction displays relative to the UCS, you can review the *Properties* panel>*View direction and model box* tab and modify its properties to define the view direction, as shown in Figure 9–19.

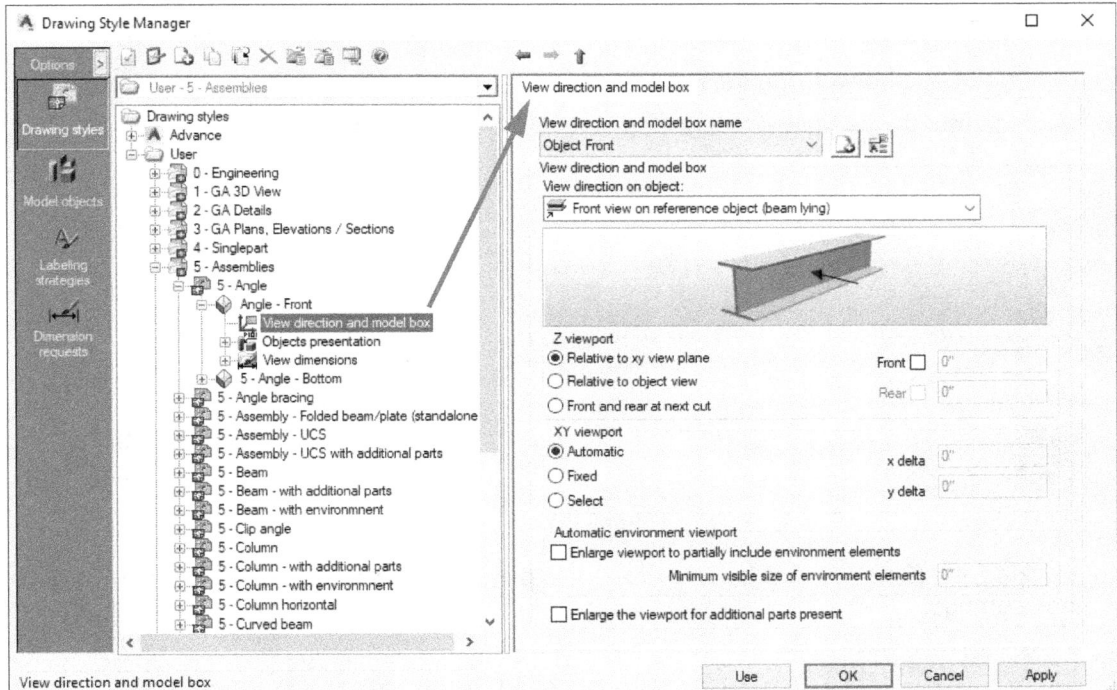

Figure 9–19

How To: Set the View Direction

1. In the tree panel, click on an object's view direction (e.g., Angle - Front or Angle - Bottom).

2. In the *Properties* panel>*View definition* tab, in the *View name* section, you can modify the existing name using ![Rename icon] (Rename) or create a new one in place of the existing one using ![New icon] (New), as shown in Figure 9–20.

3. You can modify the *View direction and model box* for the view.

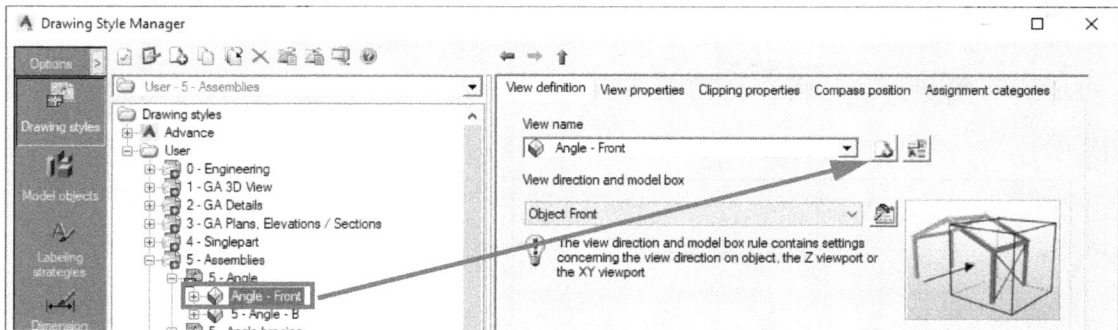

Figure 9–20

4. Clicking ![Set icon] (Set) will change your *Properties* panel to show the View direction and model box settings.

Objects Presentation

Within the model object structure, the *Objects presentation* rules will determine how the model object is displayed, its presentation, and its labeling as shown in Figure 9–21.

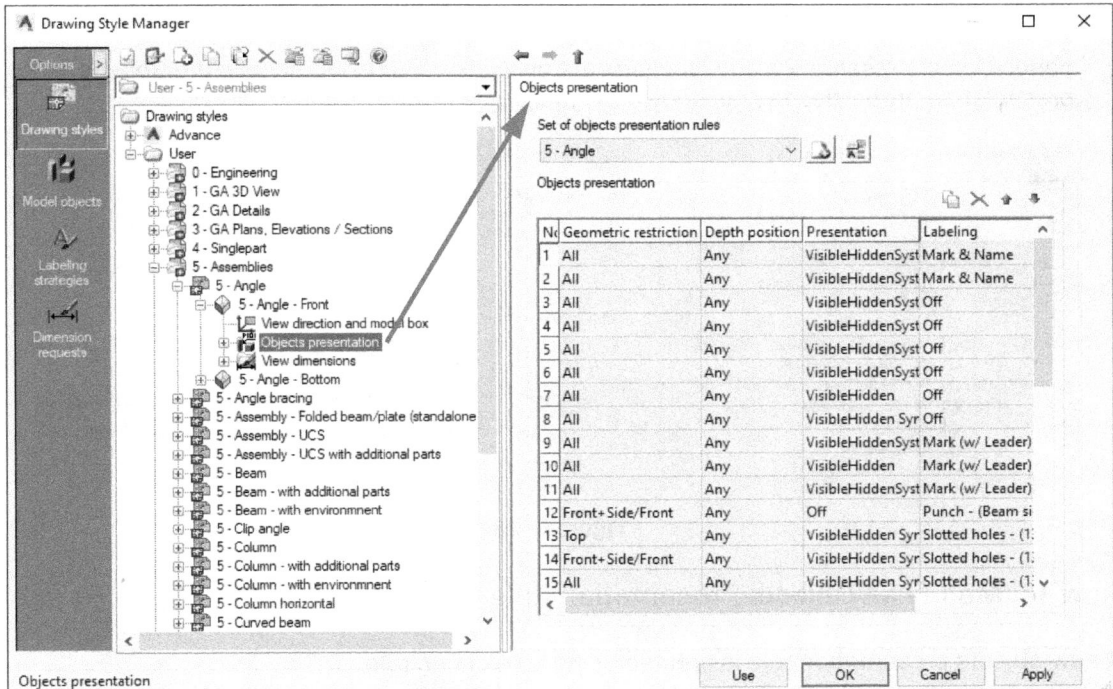

Figure 9–21

How To: Set the Objects Presentation

There are two ways to access the *Objects presentation* rules.

1. In the tree panel, expand a model object>view direction and select **Objects presentation** (as shown in Figure 9–22) to get the *Objects presentation* tab.

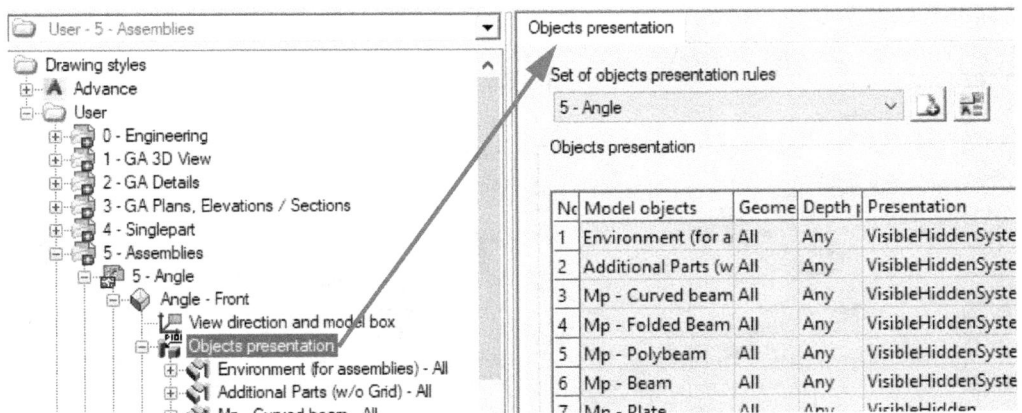

Figure 9–22

2. Alternatively, in the tree panel, click on the model object and from the *Properties* panel>*Object presentation and labeling* section, click (Set) to access the *Objects presentation* properties tab, as shown in Figure 9–23.

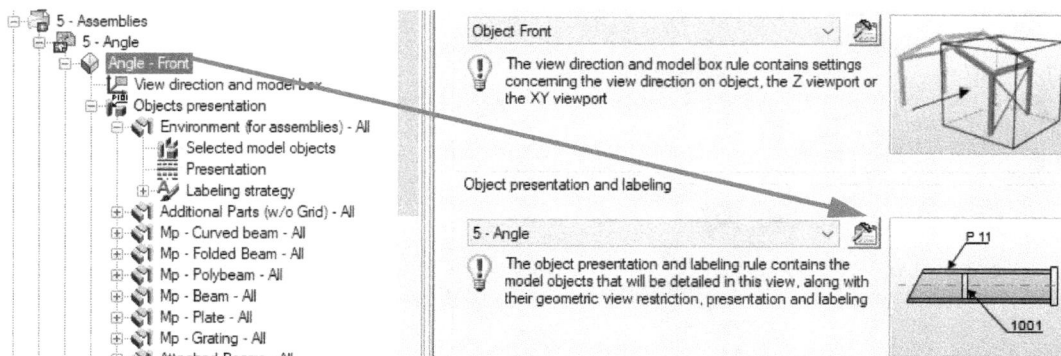

Figure 9–23

How To: Modify an Objects Presentation Rule

A new objects presentation rule is created using an existing rule.

1. Expand a model object, then expand an object view direction.

   Note: The first entry in the table cannot be deleted. Move the item down the list to delete it.

2. Select **Objects presentation**, then select a rule to use as a template.

3. Click (Add new presentation rule).

   • The new rule is added to the set of objects presentation rules, as shown in Figure 9–24.

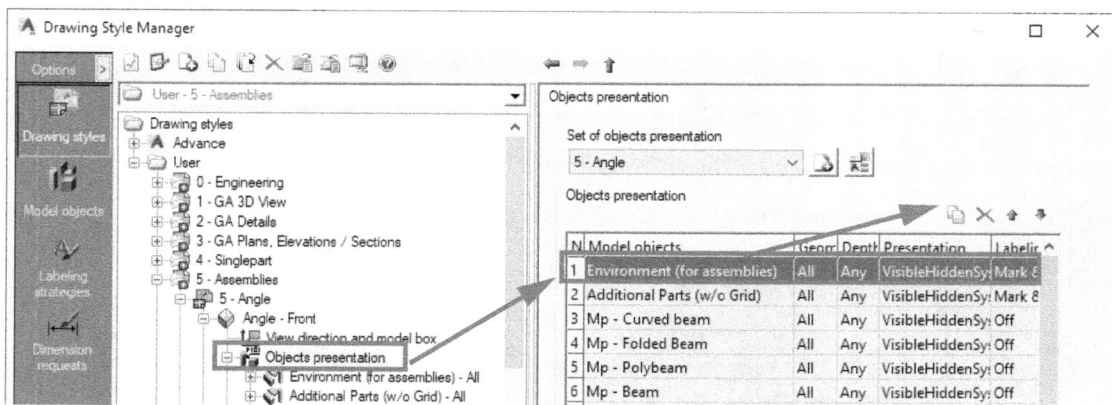

Figure 9–24

- To delete an objects presentation rule, make sure it is not first in the list (if it is, move it down), then click ✕ (Delete).

- To change the order of the list, use ⬆ (Move up selected object presentation) and ⬇ (Move down selected object presentation).

Presentation Rule

The presentation rule contains labeling settings for color and line type, as shown in Figure 9-25.

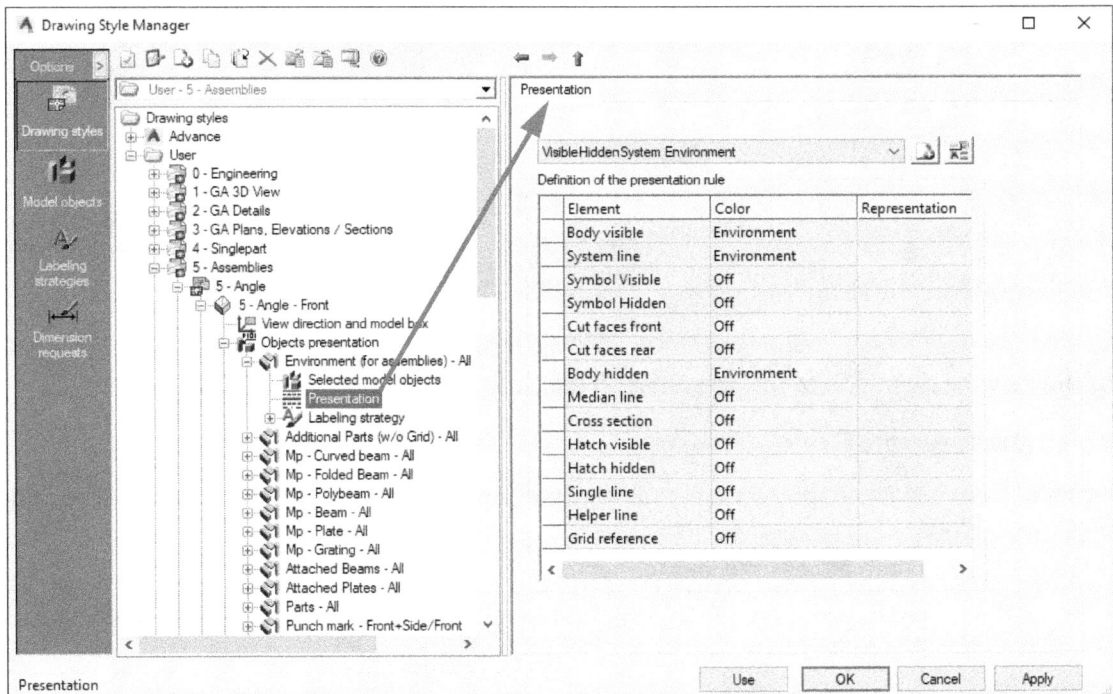

Figure 9-25

How To: Set the Presentation Rule

1. In the tree panel, expand a model object>*View direction>Objects presentation,* then select an object presentation rule.

2. In the *Properties* panel>*Objects presentation* tab>*Presentation rule* section, click (Set), as shown in Figure 9–26.

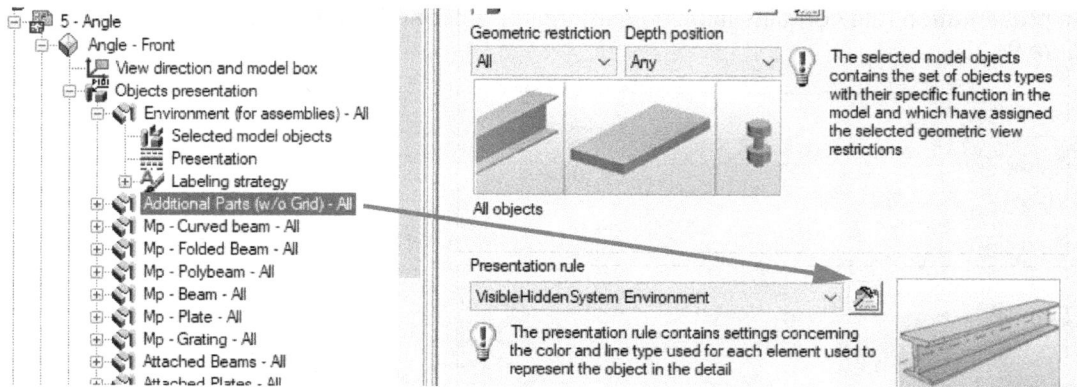

Figure 9–26

3. Review the definition of the presentation rule list.

View Dimension Style

Model objects have a set of view dimension strategies that determine the dimension type and dimension definition.

There are two ways to access the view dimension:

*   In the tree panel, expand model object>*View direction,* then select **View dimensions**, as shown in Figure 9–27.

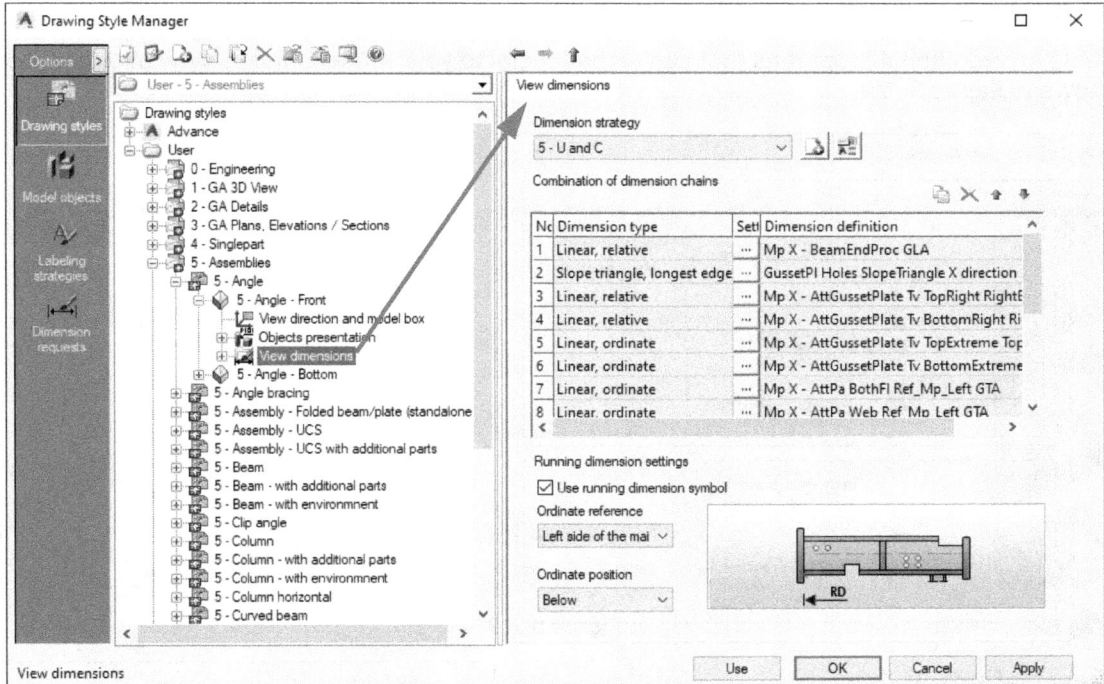

Figure 9–27

* In the tree panel, click on a view direction and from the *Properties* panel>*View dimension* section, click [Set icon] (Set), as shown in Figure 9–28.

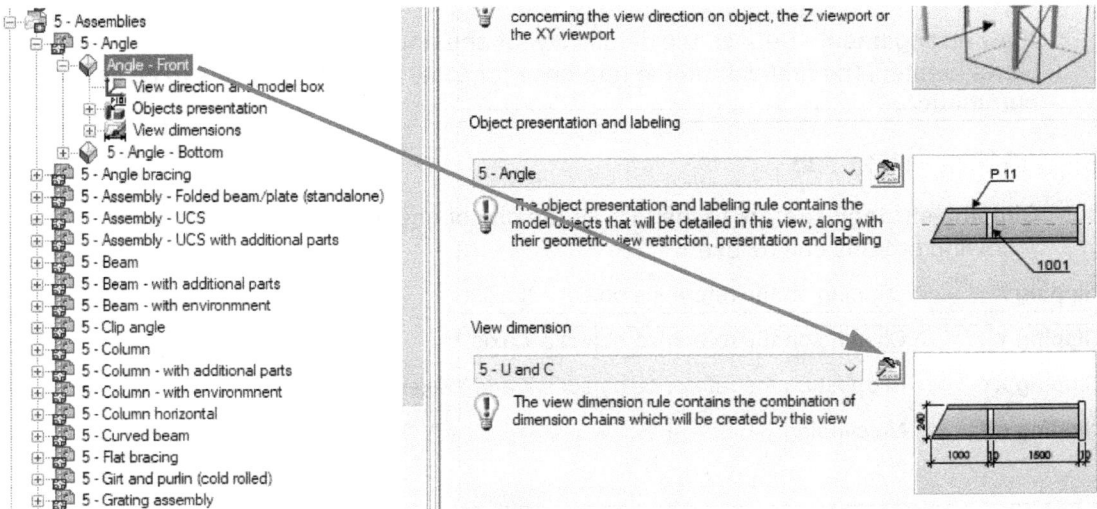

Figure 9–28

Additional Properties

There are additional properties that can be beneficial to your model object's view direction, as shown in Figure 9-29.

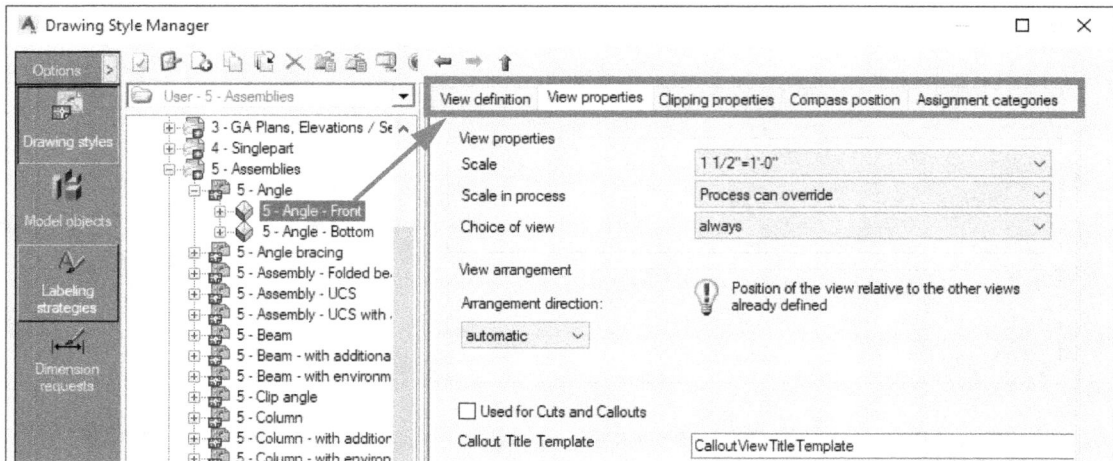

Figure 9-29

View Definition Tab

This tab gives you quick access to modify the view name, view direction and model box, object presentation and labeling, and view dimension.

View Properties Tab

- **Scale** - Set a view scale, scale in process, and choice of view.
- **View arrangement** - Defines the details layout and position of the view relative to other views created. The first view is the reference for other views so it needs to be set as **automatic**.

Clipping Properties Tab

- **Clipping strategy** - Refers to the UCS of the object and not the drawing's position. The following options can be used:

Clipping X	Clipping on the reference object's X axis.
Clipping Y	Clipping on the reference object's Y axis.
Clipping XY	Clipping on the reference object's X and Y axes.
Clipping off	No clipping.

You can define the clipping strategies by renaming or creating a new clipping strategy.

- **Minimum length to cut** - Identifies the size of the clipping. Smaller intervals mean a larger clipping.

- **How much should be kept on each side of the clipping area** - Is the excluded region's length.

- **Clipping representations in length** - Measurements that identify the distance.

Compass Position Tab

- **Compass position on assembly drawings** - Select the **Show compass** checkbox to display the compass in the drawing and set its positioning as *Upper left*, *Upper right*, *Lower left* or *Lower right*, as shown in Figure 9–30.
- **Compass representation** - Display the compass as an arrow or symbol.

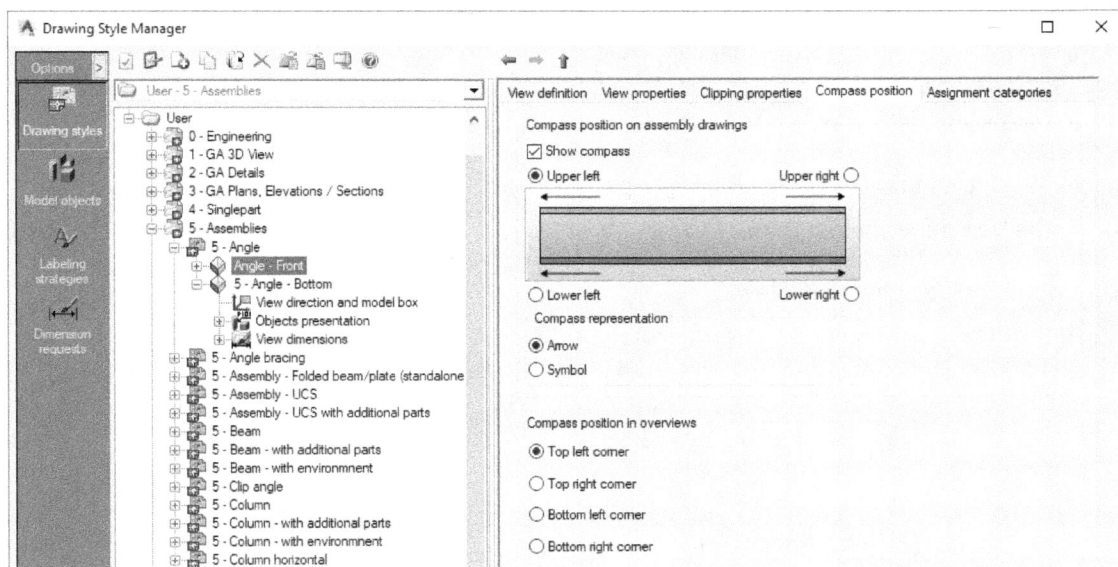

Figure 9–30

Practice 9a
Create and Edit a New Drawing Style

Practice Objectives

- Navigate the *Drawing Style Manager*.
- Create a new drawing style.
- Copy an existing drawing style and change its category.
- Copy and modify a view direction.
- Create an object presentation.
- Create a drawing style elevation view.

In this practice, you will create and copy a drawing style and change its category, then modify the view direction and create an object presentation. You will then create a drawing style elevation view.

Task 1: Create a new User drawing style.

1. In the practice files folder, open **Platform-Drawing Styles.dwg**.

2. In the *Output* tab>*Document Manager* panel, click ▥ (Drawing Style Manager).

3. In the tree panel, select **User**.

4. On the toolbar, click ↳ (New).

5. In the *New Drawing style category* dialog box, type **Class Drawing Styles** for the name, as shown in Figure 9–31.

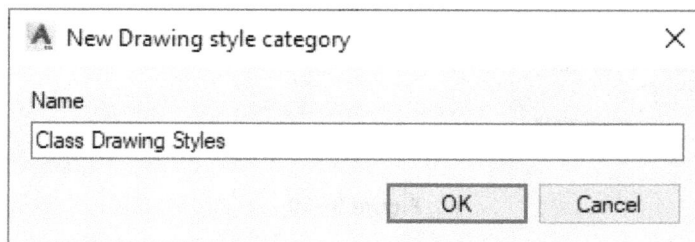

A New Drawing style category   ✕

Name

Class Drawing Styles

[OK] [Cancel]

Figure 9–31

6. Click **OK**.

7. Review the new *User* drawing style as shown in Figure 9–32.

Figure 9–32

Task 2: Copy a view direction.

1. In the *User* drawing style, expand **3 - GA Plans, Elevations / Sections** and select **3 - Elevation View - Symbol**.

2. On the toolbar, click (Copy).

3. In the *New Drawing style* dialog box, type **Class Elevation View - Symbol** for the name, as shown in Figure 9–33.

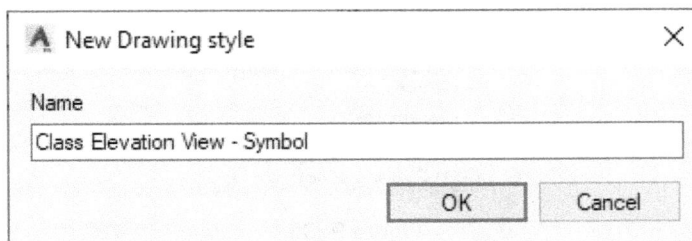

Figure 9–33

4. Click **OK** to add the view.

5. Right-click on **Class Elevation View - Symbol** and select **Properties**, as shown in Figure 9–34.

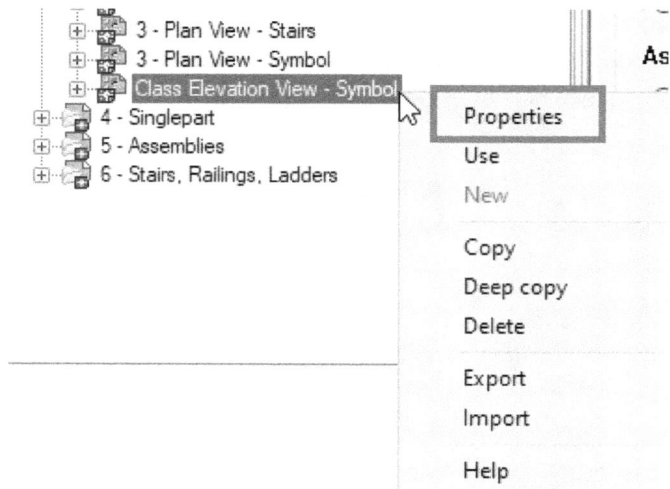

Figure 9–34

6. In the *Drawing style* dialog box, change the *Category* to **Class Drawing Styles**, as shown in Figure 9–35.

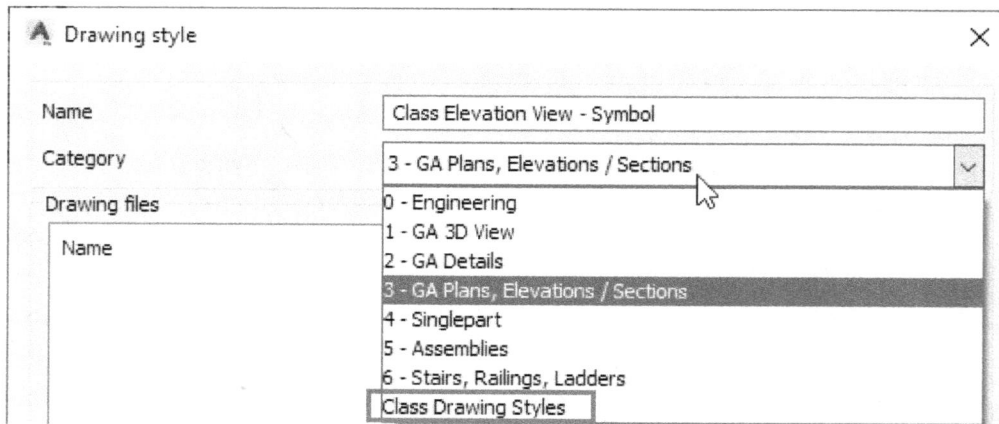

Figure 9–35

7. Click **OK**.

8. Expand **Class Drawing Styles>Elevation View - Symbol** and select **3 - Elevation View - Symbol**.

9. In the *Properties* panel, click on the *View definition* tab.

10. In the *View name* section, click ⬚ (New), as shown in Figure 9–36.

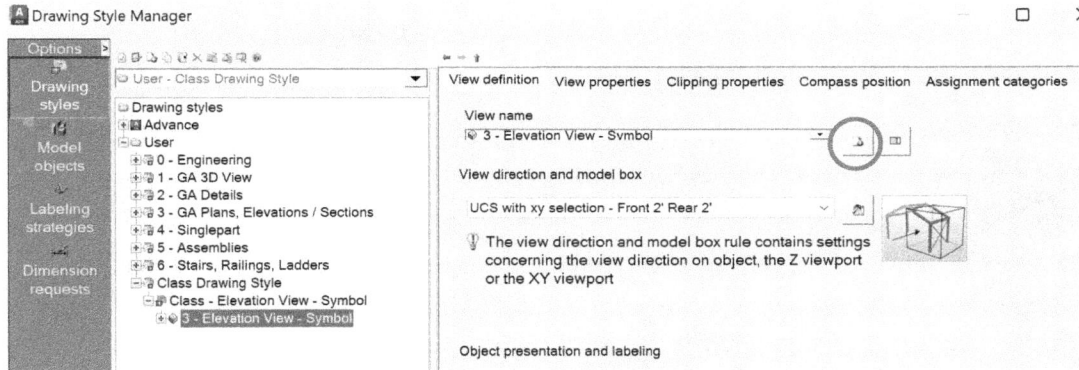

Figure 9–36

Figure 9–36

11. In the *New Drawing style* dialog box, type **Class - Elevation View - Symbol**.

12. Click **OK**.

13. In the *Properties* panel, change the *View direction and model box* to **UCS with xy selection - Front 4' Rear 4'**, as shown in Figure 9–37.

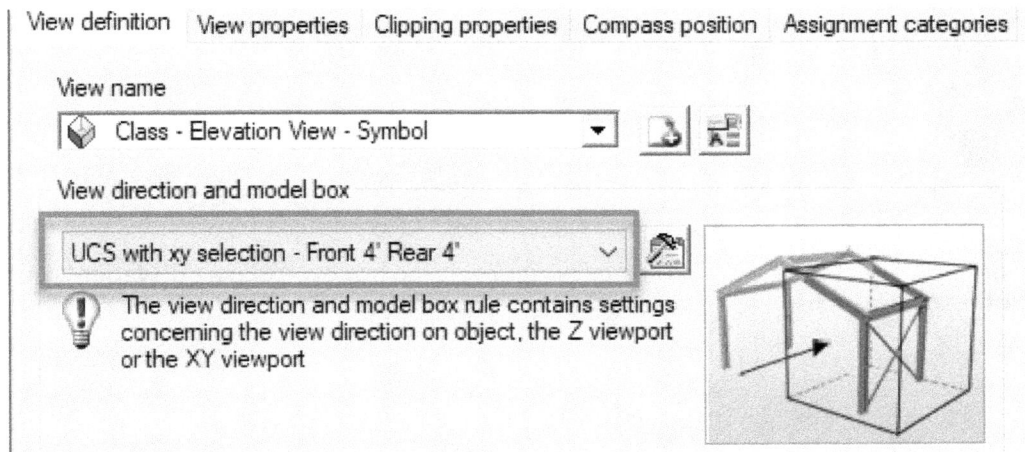

Figure 9–37

Task 3: Create an objects presentation.

1. Expand **Class - Elevation View - Symbol** and select **Objects presentation** as shown in Figure 9–38, then click 🔯 (New).

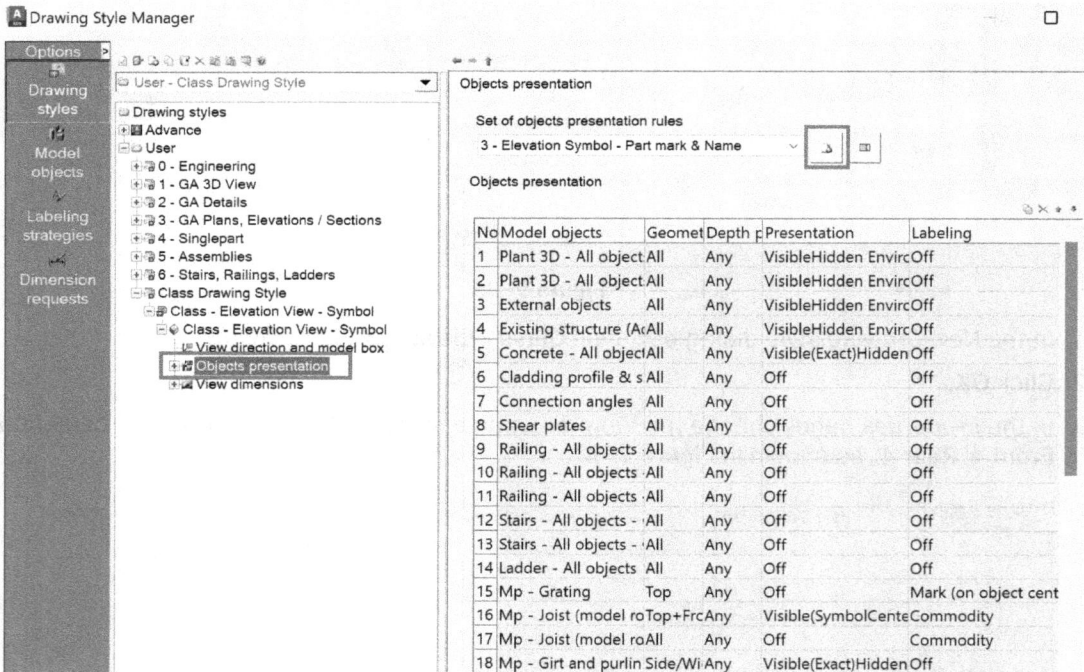

Figure 9–38

2. In the New *objects presentation and labeling rule* dialog box, type **Class - Elevation Symbol - Part mark & Name**.

3. Click **OK**.

4. Within the *Objects presentations* section>*Presentation* column, change No. *9 - 13* to **Visible(Exact)Hidden** as shown in Figure 9–39.

Objects presentation

Set of objects presentation rules

Class - Elevation Symbol - Part mark & Name

Objects presentation

No.	Model objects	Geometri	Depth po	Presentation	Labeling
1	Plant 3D - All objects - Grou	All	Any	VisibleHidden Environment	Off
2	Plant 3D - All objects - Grou	All	Any	VisibleHidden Environment	Off
3	External objects	All	Any	VisibleHidden Environment	Off
4	Existing structure (Advance	All	Any	VisibleHidden Environment	Off
5	Concrete - All objects	All	Any	Visible(Exact)HiddenCutCor	Off
6	Cladding profile & sheet	All	Any	Off	Off
7	Connection angles	All	Any	Off	Off
8	Shear plates	All	Any	Off	Off
9	Railing - All objects - Group	All	Any	Visible(Exact)Hidden	Off
10	Railing - All objects - Group	All	Any	Visible(Exact)Hidden	Off
11	Railing - All objects - Group	All	Any	Visible(Exact)Hidden	Off
12	Stairs - All objects - Group 1	All	Any	Visible(Exact)Hidden	Off
13	Stairs - All objects - Group 2	All	Any	Visible(Exact)Hidden	Off
14	Ladder - All objects	All	Any	Off	Off
15	Mp - Grating	Top	Any	Off	Mark (on o

Figure 9–39

5. Click **OK** to close the *Drawing Style Manager*.

Task 4: Add a detail section.

1. Change your UCS so that Y is up and X is to the right, as shown in Figure 9–40.

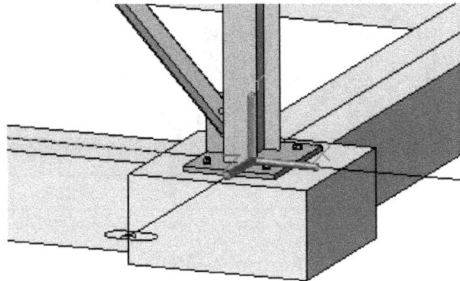

Figure 9–40

2. In the *Output* tab>*Document Manager* panel, click ⬛ (Drawing Style Manager).

3. In the tree panel, expand **Class Drawing Styles**, select **Class - Elevation View - Symbol** and click **Use**.

4. In the *Create detail* dialog box, select **Create with default settings**, as shown in Figure 9–41.

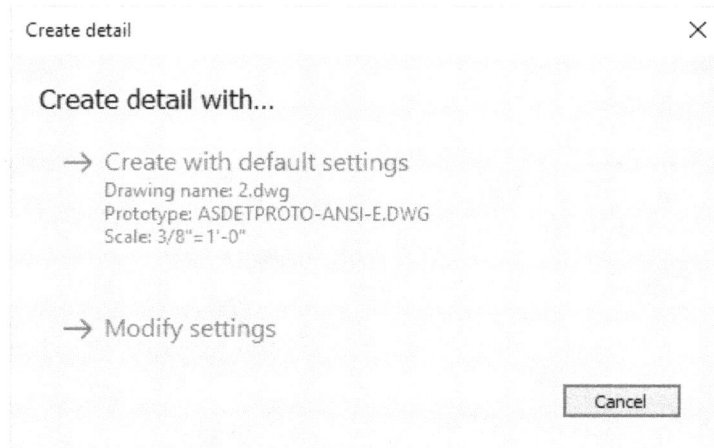

Figure 9–41

5. Draw a selection window around the end of the structure as shown in Figure 9–42.

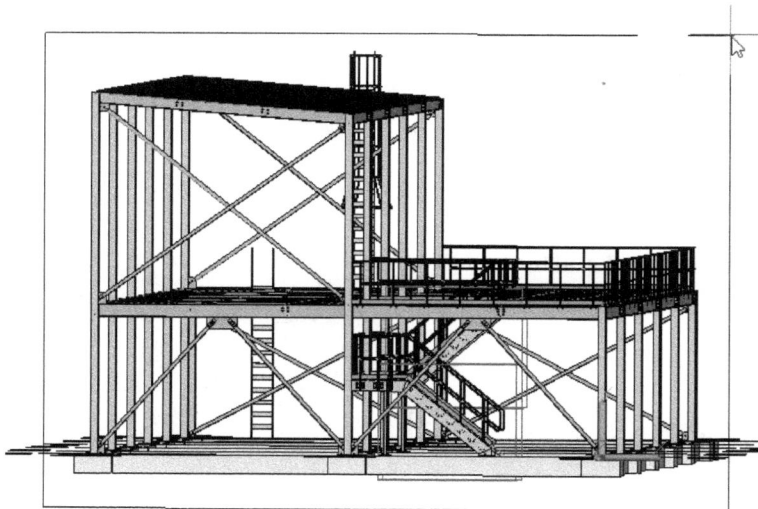

Figure 9–42

6. Click **OK** to close the *Drawing Style Manger*.

7. In the *Output* tab>*Document Manager* panel, select ⊞ (Document Manager).

8. In the *Document Manager* dialog box, expand **Details>Up to date** and select **1.dwg**.

9. Click **Open drawing**, as shown in Figure 9–43.

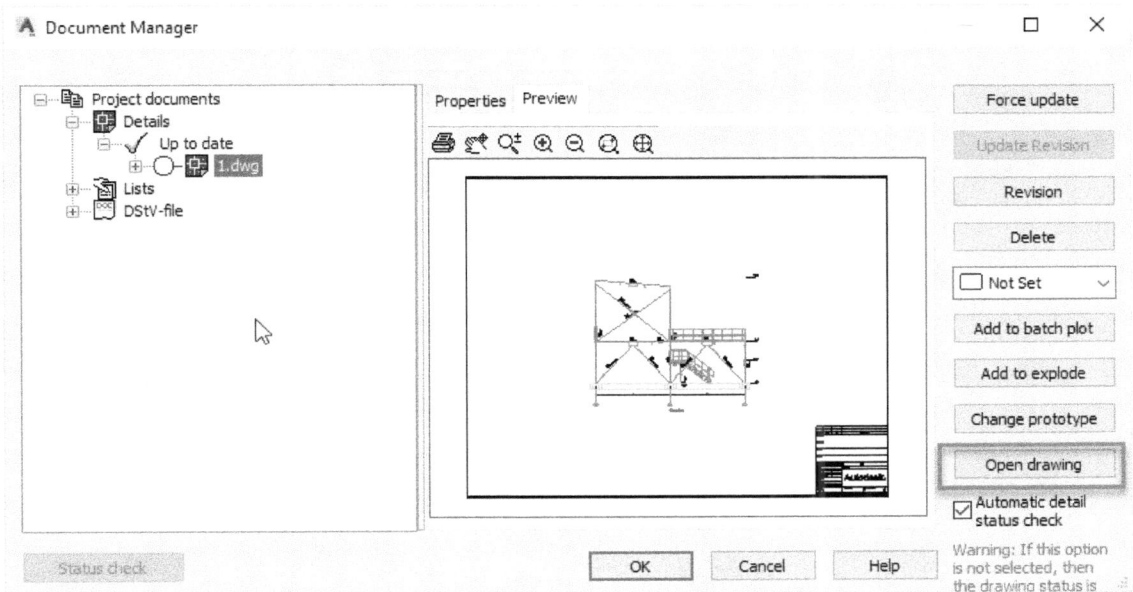

Figure 9–43

10. In order to see the new drawing style in the palette, close all palettes and the Advance Steel software.

11. Start the Advance Steel software.

12. In the *Output* tab>*Documents* panel, select ⊞ (Drawing Styles) to open the *Drawing Styles* tool palette.

13. In the *Drawing Styles* tool palette, change the *content* category from ▦ (Advance) to 👤 (User).

14. Click on the blank category at the bottom of the tool palette to show the drawing style **Class - Elevation View - Symbol**, as shown in Figure 9–44.

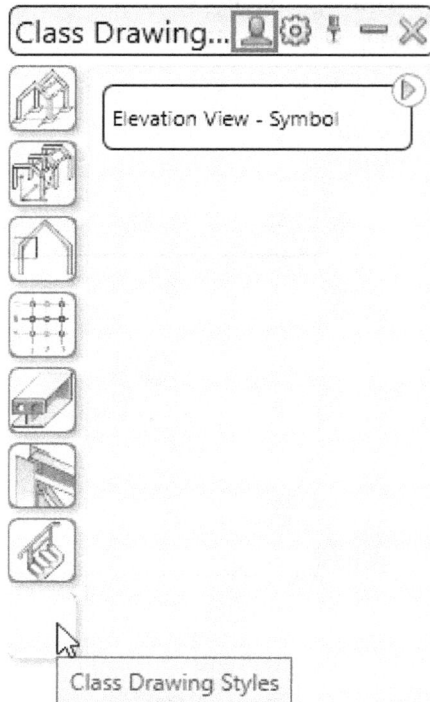

Figure 9–44

15. Save the drawing.

End of practice

Chapter Review Questions

1. The *Drawing Style Manager* interface has what tools?

    a. Drawing Styles, Properties Panel, and Toolbar Panel.

    b. Options, Toolbar, Model Objects, and Direction View.

    c. Toolbar, Tree Panel, Properties Panel, and Component Panel.

    d. Tree Panel, Model Objects, Component Panel, and Tools.

2. In the *Drawing Style Manager*, you can copy a drawing style from the *Advance* folder.

    a. True

    b. False

3. What tool would you use to create a new drawing style?

    a. Copy

    b. Deep Copy

    c. Compact

    d. New

4. What do objects presentation rules determine?

    a. How model objects are displayed and labeled.

    b. How model objects are defined and viewed.

    c. How model objects are set and labeled.

    d. How model objects are listed and displayed.

Command Summary

Button	Command	Location
	Advance	• **Drawing Style Tool Palette**
	Back	• **Drawing Style Manager:** Toolbar
	Compact	• **Drawing Style Manager:** Toolbar
	Context Help	• **Drawing Style Manager:** Toolbar
	Copy	• **Drawing Style Manager:** Toolbar
	Delete	• **Drawing Style Manager:** Toolbar
	Document Manager	• **Ribbon:** *Output* tab>*Document Manager* panel
	Drawing Styles	• **Ribbon:** *Output* tab>*Document Manager* panel
	Drawing Style Manager	• **Ribbon:** *Output* tab>*Document Manager* panel
	Export	• **Drawing Style Manager:** Toolbar
	Import	• **Drawing Style Manager:** Toolbar
	New	• **Drawing Style Manager:** Toolbar
	Properties	• **Drawing Style Manager:** Toolbar
	Rename	• **Drawing Style Manager:** *Properties* panel
	Set	• **Drawing Style Manager:** *Properties* panel
	Up	• **Drawing Style Manager:** Toolbar
	Use	• **Drawing Style Manager:** Toolbar
	User	• **Drawing Style Tool Palette**

Bill of Materials Template Editor

Bills of Materials (BOMs) are an essential part of any project and are heavily relied on by manufacturers and engineers for document control. The Advance Steel Bill of Materials Template Editor allows users to customize existing BOM template elements, including graphics, fonts, colors, token attributes and list structure, to fit your company standards.

Learning Objectives

- Work with element properties.
- Edit tokens.
- Format BOM borders.
- Understand list structure.
- Filter data with report contents.

A.1 BOM Template Editor User Interface

The BOM Template Editor has several tools that allow you to modify and edit a new or existing BOM in the template. You are able to modify and add your company logo, labels, lines, tokens, check boxes and more, as shown in Figure A-1.

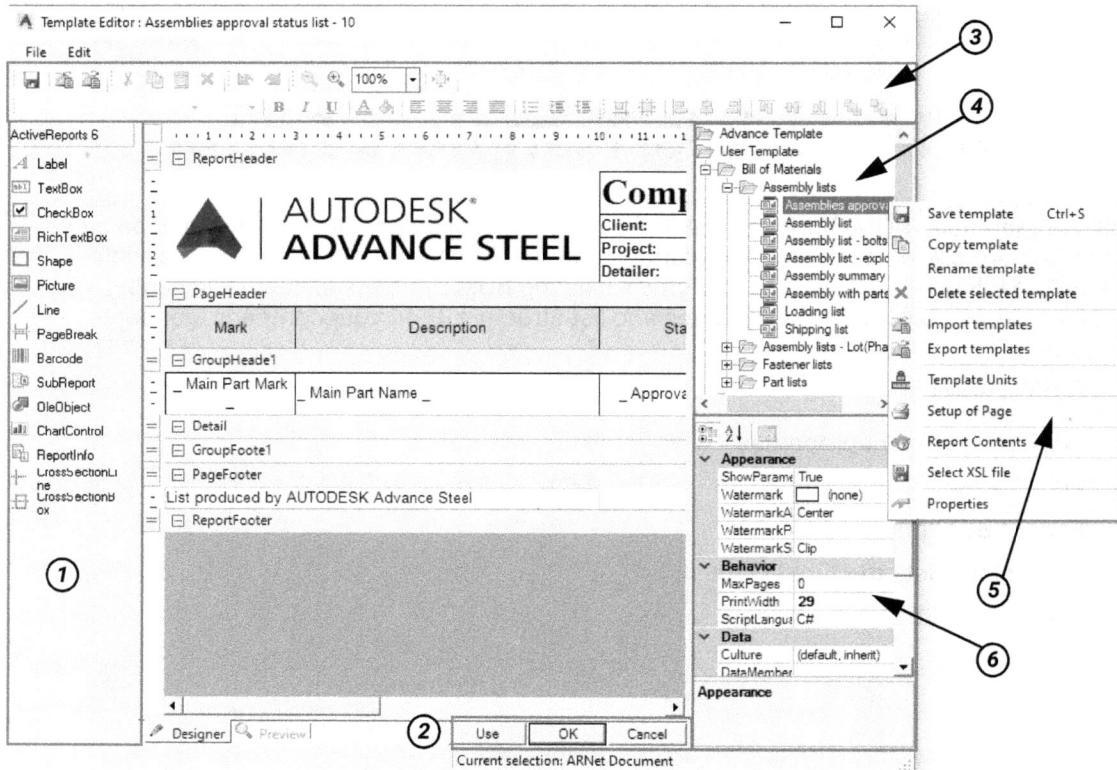

Figure A-1

1. List of controls   4. Template tree

2. Buttons for applying or canceling modifications   5. Tools for template management

3. Tools for template customization   6. Properties list

How To: Launch the BOM Template Editor

1. In the *Output* tab>*Document Manager* panel, click ▤ (BOM Editor).

2. On the Template tree>*User Template* folder, locate the template you want to modify.

   Note: You cannot edit the templates in the Advance Template folder.

3. You can edit elements and boxes in the template by selecting them and adjusting the settings in the *Properties* list, as shown in Figure A−2.

Appearance	
ShowParameterUI	True
Watermark	☐ (none)
WatermarkAlignment	Center
WatermarkPrintOnPage	
WatermarkSizeMode	Clip
Behavior	
MaxPages	0
PrintWidth	**28.6**
ScriptLanguage	C#
Data	
Culture	(default, inherit)
DataMember	
DataSource	
UserData	
Design	
TrayHeight	80
TrayLargeIcon	**True**
Misc	
ExpressionErrorMessag	
Version	6.1.2577.0

Figure A−2

Tokens

Token attributes can be modified or changed by selecting field content, then right-clicking and selecting **Field content** from the contextual menu to bring up the *Content* dialog box, as shown in Figure A–3.

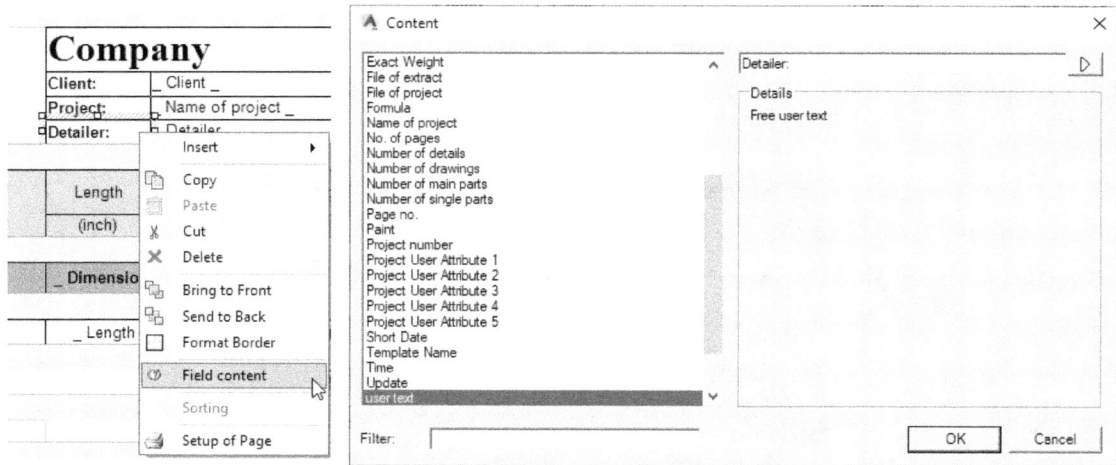

Figure A–3

Formatting

A cell's borders can be modified by selecting the content, then right-clicking and selecting **Format Border**. You can edit the border's presets, line styles and color, as shown in Figure A–4.

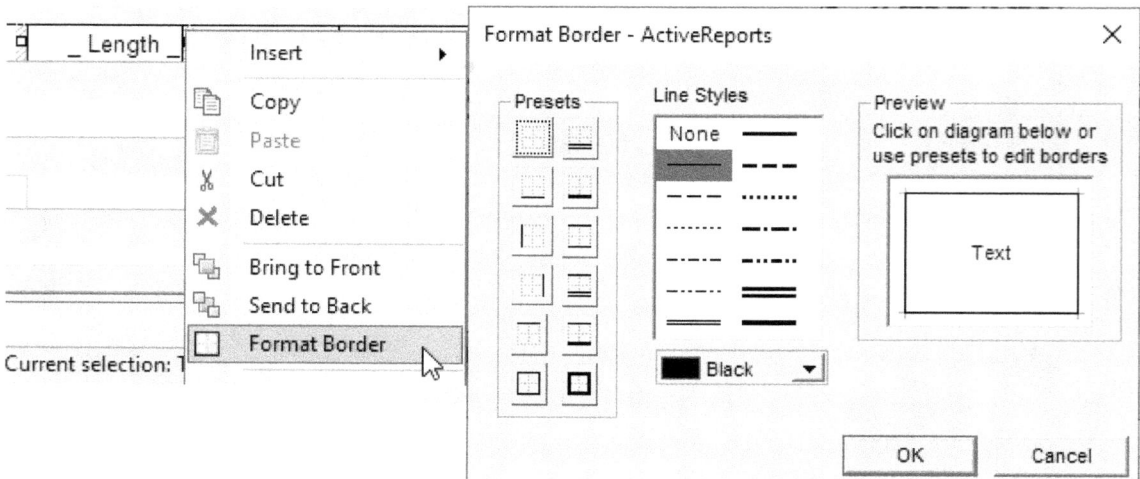

Figure A–4

You can also drag elements like lines and shapes from the list of controls on the left side of the *Template Editor* dialog box, as shown in Figure A–5.

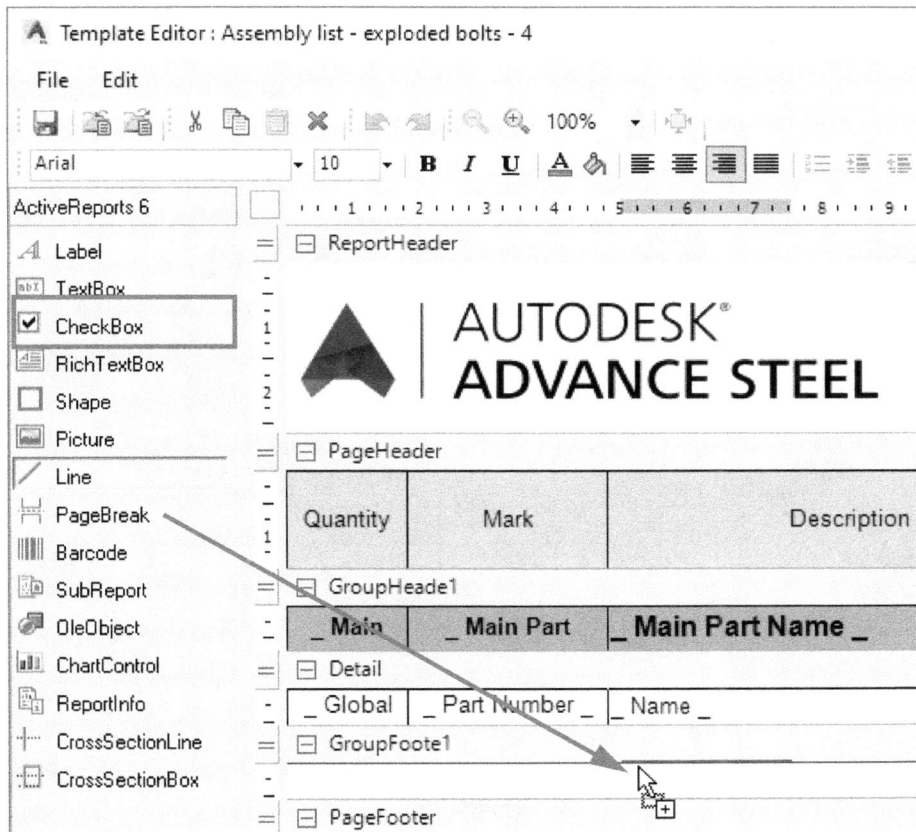

Figure A–5

You can further align, change background and foreground colors, change fonts, and adjust the content by using the toolbar at the top of the *Template Editor*, as shown in Figure A–6.

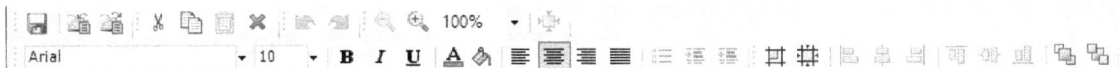

Figure A–6

List Structure

Each BOM list is divided into sections: ReportHeader, ReportFooter, PageHeader, PageFooter, GroupHeaderX and GroupFooterX, as shown in Figure A-7.

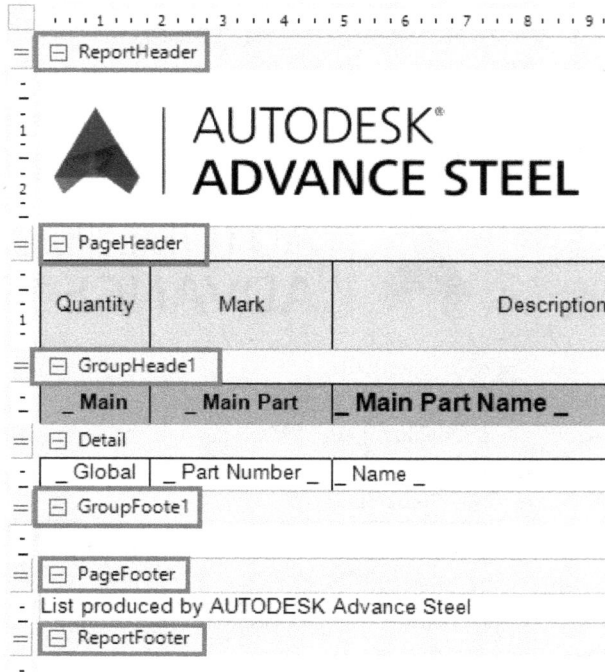

Figure A-7

The *Detail* line is the portion of the list that will show each element and its properties, as shown in Figure A-8.

Figure A-8

Report Contents

Report Contents allows you to filter the data that will be pulled from the drawing for certain objects, like beams, anchors, bolts and plates, to be included in the BOM.

How To: Edit the Report Contents

1. From the *Template* tree, select on the assembly list you want to edit.

2. Right-click and select **Report Contents** from the contextual menu, as shown in Figure A–9.

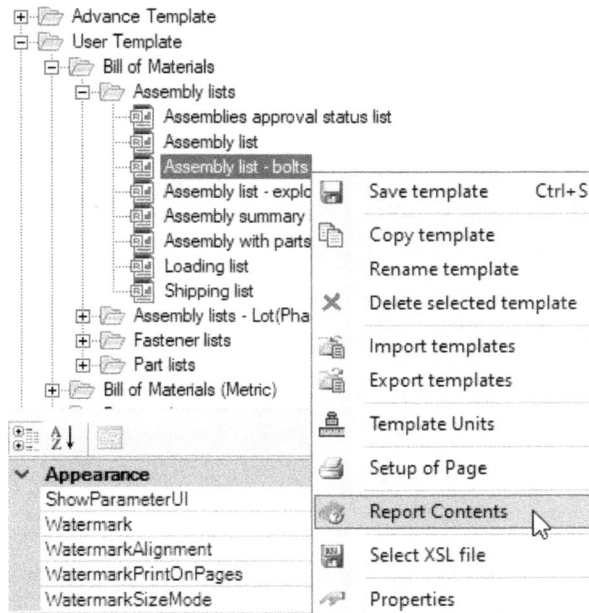

Figure A–9

3. From the list of object types in the *Report Contents* dialog box, select all the object types you want to add to the BOM, as shown in Figure A–10.

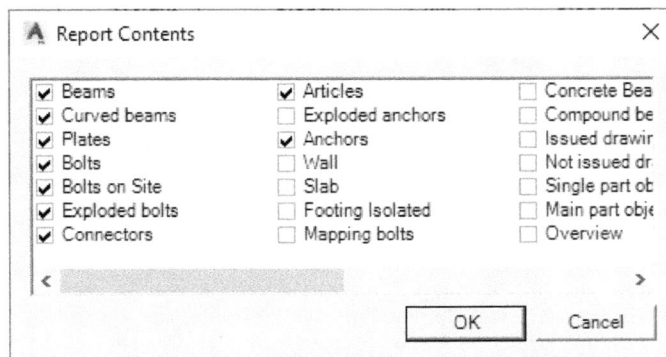

Figure A–10

4. Click **OK** and save the template.

Command Summary

Button	Command	Location
	BOM Editor	• **Ribbon:** *Output* tab>*Document Manager* panel

Management Tools

The Autodesk® Advance Steel Management Tools is a database enabling you to edit or modify drawing and element properties, preferred object sizes, and object properties; add bolt and anchor types; and define units, defaults, and shear stud types. The Management Tools is used to customize and set a substantial amount of program defaults in the database library to suit your company needs.

Learning Objectives

- Explore the *Management Tools* user interface.
- Review default values within the *Management Tools*.
- Use the *Management Tools* filter.

B.1 Management Tools Defaults

Program Defaults are accessed through the Management Tools. Defaults are the basic settings for object properties and their default values, as shown in Figure B–1.

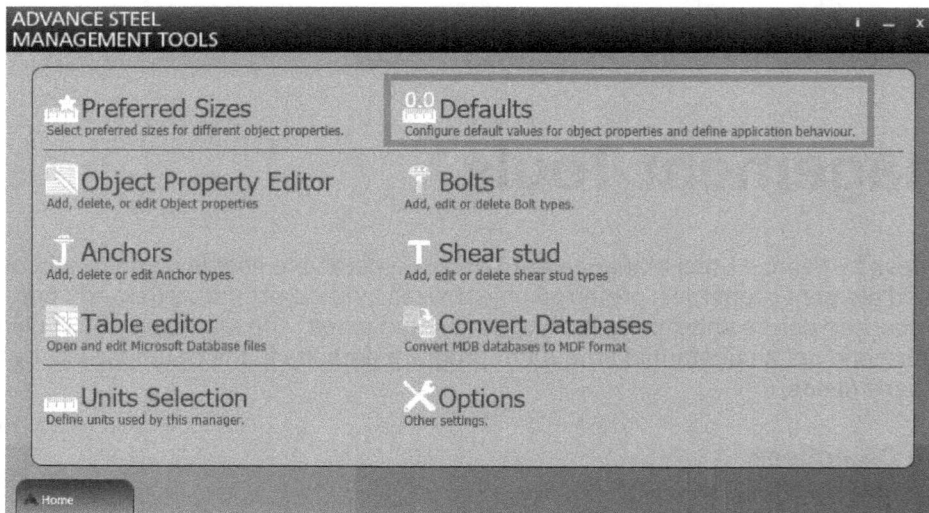

Figure B–1

How To: Access Management Tools Defaults

1. In the *Home* tab>*Settings* panel, click (Management Tools).

2. Within the Advance Steel Management Tools, select (Defaults).

3. From the Management Tools Defaults, you can expand each group to review its sub-group and settings, as shown in Figure B–2.

   Note: Hover your cursor over a setting to get a tooltip description.

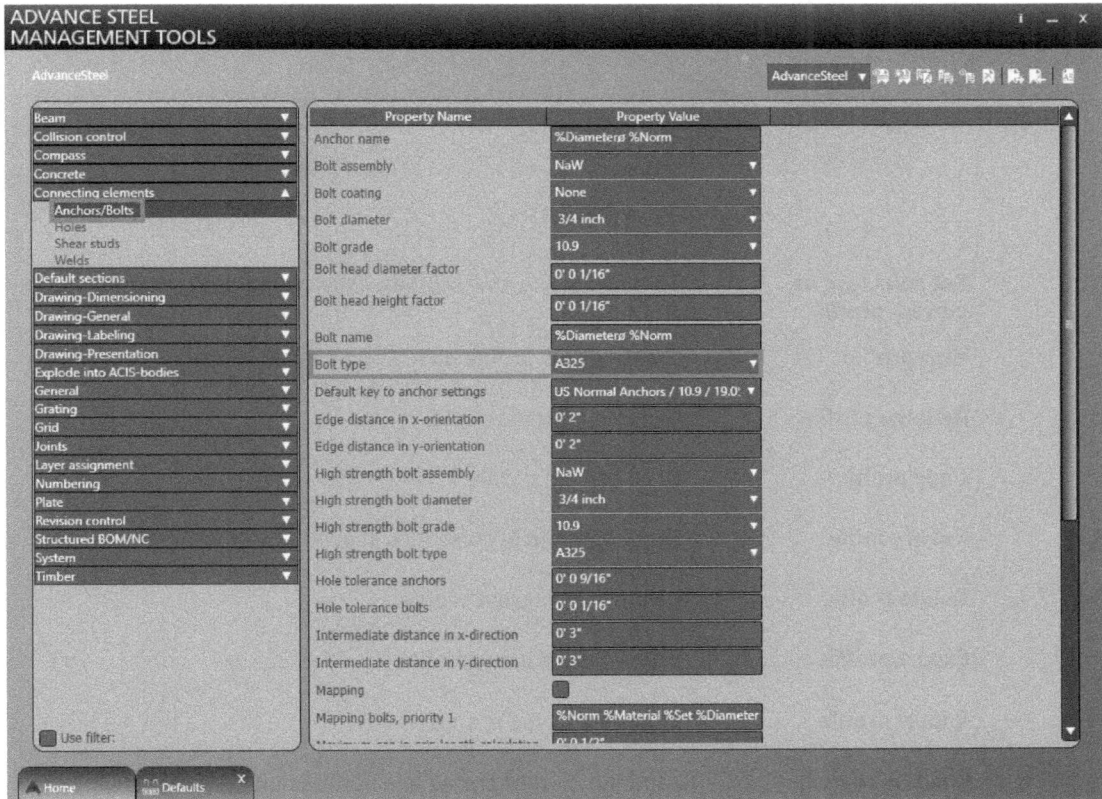

Figure B–2

4. Make the required changes to the settings.

5. From the toolbar, click (Load Settings in Advance).

Toolbar

You can save your changes to the current Advance Steel profile or you can utilize the tools on the toolbar to rename, copy, reset, delete, or create a new profile (as shown in Figure B–3).

Current Profile

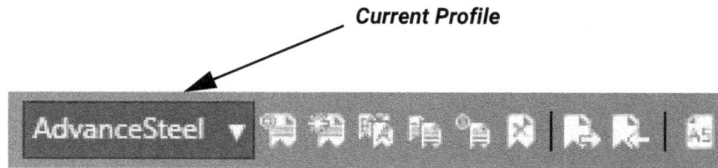

Figure B–3

	Set selection as current profile	To activate the new profile as the current one.
	New profile	Create a new profile based on an existing one.
	Rename profile	Rename current profile.
	Copy profile	Copy current profile.
	Reset profile	Reset settings in current profile.
	Delete profile	Deletes the current profile.
	Export profile	Exports the current profile.
	Import profile	Imports a saved profile.
	Load Settings in Advance	Load the current profile and settings into a drawing.

Note: *If you do not click* ⬛ *(Load Settings in Advance) from the toolbar, the changes do not load until you restart the software.*

Filter

Within the Management Tools Defaults, there are over 1,000 default settings. To help you find a setting, you can use the filter function, as shown in Figure B–4.

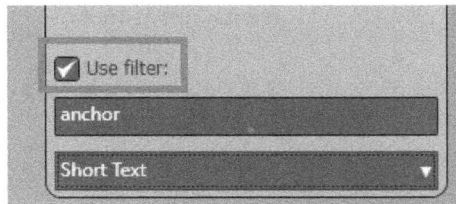

Figure B–4

How To: Use Filter

1. In the lower left corner of the Management Tools, select the **Use filter** checkbox.

   Note: Avoid using plurals when using the filter tool.

2. Enter an object name or property term.

3. Press <Enter>. The defaults are filtered to only show settings pertaining to the filter term, as shown in Figure B–5.

Figure B–5

Command Summary

Button	Command	Location
	Defaults	• **Management Tools**
	Management Tools	• **Ribbon:** *Home* tab>*Settings* panel
	Load Settings in Advance	• **Management Tools>**Toolbar

User Sections

Autodesk® Advance Steel has a database of over 100 standard cross sections, all with varying shapes and sizes. You are provided with tools that enable you to create non-standard user sections or customize existing sections that can then be added to the database and used in other projects. Creating a user section is done in three parts: using special layers to draw the section, defining key points along the polylines, and generating the section to add to the section library.

Learning Objectives

- Use layers and basic elements.
- Identify and place key points.
- Generate sections.
- Add sections to a drawing.

C.1 Layers and Basic Elements

When drawing a user section, you need to draw each section detail on a specific layer, as shown in Figure C–1. Then, you have to define key points in order to generate the user section and add it to the database. If you need to create multiple sections and sizes, you will create them all in one drawing file, but each section and size needs to be within its own frame.

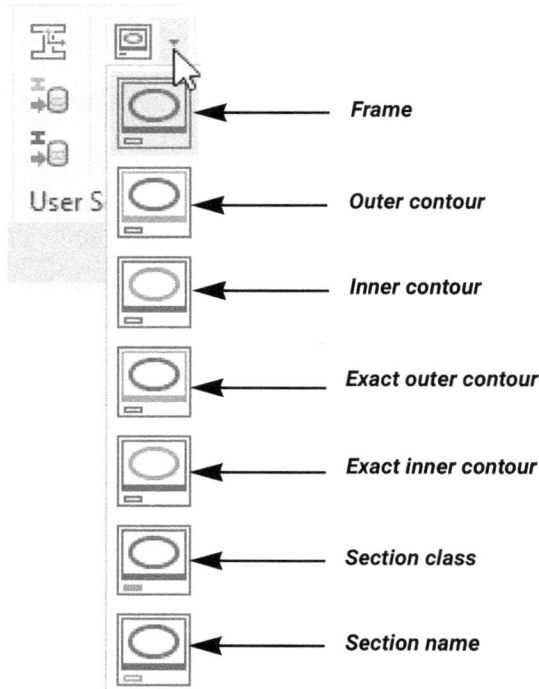

Figure C–1

How To: Start a User Section Creation

1. In the Quick Access Toolbar or Application Menu, click ☐ (New).
2. In the *Select template* dialog box, select the appropriate template and click **Open**.
3. Set your display to **Top** view.
4. Using the basic AutoCAD drawing tools, start drawing your section.

How To: Access Layers

In the *Extended Modeling* tab>*User Section* panel, expand 🔲 ▾ (Frame).

Frame

The Frame layer is the rectangular polyline that will surround the section and labels.

Outer Contour

The Outer Contour layer defines the approximate shape of the section and should not include exact details. This must be drawn using a closed polyline.

Inner Contour

The Inner Contour layer is for creating hollow surfaces. Like Outer Contour, this layer should not include exact details. This must be drawn using a closed polyline.

Exact Outer Contour

The Exact Outer Contour layer is for specific details of the section and needs to include exact detail specifications. This must be drawn using a closed polyline.

Exact Inner Contour

The Exact Inner Contour layer is for specific hollow details and should include exact detail specifications. This must be drawn using a closed polyline.

Section Class

The Section Class layer is for text associated with the object types class property.

> **Note:** *Avoid using special characters like periods and commas.*

Section Name

The Section Name layer is for text that will indicate the actual section name and size.

Creating Required Elements

Creating the required elements can be done using basic AutoCAD commands. For clarification, a section has been split into its parts so you can see each element of the section and which layer it is on, as shown in Figure C–2.

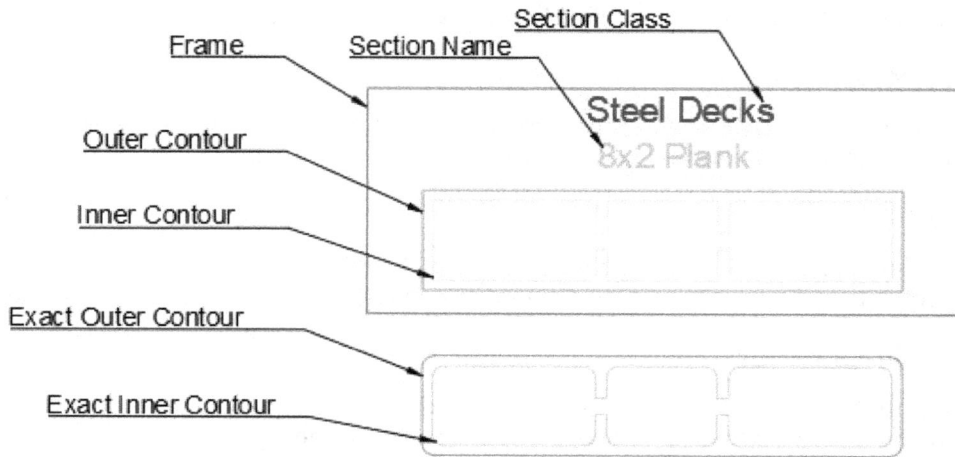

Figure C–2

Identify Key Points

Once the user section has been created on the appropriate layers, you must specify the key points for the reference axis system line and UCS coordinates.

Reference Axis

You can define the nine plausible reference axis location points for the system line on the user section.

How To: Add a Reference Axis

1. In the *Extended Modeling* tab>*User Section* panel, click ⬜ ˅ (Reference Axis left-top).
2. In the *Reference Axis* drop-down list, select the appropriate reference axis type to insert (as shown in Figure C–3), making sure to snap to the outer contour of the section.

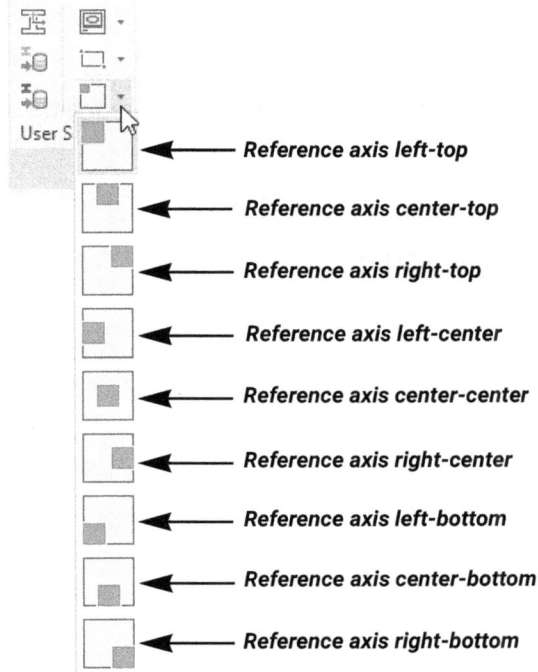

Figure C–3

- To add a reference axis, you must use snap points on the lines. If you do not have a snapping point, for example when using **Reference axis center-center**, you can add construction lines as shown in Figure C–4. If using construction lines, try not to draw them on the special layers; if you do, you must delete the construction lines after placing the reference axis or they will cause errors when you generate the section.

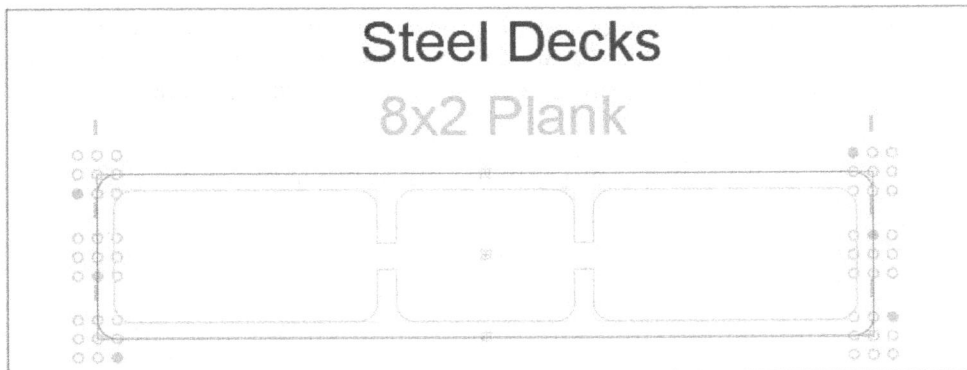

Figure C–4

Add Coordinates

On the section's surface, you need to specify the user coordinate system (UCS) for the four outer contour edges. Each coordinate is represented by two magenta arrows, both at a 90° angle, as shown in Figure C–5.

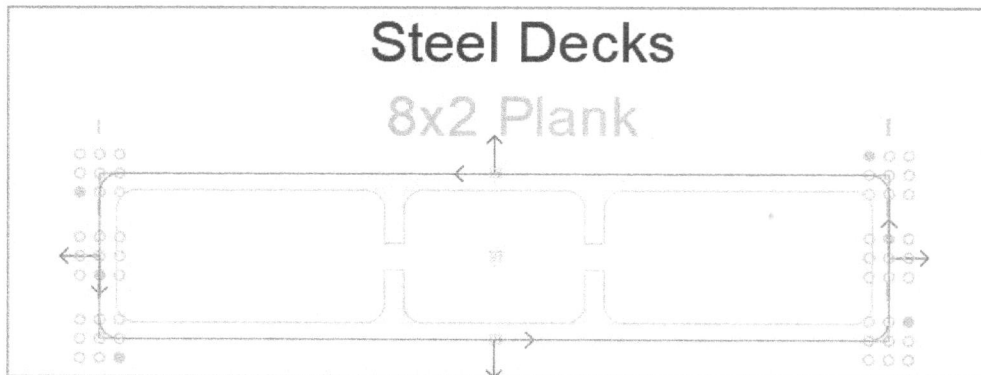

Figure C–5

How To: Add Coordinates

1. In the *Extended Modeling* tab>*User Section* panel, click ⬚ (Add Coordinates).
2. Select anywhere along the outer contours edge to define a coordinate system.

Generate Sections

When the key points and USC have been added to the section, you can generate and add the section to the Autodesk Advance Steel library. You can use either **Generate Selected Section** for one section or **Generate All Sections** for multiple sections in your drawing, as shown in Figure C–6.

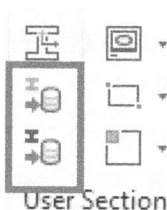

Figure C–6

How To: Generate Sections

1. In the *Extended Modeling* tab>*User Section* panel, click (Generate Selected Section) for a single section or (Generate All Sections) for multiple sections.

    Note: *If you only have one section, you can still use the **Generate All Sections** command.*

2. Select the frame around your section.

3. Press <Enter> to accept the selection.

    Note: *For multiple sections, the sections will be generated out of sequence, but this will not affect the final section creations.*

After generating the section, you will get a message that your section creation was successful, as shown in Figure C–7.

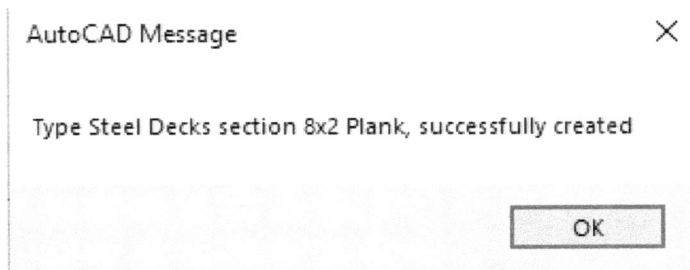

Figure C–7

4. Save your drawing in a secure location and not within the Advance Steel installed directories.

C.2 Adding User Sections

After the section is generated and automatically added to the library, it can be obtained the same way you would access a standard section, as shown in Figure C–8.

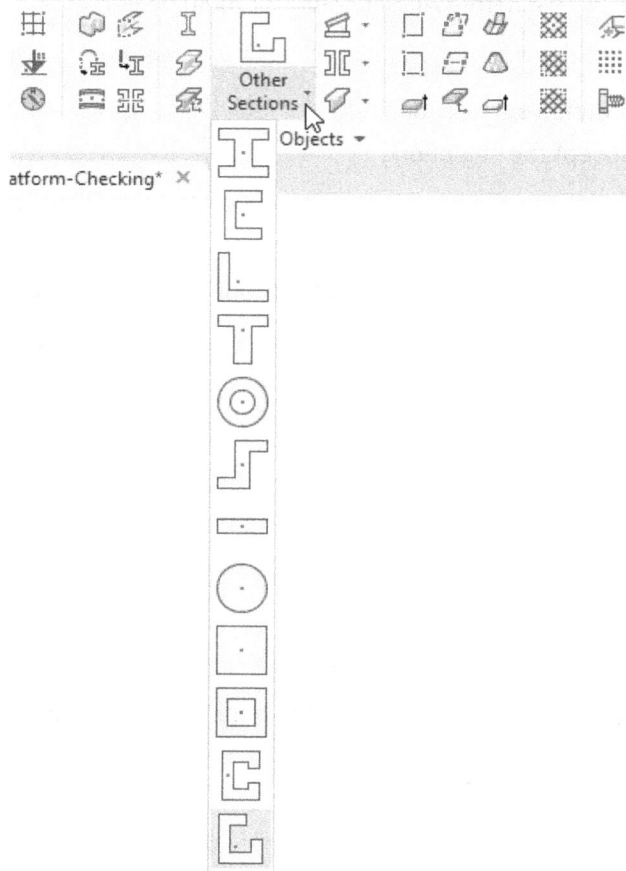

Figure C–8

How To: Add a User Section

1. In the *Home* tab>*Objects* panel, expand ⌐ (Other Sections).

2. In the drawing, specify the location of the start and end point of the system axis, then press <Enter>.

3. In the *Beam* dialog box>*Section & Material* tab, select your class name in the *Section* area, as shown in Figure C–9. The section names display in the *Section* drop-down list for selection.

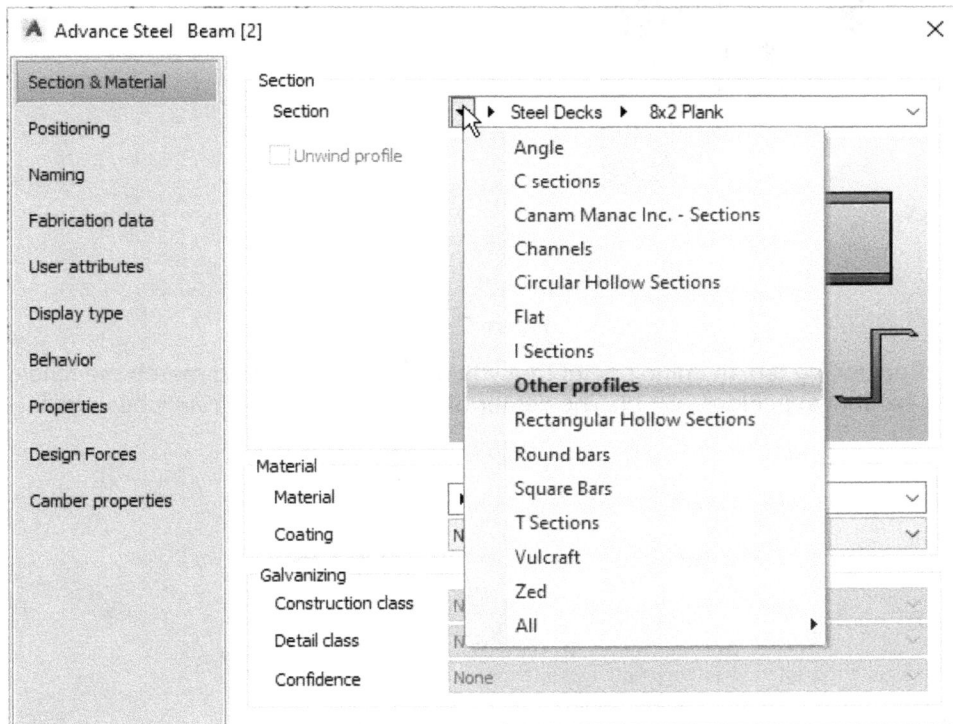

Figure C–9

4. Your section is updated in the drawing.

5. In the *Beam* dialog box>*Display type* tab, the **Standard** display type displays the simplified section shape (as shown in Figure C−10) while the **Exact** display type displays the exact contours that have been defined.

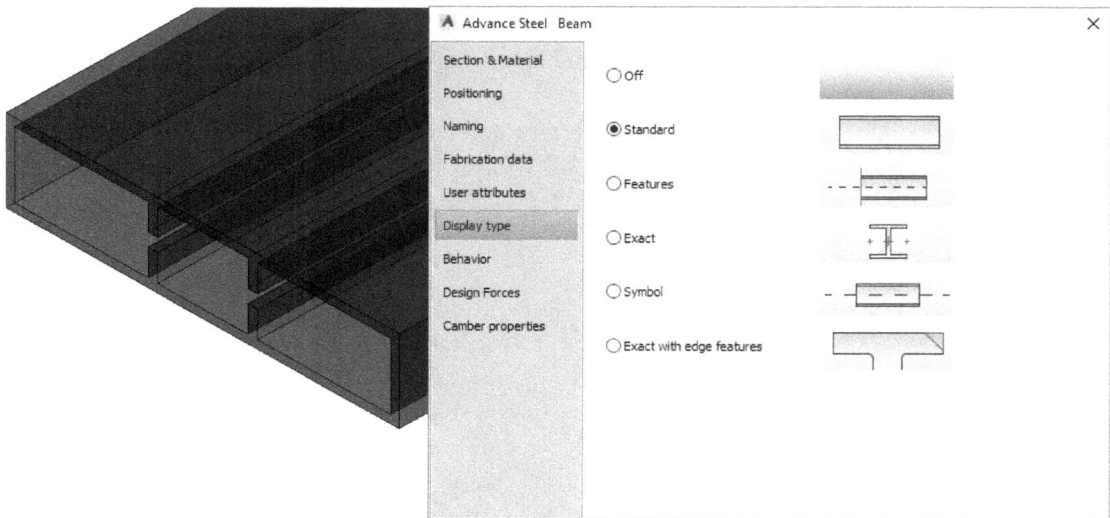

Figure C−10

6. In the *Positioning* tab, the nine points that you have defined should match the nine offset points, as shown in Figure C−11. The center of gravity point will already be calculated for you.

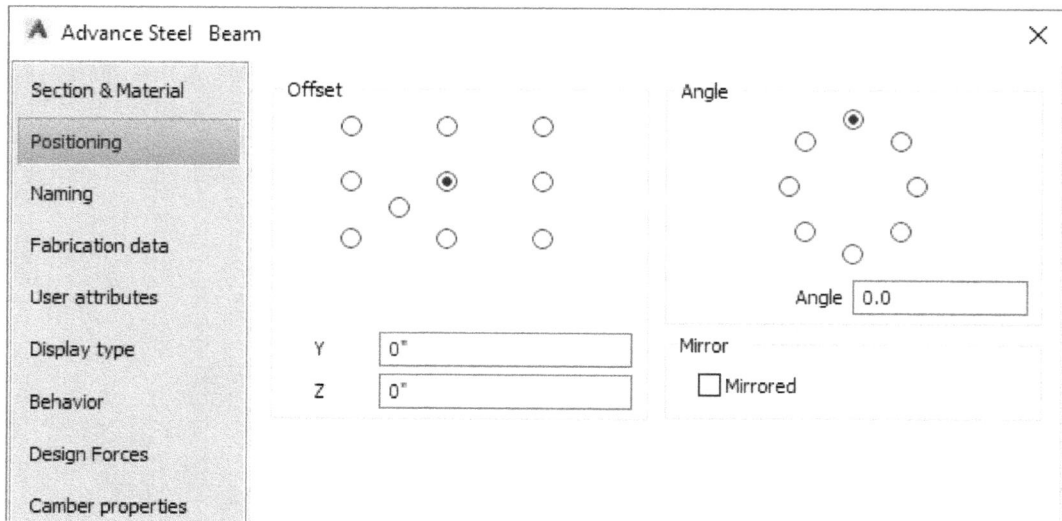

Figure C−11

Command Summary

Button	Command	Location
	Add Coordinates	• **Ribbon:** *Extended Modeling* tab>*User Section* panel
	Generate All Sections	• **Ribbon:** *Extended Modeling* tab>*User Section* panel
	Generate Selected Section	• **Ribbon:** *Extended Modeling* tab>*User Section* panel
	Other Sections	• **Ribbon:** *Home* tab>*Objects* panel>expand *Additional Settings*
	Reference Axis	• **Ribbon:** *Extended Modeling* tab>*User Section* panel> expand *Additional Settings*

Quick Connections and Quick Cuts

Autodesk® Advance Steel has **Quick Connection All**, which provides a parametric rule-based approach to create connections between structural elements and a specific node. It enables the user to create connections and set up rules and options for automatically placing connections. **Quick Cut** automatically cuts structural elements to fit with other structural elements.

Learning Objectives

- Set up quick connections.
- Apply connections automatically.
- Set up quick cuts.
- Modify beam intersections automatically.

D.1 Quick Connections

While you typically apply each connection individually so that you can verify the exact information for each joint, you can also create and use Quick Connections for repetitive connections, as shown in Figure D–1. These connections must be set up in the *Quick Connection* dialog box before you can apply them to the model.

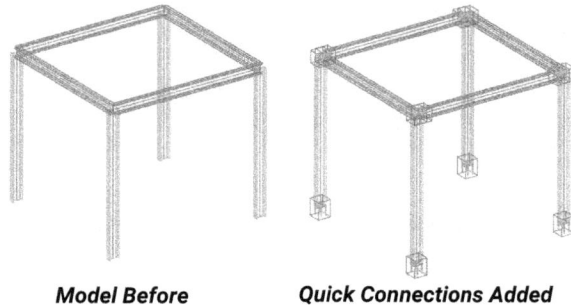

Model Before   **Quick Connections Added**

Figure D–1

How To: Create Quick Connections

1. In the *Extend Modeling* tab>*Joint Utilities* panel, click ⬚ (Configure Quick Connections). This opens the *Quick Connection* dialog box, shown in Figure D–2.

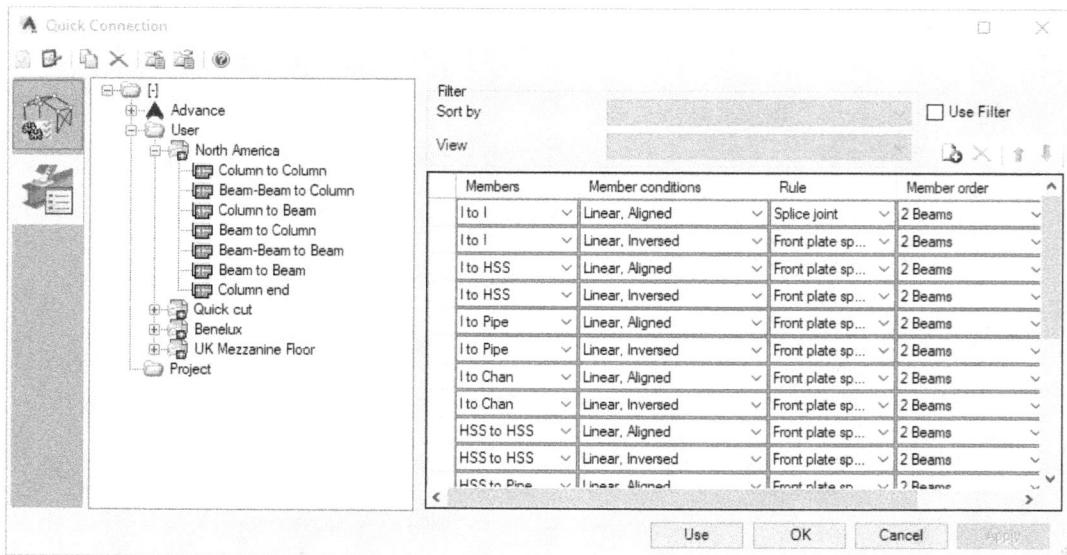

Figure D–2

2. Make changes to the types of members you want to connect and then click **OK**.

   - This may be time consuming, but it can save you time when you use standard connections.

3. In the *Extend Modeling* tab>*Joint Utilities* panel, click ⬛ (Quick Connect All). After processing, the connections are added to every unconnected joint in the model. This will apply the North America profile from the *Advance* category.

Additional Joint Utilities

⬛	**Repeat Rule**	Repeats the last rule that was used in the *Connection Vault*.
⬛	**Display**	Displays the joint boxes of selected connection objects.
⬛	**Select**	Selects **Select a joint box** and all of the other elements in the connection.
⬛	**Update**	Updates the selected joint if other elements have been changed.
⬛	**Transfer Properties**	Transfers the properties of a connection to another connection of the same type (i.e., base plate to base plate).

D.2 Quick Cuts

You can modify the intersection of all beams in a project using the **Quick Cut** command. Similar to **Quick Connect All**, this tool looks at the *Quick Connection* dialog box (shown in Figure D–3) for directions on how to modify places where beams overlap.

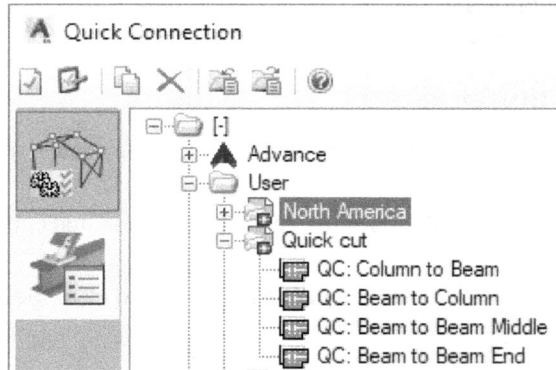

Figure D–3

- In the *Extend Modeling* tab>*Joint Utilities* panel, click ⌶ (Quick Cut). The specified cuts are created, as shown in the inset in Figure D–4.

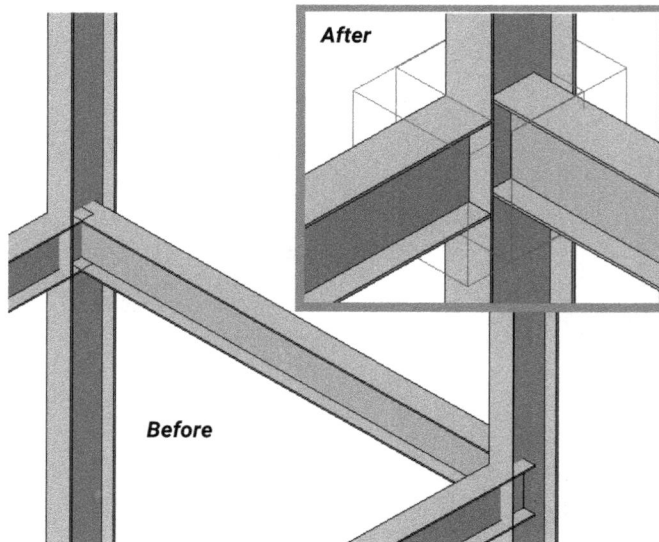

Figure D–4

Command Summary

Button	Command	Location
	Configure Quick Connections	• **Ribbon:** *Extended Modeling* tab>*Joint Utilities* panel
	Add Coordinates	• **Ribbon:** *Extended Modeling* tab>*Joint Utilities* panel
	Quick Cut	• **Ribbon:** *Extended Modeling* tab>*Joint Utilities* panel

Index

U

V

W

X

www.ingramcontent.com/pod-product-compliance
Lightning Source LLC
Chambersburg PA
CBHW060953210326
41598CB00031B/4813